ANNUAL REVIEW OF
EARTH AND
PLANETARY SCIENCES

ANNUAL REVIEW OF EARTH AND PLANETARY SCIENCES

GEORGE W. WETHERILL, *Editor*

Carnegie Institution of Washington

ARDEN L. ALBEE, *Associate Editor*

California Institute of Technology

FRANCIS G. STEHLI, *Associate Editor*

Case Western Reserve University

VOLUME 9

1981

ANNUAL REVIEWS INC. 4139 EL CAMINO WAY PALO ALTO, CALIFORNIA 94306 USA

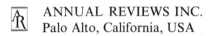

ANNUAL REVIEWS INC.
Palo Alto, California, USA

REPRINTS The conspicuous number aligned in the margin with the title of each article in this volume is a key for use in ordering reprints. Available reprints are priced at the uniform rate of $2.00 each postpaid. The minimum acceptable reprint order is 5 reprints and/or $10.00 prepaid. A quantity discount is available.

International Standard Serial Number: 0084-6597
International Standard Book Number: 0-8243-2009-3
Library of Congress Catalog Card Number: 72-82137

Annual Reviews Inc. and the Editors of its publications assume no responsibility for the statements expressed by the contributors to this Review.

PRINTED AND BOUND IN THE UNITED STATES OF AMERICA

Annual Review of Earth and Planetary Sciences
Volume 9, 1981

CONTENTS

ERRATUM

Volume 8 (*1980*)

In "Geochemistry of Evaporitic Lacustrine Deposits," by Hans P. Eugster, on page 44, the equation following the first reaction was printed incorrectly.

Replace

precipitation of calcite is governed by the relation

$$Ca^{2+} + CO_3^{2-} \rightarrow CaCO_3 \qquad K = (a_{Ca^{2+}})(a_{SO_4^{2-}})(a_{H_2O})^2.$$

by

precipitation of calcite is governed by the relation

$$Ca^{2+} + CO_3^{2-} \rightarrow CaCO_3 \qquad K = (a_{Ca^{2+}})(a_{CO_3^{2-}}).$$

SOME RELATED ARTICLES IN OTHER *ANNUAL REVIEWS*

From the *Annual Review of Astronomy and Astrophysics*, Volume 18 (1980):

*Infrared Spectroscopic Observations of the Outer Planets, Their
Statellites, and the Asteroids*, Harold P. Larson
Formation of the Terrestrial Planets, George W. Wetherwill

From the *Annual Review of Ecology and Systematics*, Volume 11 (1980):

*The Late Quaternary Vegetation History of the Southeastern United
States*, W. A. Watts

From the *Annual Review of Fluid Mechanics*, Volume 13 (1981):

Coastal Sediment Processes, Kiyoshi Horikawa
Debris Flow, Tamotsu Takahashi
Turbulence In and Above Plant Canopies, M. R. Raupach and A. S. Thom
Progress in the Modeling of Planetary Boundary Layers, O. Zeman

From the *Annual Review of Microbiology*, Volume 34 (1980):

Ore Leaching by Bacteria, D. G. Lundgren and M. Silver

Alfred O. Nier

Ann. Rev. Earth Planet. Sci. 1981. 9:1–17

SOME REMINISCENCES OF ISOTOPES, GEOCHRONOLOGY, AND MASS SPECTROMETRY

✴ 10142

Alfred O. Nier

School of Physics and Astronomy, University of Minnesota, Minneapolis, Minnesota 55455

When asked to prepare this article, describing any experiences and analyzing current trends in light of that experience, it would have been difficult to refuse the invitation—especially when one has lived through a period as exciting as the past fifty years. On a more personal note, it has been interesting and pleasurable to recall events that took place along the road to the present.

It appears that I was destined for a technical career from an early age. While in grade school in St. Paul, Minnesota, I enjoyed arithmetic more than any other subject and had an interest in building things. Each Christmas I received additional gears, pulleys, and other accessories for my Meccano set and built ever more complex devices. My closest friend had similar interests, and we regularly visited the local library to read *Popular Mechanics*, *The Boy Mechanic*, and other publications. As a student in high school my interest expanded to science, including electricity and radio, which was then developing. In those days many built their own radios, and my friends and I spent countless hours discussing the latest circuits appearing in the radio magazines. It was thus not surprising that upon my graduation from high school in 1927, my parents and I agreed that I should become an electrical engineer.

A turning point in my life came in 1928 while I was in my sophomore year at the University of Minnesota. All engineering students were required to take a year of physics. The classes were large and we had weekly quizzes. The instructor in the course, Professor Henry Erikson, who was also the chairman of the Physics Department, discovered that I had received grades of 100% on the first three tests of the year. He called

1

me to his office and urged that I consider a career in physics. To reinforce his urging he engaged me as an hourly research assistant and kept me employed throughtout my undergraduate days. I was not his only protegé. He had encouraged many others in the same way.

Life as a Graduate Student

When I graduated in electrical engineering, Professor Erikson offered me a teaching assistantship in physics which, much to his disappointment, I turned down. I had planned to work for a few years before pursuing graduate work. It was 1931 and jobs in engineering were virtually non-existent; I received no offers. Luckily I was rescued by Professor Henry Hartig of the Electrical Engineering Department, who was himself a physicist, and also interested in young people and their careers. He arranged for a teaching assistantship for me, and I worked toward a Master's degree in electrical engineering. At the time, at least at our institution, there were very few advanced courses in engineering, so I took virtually all my course work in physics and in 1933, upon receiving my Master's degree, I proceeded to work toward my PhD degree in physics, which I received in 1936.

It was an exciting time in physics. Nuclear physics was just developing, the neutron had been discovered, and new radioactive isotopes were being produced in cyclotrons almost daily. Professor John T. Tate, who was also the editor of the *Physical Review*, taught a course called Contemporary Experimental Physics, in which he discussed, among other things, the latest manuscripts that crossed his desk. He was a truly great teacher; students took his course once for credit and then came to the class the remainder of their graduate days in order to hear first hand about the exciting new developments in physics. Tate was not only an outstanding expositor, but he also had such a feeling for physics that in his classes one had the thrill of living on the frontiers of discovery.

By the late 1920s Tate had become the main adviser for graduate students in experimental physics at Minnesota. The study of the ionization and dissociation of gases by electron impact was a lively subject of investigation in physics. The laboratory had developed the technique of producing sharply collimated beams of monoenergetic electrons, so the whole subject was put on a sound quantitative basis, and Tate's graduate students made determinations of ionization potentials and cross sections which are still among the best available.

When I became a graduate student in physics, I chose Tate as my adviser but was determined to work on something other than the electron impact of gases. I became interested in electrical discharges in gases and read everything I could about the field. The study of plasma oscillations seemed an interesting topic in which I could combine my background in radio communications with investigation of basic physical phenomena. I proceeded to assemble a vacuum system and, to familiarize myself with the subject, began some modest experiments with electrical discharges at low pressures.

Tate was a very busy person. He taught eight hours of classes per week, was the adviser to at least six graduate students, and had some administrative duties at the University. In addition to his duties as editor of the *Physical Review*, he was one of the organizers of the newly formed American Institute of Physics and frequent trips—by train, in those days—were required between Minneapolis and New York City.

Tate's contacts with his graduate students were only minimal—in part because of his busy schedule, and in part because he believed graduate students should demonstrate initiative and independence if they were to assume responsible positions in the world. In any event, on one of his not-too-frequent visits to my laboratory he looked at what I was doing and suggested that I might be pursuing something "which the General Electric Company had done years ago, but had never bothered to publish."

Following this encouragement(!), I investigated other opportunities. By chance, John H. Williams had just come to Minnesota to work with Tate as a postdoctoral research assistant, and Tate suggested I might work with Williams on a new mass spectrometer being built for measuring the relative abundances of isotopes. My partnership with Williams lasted only a few months. It was 1934 and nuclear physics was emerging as the most prominent field of physics. Both Tate and Williams were anxious to participate, and Williams started a program using a 300 kV transformer and rectifier to produce energetic particles for bombardment of light-element targets. (During World War II, Williams distinguished himself at the Los Alamos Laboratory and later became an Atomic Energy Commissioner as well as a president of the American Physical Society.)

By coincidence, a large solenoid, intended for an electromagnet that was abandoned before it was completed, became available. A 180° deflection mass spectrometer was constructed which gave a mass resolution higher than previously attained. The design was based on that of a similar instrument employed earlier by Tate & Smith (1934) for the study of appearance potentials and ionization cross sections of rare gases and alkali metals. In my initial studies with the new instrument a quadrant electrometer was employed. (One who never has had to measure small currents with a quadrant electrometer cannot possibly appreciate modern electronic devices!) In the course of the work an electrometer vacuum tube, as it was called, became available so the measurement of small ion currents was greatly facilitated.

All-glass, grease and wax-free, mercury-pumped vacuum systems were hallmarks of the Minnesota Physics Laboratory at the time. The combination of a clean vacuum system, the large solenoid, and the commercial availability of electrometer vacuum tubes made possible the construction of a sensitive, relatively high resolution mass spectrometer which was not plagued by impurities that might be mistaken for rare isotopes. The stage was set for the beginning of a systematic study of the relative abundances of isotopes and for the determination of the existence or nonexistence of rare isotopes.

A quick look at argon verified that ^{38}Ar, discovered only the year before by Zeeman & de Gier (1934), indeed existed. The relative abundances of the three isotopes, ^{36}Ar, ^{38}Ar, and ^{40}Ar, were determined and low upper limits set for the existence of possible nearby isotopes such as ^{37}Ar, ^{39}Ar, ^{41}Ar, and ^{42}Ar. Potassium was a more interesting element. Its radioactivity had been known for many years, and the literature of the early 1930s contained speculations as to which isotope might be responsible, the then known ^{39}K or ^{41}K, or possibly some other as yet undiscovered isotope such as ^{42}K or ^{40}K. The problem was made to order for the new apparatus. A careful search showed that potassium indeed had a third naturally occurring isotope ^{40}K (Nier 1935), having an abundance only 1/8600 of ^{39}K, and subsequently shown by others to be responsible for potassium's radioactivity. The 1948 identification of excess ^{40}Ar in potassium minerals by L. T. Aldrich and the author demonstrated that a measurement of the relative amounts of potassium and ^{40}Ar in a mineral might be used for measuring geological ages (Aldrich & Nier 1948a).

Harvard and an Introduction to Isotope Geology

Upon receiving my PhD in 1936, I had the good fortune to receive a National Research Council Fellowship and accepted K. T. Bainbridge's invitation to come to Harvard University. Bainbridge had established an enviable reputation for his mass spectrographic studies of masses and relative abundances of isotopes. The invitation was reinforced by a grant of $5,000, from Harvard's Milton Fund, for the construction of a bigger and better mass spectrometer than I had previously used. By today's standards this would not be considered a large sum. In 1936, a depression year, and before the time of massive government grants, the amount was huge and fully adequate for the task at hand. Moreover, it was free of stipulations other than the usual understanding that the money should be spent prudently.

The two years at Harvard were truly memorable. The isotopic composition of 19 elements was carefully studied and four new isotopes, ^{36}S, ^{46}Ca, ^{48}Ca, and ^{184}Os, discovered. Of more importance, however, was an introduction to problems of geophysics and the measurement of geological age. The lead-uranium method had long been used for determining the age of a uranium mineral. Since the lead resulting from the radioactive decay of uranium was frequently contaminated by common lead, it was customary to infer the isotopic composition, and hence the amount of impurity, from the atomic weight.

The possibility of measuring the isotopic composition accurately added a new dimension to the field. Not only would a direct measurement of the isotopic composition make possible a more accurate determination of the amount of common lead impurity, but in the case of relatively pure uranium-lead samples, a determination of the ^{207}Pb/^{206}Pb abundance ratio provided an independent method of measuring the age of a uranium mineral since ^{207}Pb and ^{206}Pb are the end products of the decay of ^{235}U and ^{238}U, respectively, and the decay rates of ^{235}U and ^{238}U are drastically different.

For one interested in geochronology, Cambridge, Massachusetts, was an exciting place to be in 1936. First of all, in Harvard's Chemistry Department was Gregory P. Baxter who, together with his predecessor, T. W. Richards, had made precise measurements of the atomic weights of many elements. Included in their work were many determinations of atomic weights of both common and radiogenic lead samples, and material from the studies had been carefully preserved. There was Professor Emeritus Alfred Lane of Tufts College who lived only a few blocks from Harvard Square. Lane was chairman of the National Research Council Committee on the Measurement of Geologic Age, and he spent much of his time interacting with people in the field, distributing samples, and encouraging cooperation among workers. The prospect of having accurate isotope measurements of lead samples was a source of sheer delight to him. Working with Lane was John P. Marble, a former student of Baxter's living in Washington, DC, but making frequent trips to Cambridge. Then there was the group at MIT, Robley Evans, Clark Goodman, William Urry, Norman Keevil, and others, working on radioactivity of minerals. And finally there was K. T. Bainbridge who appreciated the importance of the field and encouraged me in every way he could.

Key to the successful exploitation of the ^{207}Pb/^{206}Pb method of measuring ages of uranium minerals was a knowledge of the ^{235}U/^{238}U isotope abundance ratio, a quantity that had not been determined. I felt that with a sample of a volatile uranium compound such as UF_6 a successful determination might be made, and Lane, armed with a $500 grant he had obtained on my behalf from the Geological Society of America, looked for a prospective provider of the then rare compound. He was not successful in his quest and to the best of my knowledge the money reverted. Fortunately, Baxter came to the rescue, providing samples of UCl_4 and UBr_4, which were volatile at elevated temperatures. The well-known ratio $^{238}U/^{235}U = 139$ (Nier 1939), now given more accurately as 137.8, resulted from the measurements.

Baxter was extremely generous in providing and personally converting lead samples from the Harvard collection to PbI_2, a form which I found best for my analyses. The atomic weight of common lead had been measured for a large number of samples by the Harvard chemists and was always found to have a value very close to 207.21. An isotopic analysis of the samples, however, now revealed that the isotope abundances varied considerably, but in such a way that the computed atomic weight was essentially constant and in agreement with the chemically determined value. It appeared from the initial as well as subsequent studies that all common lead samples could be considered as a combination of a primordial lead having isotopes ^{204}Pb, ^{206}Pb, ^{207}Pb, and ^{208}Pb, to which was added approximately equal but variable amounts of ^{206}Pb and ^{208}Pb as well as lesser amounts of ^{207}Pb (Nier 1938).

Baxter's first reaction was one of total disbelief—something had to be wrong with the mass spectrometer measurements! Then, as if I were a student in freshman quantitative analysis, he had me repeat the work, this time giving me random samples as unknowns. After about a half a dozen such "examinations" he had to agree that the isotopic analyses were correct; he had become a convert. His interest in isotopic measurements, already substantial, grew even more and he prepared many samples for me. I could jokingly remark that as a postdoctoral fellow I had acquired a Harvard full professor as a research assistant! At the time, I could perform a lead isotope analysis only every other day, most of the time being spent in cleaning and baking the mass spectrometer, the actual measurements requiring at most a few hours. Baxter taught a course a few times a week on the Radcliffe campus and passed by or through the Harvard Research Laboratory of Physics on his way to Radcliffe. Frequently he visited me both going to and returning from his Radcliffe class to see what progress had been made on the isotope front during the hour he had lectured at Radcliffe!

Among the few geologists actively interested at the time in quantitative geologic age measurements was Arthur Holmes, then at the University of Durham, England. By coincidence, at the very time the common lead isotope studies were underway, Holmes (1937) published a paper on the origin of primary lead ores and concluded that since the atomic weight of all samples studied was so nearly the same, they could not have their origin in the ordinary igneous rocks but must be derived from some deeper source. This was contrary to conventional thought and led to a spirited discussion in the literature between Holmes on the one side and L. C. Graton of Harvard and A. Knopf of Yale as well as others on the opposite side.

Graton was delighted to learn of the isotope variation study and felt that it dealt a fatal blow to Holmes' proposal as to the origin of primary lead ores. I immediately communicated my results to Holmes, and he too was pleased as he felt they substantiated points he had been trying to make! Later, Holmes made a detailed analyses of the $^{208}Pb/^{206}Pb$ ratios in lead ores which corresponded to a present day Th/U ratio of ~ 3.4, a value close to that actually observed in many igneous rocks in the crust of the earth. On this basis, Holmes revised his earlier statements concerning the primordial deep reservoir for the source of lead ores and accepted the possible derivation of Pb ores from igneous rocks and magmas. Holmes, more than any other, was central to various ideas in earth sciences in which Pb isotope studies of ores and rocks played a crucial role, for example, the construction of an absolute geologic time scale, ages of oldest crust and minerals, the age of the earth from Pb-U isotope systematics, origins and correlations of ores and rocks, and many others.

Holmes and I began to correspond regularly, a correspondence that continued until shortly before his death in 1965. We also collaborated on some work in the early 1950s. I saw him only once, in 1954, when I called on him and his wife, Doris Reynolds, also a geologist of note, and spent an afternoon with them at their cottage in Edinburgh. It was a pleasure finally to meet him and it was like seeing a distant relative you had never met before but whom you felt you knew.

The happy days at Harvard came to an end when my two-year fellowship expired and I accepted a faculty position back at the University of Minnesota. Before leaving, however, I became acquainted with Earl Gulbransen, then an instructor in chemistry at Tufts College. We decided that a study of the $^{13}C/^{12}C$ abundance ratio in different sources of carbon would be worthwhile. The mass spectrometer that had been developed could measure isotope abundance ratios to a fraction of a percent, and based on considerations of isotope fractionation by chemical equilibria there was good reason to believe that variations might exist. Fourteen different samples of carbon were investigated, and variations of up to 5 percent were found in the ratio, with limestone-derived samples having the largest amount of ^{13}C and plant-form-derived the least (Nier & Gulbransen 1939). A continuation of the work at Minnesota (Murphey & Nier 1941) confirmed the results.

1938–1945

The transition back to Minnesota was relatively smooth. John Tate had become dean of the college and no longer taught classes but had a few graduate students and continued his editorship of the *Physical Review*. Funds for research were still very meager, and Tate diverted funds from his own modest research allotment to make it possible for me to have a magnet similar to the excellent one I had at Harvard. Bainbridge let me take with me the mass spectrometer "tubes" I had built at Harvard, so with the addition of some electronic supplies, batteries, a vacuum system, and a few small items I was able to resume research in about six months. Now, however, progress was slower. Eight hours of teaching per week plus other faculty duties constituted quite a contrast from the carefree days of the NRC fellowship! Nevertheless, 15 more samples of common lead and 8 more of radiogenic lead were investigated (Nier et al 1941), the variations observed in the common lead isotopes confirming the earlier values, and isotopic data found for the radiogenic leads contributing to our list of ones for which precise ages could be given. The onset of World War II prevented additional lead isotope measurements or more careful study of the data on hand and my own participation in the field stopped. It was picked up and pursued vigorously by others when the war ended. On the basis of the prewar data, Holmes (1946) and Houtermans (1946) independently concluded that the common lead isotope variations indicated that the earth had to be of the order of 3 billion years old. Lead isotope studies were resumed by a number of laboratories, particularly at the University of Toronto. The results of the more extensive studies of ore samples by a variety of investigators (Russell & Farquhar 1960) together with those of meteorites (Patterson 1956) has led to the present accepted age of the solar system, 4.6 billion years.

In 1939 there were new ideas that engaged my interest. Clusius & Dickel (1939) had shown that practical isotope separations might be accomplished using thermal diffusion columns, and Furry et al (1939) worked out the theory of a methane-containing column which might be used to separate ^{13}C and ^{12}C. Separated ^{13}C promised to be interesting for nuclear experiments or as an isotope tracer in biological or chemical experiments (practical amounts of radioactive ^{14}C were not yet available).

Our shop built a column for me and useful amounts of methane were produced in which the $^{13}C/^{12}C$ abundance ratio was over 10 times that of naturally occurring carbon (Nier & Bardeen 1941).

Enriched ^{13}C was scarce in the world at the time and we had one of the few mass spectrometers in existence that could make precise isotope abundance measurements. I gained many new friends in the biological sciences! My botanical colleagues grew radish plants in atomspheres of ^{13}C-enriched CO_2 to study photosynthesis, and my physiological colleagues conducted metabolic studies on mice. One of the most rewarding experiences was collaboration with H. G. Wood and C. H. Werkman of the Bacteriology Department of Iowa State University in studies of the assimilation of CO_2 by heterotrophic bacteria (Wood et al 1941).

Nuclear fission of uranium was discovered in 1939 and there was some question as to which isotope was responsible. At the April meeting of the American Physical Society, John Dunning of Columbia University introduced me to Enrico Fermi, also at Columbia at that time, and they urged me to try to separate the uranium isotopes using a mass spectrometer. On October 28, 1939, Fermi wrote to me, again urging me to proceed with the experiment, stating that a determination of whether or not ^{235}U was responsible for the slow neutron fission observed was "of a considerable theoretical and possible practical interest"! The experiment was performed and ^{235}U shown to be the responsible isotope (Nier et al 1940). Of special interest here is the fact that the source of uranium for the mass spectrometer (used as an isotope separator) was some of the UBr_4 given to me by Professor Baxter of Harvard for the geological age studies.

Up to 1939 essentially all mass spectrometers employed 180° magnetic deflection for separating ions of different mass. Many of the early ones used solenoids. My Harvard instrument, and its Minnesota successor, used a two-ton electromagnet powered by a 5 kW DC generator. It seemed that it should be possible to obtain equivalent performance in some easier way and hence make it possible for more people to have mass spectrometers—and indeed it was. In the early 1930s it had been shown (Barber 1933, Stephens 1934) that a diverging bundle of ions of a given mass and velocity, if directed normally between the poles of a magnet having 'a wedge-shaped magnetic field region, would be focussed at a point provided the source of ions, the apex of the wedge, and the focal point were on a straight line. The 180° instrument is merely a special case covered by the rule. Bainbridge & Jordan (1936) at Harvard had made use of the principle of employing a 60° wedge magnet as part of their mass spectrograph built for the precision determination of atomic masses. In 1940 it was shown

(Nier 1940) that a mass spectrometer using a 60° wedge magnet had essentially as good focusing properties as the earlier 180° instruments. The saving in weight and power consumption was substantial, and the mass spectrometer became a tool within the reach of many more researchers in fields where gas or isotope analyses were required.

World War II stimulated a revolution in the application and availability of sophisticated instruments in both basic and applied research. Mass spectrometry was no exception. In 1941 only two companies manufactured mass spectrometers. The only groups that could afford to buy the instruments were the oil companies who found them ideal for analyzing the complex hydrocarbons produced in their refineries. In 1941, when it became apparent that enriched ^{235}U might be used for making atomic bombs, we had the only mass spectrometer in the world that could make isotope analyses of uranium samples, and for several years we performed such analyses for various groups in the country working on means of enriching ^{235}U. Further development of instruments was subsequently supported by contracts from the wartime Office of Scientific Research and Development (OSRD), and by the time the Manhattan Project (atomic bomb program) was underway the designs had been transferred to the General Electric Company. They produced uranium isotope analysis instruments in quantity for the Project as well as helium leak detectors, which had been developed at Minnesota, as an offshoot of the uranium isotope instrument development. From 1943 through 1945 I headed an instrument development laboratory for the Kellex Corporation, NYC, working on problems relating to the gaseous diffusion plant built in Oak Ridge, Tennessee, for the separation of ^{235}U.

The Postwar Period

When World War II ended, mass spectrometry was no longer the exclusive domain of a handful of physicists or chemists who built their own instruments for their research. As happened in other areas of instrumentation, small companies sprang up, often headed by individuals who participated in instrument development during the war. Also, there were now many more persons who had become acquainted with vacuum technology and electronic techniques. Non-physicists, who could not afford or who had special requirements beyond the commercially available mass spectrometers, still undertook the construction of instruments of their own.

During the decade following 1945 I received inquiries from many geologists and biologists who were interested in applying mass spectrometry to their research. The head of our machine shop, R. B. Thorness, who headed a development shop for me during the war and returned with me to the University after the war ended, built dozens of mass spectrometer "tubes," as we called them, for individuals all over the world (the tube was the vacuum housing and its contents, i.e. the ion source, collection system, etc). The tube, when supplemented by a magnet, a vacuum system, and electronic power supplies (then becoming available commercially), constituted a complete mass spectrometer. The do-it-yourself era faded as more companies entered the field and grants and contracts became available to help buy instruments for those who were merely interested in mass spectrometers as tools in their research.

Mass spectrometry received a tremendous boost when Paul & Steinwedel (1953) of the University of Bonn, West Germany, introduced the quadrupole mass spectrometer. Here, the beam of ions to be mass analyzed is sent axially between four rods. A proper choice of a combination of direct and radio-frequency voltages between the rods permits ions of only one mass to emerge from the space between the rods. Many of the commerical instruments in use today employ this method of mass separation rather than magnetic deflection.

The success of the Manhattan Project stimulated an enormous interest in nuclear physics and its application to peaceful purposes. A large number of scientists had gained experience in working with unusual chemicals and with radioactive isotopes. The means for making quantitative measurements of both stable and radioactive isotopes were at hand, and a new frontier was opened in geochemistry and cosmochemistry.

Especially significant was the stimulus given the field by a small group at the University of Chicago shortly after the war. Harold Urey, the co-discoverer of deuterium, and well known for his work on isotope fractionaction by chemical equilibria, had moved to Chicago. Harrison Brown and Hans Suess, also at Chicago at the time, got Urey interested in problems of the origin and evolution of the earth and solar system; Urey's profound insights and expertise in applying physical and chemical boundary conditions to understanding these fundamental problems had led to the field, now generally called cosmochemistry, in which isotope measurement played a vital role. In addition to Brown and Suess, others during this period in Chicago included Harmon Craig, Samuel Epstein, Irving Friedman, Edward Goldberg, Mark Inghram, Claire Patterson, John Reynolds, George Tilton, Gerald Wasserburg, and George Wetherill, to mention but a few. The influence this group, and their students and associates, has had on our understanding of natural processes in the solar system, if not the universe, is so broad that it is hard to describe. Isotopic variations in nature, paleotemperatures, the age of the solar system, geobiology, lunar geochemistry, and some of the conditions in the early days of the universe are but a few of the subjects touched and further developed by those who had the good fortune to be stimulated by the interest in geochemistry and cosmochemistry that prevailed at Chicago following the war.

Following the war, after an absence from scientific pursuit for five years, my own interests turned mainly to problems unrelated to geology. There were some exceptions, however. I have already mentioned the work of L. T. Aldrich, then a graduate student, in which it was demonstrated that ^{40}Ar indeed existed in potassium minerals so the decay of ^{40}K to ^{40}Ar appeared feasible as a means of measuring geological ages. In another study (Aldrich & Nier 1948b) a comprehensive study was made of the $^{3}He/^{4}He$ isotope abundance ratio in a large number of samples, and it was demonstrated that significantly large variations—over a factor of 100—were present, depending upon the source of the helium. Because of the small amounts of gas available, the mass spectrometer was run in a static mode for the first time in these argon and helium investigations.

Our main efforts for some 10 years after the war were devoted to problems largely unrelated to geophysics or geochemistry. We produced enriched ^{3}He by thermal diffusion and collaborated with individuals elsewhere who used the material in low temperature studies (this was before reactor-produced ^{3}He was released by the Atomic Energy Commission). We developed a high resolution mass spectrometer of a new design and began a systematic program of precision atomic mass determination. It was shown that the instrument could also identify molecules from their precise weight (Nier 1954), making it possible to distinguish between two molecules in a gas mixture having the same mass number, e.g. CO_2 and C_3H_8, both having masses of 44 amu. The possibility of making use of this property in identifying impurities in a gas analysis is obvious. Since the difference from integral numbers of the masses of different isotopes and elements varies from element to element, an important application of this general type of instrument has been in making determinations of molecular structure from the exact masses of compounds in question (Beynon 1960).

During this period we continued our collaboration with biological and medical colleagues, enriching ^{18}O by thermal diffusion for use as a tracer and assisting in metabolic studies. A portable mass spectrometer was developed for use in respiratory studies and for monitoring respiration during surgery.

Isotope geology was not forgotten entirely in our laboratory. John Hoffman, a graduate student, conducted a successful investigation of the ^{3}He and ^{4}He distribution in large iron meteorites (Hoffman & Nier 1960). Peter Signer, a postdoctoral fellow and later a faculty member, began a more comprehensive study of the rare gases found in meteorites (Signer & Nier 1960).

A highly significant collaboration developed between our group and Goldich and coworkers (1961) in the Geology Department at the University of Minnesota, in developing and applying the K-Ar age determination method for broad scale reconnaissance studies of the continental basement rocks in Minnesota and a large part of the US, as well as in studying detailed aspects of metamorphism of rocks. All these were exciting applications of isotopic measurements to various geological and cosmological problems, which now have evolved into highly mature fields in their own right.

Mass Spectrometry and Space Research

A person acquainted with analytical instruments, if asked in 1950 to describe a mass spectrometer, would quite properly have referred to something that filled a fair fraction of a small room. Yet when sounding rockets became available and studies of the composition of the earth's atmosphere above 100 km began, the size was miraculously reduced to smaller than a cubic foot and the weight to less than 20 pounds. Mass spectrometers, having established their power in analyzing complex gas mixtures, were natural candidates for atmospheric composition studies and the incentive to miniaturize them was enormous. Dozens of mass spectrometers have been carried aloft on sounding rockets and satellites and a great deal learned about both the neutral and ionic composition of the upper atmosphere.

One of the attractions of earth and planetary science is that it cuts across traditional disciplinary lines. While individual experts in the various fields are required, it is the syntheses of their investigations that leads to a better understanding of the universe in which we live. Nowhere was this better illustrated than in the Viking Project, the 1976 landing on and orbiting of Mars by instrumented spacecraft (Soffen & Snyder 1976).

The two Orbiters and two Landers carried sophisticated instruments for studying a variety of phenomena—imaging of the surface, measuring water vapor content of the atmosphere, thermal mapping, and studying atmospheric composition and structure, biology, organic and inorganic composition of soil, meterology, seismology, and magnetic and physical properties of the soil. Some 70 scientists, organized in teams associated with the various instruments, were involved.

I had the good fortune to be associated with two of the teams—Entry Science and Molecular Analyses. The Entry Science team made measurements in the atmosphere as the Landers descended to the surface. A retarding potential analyzer was used to study the ionosphere, a mass spectrometer to investigate the neutral composition of the upper atmosphere, and pressure and temperature sensors to determine the structure of the lower atmosphere. The Molecular Analysis team looked for volatiles in the soil (particularly organic compounds) using a mass spectrometer combined with a gas chromatograph. This team also measured the composition of the atmosphere around the Lander.

The mass spectrometers performed superbly. The Entry instrument detected N_2, O_2, Ar, CO, NO, and O in addition to the CO_2 known to be the main constituent (Nier & McElroy 1977). The Lander instrument observed Kr and Xe as well as most of the constituents seen by the Entry instrument (Owen et al 1977). It was able to determine that if organic compounds were present in the soil the amounts were extremely low (Biemann et al 1977). One of the extremely interesting results from atmospheric analyses was that whereas the $^{13}C/^{12}C$ and $^{18}O/^{16}O$ abundance ratios appear to be the same as the corresponding terrestrial values, the $^{15}N/^{14}N$ abundance ratio is about 60 percent higher than that on earth. The difference is attributed to a selective escape of ^{14}N from an atmosphere initially rich in N_2 (McElroy et al 1976).

The investigations were not without fascinating sidelights. It had been reported that when the Russian spacecraft Mars 6 had descended on Mars in 1974, the atmosphere was found to have as much as 35 percent argon. The result was not based on mass spectrometric measurements—these were lost when the spacecraft failed to return data after it reached the surface. The conclusion was reached from the abnormally high current in the sputter ion pump evacuating the mass spectrometer during the descent; this current was monitored (Istomin & Grechnev 1976). When we heard of the results in early 1975, I immediately simulated the experiment as nearly as I could in our laboratory and verified that a CO_2 mixture containing 35 percent argon would indeed give a response similar to that observed by the Russians. Their results *could* thus be correct, although they could also be due to some malfunction of their monitoring equipment. My results were reported to the Project and spread beyond. I learned later from one of my German friends that the rumor was about that I had somehow mysteriously obtained possession of one of the Russian pumps! The report of a high argon content of the atmosphere generated a host of publications theorizing as to how this could come about.

As the Viking project developed, it was plagued with serious weight and cost problems. There was a general feeling that the biology, imaging, and organic analyses experiments were more important than the others. The Entry mass spectrometer was in the less favored group. Yet on the morning of July 20, 1976, when Viking 1 descended on Mars, it had its moment of glory. Laboratory tests on the ion sputter pump maintaining the vacuum in the gas chromatograph–mass spectrometer had shown that it might have a short life if subjected to too much argon; analyses for organic compounds in the soil might not be obtained. When we received our entry data a few hours after the descent, the collective eyes of the Project were on us and there was a great sigh of relief when, after a few minutes perusal of our mass spectra, we were able to state unequivocally that the argon content of the atmosphere was less than two percent!

The Viking Project was a truly interdisciplinary effort. After the science teams were selected in early 1970, the scientists met regularly until well after the descent of the second Lander on Mars on September 4, 1976. Scientists had to learn about each others' instruments and strategies and everyone's knowledge of geophysics and geochemistry was broadened. Since a central theme of the Project was a search for life on Mars, the non-biologists had to learn about biology and how their particular experiment could contribute to an understanding of how life might have developed. When the spacecraft finally arrived at Mars and data began to become available, seminars were held almost daily. We shared the joy as pictures became available and the agonies of our biology colleagues when ambiguous results were obtained. Everyone left the Project with a broader feeling for the unity of science in planetary exploration.

The successes of the various Pioneer, Mariner, and Voyager missions demonstrated that planetary exploration can be profitably pursued. Worthwhile scientific problems can be posed and instrumentation developed for solving the problems. One of the most interesting space ventures yet to be organized and approved for funding will be a rendezvous mission to a comet. Comets are generally believed to be the most primitive objects in the solar system; a study of a comet should provide independent clues as to the chemical and physical conditions that existed when the planets were formed.

A comet is sometimes described as a dirty snowball; that is, it consists of ices in which dust is trapped. In addition to water ice, organic compounds are known to be present. It has been postulated that in early days the earth may have received organic compounds from an influx of comets. In a rendezvous mission one would have the opportunity for making *in situ* as well as remote measurements. The mission would be highly interdisciplinary and a large number of skills would be called for.

In such a program mass spectrometry would play a key role. A mass spectrometer is an extremely sensitive and versatile instrument, ideally suited to the analysis of a tenuous atmosphere of unknown composition. It can be programmed to operate in a number of different modes which make it possible to perform a number of functions. It can measure both stable compounds and free radicals, which might decompose upon striking surfaces. Hence it could measure both the so-called parent molecules near the comet nucleus and fragments that appear farther away. It can measure the speed of particles reaching it, as well as determine the direction from which they came. Finally, in the process of making analyses it also determines isotopic composition, a matter of considerable importance in the examination of so primordial a material.

Not only is an analysis of the gas emitted by the comet nucleus of interest, but so is the gas entrained in or adsorbed on the surfaces of the dust grains. Is it similar to the gas found in the atmosphere surrounding the comet or is it different? Does it have any relation to the gas found in meteorites such as carboneaceous chondrites? Is there anything unusual about the isotopic composition?

The successes of the various planetary missions—the Pioneers, Mariners, Vikings, and Voyagers—indicate that meaningful explorations of the solar system can be made. The day has arrived when automated instruments can be sent to as yet unexplored parts of the solar system and in situ experiments and measurements performed which would not have been possible in a terrestrial laboratory only a few years ago. The solar system, like the world, will grow smaller.

Literature Cited

Aldrich, L. T., Nier, A. O. 1948a. Argon 40 in potassium minerals. *Phys. Rev.* 74: 876–77

Aldrich, L. T., Nier, A. O. 1948b. The occurence of ^{3}He in natural sources of helium. *Phys. Rev.* 74: 1500–94

Bainbridge, K. T., Jordan, E. B. 1936. Mass spectrum analysis. *Phys. Rev.* 50: 282–96

Barber, N. F. 1933. Shape of an electron beam in a magnetic field. *Proc. Leeds Philos. Soc.* 2: 427–34

Beynon, J. H. 1960. *Mass Spectrometry and its Applications to Organic Chemistry.* Amsterdam: Elsevier. 640 pp.

Biemann, K., Oro, J., Toulimin, P., III, Orgel, L. E., Nier, A. O, Anderson, D. M., Simmonds, P. G., Flory, D., Diaz, A. V., Rushneck, D. R., Biller, J. E., LaFleur, A. L. 1977. The search for organic substances and inorganic volatile compounds in the surface of Mars. *J. Geophys. Res.* 82: 4641–58

Clusius, K., Dickel, G. 1939. Thermal diffusion tube for separation of isotopes. *Z. Phys. Chem. Abt. B.* 44: 397–473

Furry, W. H., Jones, R. C., Onsager, L. 1939. The theory of isotope separation by thermal diffusion. *Phys. Rev.* 55: 1083–95

Goldich, S. S., Nier, A. O., Baadsgaard, H., Hoffman, J. H., Krueger, H. W. 1961. *The Precambrian Geology and Geochronology of Minnesota.* Minneapolis: Univ. Minn. Press. 193 pp.

Hoffman, J. H., Nier, A. O. 1960. Cosmic-ray-produced helium in the Keen Mountain and Casas Grandes Meteorites. *J. Geophys. Res.* 65: 1063–68

Holmes, A. 1937. The origin of primary lead ores. *Econ. Geol.* 32: 763–82

Holmes, A. 1946. An estimate of the age of the earth. *Nature* 157: 680–84

Houtermans, F. G. 1946. Die Isotopenhäufigkeiten im natürlichen Blei und das Alter des Urans. *Naturwissenschaften* 33: 185–86

Istomin, V. G., Grechnev, K. V. 1976. Argon in the Martian atmosphere: Evidence from the Mar 6 descent module. *Icarus* 28: 155–58

McElroy, M. B., Yung, Y. L., Nier, A. O. 1976. Isotopic composition of nitrogen: Implications for the past history of Mars' atmosphere. *Science* 194: 70–72

Murphey, B. F., Nier, A. O. 1941. Variations in the relative abundances of the carbon isotopes. *Phys. Rev.* 59: 771–72

Nier, A. O. 1935. Evidence for the existence of an isotope of potassium of mass 40. *Phys. Rev.* 48: 283–84

Nier, A. O. 1938. Variations in the relative abundances of the isotopes of common lead from various sources. *J. Am. Chem. Soc.* 60: 1571–76

Nier, A. O. 1939. The isotopic constitution of uranium and the half-lives of the uranium isotopes. I. *Phys. Rev.* 55: 150–53

Nier, A. O. 1940. A mass spectrometer for routine isotope abundance measurements. *Rev. Sci. Instrum.* 11:212–16

Nier, A. O. 1954. The determination of isotopic masses and abundances by mass spectrometry. *Z. Electrochem.* 58:550–567; reprinted in 1955 in *Science* 121:737–44

Nier, A. O., Bardeen, J. 1941. The production of concentrated carbon 13 by thermal diffusion. *J. Chem. Phys.* 9:690–92

Nier, A. O., Gulbransen, E. A. 1939. Variations in the relative abundances of the carbon isotopes. *J. Am. Chem. Soc.* 61:697–98

Nier, A. O., McElroy, M. B. 1977. Composition and structure of Mars' upper atmosphere: Results from the neutral mass spectrometers on Viking 1 and 2. *J. Geophys. Res.* 82:4341–49

Nier, A. O., Dunning, J. R., Booth, E. T., Grosse, A. V. 1940. Nuclear fission of separated uranium isotopes. *Phys. Rev.* 57:546

Nier, A. O., Thompson, R. W., Murphey, B. F. 1941. The isotopic constitution of lead and the measurement of geological time. III. *Phys. Rev.* 60:112–16

Owen, T., Biemann, K., Rushneck, D. R., Biller, J. E., Howarth, D. W., LaFleur, A. L. 1977. The composition of the atmosphere at the surface of Mars. *J. Geophys. Res.* 82:4635–39

Patterson, C. 1956. Age of meteorites and the earth. *Geochim. Cosmochim. Acta* 10:230–37

Paul, W., Steinwedel, H. 1953. A new mass spectrometer without a magnetic field. *Z. Naturforsch.* 8a:448–50

Russell, R. D., Farquhar, R. M. 1960. *Lead Isotopes in Geology.* New York/London: Interscience. 243 pp.

Signer, P., Nier, A. O. 1960. The distribution of cosmic-ray-produced rare gases in iron meteorites. *J. Geophys. Res.* 65:2947–64

Soffen, G. A., Snyder, C. W. 1976. The first Viking mission to Mars. *Science* 193:759–65

Stephens, W. E. 1934. Magnetic refocussing of electron paths, *Phys. Rev.* 45:513–18

Tate, J. T., Smith, P. T. 1934. Ionization potentials and probabilities for the formation of multiply charged ions in the alkali vapors and in krypton and xenon. *Phys. Rev.* 46:773–76

Wood, H. G., Werkman, C. H., Hemmingway, A., Nier, A. O. 1941. Heavy carbon as a tracer in heterotrophic carbon dioxide assimilation. *J. Biol. Chem.* 139:365–76

Zeeman, R., de Gier, J. 1934. New isotope of argon. *Proc. K. Akad. Amsterdam 37.* 3:127–29

Ann. Rev. Earth Planet. Sci. 1981. 9:19–58

PARTICLES ABOVE THE TROPOPAUSE:

✖ 10143

Measurements and Models of Stratospheric
Aerosols, Meteoric Debris, Nacreous Clouds, and
Noctilucent Clouds[1]

Owen B. Toon and Neil H. Farlow
Space Science Division, NASA–Ames Research Center, Moffett Field,
California 94035

INTRODUCTION

Water clouds and aerosols are everyday sights in the lower atmosphere, but they are much less commonly observed at high altitudes. The violent eruption of the Krakatoa volcano in 1883 injected large quantities of volcanic ash and gas into the stratosphere. The ash particles, as well as particles created from the gas, remained in the stratosphere for several years and caused obvious worldwide optical phenomena such as colored twilights, persistent halos around the sun and moon, blue and green moons, and a dull blue sky. These optical phenomena first attracted the attention of the scientific world to the unusual clouds and aerosols at high altitude.

Stratospheric aerosols are submicron-sized particles composed mainly of sulfuric acid and are found between the tropopause and about 30 km. They were commonly observed following volcanic eruptions during the early part of the twentieth century and several studies of their properties and their possible effects on climate were conducted (Royal Society 1888, Humphreys 1940). The modern era began in 1960 when Junge and his coworkers first directly sampled stratospheric aerosols from a balloon and showed that aerosols were present in the stratosphere even during volcanically quiescent periods (Junge et al 1961). Numerous aircraft and balloon observations have been made during the past two decades. During the 1980s a vast improvement of our knowledge will be obtained from

---

[1] The US Government has the right to retain a nonexclusive royalty-free license in and to any copyright covering this paper.

19

more sophisticated direct sampling and from satellites dedicated to observing the global distribution of the aerosols (McCormick et al 1979).

Noctilucent clouds contain submicron-sized particles composed mainly of water ice. They are found during summer in a few-kilometer-thick layer at the mesopause which is near 85 km altitude. These clouds were first widely observed after the Krakatoa eruption, perhaps because the clouds were especially bright at that time, but more likely because many scientists were then interested in observing high-altitude clouds. During the century following their discovery many hypotheses were put forward to explain their seasonal existence in a narrow layer, at such high altitudes. The modern era of their study began in the late 1950s when rocket flights proved that mesopause temperatures were low enough to cause ice to condense. Numerous rocket flights were made in an attempt to capture cloud particles and the lack of success is consistent with the volatile composition of the particles.

Nacreous clouds are composed of micron-sized ice particles. The clouds occur during winter near 20 km altitude in the stratosphere. Nacreous clouds are only rarely seen and have never been directly sampled, so it might be said that the modern era of their investigation has not yet begun. Satellite observations during the 1980s may greatly improve our understanding of these clouds (McCormick et al 1979).

Meteoric debris entering the atmosphere is observed in the trails of shooting stars. The modern era of collection of meteoric debris in the atmosphere began in the 1960s when numerous attempts were made using rockets, but it is only very recently that unambiguous collections have been obtained.

In this paper we will outline the present status of our knowledge of stratospheric aerosols, meteoric debris, nacreous clouds, and noctilucent clouds. Considerable progress has been made in studies of these particles during the previous decade and it is appropriate to synthesize the information to provide a background for studies planned for the 1980s. The measurement techniques and the physical processes affecting these particles are quite similar. In addition, these particulates interact with each other. Therefore, a review that encompasses all particles seems appropriate. We will describe the observed properties of these clouds and aerosols and then discuss theoretical studies.

MEASUREMENTS

Stratospheric Aerosols

The properties of stratospheric aerosols were reviewed by Cadle & Grams (1975) and by Toon & Pollack (1976).

COMPOSITION   After two decades of measurements, most investigators believe that stratospheric aerosol particles are impure sulfuric acid droplets containing various crystalline nitrogen-sulfate compounds intermixed with traces of halogens, tropospheric constituents, and meteoric elements. Following major volcanic eruptions, ash granules are also usually present. Evidence from the investigations that led to these conclusions is both indirect and specific. For example, the identification of sulfuric acid is indirect, while detection of crystals of ammonium sulfate and ammonium peroxidisulfate by electron diffraction is specific.

The indirect evidence that supports the identification of sulfuric acid dates from the early 1960s when Junge et al (1961) determined that sulfur was the major element in stratospheric aerosols. Because of the water-soluble and hygroscopic nature of the samples, together with the independent electron diffraction identification of $(NH_4)_2SO_4$ and $(NH_4)_2S_2O_8$ by Friend et al (1961), Junge et al (1961) assumed that the sulfur must be present as a sulfate. When chemical spot tests showed insufficient NH_4^+ ions present to account for all the sulfate, they concluded that at least part of the liquid in the particles must be H_2SO_4.

During the 1960s several attempts were made to identify the composition based upon particle morphology, but such techniques are not reliable as shown by Farlow et al (1977). In an attempt to more rigorously identify the acid and other components in the aerosols, Bigg et al (1974) developed several reagent-film methods whereby particles would react with thin films of chemicals vapor-coated on the collecting surfaces. Applying these techniques, Bigg (1975) concluded that the great majority of particles were composed mainly of sulfuric acid and varying amounts of ammonium sulfate.

Rosen (1971), using a heated intake tube to his balloon-borne photoelectric particle counter, found that the aerosol particles would evaporate, reducing his counts drastically, at temperatures between 80 and 100°C. Examining vapor pressure curves for various likely substances, he found that a 75% solution of H_2SO_4 was in reasonable agreement with evaporation at the observed temperatures.

Specific measurements made on bulk samples collected with filters in the stratosphere by Cadle et al (1970), Lazrus et al (1971), and Cadle (1972) showed that most of the aerosol was composed of SO_4^- ions, with very little to moderate amounts of NH_4^+ ions, which were, generally, less than 10% of the sulfate concentrations. Again they inferred that most of the aerosol particles were H_2SO_4.

Farlow et al (1978) also used electron diffraction to specifically identify crystals of $(NH_4)_2SO_4$ and $(NH_4)_2S_2O_8$. In addition, they tentatively detected other nitrogen-sulfur compounds such as $NOHSO_4$ and

$NOHS_2O_7$. The possible presence of these chemicals suggests that interactions between the aerosol particles and trace nitrogen oxide gases may
be taking place. Finally, Cunningham et al (1979) used a new Raman
microprobe method (Etz et al 1978) to examine aerosol samples collected
at the South Pole that he believed were down-flowing stratospheric
aerosols. He identified $(NH_4)_2SO_4$ as the principal component, as did
Cadle et al (1968), who in work on similar samples also detected
$(NH_4)_2S_2O_8$.

Hayes et al (1980) have raised questions about the presence of crystalline
$(NH_4)_2SO_4$ and $(NH_4)_2S_2O_8$ in stratospheric aerosols. Evidence is
accumulating that these compounds may form within a few minutes or
hours after exposure of the samples to normal laboratory environments
where traces of NH_3 near 100 ppb are common. These researchers have
found that when the collections are protected with pure argon from the
time they are collected at flight altitudes, through insertion into the
electron microscopes in the laboratory, then no sulfate crystals are
detected. But when these specimens are removed from the microscope,
exposed in air for an hour, then reexamined, a profusion of crystals—
both $(NH_4)_2SO_4$ and $(NH_4)_2S_2O_8$—are found. Moreover, weak halos
appear when the unreacted argon-protected aerosol particles are examined
in the electron diffraction mode. These halos have diffraction d-spacings
identical to those obtained from test aerosol droplets of pure H_2SO_4
(Hayes et al 1980). Thus the researchers believe their measurements are
the first specific identification of H_2SO_4 in stratospheric aerosols.

Other chemical identifications have been made of different components
in stratospheric aerosols following volcanic explosions. Mineral elements
such as Al, Si, Na, K, Ca, Fe, Mg, Ti, and, of course, sulfur, have been
found in various proportions (Farlow et al 1973). Some of these elements
(Si, Na) as well as others (Cl, Br, Mn) have been measured quantitatively
by Cadle & Grams (1975) and Lazrus & Gandrud (1974). Moreover, the
worldwide distribution of sulfate before and after a major eruption
(Volcan de Fuego in 1974) is detailed by Lazrus et al (1979). In that
report, the fraction of sulfate estimated to be combined with NH_4^+ varies
from a negligible amount to 100%, usually, however, from about 25% to
50%.

PARTICLE SIZE   Toon & Pollack (1976) and Pinnick et al (1976) independently derived time- and space-averaged size distributions that were
consistent with the available impactor and optical measurements of size
distributions as well as with size-integrated measurements of total mass
and number. These distributions are quite similar to each other, having

unimodal, log normal shapes with maxima between a radius of 0.01 and 0.1 μm. Gras & Michael (1979) have obtained new impactor and optical data that suggest that the earlier size distributions contain too few particles between 0.1 and 0.5 μm (Figure 1). Unfortunately, Gras &

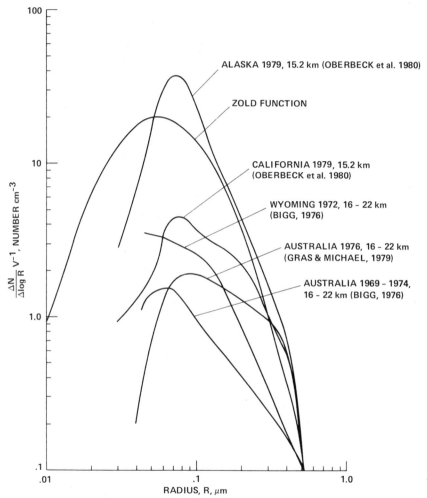

Figure 1 Particle concentration size distributions obtained by impactors flown on aircraft (V. R. Oberbeck et al, personal communication, 1980) or balloons (Bigg 1976, Gras & Michael 1979) compared to an analytic average-size distribution (Zold function; Toon & Pollack 1976). The curves of Oberbeck et al represent single measurements at the specified altitudes. The curves of Bigg et al (1976) and Gras & Michael (1979) are averages of many measurements over the altitude ranges shown.

Michael did not compare their distributions with measurements of other parameters such as mass, number, and size ratio near 0.15 μm. However, their distributions would appear to be inconsistent with these data. Further work to define the "typical" distribution is clearly warranted. Figure 1 presents a collection of various recent individual and averaged observed distributions (Bigg 1976, V. R. Oberbeck et al, personal communication, 1980, Gras & Michael 1979) together with an analytic average size distribution (Zold function) proposed earlier by Toon & Pollack (1976). Measurements near 1 μm are lacking and need to be made because some optical properties such as the infrared opacity are sensitive to these large sizes.

Individual measured size distributions are often quite different than any "typical" size distribution. For example, some are bimodal. The size

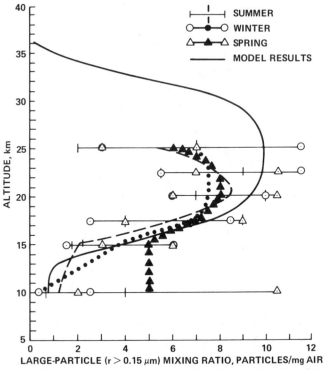

Figure 2 Observed and calculated mixing ratios of particles with radii >0.15 μm. The observations (Hofmann et al 1975) were made using a light-scattering particle counter. A large number of measurements were made over Wyoming; the mean value and a measure of the standard deviation are presented for different seasons. Model calculations from Toon et al (1979), Turco et al (1979).

distribution varies with altitude (compare Figures 2, 3, and 4) with older, larger particles being found higher in the aerosol layer. As particles with varying growth histories are mixed, bimodal distributions will result. The mixed nature of particles can be better studied by examining the area or volume size distribution because mere inflections in number size distributions are revealed as modes in these charting methods. V. R. Oberbeck et al (personal communication, 1980) have examined the maturity of stratospheric aerosols versus altitude at two locations—Alaska and California—in a preliminary study. In Alaska, they found consistent

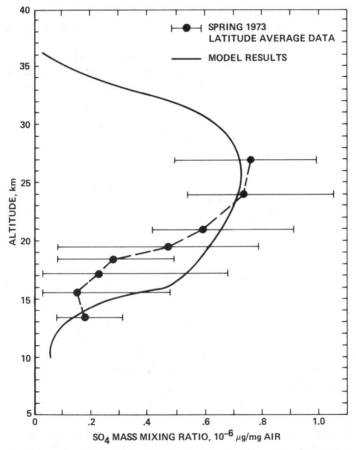

Figure 3 Observed and calculated aerosol sulfate mass mixing ratios (mg SO_4^{2-}/kg air). The observations (Lazrus & Gandrud 1974) made with filters have been averaged over latitude from pole to pole. The error bars give the extreme values found at individual latitudes. Model calculations from Toon et al (1979) and Turco et al (1979).

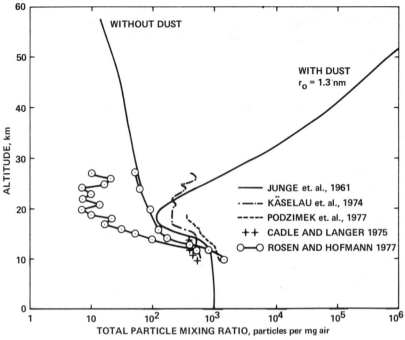

Figure 4 Observed and calculated total particle number mixing ratios (particles/mg air). The data were obtained over a period of 15 years by various investigators using condensation chambers. The large spread in the data may be partly due to real differences in aerosol abundances and partly due to different sampling techniques. Model calculations from Hunten et al (1980).

results over several days showing that a nuclei mode existed near the tropopause, mixed with a more mature one (see Figure 5). Then at higher altitudes they found the distribution modes occurred at progressively larger radii. In California they found, again, the presence of a nuclei mode at the lower altitude, but mainly a more mature one at higher altitudes.

Although data exist on the size distribution of volcanic ash after large eruptions, Toon & Pollack (1976) were unable to find any data on the size of sulfates. Observations following the Fuego eruption by Hofmann & Rosen (1977) using their balloon-borne photoelectric counter clearly showed an increase in the number of particles larger than 0.25 μm relative to those larger than 0.15 μm. Gras & Laby (1979) also reported a slight flattening of the size distribution in the size range above 0.1 μm.

Farlow et al (1979) obtained size distribution measurements in the latter part of the same time period and somewhat afterwards (1976–1977) using their impactor on a U-2 aircraft. The average mode radius they

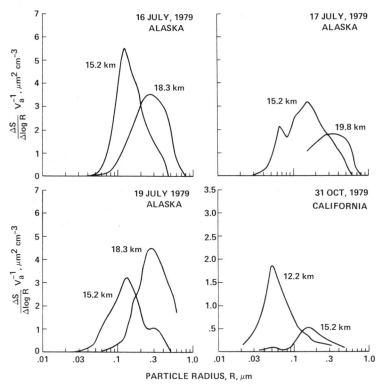

Figure 5 Particle surface area size distributions showing increasing particle maturity at higher altitudes where the distribution modes occur at larger radii. Similar aging occurs at both geographic locations shown. Distributions represent single measurements at the specified altitudes, with 95% confidence that they depict the real populations.

found was much larger than normal, approaching 0.3 μm in the altitude region 12 to 20 km.

Despite the existence of several stratospheric aerosol measuring programs during and after the eruption of Volcan de Fuego, general agreement among size distribution data was not obtained. Part of the difficulty encountered by the experimenters using the impaction method was that collecting surfaces were heavily flooded by the sudden increase in the number and size of the particles (Gras 1976, G. V. Ferry, personal communication, 1976). This prevented the acquisition of useful distributions until adjustments could be made in the sampling techniques and the sudden surge of aerosols had subsided somewhat.

CONCENTRATION   Three types of concentration measurements are commonly reported: the total number of particles; the mass of particles; and the number of particles larger than 0.15 μm radius. Figures 2, 3, and 4 present typical values of these concentrations for nonvolcanic periods as functions of altitude. The concentrations vary with latitude as discussed by Lazrus & Gandrud (1974), Rosen et al (1978), and Rosen et al (1975). Basically, the concentrations are of larger magnitude and reach their peak at higher altitude at the equator than at the poles.

Volcanic eruptions strongly perturb the mass and large particle concentrations. The maximum effect caused by Fuego, occurring about six months after the eruption, was to increase the aerosol concentration nearly an order of magnitude (Hofmann & Rosen 1977): numbers increased to a maximum near 8 cm^{-3} (80 mg^{-1} air). Gras (1976), sampling in the southern hemisphere (34° S), discovered the increase of particles flooded his collectors, yielding concentrations substantially above 2 particles cm^{-3}—the maximum value he could reliably measure under those conditions.

Aerosol concentrations steadily decreased after this initial buildup, reaching levels of around 2–3 particles >0.15 μm cm^{-3} radius two years later (Farlow et al 1979, Hofmann & Rosen 1980). Even so, these values were still about double those just before the eruption. Now, however, Hofmann & Rosen (1980) believe a minimum concentration has again been established, as before Fuego, of 1 particle >0.15 μm cm^{-3} radius. This agrees with recent measurements by V. R. Oberbeck et al (personal communication, 1980) for similar sizes where they found 1 to 2 cm^{-3}. The methods of these latter researchers provide detailed particle size distributions down to radii near 0.03 μm. When these smaller particles are taken into account, then concentrations average about 12 particles >0.03 μm cm^{-3}. Because these particle sizes are well into the Aitken nuclei range, these data suggest there are about 10 or 11 nuclei cm^{-3}, similar to the pre-Fuego nuclei measurements of Rosen et al (1974). Therefore, in summary, particle concentrations during nonvolcanic periods over the past two decades averaged around 1 cm^{-3} for larger particles (>0.15 μm radius), 10 cm^{-3} for Aitken nuclei. During volcanic activity like that at Fuego, particle concentrations are as high as about 8 cm^{-3} while nuclei numbers do not seem to change. Gras & Laby (1979), however, note that concentrations of these small particles may actually decrease following eruptions.

Another measure of stratospheric aerosol concentration is that of sulfate mass. Lazrus et al (1971) and Lazrus & Gandrud (1974, 1977) made extensive measurements using filters flown on high-altitude aircraft. Generally, their findings during non-volcanic periods before and after the Fuego eruption show SO_4^{2-} concentrations in the central part of the aerosol layer to be around 1 ppbm, while during periods strongly influenced by Fuego, the values rose to over 3 ppbm (Lazrus et al 1979). Farlow et al (1979), from their particle size distributions, have estimated the concentration of sulfate their collections might represent if they assumed the particles were 75% H_2SO_4. For the year 1976, their approximation was about 3 ppbm, a factor of three above the measurements of Lazrus & Gandrud (1977). They suggested that if the aerosol were more dilute, then their estimate of SO_4^{2-} would coincide with the filter measurements. More recently, V. R. Oberbeck et al (personal communication, 1980) have estimated 1 to 2 ppbm of SO_4^{2-} based on their size distributions obtained in Alaska in 1979. Estimates like these that are only a factor of two or three different are encouragingly close, considering the radically different techniques that are being used. The suggestion, however, that the aerosol may be more dilute than 75% H_2SO_4 needs further exploration in the 1980s.

A feature that also should be examined in more detail in this decade is the concentration of Aitken nuclei, particularly following major eruptions such as that of Mt. St. Helens. This component of stratospheric aerosols is not well measured and its role in aerosol formation and growth is an important one.

One of the more interesting aspects of the mass and large particle altitude distributions is that they suggest different altitudes for the top of the aerosol layer (Figures 2 and 3). Whether the discrepancy is due to measurement error, to a greatly decreasing particle size with altitude, or to a bias of the mass measurements by sulfurous vapors is not known.

Another measure of aerosol concentration is the optical depth. According to Toon & Pollack (1976) typical visible optical depths during volcanically quiescent periods are about 0.005. Following large volcanic explosions optical depths as large as several tenths are found (Pollack et al 1976). Satellite observations will greatly improve our knowledge of optical depths during the 1980s.

Noctilucent Clouds

Observational studies of noctilucent clouds prior to 1970 have been reviewed by Fogle & Haurwitz (1966) and Bronshten & Grishin (1976).

DISTRIBUTION   Noctilucent clouds have been reported poleward of 45° N and poleward of 50° S. The clouds can only be seen from the ground when the sun is more than 6° below the horizon, so that the lower atmosphere is in darkness, and when the sun is less than 16° below the horizon so that the 80 km altitude level is still in sunlight. These geometric constraints make it impossible to see the clouds from the ground at latitudes poleward of about 70°. However, satellite observations confirm that the clouds extend to the poles where they obtain their greatest optical thickness (Donahue et al 1972). The lower latitude clouds seem to be patchy outbreaks of this dense, uniform polar haze.

The distribution of noctilucent clouds over long time periods is independent of longitude. Individual displays can be limited to an area of a few tens of thousands of km^2. In some cases the clouds have been observed over an area as large as 3.5×10^6 km^2 or the whole of Northern Canada. Displays have been observed simultaneously in Canada, Europe, and the USSR.

The altitude distribution of noctilucent clouds is sharply peaked at 83 km altitude. However, clouds are found over a range of altitudes from 73 to 95 km. The thickness of the clouds is difficult to determine but it appears to be a few kilometers or less. The large range of cloud altitude is partly due to multiple cloud layers and partly due to the wave motions at the mesopause whose amplitude can exceed 5 km.

The morphology of noctilucent cloud displays is rather complex and a classification scheme has been devised which includes veils, whorls, bands, and billows. These structures are clearly related to the dynamics at the mesopause.

Noctilucent clouds are only observed during the summer time. Poleward of 80° N, Donahue et al (1972) observed the clouds on nearly 100% of their satellite observations during one 15-day period in 1969. Clouds were seen over the South Pole in the summer of 1969 and again over the North Pole in the summer of 1970. During July, at 60° N, noctilucent clouds were seen on 75% of the clear nights from a dense set of North American observing sites. At any one location in the Soviet Union noctilucent clouds were not seen more than ten nights per year during the intensive observing period from 1957–1958. However, these sightings all occurred within a few months so the frequency of occurrence during the summer was rather high at some sites. Although the time of year of maximum occurrence seems to shift with latitude this effect is probably due to observational constraints.

The time of year when noctilucent clouds occur and the altitude at which they are seen coincide with the time and location of the lowest temperatures found in the atmosphere. Theon et al (1967) measured mesospheric temperatures during a noctilucent cloud display and found them to be as low as 130° K.

When noctilucent clouds were first reported after the Krakatoa eruption, it was believed that they were products of volcanic explosions. However, by the 1930s it was clear that the clouds were seen when no eruptions occurred, and that large eruptions, such as that of Katmai in 1912, were not always followed by displays. Likewise, correlations between cloud displays and meteor showers have not stood the test of time.

COMPOSITION   Noctilucent clouds are believed to be composed primarily of water ice because water vapor condensation provides a natural explanation for the observed seasonal, latitudinal, and vertical distribution of noctilucent clouds. The low temperatures at the mesopause found by Theon et al (1967), together with calculations of the water vapor at the mesopause (Reid 1975), indicate that water can be supersaturated at the mesopause. Temperatures low enough to supersaturate water occur only at the mesopause, at high latitudes, during summer. Simultaneous measurements of cloud brightness and polarization can be explained using the optical constants of ice, but not using those of meteoritic materials such as iron and nickel (Tozer & Beeson 1974). Two major questions about the composition of noctilucent clouds persist: What is the crystal habit of the ice? What role do ions and meteoric debris play as nuclei for the ice particles?

Ice has three different forms at low pressures and temperatures—hexagonal, cubic, and amorphous (Hobbs 1974). The temperature boundaries between these phases are poorly defined but it is most likely that noctilucent clouds are composed of amorphous or cubic ice. The heats of formation of cubic and amorphous ice are small compared with the heat of formation of hexagonal ice, which suggests that the vapor pressures of the various forms of ice differ very little. However, the shape of the crystals is significant for radiative transfer studies and for calculations of particle sedimentation velocities. The crystal habit of submicron particles of ice was observed by Fernández-Moran (1960) and found to be cubic. Amorphous ice is likewise equidimensional in shape. There is no laboratory support for strongly irregular shapes.

The homogeneous nucleation of ice occurs only when the water vapor pressure greatly exceeds the vapor pressure of ice. However, ice nucleation on particles or on ions occurs at low supersaturations. Although meteoric debris is not a particularly good ice nucleus (Pruppacher & Klett 1978), nucleation on meteoric debris would nevertheless be greatly preferred over homogeneous nucleation. As we shall discuss, collections of meteoric particles of size larger than 100 Å show that the number of such meteoric particles at noctilucent cloud altitudes is very much smaller than the number of noctilucent cloud particles. Hence, such large particles are not the nucleation centers for the clouds. However, modern theories of meteor ablation find that most particles should be on the order of 10 Å in size and the total number of these particles ($\approx 10^3$ cm^{-3}) is greatly in excess of the estimated number of noctilucent cloud particles (Hunten et al 1980). These meteoric particles originate from a relatively steady influx of interplanetary debris, rather than from sporadic meteor showers.

There are also large numbers of ions at the mesopause that could serve as nuclei. Witt (1969) concluded that ions are the preferred nuclei because he attempted to capture meteoric debris imbedded in noctilucent cloud particles and failed to find any debris even though he could detect meteoric particles smaller than 50 Å. Although ions may serve as nuclei it appears premature to rule out a significant role for meteoric particles because their anticipated size is small. Further work in this area is required.

PARTICLE SIZE   The particle size distribution can be determined only indirectly from observations of color, brightness, and linear polarization. Some early studies attempted to deduce the particle size from the silvery blue color typical of the clouds. Gadsden (1975) emphasizes the importance for cloud color of the passage of sunlight through the ozone layer before the light reaches the clouds. Ozone absorbs in the red, which would cause the clouds to appear blue irrespective of their particle size. Hence, color observations involving wavelengths in the ozone bands appear unreliable as indicators of size. Bronshten & Grishin (1976) summarize the measurements of size based on the phase angle dependence of the brightness and polarization prior to 1961. Early estimates indicated radii of about 0.5 μm while later estimates based upon a wider range of angles between the sun and the observer suggested radii close to 0.1 μm. Throughout the 1960s a radius value of 0.1 μm continued to be favored by groundbased observers. Rocket and satellite measurements made since 1970 suggest a typical radius of less than 0.1 μm.

Tozer & Beeson (1974) measured brightness and polarization at 5400 and 4100 Å. From theoretical models incorporating various possible size distributions they found that the particle size increased with the brightness of the display, but even in the brightest display 98% of the particles were smaller than 0.13 μm. Since the particles were so much smaller than the wavelength of the light, a precise determination of particle size was not possible, but a typical radius seemed to be about 0.05 μm.

Witt et al (1976) made measurements of the brightness and polarization at 5360 Å during a weak display. The measurements were consistent with radii smaller than 0.05 μm.

Hummel & Olivero (1976) analyzed the two-color satellite observations by Donahue et al (1972) and deduced that the particle sizes must be less than 0.1 μm. These results are significant because the clouds were optically very thick and Tozer & Beeson's (1974) results would suggest that such bright displays should contain the largest particles.

Gadsden (1978) has criticized the results of Hummel & Olivero (1976) because the color ratio they analyzed had a large noise level due to non-uniformity in the scanning mode of the satellite. Although Tozer & Beeson (1974) and Witt et al (1976) observed weak displays that required only small particle number densities, the bright cloud observations analyzed by Hummel & Olivero (1976) imply high particle concentrations. Gadsden (1978) points out that the optical cross section decreases rapidly with particle size. Therefore the total mass of water in a cloud composed of small particles is very much higher than the mass of water in an equally bright cloud composed of large particles. If there were 20 particles cm^{-3} of radius 0.13 μm over a 3 km layer (Reid 1975), the cloud observed by Donahue et al (1972) would contain the entire column content of water above 76 km altitude at a water content of 3 ppm. If the particle radius were reduced to 0.09 μm the required mass of water would be doubled. These results suggest that very bright clouds composed of small particles may be difficult to obtain. Clearly, further observational studies of the particle sizes in bright clouds and their variations with brightness are desirable.

CONCENTRATION   Estimates of the concentration of particles depend upon knowledge of the particle size. Only Tozer & Beeson (1974) and Witt et al (1976) found both size and number in a consistent manner. Witt et al obtained 0.3 particles cm^{-3} while Tozer & Beeson (1974), who observed a brighter display, suggested there were 2 to 30 cm^{-3}.

Studies of particle concentrations prior to 1960, based upon photographic exposure times and simple calculations, do not appear reliable. Donahue et al (1972) assumed monodispersed 0.13 μm particles and suggested that their satellite observations were consistent with 15 to 40 particles cm^{-3} depending upon whether the clouds were 5 or 1 km thick. According to Hummel (1977), Donahue et al used an incorrect scattering cross section so that the corrected concentrations should be 75 to 255 cm^{-3}. Fogle & Rees (1972) observed a weak cloud display and by assuming monodispersed 0.13 μm particles obtained a concentration of about 1 cm^{-3}.

These results suggest a highly variable concentration. Tozer & Beeson (1974) find that brighter clouds may have larger particles, but not necessarily more of them. However, the interplay between increased size and increased number as the brightness increases is not clearly determined by current measurements.

Another measure of the concentration of the particles is their optical depth. The vertical optical depth of noctilucent clouds is so low that they do not measurably extinguish star light so the optical depth cannot be found directly. All measurements of optical depth have been based upon the brightness of scattered light at some angle of observation. If one knows the angular scattering function then the optical depth can be found. Unfortunately, the angular scattering function is not known. The various optical depths quoted in the literature are poorly defined because they usually refer to the product of the optical depth and the phase function at some angle. Fiocco & Grams (1969) determined an optical depth for backscattered laser light of about 10^{-4}. Donahue et al (1972) suggested an optical depth of 4×10^{-5} to 7×10^{-5} for the polar clouds seen from a satellite. However, Hummel & Olivero (1976) state that Donahue et al (1972) did not remove the phase function and that the true optical depth is about 2.7×10^{-3}. Hummel & Olivero deduced this optical depth by assuming the particles were of size 0.13 μm, an assumption that they later stated was incorrect. The optical depth is obviously strongly variable and the available values in the literature give rough estimates. However, reliable estimates of true optical depths will require further measurements and more sophisticated analyses.

Nacreous Clouds

The properties of nacreous clouds have been reviewed by Hesstvedt (1969), Stanford (1977), and Stanford & Davis (1974).

DISTRIBUTION   Nacreous clouds have been seen at high latitudes in both hemispheres, but they have been most frequently reported over Scandinavia. Their altitudes range from 17 to 31 km, with a mean of 23 km. Mother-of-pearl clouds are typically shaped like lenticular clouds and may form in stratospheric air lifted adiabatically over the crests of topographically induced tropospheric lee waves. The brilliantly colored clouds remain stationary because the lee waves are stationary.

Nacreous clouds are only observed during the winter months and are quite rare, being reported on average only a few times per year during the previous century (Stanford & Davis 1974). The rarity of observations may be partly caused by the difficulty of observing clouds during the winter months when extended twilight conditions without tropospheric clouds do not often occur. The rarity of sightings also may be due to the infrequent occurrence of stratospheric temperatures low enough to supersaturate water. Stanford (1974a) made an extensive search of available stratospheric temperature measurements during the winter at latitudes where twilight conditions would allow clouds to be observed. In 10^5 radiosonde reports over 37 winter months he found only one case in the Northern Hemisphere with a temperature low enough ($-88°$C) to saturate 3 ppm of water vapor at 23 km altitude. Stanford (1973, 1977) pointed out that temperatures in the Antarctic stratosphere are commonly as low as $-90°C$ so that much more persistent clouds might be expected there. During the winters of 1950 and 1951 persistent widespread stratospheric clouds were observed in Antarctica. These clouds were optically thin and did not noticeably extinguish star light, but could only be seen with the proper twilight conditions.

Nacreous clouds in the form of lee waves may remain stationary for several hours. Although they are typically not reported on consecutive days, there have been periods when clouds were seen several days in a row. Sightings of the clouds are not correlated with volcanic eruptions (Stanford 1974b). During the period from about 1900 to 1930 no clouds were reported despite attempts by experienced observers to see them.

COMPOSITION   When nacreous clouds were first observed their composition was a mystery, but temperature soundings of the stratosphere since the mid-1930s and more recent water vapor measurements have made it clear that the clouds are mainly composed of water ice. The major compositional question is whether or not stratospheric aerosols are a significant contaminant within the nacreous cloud particles. It is thought that nacreous cloud particles form on stratosphere aerosols and therefore small particles would have a significant sulfate concentration. However, the estimated size of visible nacreous cloud particles is so much larger than that of stratospheric aerosols that the mass of the stratospheric aerosols may be negligible.

Hallett & Lewis (1967) suggested that sulfuric acid particles play an important role in the shape of nacreous clouds. Nacreous clouds have colorless trails behind them much like those behind lenticular clouds. Lenticular clouds have these trails because of supercooling of ice in the body of the clouds and freezing in the trails. Hallett & Lewis proposed that small nacreous cloud particles at the forming edge of the cloud are supercooled sulfuric acid-water droplets. Strong sulfuric acid solutions are known to supercool very easily. As the particles grow they become more dilute until they finally freeze in the main body of the cloud. Once formed the frozen ice crystals could persist at lower vapor pressures than liquid drops to form the trailing wakes of the clouds.

PARTICLE SIZE AND CONCENTRATION   The most distinctive optical property of nacreous clouds is their brilliant coloration which takes the same shape as the cloud. Hesstvedt (1962, 1969) has pointed out that although the coloration is partly due to the angular dependence of diffraction by nearly monodispersed particles, the angular size of many clouds is too small for the different colors to be due purely to the change in viewing angle. Rather he proposes that the particle size, which is uniform at any one location in the cloud, varies strongly across the cloud with the smallest size being found toward the outer edges of the cloud. Such a distribution also explains why the colors take the same shape as does the cloud.

Hallett & Lewis (1967) briefly commented upon the particle sizes in nacreous clouds and suggested that both the angular dependence of cloud color and the high polarization of the light were consistent with particle sizes near 1 μm. However, detailed optical analyses of particle size are lacking.

Although there is no direct evidence, it is generally believed that nacreous clouds form by deposition of H_2O on preexisting stratospheric aerosol particles. Hence, the number concentration of nacreous cloud particles should be equal to or less than that of the stratospheric aerosols, about 5–20 cm^{-3} at 20 km. The radius of the particles depends upon the cube root of the mass of water condensed and upon the cube root of the inverse of the number of cloud particles. An approximate maximum radius of about 4 μm may be found by assuming that 3 ppm of water condenses to form 1 particle cm^{-3} at 20 km.

Meteoric Dust

Each day about 40 metric tons of interplanetary meteoric debris enter the atmosphere (Hughes 1978). A similar mass may enter from sporadic meteor showers during short time intervals. The mass median weight of the particles is about 10 μg, which corresponds to a radius of about 100 μm. A large fraction of the incoming mass is vaporized upon entering the atmosphere (Hunten et al 1980). Few measurements have been made of meteoric particles within the atmosphere, but three types of particles are expected. Small bodies—micrometeorites—are slowed to terminal velocity in the upper atmosphere without reaching high enough temperature to ablate. The size of the largest micrometeorite depends upon the cube of the incoming particle velocity and is about 100 μm at 12 km s^{-1}, 25 μm at 20 km s^{-1}, and 3 μm at 40 km s^{-1}. Most particles enter at speeds near 14 km s^{-1} (Hunten et al 1980). Particles of slightly larger size than the micrometeorites dominate the incoming mass and these ablate completely upon entering the atmosphere. Their smoke trails may then recondense into very small particles. Some of these larger particles fragment on entering the atmosphere, but most of the fragments are thought to be larger than micrometeorites and therefore probably ablate independently (Hughes 1978, Hawkes & Jones 1975). Very large meteors—the largest of which compose the classical meteorites—do not ablate completely but have enough residual mass to reach the ground. Because of their large size and high fall velocities there are so few of these particles in the atmosphere that they are of little interest.

STRATOSPHERIC COLLECTIONS   Only micrometeorites are large enough to be easily captured in the stratosphere. Esat et al (1979) and Ganapathy & Brownlee (1979) have identified meteoric particles in collections made in the stratosphere for them by NASA researchers using U-2 aircraft. Brownlee et al (1976) found that 90% of the stratospheric particles with radii between 1.5 and 4 μm were aluminum oxide spheres probably produced from the exhaust of solid fuel rocket engines. Many of the remaining particles and most of these with radii between 4 μm and 19 μm were judged to be meteoric because their elemental composition closely resembled carbonaceous chondrite meteorites and because the particles contained large quantities of solar-wind-implanted helium (Brownlee 1978). Only about 150 particles were captured in a sampling volume of 1.4×10^5 m^3.

Attempts to collect smaller, more numerous micrometeorites have been controversial because of the difficulty of preventing terrestrial contamination, and because smaller particles cannot be individually analyzed to determine their composition. The stratospheric mass budget during volcanically quiescent periods involves about 3×10^5 metric tons per year of terrestrial material (Turco et al 1980b). The meteor influx is only about 1.6×10^4 tons per year. Hence the meteoric debris should only constitute a small mass fraction of the aerosols below 30 km. Shedlovsky & Paisley (1966) analyzed the Fe/Na ratio of aerosols collected at 20 km altitude and found the ratio to be similar to that found in crustal rock. The measurement uncertainty placed a 10% upper limit on the chondritic meteoric contribution to the iron abundance and an upper limit of 9×10^4 tons on the annual influx of submicron meteoric material.

Bigg et al (1971) reported the collection of particles in the 30 to 40 km altitude region in which the stratospheric aerosols should be depleted. They found solid particles, which originally had volatile coatings, whose morphology seemed to suggest a meteoric origin. However, Farlow et al (1973), who determined the elemental composition of these particles, found similar particles throughout the lower stratosphere, suggesting that they were of terrestrial origin.

Particles formed from the ablation debris of meteors, which probably constitute most of the meteor mass, are too small to capture directly in the stratosphere. However, model studies suggest that such particles may have very high number concentrations. The particles might be detected using condensation nuclei counters to search for a large increase in number with increasing altitude. Unfortunately the counters do not count all particles, but only those larger than a certain size, which is normally on the order of 50 to 100 Å. Hence the ability of the counters to detect small meteoric ablation debris is not well established. Hunten et al (1980) suggest that the large differences in measured values among the various counters is due to their varying ability to detect the debris. Measurements by Käselau et al (1974) strongly suggest a large number of meteoric particles in the upper stratosphere. More comprehensive measurements of condensation nuclei by Rosen et al (1978) are consistent with a meteoric source, but do not strongly reveal it perhaps because the measurements do not extend to high enough altitude. Clearly further measurements are required.

HIGH ALTITUDE COLLECTIONS   Above the stratospheric aerosol layer contamination problems are lessened but sampling must be done from short-duration rocket flights, or remotely. Only particles in the micrometeorite range have been studied.

Lidar returns from upper atmospheric dust layers have been discussed by Poultney (1972). There is little evidence for persistent dust layers, but several groups have recorded transient dust layers associated with cometary influxes. Kent et al (1971) observed enhanced lidar backscattering at 60 to 75 km and then three days later enhanced scattering near 40 km following the passage of dust from comet Bennett. Clemesha & Nakamura (1972) and Clemesha & Simonich (1978) observed an aerosol layer near 47 km which persisted for several days and then found enhanced lidar backscattering from 40 to 70 km for several months afterward. The 47 km layer fell at a rate consistent with 1 μm particles and the backscattering was consistent with 10^4 particles m^{-3} with radii greater than 0.25 μm.

Fiocco & Grams (1969) reported persistent high latitude dust layers during the summer at altitudes of 60 to 70 km. These observations were barely above the noise level and were made with one of the first operating lidars. Since the dust layers would be of significance for noctilucent clouds, repeated measurements would be valuable.

Numerous attempts have been made to collect particles at high altitude from rockets. Unfortunately, the particles collected are too small for compositional analyses and the possibility of contamination is difficult to eliminate. Two groups of experimental results have been reported. In the first, the number of micrometeorites found was small and was consistent with the most probable flux of incoming micrometeorites deduced from spacecraft measurements (Farlow & Ferry 1972, McDonnell 1978). The other group of measurements found very much larger numbers of particles (Lindblad et al 1973).

Table 1   Number of meteoric particles per m^3

Altitude (km)	Radius range (μm)			
	.01–.02	.02–.05	.05–.1	.1–.2
60	3×10^6	10^5	80	.2
80	2×10^4	77	.08	.01
90	10	.08	.01	.002

Farlow & Ferry (1972) reviewed measurements from a large number of rocket flights and found that the number of particles larger than 0.02 μm radius captured in the mesosphere is less than a few hundred per m^3, which is the measurement limit set by very high quality contamination control. They were unable to detect any enhancement of particles following meteor showers or the passage of Earth through cometary dust trails. Likewise, dust was not found to be enhanced at any latitude.

Lindblad et al (1973) discussed measurements from a collection of rocket flights employing a variety of sensors, including acoustic detectors, plasma analysis, and direct particle sampling experiments. The results from these sensors differed from each other by about 3 orders of magnitude and the lowest values were 2 or 3 orders of magnitude larger than the satellite flux values or the values found by Farlow & Ferry (1972). Possibly some of the discrepancy is due to altitude and particle size collection differences. Model results suggest the micrometeorite distribution in the atmosphere is a very strong function of size and altitude (Table 1). However, much of the difference remains unresolved and may be due to contamination problems.

NOCTILUCENT CLOUD COLLECTIONS   Several attempts have been made to capture meteoric particles during noctilucent cloud displays. Farlow et al (1970) captured 6000 particles m^{-3} larger than 0.02 μm radius over a wide altitude range. These may have been collected below the cloud. Hallgren et al (1973) captured particles over a narrower altitude range during a cloud display and found 1 to 5×10^3 particles m^{-3} whose average diameter was 0.1 μm or larger. These number densities are 10^{-2} to 10^{-3} of those inferred for cloud particles from optical studies of noctilucent cloud displays, which suggests that meteoric debris does not play a significant direct role in forming clouds. Witt (1969) also found meteoric debris was able to account for only 10^{-2} to 10^{-3} of the number of cloud particles, but he sampled meteoric debris as small as 50 Å.

Table 2 Rates of physical processes

Cloud type	Nucleation	Coagulation	Growth or evaporation	Sedimentation	Dynamical transport
Stratospheric aerosols	Occurs on tropospheric particles, meteoric dust, ions and radicals.	Near tropopause $N \simeq 100$ cm^{-3}, $A_0 \simeq 0.05$ μm, $\tau_c \simeq 2$ months. Controls particle numbers.	If $n_g = 2n_v$, $n_v = 10^3$ cm^{-3}, $A_0 = 0.1$ μm, $\tau_g \simeq 2 \times 10^{10}$ s. Hence $S \gg 2$. Growth rate controls mass and is dominated by ability of chemical processes to keep $S \gg 2$. Evaporation occurs above 30 km and establishes layer top.	$l = 5$ km, $A_0 = 0.1$ μm, $\eta_g = 4 \times 10^{18}$ cm^{-3}, $H = 6$ km, $Z_0 = 14$ km, $\tau_f \simeq 7$ years. Not important for $a < 0.1$ μm	Controls lifetime of small particles. τ_T is month to few years.
Noctilucent clouds	Occurs on meteoric dust and on ions. Controls particle numbers.	$N \simeq 1 - 100$ cm^{-3}. $A_0 \simeq 0.1$ μm, $\tau_c \gtrsim 1$ month. Not important process.	Observed H_2O gives $n_g = 10^9$ cm^{-3}, $A_0 = 0.1$ μm $\tau_g \simeq 5$ hours. Growth rate controls mass and $S \gg 1$. Evaporation determines layer thickness and location.	$l = 1$ km, $A_0 = 0.1$ μm $N_g = 4 \times 10^{14}$ cm^{-3}, $H = 6$ km, $Z_0 = 80$ km, $\tau_f \simeq 2$ hours. Controls lifetime of cloud particles.	Temperature anomalies with transport control lifetime of clouds. Turbulence may be important to cloud particle lifetimes.
Nacreous clouds	Occurs on stratospheric particles. Controls particle numbers.	$N \simeq 10$ cm^{-3}, $A_0 \simeq 1$ μm, $\tau_c \simeq 4$ months. Not important process.	Observed H_2O gives $n_g \simeq 4 \times 10^{12}$ cm^{-3}, n_v is small, $A_0 = 1$ μm, then $\tau_g \simeq 50$ s. Hence $S \simeq 1$. Growth rate controls mass. Evaporation determines layer thickness and location.	$l = 1$ km, $A_0 = 1$ μm, $n_g = 10^{18}$ cm^{-3}, $H = 6$ km, $Z_0 = 24$ km, $\tau_f \simeq 1$ month. Not important except for very long-lived clouds.	Temperature anomalies associated with transport controls lifetime of clouds. Particles removed from clouds by winds.
Meteoric debris	Occurs homogeneously.	$N \simeq 10^3$ to 10^4 cm^{-3}, $A_0 \simeq 10^{-3}$ μm, $\tau_c \simeq$ few days to 1 month. Controls particle numbers and particle size.	Process does not occur due to low vapor pressure.	$l/H \gg 1$, $H = 6$ km, $Z_0 = 30$ km, $n_g = 4 \times 10^{17}$, $A_0 = 10^{-2}$ μm, $\tau_f \simeq 15$ years; $n_g = 6 \times 10^{15}$, $Z_0 = 60$ km, $\tau_f = 3$ months. Not important except for large particles at high altitudes.	Controls lifetime of debris for most sizes.

MODELS

Numerical models of the formation, growth, and evolution of stratospheric aerosols, noctilucent clouds, nacreous clouds, and meteoric debris have been constructed. The numerical models all contain similar physical processes, which we shall describe below. Then we will present the results of available model studies and compare them with observations.

Physical Processes

Twomey (1977) and Pruppacher & Klett (1978) provide good reviews of the physical process affecting aerosols and clouds. Table 2 summarizes the rates of the various processes for the four types of particles above the tropopause.

NUCLEATION   Nucleation is the basic process that first creates particles and establishes their numbers. Particles form by collisions between gas molecules and molecular clusters. For involatile substances such as meteoric ablation debris, it is believed that every collision between two molecules or clusters results in a stable particle (Hunten et al 1980). However, clusters of volatile substances are metastable. The energy gained by adding a molecule to the cluster can be too small to supply the energy required to support the surface area of the enlarged cluster unless the cluster is larger than a "critical" size. Predicting the abundance of critical size clusters is very difficult because it depends exponentially on poorly known quantities such as surface free energy. If gas molecules collide with other molecules on a preexisting aerosol surface, then the cluster size can be as large as the preexisting aerosol and it is possible that no barrier to nucleation exists. For this reason, volatile aerosols often form on preexisting particles and the number concentration of the new particles is equal to or less than that of the preexisting ones. Stratospheric aerosols may nucleate on particles that originated in the troposphere or on meteoric debris (Toon et al 1979, Turco et al 1980a, Junge et al 1961). Nacreous clouds probably form on stratospheric aerosols and noctilucent clouds may form on meteoric debris (Hesstvedt 1969).

Many theorists believe that nucleation on ions or radicals may be important. This type of nucleation is favored because of the enhanced binding energy of radicals and ion clusters compared with that of pure materials. Witt (1969) favors ion nucleation of noctilucent clouds, because of the lack of evidence of meteoric particles in the clouds. Friend et al (1980) propose that radical precursors of H_2SO_4 nucleate whenever they collide and form stratospheric aerosols. Experimentally, it is difficult to distinguish the various possible types of nucleation and this area is a good one for future work.

COAGULATION   Brownian motion causes particles to collide with each other. Collisions usually result in the sticking together of particles, which reduces the particle number and creates larger particles. When coagulation is important, the observed number of particles is usually determined by a balance between the mean residence time of the particles and the rate of loss by coagulation. The number production rate by nucleation is not very significant if the coagulation rate is high because coagulation proceeds much more rapidly when larger numbers of particles are present so that the final number of particles is nearly independent of the initial number.

The coagulation rate for monodispersed particles whose size is smaller than that of the mean free path of the gas, as is generally the case for particles above the tropopause, yields an "e" folding time of (Hamill et al 1977)

$$\frac{1}{\tau_C} = \frac{1}{N}\frac{dN}{dt} = 4\left(\frac{3k_B T}{\rho}\right)^{1/2} N A_0^{1/2} \simeq 10^{-6}\, N A_0^{1/2} \text{ cm}^{5/2} \text{ s}^{-1}$$

where k_B is Boltzman's constant, T is temperature, ρ is the particle density, N is the number density of particles, and A_0 is a typical particle radius. Table 2 presents typical values of τ_C which show that the coagulation time is much longer than the lifetimes of nacreous or noctilucent clouds. Coagulation proceeds very rapidly for small stratospheric aerosols, and their declining mixing ratio with altitude (Figure 4) is due to the increasing residence time of the particles with altitude so that coagulation removes more and more particles. Coagulation is not very important for large stratospheric aerosols which do not grow by coagulation (Toon et al 1979). Coagulation of meteoric dust particles is the only mechanism by which they can grow (Hunten et al 1980).

CONDENSATIONAL GROWTH   Condensation and evaporation occur when molecules enter or leave a volatile aerosol. Both processes continually occur and whether the particle as a whole expands or shrinks depends upon whether the partial pressure of the gas phase molecules exceeds the vapor pressure (condensation) or is less than the vapor pressure (evaporation). This balance is very sensitive to temperature because of the strong dependence of vapor pressure on temperature. Stratospheric sulfuric acid aerosols are not found above about 30 km because the high temperatures at those altitudes cause them to evaporate. Similarly, noctilucent clouds are restricted to the narrow zone of low temperatures near the mesopause.

The growth rate for particles smaller than the mean free path of air is (Hamill et al 1977)

$$\frac{1}{\tau_g} = \frac{1}{A_0}\frac{da}{dt} = \frac{m_p}{4\rho}\left(\frac{8k_BT}{\pi m_p}\right)^{1/2}\frac{n_g - n_v}{A_0} \simeq 5 \times 10^{-19} \text{ cm}^4 \text{ s}^{-1}\frac{n_g - n_v}{A_0}$$

where a is the radius of the aerosol, m_p is the mass of a molecule of the aerosol material, n_v is the number concentration of molecules at the vapor pressure, and n_g is the number concentration of aerosol molecules in the gas phase. Often growth rates are expressed in terms of the supersaturation $S = n_g/n_v$. For sulfuric acid another factor is needed in the growth rate to account for the fact that both acid and water molecules are added to growing aerosols (Hamill et al 1977). The numerical values for the leading term in τ_g for sulfuric acid and ice, however, are identical to within a factor of two. Table 2 shows that the growth rates for various types of particles differ by large factors. As we will discuss, S is much greater than unity for sulfuric acid particles and for noctilucent clouds. This is an important conclusion because if $S \simeq 1$, minor impurities, which slightly affect the vapor pressure, determine the value of $S - 1$ and control the growth rate. Calculating the properties of clouds with $S \simeq 1$ is quite difficult and accounts for many of the problems in tropospheric cloud physics.

For sulfuric acid, the vapor pressure can be easily calculated, but the ambient number of acid molecules cannot be easily found. If the number of gas molecules were close to the vapor pressure, then growth would not have time to occur. Chemical models (Turco et al 1979) show that photochemical processes are capable of maintaining $S \simeq 10^2$ so that growth can occur near 20 km. At 35 km the temperature has increased and the vapor pressure is about 10^8 cm^{-3}. Chemical processes cannot supply such large number densities, hence the particles evaporate. A 0.1 μm particle placed at 35 km could evaporate in only 2×10^2 seconds. This process creates the top of the stratospheric aerosol layer.

Noctilucent cloud particles would require many hours to grow to observed sizes if the ambient water vapor were close to the vapor pressure at 130 K. However, the amount of water present in the stratosphere (3 ppm) and presumed to be present in the mesosphere is sufficient to allow the particles to grow. At common noctilucent cloud temperatures, S must be much greater than unity. Evaporation controls the layer thickness because slightly higher temperatures would require more water than is present in order to achieve $S > 1$.

Nacreous clouds potentially grow very rapidly. Their sizes must be limited by short lifetimes due to transport and by the growth rate's being reduced due to S being close to unity. The rapid growth rate and a value of S close to one may account for the narrow particle size distribution and hence the brilliant colors of these clouds.

SEDIMENTATION   Gravitational sedimentation is not as important a process above the tropopause as might be suspected. The time to fall a typical distance l, through an atmosphere whose density varies exponentially with height, to reach altitude Z_0 for a particle smaller than the mean free path of air molecules is

$$\tau_f = H/V(Z_0)(1 - \exp[-l/H])$$

$$\simeq 2 \times 10^{-21} \text{ cm}^3 \text{ s} \frac{H}{A_0} n_g(Z_0)(1 - \exp[-l/H]).$$

Here H is the scale height of the atmosphere, $V(Z_0)$ is the fall velocity at Z_0, and $n_g(Z_0)$ is the gas density at Z_0. The expression for $V(Z_0)$ is given by Hamill et al (1977). Table 2 shows that fall times are generally quite long.

Stratospheric aerosols smaller than 0.1 μm do not fall out of the stratosphere. Particles larger than 0.1 μm have fall times between 30 km and 25 km that are comparable to the observed residence time at those altitudes so such large particles are affected by sedimentation near the top of the layer.

Nacreous clouds are generally short-lived and not affected by sedimentation. Stanford (1973) has argued that persistent Antarctic clouds, forming at moderately low altitudes with large particle sizes, might have a significant sedimentation transport across the tropopause.

Noctilucent clouds have a significant sedimentation sink, and sedimentation is one of the main reasons that noctilucent cloud particles do not reach sizes much above 0.1 μm.

Meteoric ablation debris with size less than 0.01 μm is not affected by sedimentation except above about 60 km. Even 1 μm–sized micrometeorites will require one month to fall to the 30 km level. Although sedimentation is quite important for micrometeorites, most ablation debris is not strongly affected by sedimentation except at the highest altitudes.

DYNAMICAL TRANSPORT   Atmospheric motions are of great importance to the distribution of the aerosols above the tropopause. Unfortunately, it is quite difficult to calculate the wind velocity fields and little work has been done on the relation between aerosols and meteorology.

The transport of stratospheric aerosols can be monitored by studying the decay of small volcanic perturbations (Hofmann & Rosen 1980) and the spread of radioactive debris that attaches itself to the aerosols

(Reiter 1975). Both types of studies show that particles just above the tropopause are removed more rapidly than those at higher altitudes. Particles at 20 km are removed from the stratosphere in one to two years. These removal times are consistent with those of gases, which suggests that sedimentation plays only a small role in removing stratospheric aerosols.

Individual nacreous cloud particles in lens-shaped clouds are rapidly removed from the region of their formation by horizontal winds. For example, Hesstvedt's (1960) models of nacreous clouds suggest that the particles are swept through the cloud in about 10^3 seconds. The nacreous clouds themselves do not move with the wind. The length of their existence is controlled by the duration of meteorological conditions that maintain large amplitude temperature oscillations in lee waves. Observations suggest that these conditions typically exist for several hours. The observed persistent Antarctic clouds may have their lifetime controlled by long time-scale dynamic and radiative processes that maintain the low Antarctic winter temperatures. The simplest possibility is that the seasonal temperature cycle determines the time between cloud formation and evaporation, but further observations of these clouds are needed to confirm the absence of transient warming events.

Noctilucent clouds have complex morphologies, and are transient, which suggests that dynamical processes are quite important in determining the lifetimes of the clouds. The typical turbulent mixing rates assumed to occur at the mesopause on the basis of studies of gaseous materials imply particle lifetimes of only a few hours, which are comparable to the lifetimes due to sedimentation. Hence, both atmospheric motions and sedimentation may be significant processes.

Meteoric debris, with submicron size, has not been observed in the atmosphere, so no direct information is available concerning the importance of dynamics to its distribution. However, sedimentation velocities of the debris are so small that atmospheric motions must control its distribution.

Stratospheric Aerosol Models

Table 2 shows that stratospheric aerosols are quite complicated. All the physical processes listed control some aspect of their distribution and the aerosols are created by involved chemical reactions. We shall not review the chemistry creating the aerosols. Volcanic injections continually provide fresh supplies of SO_2 that is eventually converted into aerosols. In addition, carbonyl sulfide, which may have a significant anthropogenic source, supplies enough sulfate to maintain the layer during the substantial periods when volcanic eruptions do not occur (Turco et al 1980b).

Models of stratospheric aerosols were recently reviewed by Whitten et al (1980). The complexity of these models is illustrated by Figure 6, which shows the processes included by Turco et al (1980a). Toon et al (1979) illustrate the sensitivity of the various observed properties of the

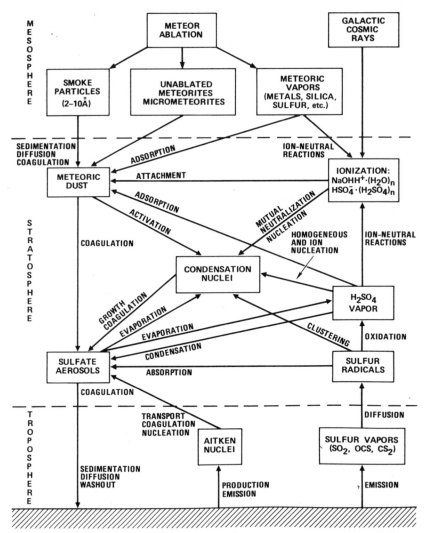

Figure 6 A schematic outline of the physical and chemical processes included in a model of stratospheric aerosols (Turco et al 1980a). Meteors, ions, radicals, and tropospheric aerosols all serve as nucleation sources for the aerosols, which grow, evaporate, coagulate, fall, and diffuse vertically.

aerosols to changes in atmospheric parameters and to the physical processes controlling the model particles. At present three fundamental questions remain to be resolved: What is the relative importance of the various possible nucleation processes? What is the contribution of materials other than sulfuric acid to the aerosol layer? What physical and chemical processes control the layer at its upper boundary near 30 km?

The stratosphere is a particle sink for the total number of particles, as illustrated by the observed declining mixing ratio with altitude in Figure 4. Figure 4 demonstrates that the observations are consistent with models in which tropospheric particles cross the tropopause in large numbers and then are reduced in number by coagulation as they are slowly transported to higher altitudes by winds. Above 25 km meteoric dust may provide an additional supply of particles. The large spread in the data precludes the possibility of confirming the meteoric source. Rosen et al (1978) show that measurements by a single instrument produce similar values at a wide range of latitudes, and when flown together separate instruments also agree. Hence, the source of the divergent measurements is not clear. Turco et al (1980a) show that the number of particles can be a strong function of the smallest particle size measured when meteoric debris is present, which may account for part of the range of observations.

Although the stratospheric particles may form upon particles uplifted across the tropopause, it is also possible that nucleation occurs in the lower stratosphere. At higher altitudes these small particles would coagulate, which might produce a mixing ratio profile similar to that observed. The question of whether or not nucleation occurs may be answered by future measurements of the size distribution because small particles would only be expected in a source region.

The number of particles in the stratosphere is controlled by particles smaller than 0.1 μm, but most measurements such as those of mass and size distribution are based upon particles of size greater than 0.1 μm. These particles form a distinct layer in the stratosphere (Figures 2 and 3). The layer structure is caused by chemical production of aerosol mass in the stratosphere, loss of particles by evaporation above 30 km, and loss of particles across the tropopause near 15 km. The supply of sulfur gases and the particle residence time are the principal factors controlling the mass, the number of particles larger than 0.1 μm, and the particle size distribution above 0.1 μm (Toon et al 1979).

Models have been quite successful in reproducing the aerosol features for sizes larger than 0.1 μm (see Figures 2 and 3). The principal difficulty is in understanding the precise reason for the location of the top of the aerosol layer. The observational evidence is confusing because the mixing ratio of large particles declines near 25 km (Figure 2) while the mass mixing ratio declines well above 30 km (Figure 3). Observations of acid vapor above 30 km suggest that the mass mixing ratio measurements are not biased by picking up acid vapor (Turco et al 1980a). The resolution of these problems may involve meteoric dust. In current models small particles begin to evaporate near 30 km due to the Kelvin effect supplying vapor to large particles which in turn evaporate at much higher altitudes. If meteoric debris is dissolved in the acid particles, it may alter the vapor pressure and reduce the Kelvin effect. Smaller particles would then contain most of the sulfate mass, the large particles would be depleted, and little acid would be added to the vapor phase. Although this specific scenario has not been modeled, Turco et al (1980a) presented a model in which all evaporation was suppressed and found improved agreement with the acid vapor concentration above 30 km.

Noctilucent Cloud Models

Table 2 shows that coagulation is not significant in noctilucent clouds and therefore the number density is controlled by nucleation. The cloud particle size is determined by the growth that occurs before sedimentation and winds remove the particles to the high temperature evaporation regions surrounding the clouds. Several numerical noctilucent cloud models have been constructed. The outstanding problems are to determine whether nucleation occurs on ions or on meteoric debris, and to determine the conditions that allow particles to grow to the observed sizes.

The most recent and complete model is that of Reid (1975), who also reviewed previous models. Reid calculated the water vapor altitude profile and found that ample water was available at low altitudes to be transported to higher altitudes against photochemical water vapor losses and create supersaturated conditions at mesopause temperatures of 130 K. Reid did not directly calculate the nucleation rate, but simply assumed that a fixed number of very small particles was generated at the cloud top altitude and grew as they fell to lower altitudes. Reid's model was unable to produce spherical ice particles larger than about 0.08 μm radius. In order to obtain larger particles he hypothesized that the particles were strongly irregular in shape. Irregular particles fall more slowly than spherical ones and therefore grow to larger sizes.

As we discussed earlier, laboratory studies imply that noctilucent cloud particles may be nonspherical but they are probably equidimensional. There is also no reason to believe that irregular particles would become oriented. Large, irregular ice crystals in the lower atmosphere do become oriented by atmospheric drag (Pruppacher & Klett 1978) but the same physical processes do not apply to noctilucent cloud particles. Hence there is little reason to think that noctilucent cloud particles fall appreciably more slowly than spheres.

R. P. Turco et al (private communication, 1980) have improved upon Reid's (1975) work in three ways. They utilize Theon et al's (1967) observed temperature profile which produces a wider cold region than the one used by Reid. Turco et al include dynamic mixing of the particles. Random mixing processes occurring at the rate Reid (1975) assumed to apply to water vapor will also be important in transporting particles. Some particles about to fall out of the cloud bottom will be stirred back to the cloud top and therefore will grow larger. Finally, Turco et al calculate the size distribution of the cloud particles, which allows particles of various sizes to be treated individually and properly mixed by motions.

Figures 7 and 8 present results of Turco et al's model for cubic ice particles that nucleate on meteoric debris. A large number of very small

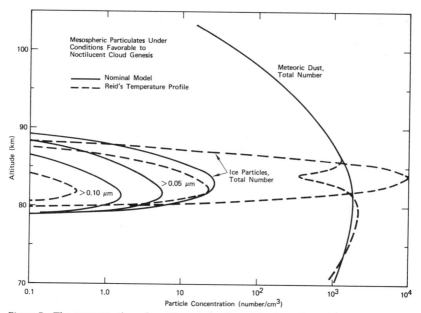

Figure 7 The concentration of noctilucent cloud particles in various size ranges and the concentration of meteoric dust particles as functions of altitude (R. P. Turco et al, private communication, 1980).

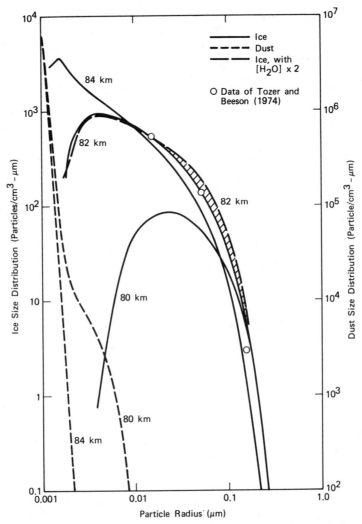

Figure 8   The size distribution of ice and dust particles at several different altitudes within the noctilucent clouds. Also shown at 82 km is the effect of increasing the water vapor abundance at lower altitude from 3 to 6 ppm (R. P. Turco et al, private communication, 1980).

meteoric particles are present, which have a steep size distribution (Hunten et al 1980). The high supersaturation of water, due to the low mesopause temperatures, causes nucleation to occur on the largest of these particles. Once nucleated these particles consume water vapor and grow, which prevents the supersaturation from becoming any higher. The number of particles larger than 0.1 μm is similar to that suggested by observations, and the size distributions resemble those of Tozer & Beeson (1974). The breadth of the computed size distribution, which is characteristic of particles grown moderately slowly in a highly supersaturated environment, suggests that the assumption of nearly monodispersed particles, made for convenience in some optical analyses of the clouds, is not valid. Although the clouds markedly affect the water vapor abundance, they do not have long enough residence times to reduce the ambient water vapor to the vapor pressure of ice. Turco et al also found that ions were nucleated in a model using slightly lower temperatures, and model results do not clearly distinguish between ion and meteoric nucleation mechanisms.

Nacreous Cloud Models

Table 2 suggests that nacreous clouds are controlled by only a few physical processes. Rather than nucleate, they form by the growth of ambient sulfuric acid droplets as temperatures approach those required to saturate ice crystals. Therefore, the number density is equal to that of stratospheric aerosols. Coagulation and sedimentation play little role in the history of the clouds unless they are very long-lived. The particles have rapid growth rates and simply continue to grow until they completely deplete the ambient water vapor so that the supersaturation is unity or until winds carry them into a region of higher temperatures, where they evaporate.

Stanford (1973) briefly discussed the physics of persistent Antarctic clouds. The temperatures are consistently so low that most of the stratospheric water would be condensed onto stratospheric aerosols, and ice particles a few microns in radius would result. These particles would fall toward the tropopause. Whether they reach the tropopause before encountering higher temperatures and evaporating is not known, but Stanford (1973) thought that ice particle sedimentation across the tropopause might be significant as a stratospheric water sink.

It is interesting to speculate on the size distribution of these particles. The very low supersaturation implies that small, pure ice particles could not be present in the cloud because of their higher vapor pressures due to the Kelvin effect. However, the Kelvin effect is not important for sizes as large as 1 μm so that a distribution of ice particle sizes could occur. Small acid-water droplets might be stable because of the water vapor pressure lowering due to the acid. As the particles grew they would add water until they became dilute enough for their vapor pressure to equal that over the ice. If they froze before reaching the ice vapor pressure, they could continue to grow into ice particles. However, if the acid particles reached the ice vapor pressure before freezing they would be stabilized at some fixed size. Adding more water would cause them to have higher vapor pressures than the ambient pressure and they would evaporate back to the stable size. Observations of the size of particles in persistent clouds would be an interesting test of these ideas.

Hesstvedt (1969, 1960) examined the dynamically complex case of nacreous clouds forming in stratospheric lee waves. He concluded that nacreous clouds form on stratospheric aerosols, and that their growth history is controlled by the adiabatic heating and cooling of the wave. The particle sizes were found to be a function of position within the clouds. Particles grew as they approached the crest of the wave, continued to grow during the descending phase until the supersaturation point was crossed, and then began to evaporate. The changing particle size in various parts of the cloud accounts for the distribution of colors in these clouds. Hesstvedt found that the particles would grow several microns during the few hundred second period that they remained within the cloud. The time the particles spent within the cloud was determined by the mean horizontal wind speed (50 m s^{-1}) and the typical cloud size (20 km).

Whether or not the ice particles actually contain an appreciable fraction of the atmospheric water depends upon the details of the calculations. At 20 km, 1 ppm of water is equivalent to 5 particles cm^{-3} of size 1.5 μm. Hesstvedt (1962) found such sized particles in his results. The narrow size distribution of the particles was explained by Hesstvedt (1960) as being due to the rapid growth rate of the particles. The Kelvin effect probably plays no important role in narrowing the size distribution.

No numerical models have been made that explicitly consider the interaction of the ice particles with stratospheric sulfuric acid particles. The stratospheric aerosols should begin to grow before ice is supersaturated. Subvisible dilute-acid clouds may occur commonly in the stratosphere at temperatures above the saturation point of ice.

Meteoric Debris Models

Table 2 shows that meteoric debris is controlled by nucleation, coagulation, sedimentation, and transport. Hunten et al (1980) modeled all these interactions above 30 km and their work was extended into the stratosphere by Turco et al (1980a). Hunten et al (1980) assumed that 44 metric tons of meteoric particles with a mean mass of 10 μg (100 μm radius) entered the Earth's atmosphere each day with a mean velocity of about 14.5 km s^{-1}. Particles smaller than 1 μg do not ablate but remain as micrometeorites. Figure 9 shows that at 30 km altitude the calculated number of micrometeorites of a few microns radius agrees well with the collections made by Brownlee (1978) at slightly lower altitude.

About two-thirds of the incoming meteor mass in Hunten et al's model ablates near 90 km and then recondenses to form large numbers of mole-

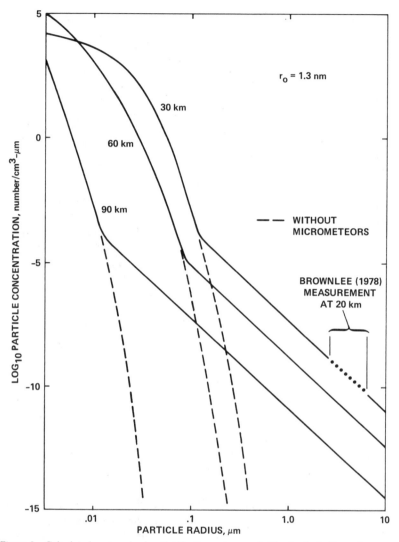

Figure 9 Calculated meteroric size distributions at several altitudes including micrometeorites (Hunten et al 1980). The particle size increases with decreasing altitude due to coagulation and above 1 μm due to the decreased sedimentation velocity.

cule-sized particles. The number of particles, as illustrated by Figure 10,
depends upon the initial particle size, which cannot be calculated in the
meteor smoke trail. The most probable size in the smoke trail was thought
to be less than 1 nm so more than 10^3 particles cm^{-3} could be present
near 90 km. Coagulation rapidly reduces these large concentrations and

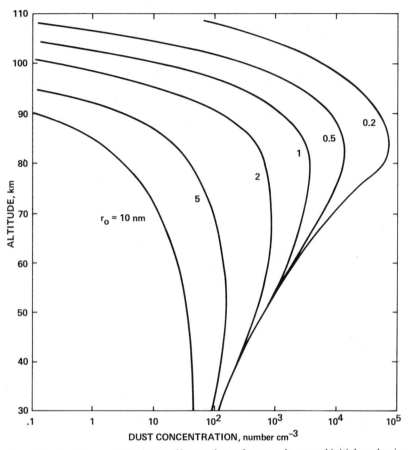

Figure 10   Particle concentration profiles are shown for several assumed initial smoke sizes with the total dust mass input held fixed (Hunten et al 1980).

at low altitudes the number of particles is nearly independent of the size of the particles in the trail. Coagulation causes the particle size to increase with decreasing altitude as is shown by Figure 9. Hunten et al (1980) pointed out that the large numbers of small meteoric particles have a large enough surface area to affect some slow chemical reactions in the stratosphere.

Table 1 presents the number of particles in various size ranges as a function of altitude for an assumed smoke size of 1.3 nm. The particle numbers increase rapidly with decreasing size as one approaches the size of the initial smoke debris. The numbers also increase rapidly with decreasing altitude because at lower altitudes more particles have coagulated from smaller sizes. The strong dependence of concentration on size and

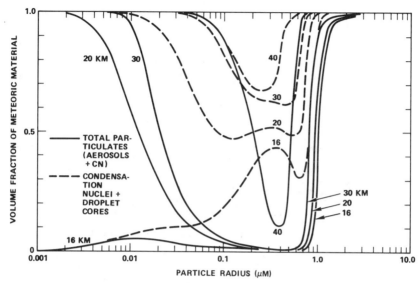

Figure 11   The volume fraction of meteoric material included within stratospheric aerosols of various sizes, at various altitudes. The solid curve refers to all stratospheric aerosols and the dashed curve refers to the fraction of the aerosols that are not sulfuric acid, but instead originate as tropospheric particles of unspecified composition (Turco et al 1980a).

altitude may account for some of the discrepancies in rocket collections of particles.

Turco et al (1980a) presented models of the interactions of meteoric dust and stratospheric aerosols. Meteoric particles are significant for large (>1 μm) and small (<0.01 μm) sizes, but as discussed earlier they are probably only a few percent of the aerosol's mass below 30 km. Figure 11 illustrates the fraction of all particles and the fraction of non-H_2SO_4 particles in the stratosphere that might be composed of meteoric material.

CONCLUSIONS

During the decade of the 1970s considerable progress was made in observing and modeling the particles above the tropopause. However, many difficult questions remain. The mode of nucleation of noctilucent clouds and stratospheric aerosols is uncertain. The size of particles in noctilucent and nacreous clouds and in meteoric debris is not precisely known. The processes that control the location of the top of the stratospheric aerosol layer are ill defined. Basic information about the geographic and temporal distributions of all the aerosols is incomplete. The interactions between the various particles above the tropopause are poorly understood.

Further progress in the 1980s may come from satellite observations of the geographic distribution of stratospheric aerosols, nacreous clouds, and possibly noctilucent clouds (McCormick et al 1979). In addition, aircraft sampling of stratospheric aerosols will add greatly to our knowledge. From coordinated studies of the aerosols the answers to many of our questions may be forthcoming.

Literature Cited

Bigg, E. K. 1975. Stratospheric particles. *J. Atmos. Sci.* 32:910–17

Bigg, E. K. 1976. Size distributions of stratospheric aerosols and their variations with altitude and time. *J. Atmos. Sci.* 33; 1080–86

Bigg, E. K., Kviz, Z., Thompson, W. J. 1971. Electron microscope photographs of extraterrestrial particles. *Tellus* 23:247–60

Bigg, E. K., Ono, A., Williams, J. A. 1974. Chemical tests for individual submicron aerosol particles. *Atmo Environ.* 8:1–13

Brownlee, D. E. 1978. Microparticle studies by sampling techniques. In *Cosmic Dust*, ed. J. A. M. McDonnell, Chap. 5. New York: Wiley

Brownlee, D. E., Ferry, G. V., Tomandl, D. 1976. Stratospheric aluminum oxide. *Science* 191:1270–71

Bronshten, V. A., Grishin, N. I. 1976. *Noctilucent Clouds*. Jerusalem: Israel Program for Scientific Translations. 237 pp.

Cadle, R. P. 1972. Composition of the stratospheric 'sulfate' layer. 1972. *EOS, Trans. Am. Geophys. Union* 53:812–20

Cadle, R. D., Grams, G. E. 1975. Stratospheric aerosol particles and their optical properties. *Rev. Geophys. Space Sci.* 13:475–501

Cadle, R. D., Langer, G. 1975. Stratospheric Aitken particles near the tropopause. *Geophys. Res. Lett.* 2:329–32

Cadle, R. D., Fischer, W. H., Frank, E. R., Lodge, J. P. Jr. 1968. Particles in the Antarctic atmosphere. *J. Atmos. Sci.* 25:100–3

Cadle, R. D., Lazrus, A. L., Pollack, W. H., Shedlovsky, J. P. 1970. The chemical composition of aerosol particles in the tropical stratosphere. *Proc. Symp. on Tropical Meteorology*, Honolulu, Hawaii, KIV:1–7

Clemesha, B. R., Nakamura, Y. 1972. Dust in the upper atmosphere. *Nature* 237:328–29

Clemesha, B. R., Simonich, D. M. 1978. Stratospheric dust measurements, 1970–1977. *J. Geophys. Res.* 83:2403–8

Cunningham, W. C., Etz, E. S., Zoller, W. H. 1979. Raman microprobe characterization of South Pole aerosol. *Microbeam Analysis*, ed. D. E. Newbury, pp. 148–53. San Francisco Press.

Donahue, T. M., Guenther, B., Blaumont, J. B. 1972. Noctilucent clouds in daytime: circumpolar particulate layers near the summer mesopause. *J. Atmos. Sci.* 29:1205–9

Esat, T. M., Brownlee, D. E., Papanastassiou, D. A., Wasserburg, G. J. 1979. Magnesium isotopic composition of inter-planetary dust particles. *Science* 206; 190–97

Etz, E. S., Rosasco, G. J., Blaha, J. J. 1978. Observation of the Raman effect from small, single particles: its use in the chemical identification of airborne particulates. *Environmental Pollutants*, ed. T. Y. Toribara, J. E. Coleman, B. E. Dahnelse, I. Feldman, pp. 413–456. New York: Plenum.

Farlow, N. H., Ferry, G. V. 1972. Cosmic dust in the mesosphere. *Space Res.* 12:369–80

Farlow, N. H., Ferry, G. V., Blanchard, M. B. 1970. Examination of surfaces exposed to a noctilucent cloud, August 1, 1968. *J. Geophys. Res.* 75:6736–50

Farlow, N. H., Ferry, G. V., Lem, H. Y. 1973. Analysis of individual particles collected from the stratosphere. *Space Res.* 13:1153–57

Farlow, N. H., Hayes, D. M., Lem, H. Y. 1977. Stratospheric aerosols: Undissolved granules and physical state. *J. Geophys. Res.* 82:4021–29

Farlow, N. H., Snetsinger, K. G., Hayes, D. M., Lem, H. Y., Tooper, B. M. 1978. Nitrogen-sulfur compounds in stratospheric aerosols. *J. Geophys. Res.* 83:6207–11

Farlow, N. H., Ferry, G. V., Lem, H. Y., Hayes, D. M. 1979. Latitudinal variations of stratospheric aerosols. *J. Geophys. Res.* 84:733–43

Fernández-Moran, H. 1960. Low temperature preparation techniques for electron microscopy of biological specimens based on rapid freezing with liquid helium II. *Ann. NY Acad. Sci.* 85:689–713

Fiocco, G., Grams, G. 1969. Optical radar observations of mesospheric aerosols in

Norway during the summer 1966. *J. Geophys. Res.* 74:2453–58

Fogle, B., Haurwitz, B. 1966. Noctilucent clouds. *Space Sci. Rev.* 6:279–340

Fogle, B., Rees, M. H. 1972. Spectral measurements of noctilucent clouds. *J. Geophys. Res.* 77:720–25

Friend, J. P., Feeley, H. W., Krey, P. W., Spar, J., Walton, A. 1961. The high altitude sampling program. Vol. 5: Supplementary HASP Studies. Final Rep. Contract DA-29-044-XZ-609, Defense Atomic Support Agency, Washington DC

Friend, J. P., Barnes, R. A., Vasta, R. M. 1980. Nucleation by free radicals from the photodissociation of sulfur dioxide in air. *J. Appl. Meteor.* In press.

Gadsden, M. 1975. Observations of the colour and polarization of noctilucent clouds. *Ann. Geophys.* 31:507–16

Gadsden, M. 1978. The sizes of particles in noctilucent clouds: Implications for mesospheric water vapor. *J. Geophys. Res.* 83:1155–56

Ganapathy, R., Brownlee, D. E. 1979. Interplanetary dust: trace element analysis of individual particles by neutron activation. *Science* 206:1075–77

Gras, J. L. 1976. Southern hemisphere midlatitude aerosol after the 1974 Fuego eruption. *Geophys. Res. Lett.* 3:533–36

Gras, J. L., Laby, J. E. 1979. Southern hemisphere stratospheric aerosol measurements 2. Time variations and the 1974–1975 aerosol events. *J. Geophys. Res.* 84:303–7

Gras, J. L., Michael, C. G. 1979. Measurement of the stratospheric aerosol particle size distribution. *J. Appl. Meteorol.* 18:855–60

Hallett, J., Lewis, R. E. J. 1967. Mother-of-pearl clouds. *Weather* 22:56–65

Hallgren, D. S., Schmalberger, D. C., Hemenway, C. L. 1973. Noctilucent cloud sampling by a multi-experiment payload. *Space Res.* 13:1105–12

Hamill, P., Toon, O. B., Kiang, C. S. 1977. Microphysical processes affecting stratospheric aerosol particles. *J. Atmos. Sci.* 34:1104–19

Hayes, D. M., Ferry, G. V., Oberbeck, V. R., Farlow, N. H. 1980. *Reactivity of stratospheric aerosols in laboratory environments.* Paper presented at AGU Spring Mtg., Toronto, Canada, May 22–27

Hawkes, R. L., Jones, J. 1975. A quantitative model for the ablation of dust ball meteors. *Mon. Not. R. Astron. Soc.* 173:339–56

Hesstvedt, E. 1960. On the physics of mother of pearl clouds. *Geophys. Publ.* 20:1–32

Hesstvedt, E. 1962. A two-dimensional model of mother-of-pearl clouds. *Tellus* 14:297–300

Hesstvedt, E. 1969. The physics of nacreous and noctilucent clouds. In *Stratospheric Circulation*, ed. W. L. Webb, 4:207–217. London & New York:Academic

Hobbs, P. V. 1974. *Ice Physics.* Oxford: Clarendon. 836 pp.

Hofmann, D. J., Rosen, J. W. 1977. Balloon observations of the time development of the stratospheric aerosol event of 1974–1975. *J. Geophys. Res.* 82:1435–40

Hofmann, D. J., Rosen, J. M. 1980. On the background stratospheric aerosol layer. *J. Atmos. Sci.* In press

Hofmann, D. J., Rosen, J. M., Pepin, T. J., Pinnick, R. G. 1975. Stratospheric aerosol measurements I: Time variations at northern midlatitudes. *J. Atmos. Sci.* 32:1446–56

Hughes, D. W. 1978. Meteors. In *Cosmic Dust*, ed. J. A. M. McDonnell, Chap. 3. New York: Wiley

Hummel, J. R. 1977. Contribution to polar albedo from a mesospheric aerosol layer. *J. Geophys. Res.* 82:1893–1900

Hummel, J. R., Olivero, J. J. 1976. Satellite observation of the mesospheric scattering layer and implied climatic consequences. *J. Geophys. Res.* 81:3177–78

Humphreys, W. J. 1940. *Physics of the Air.* New York: McGraw-Hill

Hunten, D. M., Turco, R. P., Toon, O. B. 1980. Smoke and dust particles of meteoric origin in the mesosphere and stratosphere. *J. Atmos. Sci.* 37:1342–57

Junge, C. E., Chagnon, C. W., Manson, J. E. 1961. Stratospheric aerosols. *J. Meteorol.* 18:81–108

Käselau, K. H., Fabian, P., Rohrs, H. 1974. Measurements of aerosol concentration up to a height of 27 km. *Pure Appl. Geophys.* 112:877–85

Kent, G. S., Sandford, M. E. W., Keensliside, W. 1971. Laser radar observations of dust from comet Bennett. *J. Atmos. Terr. Phys.* 33:1257–62

Lazrus, A. L., Gandrud, B. W. 1974. Stratospheric sulfate aerosol. *J. Geophys. Res.* 79:3424–31

Lazrus, A. L., Gandrud, B. W. 1977. Stratospheric sulfate at high altitudes. *Geophys. Res. Lett.* 4:521–22

Lazrus, A. L., Gandrud, B. W., Cadle, R. D. 1971. Chemical composition of air filtration samples of the stratospheric sulfate layer. *J. Geophys. Res.* 76:8083–88

Lazrus, A. L., Cadle, R. D., Gandrud, B. W., Greenberg, J. P. 1979. Sulfur and halogen chemistry of the stratosphere and of volcanic eruption plumes. *J. Geophys. Res.* 84:7869–75

Lindblad, B. A., Arinder, G., Wesel, T.

1973. Continued rocket observations of micrometeorites. *Space Res.* 13:1113–20

McCormick, M. P., Hamill, P., Pepin, T. J., Chu, W. P., Swissler, T. J., McMaster, L. R. 1979. Satellite studies of the stratospheric aerosols. *Bull. Am. Meteorol. Soc.* 60:1038–47

McDonnell, J. A. M. 1978. Microparticle studies by space instrumentation. In *Cosmic Dust*, ed. J. A. M. McDonnell, Chap. 6. New York: Wiley

Pinnick, R. G., Rosen, J. M., Hofmann, D. J. 1976. Stratospheric aerosol measurements III: Optical model calculations. *J. Atmos. Sci.* 33:304–14

Podzimek, J., Sedlacek, W. A., Haberl, J. B. 1977. Aitken nuclei measurements in the lower stratosphere. *Tellus* 29:116–27

Pollack, J. B., Toon, O. B., Sagan, C., Summers, A., Baldwin, B., Van Camp, W. 1976. Volcanic explosions and climatic change: A theoretical assessment. *J. Geophys. Res.* 81:1071–83

Poultney, S. K. 1972. Laser radar studies of upper atmospheric dust layers and the relation of temporary increases in dust to cometary micrometeoroid streams. *Space Res.* 12:403–21

Pruppacher, H. R., Klett, J. D. 1978. *Microphysics of Clouds and Precipitation*. Dordrecht, Holland: Reidel 714 pp.

Reid, G. C. 1975. Ice clouds at the summer polar mesopause. *J. Atmos. Sci.* 32:523–35

Reiter, E. R. 1975. Stratospheric-tropospheric exchange processes. *Rev. Geophys. Space Phys.* 13:459–74

Rosen, J. M. 1971. The boiling point of stratospheric aerosols. *J. Appl. Meteorol.* 10:1044–45

Rosen, J. M., Hofmann, D. J. 1977. Balloon-borne measurements of condensation nuclei. *J. Appl. Meteorol.* 16:56–62

Rosen, J. M., Pinnick, R. G., Hall, R. 1974. Recent measurements of condensation nuclei in the stratosphere. *Proc. 3rd Conf. of the Climatic Assessment Prog., Rep. DOT-TSC-OST-74-15.* Washington DC: Dept. Transp.

Rosen, J. M., Hofmann, D. J., Laby, J. 1975. Stratospheric aerosol measurements II: The world wide distribution. *J. Atmos. Sci.* 32:1457–62

Rosen, J. M., Hofmann, D. J., Käselau, K. H. 1978. Vertical profiles of condensation nuclei. *J. Appl. Meteorol.* 17:1737–40

Royal Society. 1888. *The Eruption of Krakatoa and Subsequent Phenomena*. London: Harrison & Trubner

Shedlovsky, J. P., Paisley, S. 1966. On the meteoritic component of stratospheric aerosols. *Tellus* 18:499–503

Stanford, J. L. 1973. Possible sink for stratospheric water vapor at the winter Antarctic pole. *J. Atmos. Sci.* 30:1431–36

Stanford, J. L. 1974a. Stratospheric water-vapor limits inferred from upper-air observations. Part 1 northern hemisphere. *Bull. Am. Meteorol. Soc.* 55:194–212

Stanford, J. L. 1974b. Possible long-term variations in stratospheric water-vapor content. *Weather* 29:107–12

Stanford, J. L. 1977. On the nature of persistent stratospheric clouds in the Antarctic. *Tellus* 29:530–34

Stanford, J. L., Davis, J. S. 1974. A century of stratospheric cloud reports: 1870–1972. *Bull. Am. Meteorol. Soc.* 55:213–19

Theon, J. S., Nordberg, W., Katchen, L. B., Horvath, J. J., 1967. Some observations on the thermal behavior of the mesophere. *J. Atmos. Sci.* 24:428–38

Toon, O. B., Pollack, J. B. 1976. A global average model of atmospheric aerosols for radiative transfer calculations, *J. Appl. Meteorol.* 15:226–46

Toon, O. B., Turco, R. P., Hamill, P., Kiang, C. S., Whitten, R. C. 1979. A one-dimensional model describing aerosol formation and evolution in the stratosphere. II—Sensitivity studies and comparison with observations. *J. Atmos. Sci.* 36:718–36

Tozer, W. F., Beeson, D. E. 1974. Optical model of noctilucent clouds based on polarimetric measurement from two sounding rocket campaigns. *J. Geophys. Res.* 79:5607–12

Turco, R. P., Hamill, P., Toon, O. B., Whitten, R. C., Kiang, C. S. 1979. A one-dimensional model describing aerosol formation and evolution in the stratosphere. I — Physical processes and numerical analogs. *J. Atmos. Sci.* 36:699–717

Turco, R. P., Toon, O. B., Hamill, P., Whitten, R. C. 1980a. Effects of meteoric debris on stratospheric aerosols and gases. *J. Atmos. Sci.* In press

Turco, R. P., Whitten, R. C., Toon, O. B., Pollack, J. B., Hamill, P. 1980b. OCS, stratospheric aerosols and climate. *Nature* 283:283–86

Twomey, S. 1977. *Atmospheric Aerosols*. New York: Elsevier, 302 pp.

Whitten, R. C., Toon, O. B., Turco, R. P. 1980. The stratospheric sulfate aerosol layer: processes, models, observations and simulations. *Pure Appl. Geophys.* 118:86–127

Witt, G. 1969. The nature of noctilucent clouds. *Space Res.* 9:157–69

Witt, G., Dye, J. E., Wilhelm, N. 1976. Rocket-borne measurements of scattered sunlight in the mesosphere. *J. Atmos. Terres. Phys.* 83:1155–56

Ann. Rev. Earth Planet. Sci. 1981. 9:59–80

FORM, FUNCTION, AND ARCHITECTURE OF OSTRACODE SHELLS

❋ 10144

Richard H. Benson

Department of Paleobiology, National Museum of Natural History, Smithsonian Institution, Washington, DC 20560

INTRODUCTION

Ostracodes are microscopic crustaceans that live on the sea bottom, in lakes, ponds, and estuaries, and are frequently found as fossils in sedimentary rocks ranging in age from the Cambrian to the present. They are improbable animals, at least in their appearance, which invaded a world too small for most of the organ machinery developed by their ancestors. So far as we know no other animal ever evolved from the ostracode. Nevertheless, in its struggle to survive, thousands of different species of ostracodes evolved to mark geologic history with attempts to develop suitable designs to fit the needs of environmental change. A handful of sand or a piece of drill core taken from the sea or ocean floor may have hundreds of specimens of ostracode shells.

Some ostracodologists (there are three or four hundred in the world) study these curious fossil remains for very practical reasons. While some are interested in dating rocks, especially pre-Cretaceous strata, others may be assisting other micropaleontologists to reconstruct the ecological conditions of past environments. They consider the design of each species to be uniquely adapted to the time and circumstances of its origin. The problem has been to develop a body of theory to explain how this occurs.

For many years the best that could be hoped for was to correlate species, with unique taxonomic characters or traits, with other similar ones where the age was known or the environment could be inferred from accompanying geologic data. Many of these characters were based on form analogy without knowledge of functional structure. A few ostracodologists (Liebau 1977, Benson 1972, 1974, 1975, Peypouquet 1977), risking the dangers of strained and mistaken analogy, set out in search of a set of

59

0084-6597/81/0515-0059$01.00

unifying principles in terms of static-frame and metabolic reaction that could explain the variation in morphology through recurrent patterns of structure and function.

The design of the skeleton, or exoskeleton (carapace) in the case of the ostracode, seems to be a solution to an engineering construction problem defined as a compromise between design and materials to fit the needs of a changing environment. Furthermore, because the carapace of the ostracode can enclose its softer parts and appendages, the design of the carapace represents a direct interface between the animal and surrounding stresses.

Does this exoskeleton react to these stresses according to reasonably simple mechanical principles? Are the differences between species identifiable as adjustments of architectural structure which in turn reflect underlying biological needs?

As D'Arcy Thompson said (1963), the accomplished fact of organic morphology is its own diagram of forces, if the function of the framework of operating stresses can just be determined. Similar to the recipient of a message in communication theory, we have the answer among the noise and attenuation, if we can just determine what form it takes. Then maybe we can guess at the question of causal selection with greater confidence.

Every part of science and mathematics is engaged to some degree in morphology, the identification of order and structure in form. In those cases where the analogy between traditional mathematical solutions and natural order is reasonably direct, organic morphology has succeeded in resolving the general relationships between proportion in animal form and problems of classification and genetic proximity. Principles of homology, similarity through enumeration of parts, definition of symmetry, and comparative deformation (allometry) are but a few of the analogical tools of the organic morphologist (Cain 1959). However, the problem of establishing causes in the explanation of why organic form mechanically varies is not often resolved in whole-animal or organ system morphology through the mathematical route of similitude, so powerful in the physical sciences.

It is curious that although organic skeletal form is architecturally efficient, and often copied by architects (Torroja 1967, Fuller 1965), the inquiring concepts of architecture are seldom applied above the molecular level to the practical problems of organ system morphology (for two notable exceptions see Hertel 1966, Wainwright et al 1976). The purpose of this review is to examine a few simple concepts borrowed from engineering architecture, especially membrane mechanics, or just architecture, to see how they can be used to understand morphological variation in the skeleton of a particular crustacean of microscopic dimensions. The state of the art is still mostly descriptive and comparative so that at this point the reader must excuse the absence of reference to many equations. Regrettably we are still trying to sort out our assumptions, some of which are discussed here.

THE OSTRACODE

Adult ostracodes range in size from half a millimeter to one and a half millimeters in length. There are larger pelagic ostracodes, but these do not have a heavily scleritized or calcified carapace capable of being preserved in the fossil record. This calcified carapace, its behavior in a fluid medium, and the way in which it is made, is of special interest. The most elaborate architecture is best developed among the "ornate" podocopid ostracodes.

Unless one has seen an ostracode, it is very difficult to imagine (see Figures 1–5; for basic references see van Morkhoven 1963, Pokorný 1978). At first it looks like a tiny shrimp living in the two valves of a clam shell (therefore, the German name "muschelkrebs"). The valves are equivalent to the head and thorax exoskeletal parts of the crab and lobster which are created for support and protection by hardening of the skin. Unlike the valves of a mollusk, which increase in size by adding to the margins, this exoskeleton is created anew and discarded numerous times during periodic growth.

The ostracode carapace evolved to envelope and encapsulate the rest of the animal as it diminished in size, in a way similar to the phyllopods and cladocerans. It is as though the animal shrank during its evolution to the point that its body was drawn up into its skull, which was also shrinking, but at a less rapid rate. In order for the rest of the body, the legs, feeding and reproductive parts (also covered by the exoskeleton, but very poorly scleritized), to have access to the external world, the "skull" or carapace became divided to be joined by a hinge along the dorsum, where it is also attached to the rest of the body (Figure 1 A–D).

The carapace is impregnated with massive layers of calcite during the molting and consequent growth process (unlike phyllopods and cladocerans). The result is a heavily armored animal. There have existed through geologic time as many as twenty to thirty thousand species, all representing experiments in carapace design. As unlikely as this basic body plan may seem, it has obviously been very successful.

A few more biological facts are necessary before focusing on the size and variations of carapace design of the ostracode. Except for a few species that are swimmers, the remainder crawl on or burrow in the bottom of their respective oceans (living as deep as 5,000 meters), seas, estuaries, or shallow ponds. In fresh water they may hatch and die in three months, or in the sea in two years. They feed on almost anything.

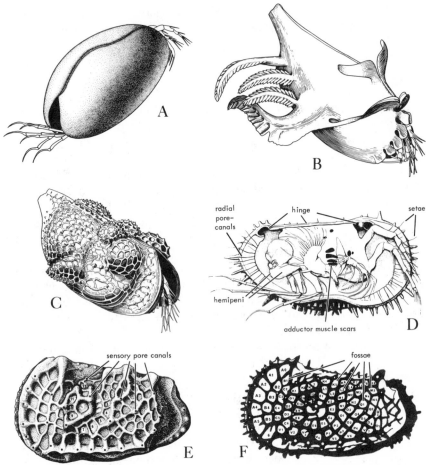

Figure 1 Variations in ostracode form. *A*, a smooth monocoque macrocyprid. *B*, an alate *Pedicythere*. *C*, a compound spherical bythoceratinid. *D*, demi-schematic of the internal anatomy of a deep-sea cytherid. *E*, a reticulate *Agrenocythere* species with the sensory canals and the central part of the histological pattern coded. *F*, the histological or reticular pattern of a second *Agrenocythere* species.

Some are parasitic, some attack much larger animals *en masse*, some are principally algal or diatom grazers. In fresh water some species are parthenogenetic (self-fertilizing) females, but most exhibit sexual reproduction with the male slightly more elongate to accommodate two large hemipeni. The sperm of the male is longer than that of an elephant, sometimes requiring it to be coiled up inside and ejaculated by a special pump mechanism.

Although some females lay eggs, most that live in turbulent waters carry these eggs and the developing young within the enveloping carapace, through one or two molt stages, after which they emerge as miniature adults. They grow through a series of nine instar or molting (ecdysis) stages, each time shedding and recreating the larger carapace at what must be a considerable metabolic effort. The animal has eyes, if it lives in depths less than six to eight hundred meters, which peer through the somewhat translucent carapace by way of "lucid spots" or by way of lenses set in the carapace as periscopic eye tubercules. It also has special glands near its head that are either used to spin small webs or attachment byssae (thread-like anchors), or are used to secrete poison into a victim. Some, especially the pelagic myodocopids, bioluminesce.

It is notable that among its crustacean attributes, the ostracode has lost its arterial circulation system to a simpler open "crankcase" model. All ostracodes may have had hearts 400 million years ago, but diffusion and "splashing" now suffices to circulate body fluids in all but the largest varieties. The respiratory system is also now very simplified with two maxillary "fans" that force oxygen-rich water currents past sensitive or receptive surfaces of the body or inner carapace. The digestive system is a straight gut with possibly two enlargements. Only the reproductive system seems to have kept pace in its relative complexity with the rate of carapace modification. As a consequence one either studies the male reproductive organs, which may occupy up to 30% of the volume of the animal, or the carapace, in order to understand the diversity of evolutionary modification and classification among ostracodes. Only the carapace is preserved in the fossil record.

Presumably the ancestors of the ostracode were much larger and had the normal complement of crustacean organ systems. They, like many modern arthropods, were probably entirely encased in a skeleton composed of an articulated system of thin imperfectly scleritized chitin tubes or tube segments capable of responding to external stress and stimuli by reactions at joints and through flexure or bending in the tough wall structure (Manton 1959). But the diminutive ostracodes that we know from the beginning of their fossil record have lived within the constraints of two heavy, rigid, uniformly stressed, and static systems of unyielding protective armor. No other animal spends so much relative effort in terms of mass of material to repeatedly armor itself.

THE CARAPACE

The two valves of the carapace (Figure 2) are joined at the dorsum by a simple chitinous or complex calcite hinge, and are closed by a strong set

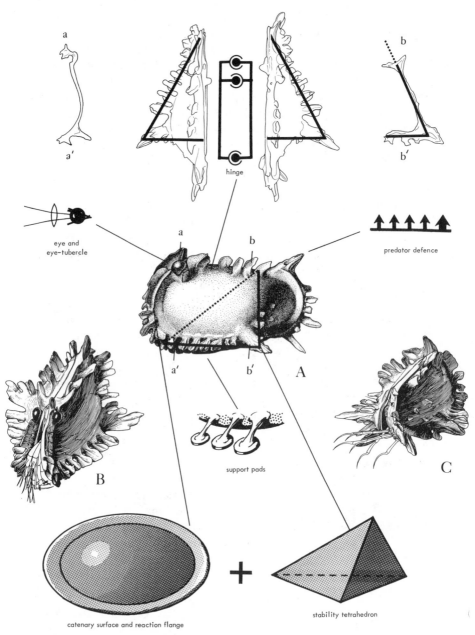

Figure 2 The morphological analysis of *Pterygocythereis ceratoptera* leading to the identification of underlying functions and implicit form.

of adductor muscles that connect across its midsection and through the rest of the animal suspended inside (Figure 3B). Except for the hinge, the rest of the margin of each valve is open or free. In the normal position of activity, the valves, with the appendages extended, are open with a gape of 15° to 20°. This gape seems to be maintained by tension at the hinge or by hydrostatic pressure within the body. Tiny setae are aligned along the gape and extend through the flange of carapace to join a sensory network. Numerous setae penetrate the open outer wall of the carapace to complete, with the eyes, the obvious part of the sensory system capable of receiving information from the outer world. When danger is perceived the appendages are drawn inward and the valves are shut forming a tight little protective box.

The carapace is formed quickly during times of molting by a fixed system of dermal cells whose pattern tends to be nearly the same among quite different species (Liebau 1977, Benson 1972). These same cells produce a smooth membraneous monocoque shell in one species (Figure 1A) or a coarsely reticulate, alate, or heavily ridged frame in another (Figure 1F). The pattern of pore canals, those through which the tactile nervous system is presumed to operate, seems to be even more conservative in its pattern of distribution over the carapace (Figure 1E; Benson 1976), whereas the particular structural pattern of the carapace may vary depending on its previous environmental and adaptive history (Figures 4 and 5; Benson 1975). So there seem to be at least three organ systems of architectural importance operating within the carapace (the sensory field, the histological dermal-cell field, and the static-frame working structure) plus the effects of accommodation of the ocular, respiratory, reproductive, and especially the closing muscular systems. The dermal-cell and sensory fields respond to the carapace shape and to the needs of the working structure, all in the context of general environmental stress, all as inertial subsets of reacting adaptive systems.

In the simplest of known ostracode carapace designs, the valves approach the shape of two elongate and truncate domes joined by a very short hinge at the dorsum and a few centrally located adductors at the crown (Figure 3). The rise of the dome compared to its span is a ratio of about 1 to 3. In the species of cyprids, which are fresh-water planktonic species, the valve approaches a nearly perfect membrane shell structure (in the engineering sense), assuming a catenary form over nearly all of its cross section taken in any direction (Figure 3C). The centralized adductor attachment must effect minimal pull when closed. The thrust is resolved at the margin by the narrow inward folding of the shell forming a flange and increasing the proportion of tension-resistant chitin (Figure 3D). This animal represents one end of a spectrum of designs. Other forms exhibit modifications of the carapace to allow for a much more forceful closure.

The adductor muscles that pull the valves tightly closed are attached normal to the carapace by chitinous tendrils extending into discrete keystone-shaped calcite prisms (called "muscle-scars") which are fused and imbedded in the shell wall. In imperfectly calcified specimens, such as might be found in cold or deep-water animals, lines of radiating wrinkles or "spotwelds" of calcite can be seen at the junction of the muscle-scars and the shell indicating the presence of stress and compensating strengthening during the construction of the shell. The adductors form the principal source of stress for which the design of the valves must compensate, especially at the very sensitive time of molting and shell formation.

The shell wall of the carapace is generally composed of many layers of calcite (Sylvester-Bradley & Benson 1971) with the C-axis (principle crystal axis) oriented normal to the surface. These layers are in turn separated by very thin sheets or stringers of chitin, which may thicken near the free margin of the carapace and carry through its hinge or join with the almost calcite-free exoskeleton of the rest of the ostracode body. The result is a wall made up of many calcite "bricks," which have little tensile strength but up to ten times greater compressive strength, and intermediate small amounts of chitin, which has great tensile strength and almost no compressive strength. The result is a structure not too dissimilar from reinforced concrete! As the design changes from one of a simple membrane shell to that of a highly complicated ridged structure, the layering of the shell follows the new structure in such a way as to indicate a conformity or congruence between the inner structure and the general design. This observation supports the contention that the various shell "ornament" or sculpture is actually doing mechanical, that is static-frame, work through the alignment of the strength of the construction material.

STRUCTURAL MECHANICS

The structural problem of the ostracode carapace is to encapsulate the animal beyond its distal-most, softer regions, and yet remain divided into two parts for appendage extension and access for feeding, respiration, and reproduction. At the same time the carapace must satisfy the needs of positional stability; that is, it must be able to react favorably to the movement of the surrounding medium. A single valve must economically, yet with strength, span a space larger than the greatest dimension of the softer remainder of the animal. Much of this space remains unoccupied by any hydrostatic or other compression-resistant support. In fact, the structure itself provides support. Furthermore, this structure must be constructed all in one stage or in repeated independent stages, with minimal assisting temporary construction support.

The most economical open structure for spanning and enclosing a considerable area is the dome, a derivative of the catenary,[1] or its modification, which has the capacity to react latitudinally as well as in the planes of the arch-shaped meridianal sections. A simple sphere encloses the maximum volume with the minimum surface; however, when the sphere is divided, the margins are weakened and subject to bending. When it rests on a plane, the stress of a sphere is no longer uniform, as the funicular surface of equilibrium departs from the mass of the wall. The positional instability (tendency to roll) of a sphere approaches the infinite, yet the number of optional reaction pathways through a given point on a sphere is also infinite and represents a high aspect of redundancy in its structure. By redundancy, it is meant that failure of the whole structure may be prevented by the availability of other reaction pathways.

The section of a dome formed of catenary meridional sections taken at variance to a direction normal to the crown represents a departure from the most efficient pathways of reaction from the static optimum (see Figure 3 for an example of a nearly perfect dome in the genus *Metapolycope*). If unexpected external stress were introduced away from the crown, bending moments would result in proportion to the force, of course, but also in proportion to departure from a conic projection of which the margin of the shell must be a section. Some ostracodes develop cone extensions for support (Figure 1C), while others become modified tetrahedrons (Figures 2 and 4D).

The valves of many benthic ostracodes are modified to form a strong union along the hinge, provide ventral stability, and support the free margins against bending and unresolved thrust. The elongate dome, modified vault structure, and shell-frame all are used to satisfy these requirements. Their efficient cross section approaches some portion of a catenary arch, depending on the orientation and the interference of other structures (Figures 2 and 4A). The introduction of supplementary ridges for directional strength tends to deform or divide the catenary into several such membranous sections. Some designs are wholly dominated by such ridges and the membrane sections disappear.

A major structural problem posed by the catenary is the resolution of the thrust that would tend to cause bending near the free margin. This

---

[1] A hypothetical surface within a mass uniformly loaded so that a section of it describes a catenary.

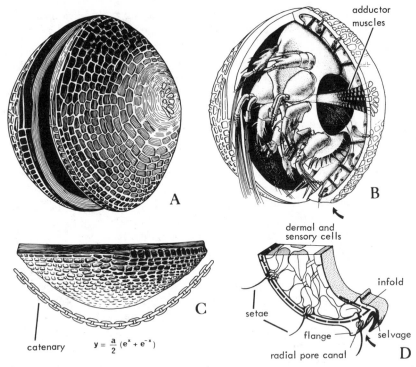

Figure 3 *Metapolycope hartmanni* and its shell architecture. *A*, the carapace. *B*, a sectional view of the shell with the body included showing the adductor muscles, marginal infolds, flanges, and wall tissue. *C*, the coincidence of the catenary (chain) with the shell curvature. *D*, the shell margin (right valve) with the reversal of the infold and the added chitinous reaction rim (selvage) indicated.

may be of considerable magnitude in forms of low rise relative to the span. Because it is not practical to join the margins with cross-tension-resisting members, a tension ring must be employed.

Calcite composing the shell has considerable compressive but little tensile strength. Either calcite mass or chitin, which has tensile but little compressive strength, must be added near the margin to resist a high bending moment. Close examination of the shell margin at the infold (Figure 3D) shows that the calcite laths (the parallel layers of calcite crystals) of the shell wall continue across the so-called zone of fusion, at one time mistakenly thought to unite two separate wall structures as the "inner" and "outer" lamella. Strength is continued through a change of direction and shape, plus the addition of mass, usually with only minor alterations in composition.

In very thin shells, the possibility of bending caused by the thrust of the catenary form is increased, and accessory stiffening structures or a considerable increase in mass may be required to maintain the shape and prevent buckling. There are several solutions to prevent bending and to absorb or redirect the thrust, within the shell and infold, or on the outside of the shell. These outside structures, which stiffen and strengthen the margins, may eliminate the primary strength purpose of the infold, and this function of the infold may become vestigial (fused).

As the ostracode design departs from the fundamental catenary or membrane surface, and the number of ridges, folds, or other framing members increases, the possibility of structural redundancy decreases. For example, in a smooth cyprid any "great circle" section taken through the carapace tends to be a catenary. The introduction of each lateral ridge reduces the options of direction and distance across which a catenary route can convey stress through a membrane surface until a tetrahedronal or arch-beam form is approached (Figures 4B and C) which has only six structural or load-bearing trusses and four surfaces with no options and zero redundancy. Intermediate between these extremes are the reticulate or box-frame designs (Figure 4C) which have varying degrees of redundancy offering the greatest flexibility of usage of truss on which to build enlarged frame systems.

Redundancy, or the presence of alternate reaction pathways, gives the paleontologist the chance to recognize tradeoffs and replacements of some structural members by the emphasis of others as the adaptive evolution of the frame system proceeds. Sometimes there is a decrease in descriptive uncertainty (negentropy) that begins to approach a more determinant mode in design. One can see this principle at work in the evolution of the reticulate ostracode *Oblitacythereis* (Benson 1977), which began as a highly determinant form (*Paleocosta*) in the Paleogene of North Africa. It became indeterminant in the Miocene, and returned as a determinant form in the Pliocene after the Messinian salinity crisis. The pathways of reaction in the box-frame pattern became increasingly redundant in the warm Miocene up to the salinity crisis. The earlier regularity of the reticular pattern was destroyed. The allopatric form coming to the Mediterranean from the cold Atlantic during the Pliocene was once again a structurally highly ordered design. It used different elements of the box-frame than its determinant ancestor.

The structural systems described above which are always changing are probably never completely in equilibrium with the environmental systems that forced them into being. A lag is perpetuated genetically, especially if the structure represents a minor metabolic taxation for its construction. For purposes of analysis it is best to assume that most realized structures

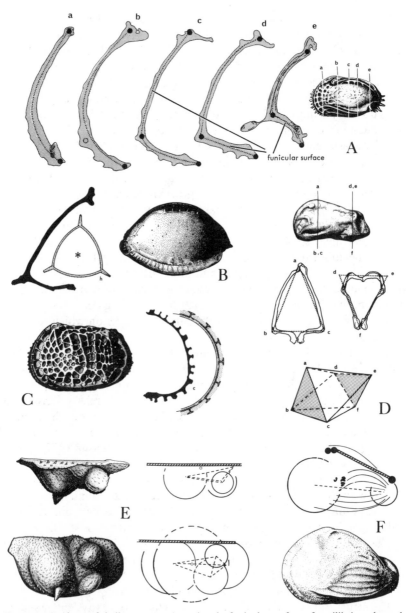

Figure 4 Analyses of shell structure. *A*, tracing the funicular surface of equilibrium through sections of a valve with two ridges from the anterior (*a*) to the posterior (*e*) showing the introduction of the lower ridge (*b–c*) and the ultimate catenary reversal (*e*). *B, C*, arch-beam and box-frame architectural shell types. *D*, a reversed tetrahedronal form. *E*, a compound-spherical type with an implicit tetrahedronal center-of-form frame. *F*, an example of the resolution of forces between two centers of form through a stress field.

exist as stages of transition in design somewhere between their potential and last stable functional optimum. The difference between the forms of ostracode that are actually found and the "best design" one might conceive usually results in the recognition of a deformed state of the ideal type. Yet the reality of stable forms, which are the majority of the species found in the fossil record, suggests that transition between types is short-lived. Only among the complex reticulate and ridged ostracodes are there presently evident redundancies and structural alternatives suitable for study. As the shape of a portion of the shell is deformed from its prior state the realignment of truss or surface elements can be observed. A study of this "deformation" in terms of a history of environmental stress offers a promising area of inquiry.

CLASSES OF ARCHITECTURAL TYPES AMONG CARAPACE "MORPHOTYPES"

Up to this point I have followed the argument that morphologic stability in carapace form should exist for structural reasons. Now it is necessary to demonstrate that the limited number of form options existing in the fossil record can be classified according to architectural principles.

In the first attempts to classify ostracode form by structural analogy (Benson 1974) aircraft fuselage and submersible hull design was considered as a source of potential analogues. The effects of gravity are cancelled out by the thrust of forward movement and the static-frame or lateral stress is in part carried in the "skin," a structural counterpart to the exoskeleton. This led to many insights into how material is saved in the secondary frames used to prevent buckling. However, these structures (thin elastic shells) typically use material that is almost equally resistant to tension and compression. It was through the examination of concrete, especially reinforced concrete (ferrocement), structural design that further progress was made. The following morphotypes, which are related to changing energy levels in the ocean, contain the majority of architectural variations among ostracode carapaces (see Figure 4 and Figure 5 for some of these).

Monocoque shell frames transmit most of their load in a thin-walled smooth calcified "skin."[2] *Metapolycope* is a good example of this type as it is almost completely undeformed although the example given here (Figure 3) is mildly, though uniformly reticulate. *Macrocypris* (Figure 1*A*) is an elongate or attenuate example of a deformed type. These species may be swimmers, burrowers, or live in interstitial waters of sediments.

---

[2] The term "monocoque" is derived from a class of aircraft structures (Nye et al 1940).

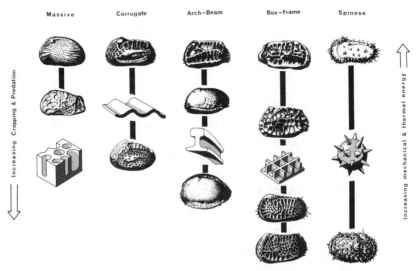

Figure 5   A general schematic of structural morphotype distribution relative to depth in the sea (after Benson 1975).

External mechanical forces are weak, uniformly distributed, and apart from any refractive interface. Streamlining may be important. The thrust, or outwardly resolved force, formed by the adductor muscles, is absorbed within the shell along the latitudinal parallels from the meridian or arch lines of compressive force to be ultimately dissipated in the elastic band of chitin at the margin. The thin shell can carry loads efficiently if the external stress is distributed uniformly. This shell design is not able to withstand substantial impact unless the shell wall becomes very thick. Most of the thin-walled varieties of this design live in habitats with uniform mechanical stresses.

More mass is required to increase the potential for strength and positional stability (as ballast) in the zone of a mechanically active water-sediment interface. This mass is added by an enlargement of the calcite crystals in the layering of the carapace wall. *Massive* ostracodes may evolve from almost any phyletic lineage or other structural morphotype. All of the massive forms share the same characteristic of strength through added shell-wall thickness and partial obliteration of other structural properties inherited from their past. *Cythere lutea* and *Hemicythere villosa*, which also become narrower in width, are examples. This design is more capable of anticipating variable loads by utilizing material to resist shearing stress or buckling. Historically the general increase in mass seems to have been an easy solution to the increased need for strength, but this design solution is metabolically expensive.

A subclass of monocoque designs could be the *compound subspherical* type (Figure 1C) in which the shell consists of a series of intersecting domes and in some cases cones that resolve the stress field through a succession of space-forming and support structures. The center of form of these individual substructures tends to form a tetrahedronal space frame. This group of ostracodes is well known in the Paleozoic and has living marine examples especially among the monoceratinids.

Arch-beam ostracodes translate much of their strength through a few curved ridges or flange (velate) structures. These lie outboard from the principle curvature and often smoother thin remainder of the shell in order to increase the moment selectively (Figure 4B). This structural morphotype has positional stability and strength along the bottom surface (venter) combined with some configuration of a beam or arch system to transmit force along the hinge or over the lateral surface. They can be very strong with thick or very delicate shells. *Aurila* and *Eocytheropteron* are two examples of the stronger type, although the same principle is approached in delicate genera like *Cytheropteron* and *Cativella*. This type also may become tetrahedronal as the arch components are straightened. Some become winged or alate as the venter is extended farther begging the question of whether added lift with movement is combined with positional stability at rest.

The *box-frame* type (Figure 4C) is the strongly reticulate ostracode whose surface is enhanced by a system of intersecting partitions (murae). Particular members may be emphasized like individual I-beams to dominate the remainder of the system of mural struts. *Bradleya* is of this type, which has a truss bridge to carry stress from the adductor to the radiating anterior margin. Shallow, smaller species are usually simpler with fewer mural struts than deep, larger species. This can be a very strong structural system yet economical in mass. It can increase its strength further by adding mass away from the center of form. The partitions or murae form at the junctions of the dermal cells of the histological field that construct the carapace and whose patterns of distribution are phyletically conservative. Some reticulate species actually become smooth in external appearance as the top flanges of the strut members broaden and join to form a two-layered light sandwich wall structure.

Corrugate or plicate ostracodes (Figure 5) differentially increase their area-to-volume ratio rapidly during growth, more rapidly than does a monocoque shell, thereby increasing strength and stiffness by introducing folds to a relatively thin wall. Placing mass symmetrically away from the former surface of geometric stability, as with the arch-beam and box-frame designs, the reaction moment (moment of inertia) can increase and still use a minimum of mass. This transformation gives mechanical advantage (resistance to longitudinal stress) by creating distance between that part of the wall under tension and that part under compression. The result is similar to that of the beam, but with less concentration of mass. This design represents a primitive solution in terms of ostracode growth. Construction requires only that the area of the shell be slightly increased over its prior condition by an accelerated growth rate. Its evolution is simple, but its reaction is less predictable. Some bending can only be prevented by development of cross-structural members. *Veenia, Procythereis,* and many other late Mesozoic cytheracean ostracodes fit this category.

Spinosity is a special structural attribute. It has functional importance, but it is not statically determinate nor geometrically stable. *Spinose* ostracodes usually tend to become larger and more spinose with depth (Figure 5). The function of spines is presumed to be to ward off predators or to extend sensory setae. The number of the spines, but not their length, seems to correlate directly with changes in depth. The positions of the major spines are not random but appear to be genetically fixed in the positions of the major normal pore canals present in all morphotypes. An increase in spinosity with depth may be the result of an increase of potential predator selection over mechanical selection, but this is speculation.

Other classes might include *compound-tetrahedronal* forms like the compound-spherical forms. These simplest geometric shapes, often found in joined sets, tend to reinforce one another as parts of another architectural type or to become dominant on their own. Structural morphotypes intergrade, and it is conceivable that a large ostracode taxon may find stable structural solutions in all of these types.

SIZE AS A FACTOR

Ostracodes range in length from about 200 micrometers to about 2,000 micrometers. The eggs are in the order of about 100 micrometers in diameter. Most modern adult ostracodes in shallow marine waters are about 900 μm and those in estuarine and the deep sea are about 1,200 with some growing to 1,600 μm. A few fresh-water species approach 2 μm. The instars increase in length with molting by a factor of 1.24 to double the volume each time. Because of the difference in shapes, the factor of increase in area will not remain constant among species as closely as that of the volume.

The average density of an ostracode is estimated to be between 1.10 and 1.15 with the density of sea water being about 1.025 to 1.030. The differences in viscosity in sea water due to pressure or temperature is not considered significant to the variation in ostracode shape, although not enough is known about this factor.

According to Hjulström (1939) a sediment particle the size of an ostracode (medium to coarse sand) would be unstable in suspension in current velocities above 1 cm/s in the smallest range and 8–9 cm/s in the largest sizes. These velocities are well within the ranges that might be expected in the benthic habitats of many marine ostracodes. Instability in this sense means that a spherical animal in turbulence would not resist the current to come to the bottom were it already in suspension. Of course, few benthic ostracodes are spherical, and most stay on or very close to the bottom. The problem may be to remain in a place of choice on the bottom as the current velocity increases. The thresholds of positional stability of sediment particle size are significantly lowered and many approach those of the ostracode as the shape of the particle departs from spherical (Zeigler & Gill 1959) or in morphological terms becomes more complex. The ability of the currents to sort or cause ostracode positional instability becomes a real factor in some habitats and differences in shape and surface roughness (surface tension effects considered but seldom measured in particle behavior) may become aerodynamically significant.

This is to say that as the ostracode molts to the next size and usually also to a greater state of complexity in form, it tends to become increasingly and perhaps suddenly unstable in a moving fluid medium relative to the bottom. Because it is generally not a swimmer and as some fluid motion is always likely to be present, there exists a problem of positional and orientational stability, if not the threat of breakage in extreme cases. This worst case condition seems to be rare. Examples of exoskeletal failure because of partial breakage are known, but not common. Undoubtedly the spherical and small ostracode has the least potential for positional benthic stability and the tetrahedronal and large ostracode approaches the greatest. What seems to be a liability for the small and more nearly equidimensional molt may in fact aid in its dispersal.

The possibility that some ostracodes (Figure 1B) are constructed to "hang glide" through local turbulence in boundary layers of weak currents needs at least to be considered. The more permanent suspension of other forms at a water-ooze interface without much bearing strength by an enlargement of the venter to create a "snow-shoe" design (Figure 4B) is even more feasible. Other general aspects of design with size are at present very speculative.

EVOLUTION OF STRUCTURAL TYPES

Thus, as a result of a hierarchy of functional responses and the presence of potentially equal series of operating mechanical reactions, there may be a definable number of ideal architectural solutions or structural systems available as options for the evolution of a successful carapace design. Convergence of form among ancestral stocks with differing recombinations of structural systems has been commonplace throughout the history of ostracodes. Arch-beam designs are present even in Paleozoic velate beyrichaceans. There are also corrugate quadrijugatorids, compound-spherical zygobolbids, box-frame kirkbiaceans, and so on. These same classes are found in Cenozoic and living forms. The fact that there are several ways to achieve strength besides just an increase in mass has allowed the diversity of ostracode form, but within a limited number of options.

How are these recombinations achieved? I suggest that changes in depth in the habitats of benthic ostracode species, which subjects them to changes in mechanical and predator selection pressures, may be responsible. This is to say that when either of these pressures is great and predictable, the morphologic variability as well as the geometric complexity of the design decreases (a decreased number of elements or descriptors). The morphologic choices are locked in as structural responses. With relaxation of these pressures, the inherent morphologic patterns of the carapace, those that control the basic pore and reticular patterns and tend to be temporarily geometrically more complex, reassert themselves. These more complex, perhaps mixed structures, while weaker in resisting mechanical pressures, provide the "roughness," or pseudospinosity necessary to ward off predators. Defensive structures such as spinosity become extremely exaggerated where severe predation occurs such as in the deep sea. If almost no stress is present, or if it tends to be dynamically uniform, the smooth thin shell would result leaving only the traces of the basic carapace patterns. As the "complex" forms reinvade regions of high selection pressure, they may not reappear in the same structural mode as before. Variation in *Bradleya* from box-frame to arch-beam in geographically isolated shallow shelves during the Neogene tends to follow this differential structural selection.

THE PROBLEM OF THE BIO-ENGINEER

It is difficult to experiment with engineering concepts in organic form, especially in animals of a scale very different from practical experience and which live in environments difficult to observe. The animal design is an accomplished fact, but it is seldom clean. The abundant fossil record of the ostracode is a long history of successful but less than perfect attempts to bring lasting stability between changing conditions and the inertial resistance of good designs. There is always the problem of conceiving and separating the design signal implicit in the ideal function of the animal, which is never actually knowable, and the noise or attenuation implicit in the indeterminant results in hand. One must take the position that within the context of the limitations of the basic arthropod design (that is, the constraints of an exoskeleton, accommodation of complex organ systems, etc, and the condition of bivalvedness), there is a limited number of actual designs, however frequently repeated, compared to all of those possible. This repeatability indicates an adherance to structural relation-ships yet to be defined, the ones we are searching for. At some level, the study of design and function of evolved structure begins to merge with that of the engineers' knowledge of structure.

If by analogy to architectural experience one can find geometric sim-ilarity, it is reasonable to propose first approximations of similitude, the concurrence between form (shape) and structure (a framework of reaction pathways). Among ostracode carapaces some of this geometric similarity is obvious and can be proposed directly, such as truss frames or the pres-ence of load-bearing surfaces described by regression approaching the catenary. In still other geometries, too complex or deformed, only the deformation or distortion of the original *Bauplan*[3] can be described.

Breaking the ostracode valves under controlled conditions can offer insight to design, for example by increasing the stress in the direction of and in the region of the adductor muscles. Most of the failure occurs along predictable pathways of static thrust. Also, lines of stress found in the calcification around particular structural features, which were formed before final calcification froze the design, help to show the magnitude and direction of local forces.

The ostracode grows by inflating itself by hydrostatic pressure to double its volume to the next stage of its design development, before it begins to harden its outer form as a static structure. The surfaces formed under inflated membrane stress or tension are easily converted to accommodate equal amounts of subsequent potential compression. However, the com-plexity of form may increase rapidly to the level where one can only conclude that some of the finished structures could only have resulted from highly controlled deflation around preselected lines of strength. The way in which the design is carried through the construction phases of the shell is still poorly understood, but studies of poorly calcified forms (Kornicker 1975) show that the major framing is anticipated early in development.

---

[3] A German concept of implicit order in organic form.

Within vertebrate animal skeletons the constituent construction elements (trabeculae, or elemental mineral crystal structures) tend to enlarge or diminish along lines of stress (Wolff's Law), presumably directed by piezo-electric forces. This same effect in calcite-enriched crustacean exoskeletons provides a theoretical basis to explain why exoskeletal design and consequent morphology should concur. That nonfunctioning elements of carapace structure exist seems also to be evident. Caution must be exercised, however, in judging structures to be nonfunctional as increased mass alone may allow ballast as well as strength.

The ostracode must be an efficient engineer; that is to say, the potential design, carried and controlled in the DNA-RNA genetic train, is the residue of historical experience. It must succeed (or fail) almost instantaneously with construction. Unlike many other organisms, the growth is not by degrees with multiple incremental options. It must work by anticipation of its success within very narrow limits. The efforts of this engineer, therefore, are likely to be much more deterministic and for our purposes easier to study.

The present thesis assumes that if a particular configuration of carapace form existed and was stable for a significant period, and was repeated in the fossil record, it represented an effective design or diagram of forces reacting to structural needs. This relatively restricted functional hypothesis attempts to reduce the accidental content of the exploratory arguments in morphology.

CONCLUSION

In recent years, mechanical functional morphology has met with limited success in the study of fossils (particularly in the view of Gould 1980), especially those invertebrate forms that are outside of our normal experience through comparison with living, familiar species. Cuvier (1825), the founder of the science of comparative anatomy, was one of the first to demonstrate consistancy in organic design by dramatically predicting bone structure of half-exhumed specimens in unfamiliar fossil vertebrate mammal or reptile species. But the general structure of these animals strongly reflected familiar functions. The power of plausible inference becomes somewhat more strained with forms of microscopic proportion living in areas or times where observation is difficult or impossible.

No one doubts that organisms are well designed. The purpose of the study of organic or skeletal architecture should not be to prove that they are. A skeletal system is the product of history, of former attempts to find viable solutions to survival within the limits imposed by selection and history. It is unlikely that any one example of skeletal structure is ever completely in equilibrium with its contemporary environmental stresses. As the skeletal form necessarily carries with it a record of its growth or construction, or the limits thereof, it also carries with it a record of structural change reflecting the general design of its kind. If one can imagine a purer solution within these constraints, it is also possible to envisage the amount of deformation away from this plan, the difference that reality has forced this solution to become. First, however, one must be able to recognize the total or integrated design or a series of potentially stable stages of such designs, those that have reached the most direct and orderly response to functional needs of the whole animal. The answers are in the fossil record. Are we asking the right questions?

If there have been mistakes in recent years in speculation about the functions of individual morphologic traits, they are probably due to a misplaced emphasis on control of inheritance suggested by particulate genetic theory. A greater emphasis is needed on the principles of architecture or order that govern the effectiveness of the whole. This report reflects a search in that direction. As Riedel (1978) has suggested, if order does not appear to occur in natural systems, we are forced to invent it. When induction through the use of old morphologic descriptors fails, we must invent a new system of morphologic descriptors. As Diogenes showed us, descriptors with implied static limits can not be used to demonstrate change. Evolution is a fact. It is the single pieces of evidence, scattered through the sands of time, that can be illusory.

ACKNOWLEDGMENTS

Larry Isham, Manuela Farfanti, and the author drew the illustrations. Carlita Sanford typed and proofread the manuscript. Alan H. Cheetham, Porter M. Kier, and Stephen J. Culver reviewed the final paper. Louis S. Kornicker helped with discussion of the metacopid ostracodes. Many thanks to all.

Literature Cited

Benson, R. H. 1972. The Bradleya Problem, with descriptions of two new psychrospheric ostracode genera, *Agrenocythere* and *Poseidonamicus* (Ostracoda: Crustacea). *Smithsonian Contrib. Paleobiol.* 12:1–138

Benson, R. H. 1974. The role of ornamentation in the design and function of the ostracode carapace. *Geoscience and Man*, 4:47–57. Baton Rouge, La: La. State Univ. Press

Benson, R. H. 1975. Morphologic stability in Ostracoda. In *Biology and Paleobiology of Ostracoda*, ed. F. M. Swain. *Bull. Am. Paleontol.* 65(282): 13–46

Benson, R. H. 1976. The Evolution of the Ostracode *Costa* analyzed by "Theta-Rho Difference." *Abh. Verh. naturwiss. Ver. Hamburg.* 18/19:127–139

Benson, R. H. 1977. Evolution of *Oblitocythereis* from *Paleocosta* (Ostracoda: Trachyleberididae) during the Cenozoic in the Mediterranean and Atlantic. *Smithsonian Contrib. Paleobiol.* 33:1–47

Cain, A. J. 1959. Function and taxonomic importance." *System. Assoc. Publ.* 3:5–19

Cuvier, G. 1825. *Recherches sur les Ossemens Fossiles.* Paris: A. Belin. 340 pp.

Fuller, R. B. 1965. Conceptuality of fundamental structures. In *Structure in Art and in Science*, ed. G. Kepes, 6:66–88. New York:George Braziller. 189 pp.

Gould, S. J. 1980. The promise of paleobiology as a nomothetic, evolutionary discipline. *Paleobiol.* 6(1):96–118

Hertel, H. 1966. *Structure, Form and Movement.* New York:Reinhold. 251 pp.

Hjulström, F. 1939. Transportation of detritus by moving water. In *Recent Marine Sediments*, ed. P. Trask, pp. 5–31. Tulsa: Am. Assoc. Petrol. Geol.

Kornicker, L. S. 1975. Antarctic Ostracoda (Myodocopina), Pt. 2. *Smithsonian Contrib. Zoology* 163:1–720

Liebau, A. 1977. *Homologous Sculpture Patterns in Trachyleberididae and Related Ostracodes* (translated from *Homologe Skulpturmuster bei Trachyleberididen und verwandten ostracoden*, dissertation, Berlin, 1971) for the Smithsonian Inst., Washington, DC by Nolit Publ. Co., Belgrade, Yugoslavia. 93 pp.

Manton, S. M. 1959. Functional morphophology and taxonomic problems of Arthropoda. In *Function and Taxonomic Importance*, ed. A. J. Cain. *System. Assoc. Publ.* 3:23–32

Nye, W. L., Hamilton, D., Eames, J. P. 1940. *Procedure Handbook for Aircraft Stress Analysis.* San Francisco: Aviation Press. 334 pp.

Peypouquet, J-P. 1977. *Les Ostracodes et la Connaissance des Paléomilieux profonds. Application au Cénozoique de l'Atlantique Nord-Oriental.* Thèse de doctorat d'état des sciences, Univ. Bordeaux I. No. 552. 448 pp.

Pokorńy, V. 1978. Ostracodes. In *Introduction to Marine Micropaleontology*, ed. B. U. Haq, A. Boersma, 4:109–149. New York & Oxford: Elsevier. 376 pp.

Riedel, R. 1978. *Order in Living Organisms* (translated by R. P. S. Jefferies). New York:Wiley. 313 pp.

Sylvester-Bradley, P. C., Benson, R. H. 1971. Terminology for surface features in ornate ostracodes. *Lethaia* 4:249–86

Thompson, D'Arcy W. 1963. *On Growth and Form.* Vols. 1,2. Cambridge Univ. Press. 1116 pp. 2nd ed.

Torroja, E. 1967. *Philosophy of Structures.* Berkley & Los Angeles: Univ. Calif. Press. 366 pp.

van Morkhoven, F. P. C. M. 1962–1963. *Post-Paleozoic Ostracoda, Their morphology, taxonomy and economic use.* Vols. 1,2. Amsterdam: Elsevier. 104 pp., 478 pp.

Wainwright, S. A., Biggs, W. D., Currey, J. D., Gosline, J. M. 1976. *Mechanical Design in Organisms.* New York:Wiley. 423 pp.

Zeigler, J. M., Gill, B. 1959. Tables and graphs for the settling velocity of quartz in water, above the range of Stokes' Law. *Woods Hole Oceanogr. Inst. Publ., Ref. No. 59–36.* Woods Hole, Mass. 13 pp., + tables and graphs

Ann. Rev. Earth Planet. Sci. 1981. 9:81–111

MECHANICS OF MOTION ON MAJOR FAULTS[1]

✺ 10145

Gerald M. Mavko

US Geological Survey, Menlo Park, California 94025

INTRODUCTION

Major faults are most simply viewed as the boundaries between litho-spheric plates, across which relative plate motion is accommodated. On a global scale, these plate boundaries appear as simple zones of infinitesimal width; when averaged over thousands of years, the displacement rates are approximately steady. The simplicity disappears, however, when one looks in more detail. The surface fault trace is never a smooth break. The zone of concentrated strain may vary from a few meters to tens of kilometers wide. Furthermore, fault motion during the time scale of scientific observation is seldom simple. A section of a fault may exhibit a combination of nearly steady fault slip, episodic slip, minor seismicity, and large damaging earthquakes.

In spite of the complexity, a great deal of progress has been made toward understanding the mechanics of fault motion, primarily because of many careful field and laboratory observations and a few clever models. This paper reviews some of our current understanding of major earthquake cycles in terms of large scale fault models. The emphasis is on observable quasi-static deformation including the process of strain accumulation and the coseismic changes in static stress and strain. Several other reviews have recently been published on related topics, including mechanisms of earth-quake instability and rupture (Dieterich 1974, Stuart 1978, 1979, Freund 1979), fracture mechanics applied to the crust (Rudnicki 1980), earthquake-related crustal deformation (Thatcher 1979), and rock properties (Kirby 1977, Tullis 1979, Logan 1979).

---

[1] The US Government has the right to retain a nonexclusive, royalty-free license in and to any copyright covering this paper.

ELASTIC REBOUND

Most theories concerning earthquakes are based on elastic rebound—the idea that elastic strain energy is gradually stored in the earth and is abruptly released during episodes of failure known as earthquakes. Comparisons of the accumulation of deformation at the earth's surface before large earthquakes with the rapid deformation during earthquakes show that they often approximately cancel except for a net rigid block translation of one side of the fault past the other. This led to the idea of a rebound (Reid 1910).

In the context of plate tectonics, the process of strain buildup and release at major plate boundaries repeats itself in a roughly cyclic fashion. The driving mechanism for the earthquake cycle is the relative plate motion across the common boundary (Andrews 1978). Whether or not strain accumulates, and the way it is released, depends on the nature of slip on the boundary.

We can describe a major cycle in terms of four time phases relative to the earthquake (Mescherikov 1968, Lensen 1970, Scholz 1972). In the *interseismic* or *strain accumulation* phase, the average fault slip on the plate boundary is slower than the long term average plate rate far from the fault. A simple geometric deficiency of slip accumulates causing strain energy to be stored in the plates on both sides of the fault. The *coseismic* phase is the period of several seconds to minutes during which rapid fault slip occurs, generating seismic waves. Most of the slip deficiency is recovered; stored elastic strain energy is converted into heat and waves (kinetic energy). The coseismic phase may or may not be preceded by a *preseismic* phase. This is a period of incipient strain release characterized by higher strain rates than occur during the strain accumulation phase. Rapid changes of any sort during this period might be interpreted as precursors. Finally the *postseismic* phase is a period of transient adjustment following the rapid earthquake movement. This adjustment may take place through aseismic creep, aftershocks, or viscoelastic relaxation.

Elastic rebound is also involved in the phenomenon known as fault creep. At several sites on the San Andreas and Calaveras faults in California, creep occurs in discrete aseismic events lasting up to several hours and separated by periods of little or no slip (Nason 1973). Creep events, like earthquakes, are episodes of strain release (C.-Y. King et al 1973, Goulty & Gilman 1978), although the amount and rate of slip are at least an order of magnitude less for a creep event than the amount and rate of slip for an earthquake with the same rupture length.

The short term unsteady slip associated with earthquakes and creep events is only a small perturbation superimposed on long term plate motion. In fact most strike-slip earthquakes are confined to only the upper-most 10–15% of a typical continental lithospheric thickness (although major thrusts include a larger fraction). Nevertheless, the unsteady slip excites transient deformation over a broad range of time scales, which forms the basis for most geophysical study of faulting.

COSEISMIC ELASTIC FIELDS

The best constrained portion of a rebound cycle is the rapid coseismic or strain release part. In general, we can determine from seismic and geodetic data the approximate area and orientation of the fault plane, the average slip, and the average stress change. This is possible primarily because the short term response of the earth to rapid fault slip is elastic. Therefore abrupt changes in strain due to earthquakes are insensitive to uncertainties in plate thickness and the inelastic rheologies responsible for long term plate motion. In this section we review some first order features of coseismic fault slip that are inferred from analysis of these elastic fields.

Dislocation and Crack Models of Faulting

Theoretical approaches to computing the static elastic fields due to faulting generally fall into two types: crack models and dislocation models. Crack models are based on a prescribed stress change on the fault plane (e.g. Starr 1928, Muskhelishvili 1953, Eshelby 1957, Knopoff 1958, Sneddon & Lowengrub 1969, Segall & Pollard 1980); dislocation models are based on a prescribed fault slip (Steketee 1958a,b). An advantage of the dislocation approach is the ability to compute the stress and displacement fields due to well-defined, arbitrarily shaped faults with arbitrary slip distributions. The deformation from complex slip distributions is constructed by linear superposition of simple slip solutions. The crack problem, on the other hand, involves mixed stress and slip boundary conditions in the plane of the fault and is generally more difficult to treat mathematically. Of course, both stress and slip changes accompany faulting, and the two descriptions are equivalent. Applications of this work to faulting have also been reviewed by Chinnery (1967) and Mavko (1978).

The most useful approach to modeling the displacement fields associated with three-dimensional faults is the dislocation formalism developed by Steketee (1958a,b). Steketee showed that if we approximate a fault as a discrete surface S of discontinuity (or dislocation) in an otherwise elastic half space, the resulting vector displacement U_k everywhere in the medium is given by an integral over the fault surface of point nuclei of strain τ_{ij}^k (Love 1944) multiplied by the local value of slip ΔU_i.

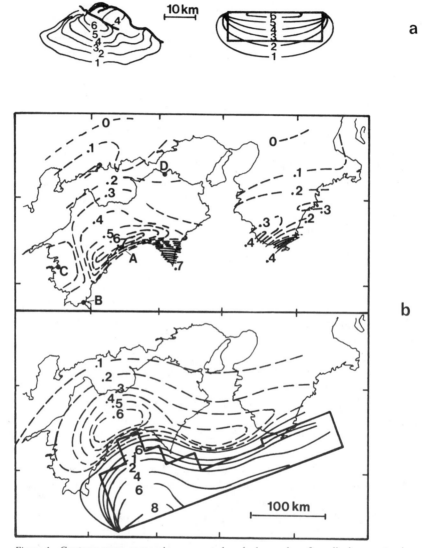

Figure 1 Contour maps comparing computed and observed surface displacement using rectangular fault models. (*a*) Observed (left) and computed (right) subsidence, in meters, associated with 1959 Hebgen Lake, Montana, earthquake (after Savage & Hastie 1966). Heavy rectangle shows the surface projection of the (normal) fault, which dips 54° S. (*b*) Observed (top) and computed (bottom) vertical displacement, in meters, associated with 1946 Nankaido, Japan, earthquake (after Fitch & Scholz 1971). Uplift is shown by solid contours; subsidence, dashed. Heavy polygon shows the surface projection of the (thrust) fault model, composed of six rectangular surfaces.

$$U_k = \frac{1}{8\pi\mu} \iint\limits_{S} \Delta U_i \tau_{ij}^k \nu_j \, dS. \tag{1}$$

In Equation (1), μ is the elastic shear modulus and ν_j are the direction cosines of the normals to dS. The τ_{ij}^k are the displacements from a nucleus of strain in a half space and have been given in analytical form by Mindlin & Cheng (1950) and Maruyama (1964). Steketee's expression (1) is based on the concept of a dislocation surface composed of infinitesimal elements dS. Strains and stresses in the medium are obtained from the derivatives of (1) and Hooke's law.

A simple application of the formula (1) is the evaluation of displacements associated with uniform slip over a rectangular slip plane. The integral has been evaluated analytically and compared with observations for vertical strike-slip faults by Chinnery (1961, 1963, 1964, 1965) and for a variety of fault models, including dip-slip faults, by Maruyama (1964), Press (1965), and Savage & Hastie (1966). In fact, a rectangular fault with uniform slip is the most commonly used geodetic model of faulting. Examples of computed surface displacement for rectangular strike-slip and dip-slip faults are illustrated in Figures 1, 3, and 4.

A problem with uniform slip models is that they predict stress singularities around the edges of the fault. Furthermore, uniform slip is sometimes not sufficient to explain complicated surface deformation. Nonuniform slip on a three-dimensional fault requires numerical integration of Equation (1). Chinnery & Petrak (1967), for example, have evaluated the stress and displacements for a Gaussian distribution of slip over a roughly rectangular surface. However, in practice, strain fields from nonuniform slip are most often calculated by piecing together a finite number of rectangular fault patches, each with uniform slip (Fitch & Scholz 1971, Thatcher 1975, Dunbar 1977, Savage et al 1979).

Effects of variable slip and stress drop can be studied more easily in two dimensions. In two dimensions a very long fault (length $\gg$ depth) can be modeled as a distribution of elastic dislocation lines, screw dislocations when slip is parallel to the long fault dimension and edge dislocations when perpendicular (Weertman 1964, Bilby & Eshelby 1968, Mavko & Nur 1978). In contrast to Steketee's formalism, constructed from infinitesimal dislocation surfaces, each dislocation line marks the edge of a semi-infinite plane of slip. Variable slip $U(x)$, where x is the in-plane coordinate parallel to the fault (perpendicular to the dislocation line), is described by the dislocation density function $B(x) = -\partial U/\partial x$ where $B(x) \, dx$ represents the total length of Burgers vectors of the infinitesimal dislocations lying between x and $x + dx$.

The displacement field from a single dislocation in an infinite medium has a simple form (Weertman & Weertman 1964). For a screw dislocation lying along the z-axis with slip b parallel to the z direction, the only nonzero displacements are in the z direction:

$$U_z = \frac{b}{2\pi}\tan^{-1}(y/x). \tag{2}$$

For an edge dislocation lying along the z axis with slip b parallel to the x direction, the displacements are as follows:

$$U_x = \frac{-b}{2\pi}\left[\tan^{-1}(y/x) + \frac{\lambda + \mu}{\lambda + 2\mu}\frac{xy}{x^2 + y^2}\right],$$

$$U_y = \frac{-b}{2\pi}\left[\frac{-\mu}{2(\lambda + 2\mu)}\log\left(\frac{x^2 + y^2}{c}\right) + \frac{\lambda + \mu}{\lambda + 2\mu}\frac{y^2}{x^2 + y^2}\right], \tag{3}$$

$$U_z = 0.$$

Here λ is Lamé's coefficient and c is an arbitrary constant. The slip due to both types of dislocation is uniform over the half plane $y = 0$, $x > 0$. Because the material is linear and the elementary solutions (2) and (3) are invariant under spatial translation in an infinite medium the displacements due to variable slip are given by the convolution (Bracewell 1965) of (2) or (3) (with $b = 1$) with the distribution $B(x)$ (Canales 1975, Mavko & Nur 1978, Stuart & Mavko 1979).

The stress change in the plane of the fault is related to the slip through the Hilbert transform:

$$\sigma = \frac{\mu}{2\pi\alpha}\int_{-\infty}^{\infty}\frac{B(x')dx'}{x - x'}. \tag{4}$$

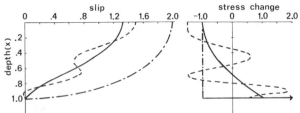

Figure 2 Three different stress-slip pairs for two-dimensional faults computed from Equation (4). Depth (x) can be interpreted as either distance from the center of a deeply buried fault, normalized by fault half width W, or actual depth, normalized by W, for vertical strike-slip faults breaking the surface. Stress is in arbitrary units τ. Displacement is in units of $\tau W\alpha/\mu$. Dashed and solid curves represent two heterogeneous fault models, both having the same moment. Dot-dashed curves represent the more familiar uniform stress drop model.

Here $\alpha = 1$ for screw dislocations, $\alpha = (1 - v)$ for edge dislocations, and σ is the component of shear stress in the direction of slip. The transform (4) can be evaluated numerically using the fast Fourier transform (Claerbout 1976, Mavko & Nur 1978, Stuart & Mavko 1979). Also, a large number of analytic transform pairs are given by Erdelyi et al (1954). Mavko & Nur (1978) outline a simple analytic procedure for inverting (4) for slip, given an arbitrary stress change σ expanded as a polynomial. Several stress-slip pairs for two-dimensional faults are shown in Figure 2.

A convenient feature of solutions constructed with screw dislocations is that a plane perpendicular to the fault (parallel to the dislocation lines) is traction-free whenever the slip is symmetric about the plane. Hence, the solution for a vertical strike-slip fault intersecting the free surface is just half of a full space solution constructed by mirroring the problem across the free surface (the method of images).

Geodetic Depth

A common feature of all static coseismic strain fields is that they rapidly decrease with distance from the fault, within several 10's of kilometers for strike-slip earthquakes and 100 km or so for major thrusts. Data showing the falloff of displacement with distance from the Nankai Trough (1946 Nankaido, Japan, thrust earthquake) and the Gomura Fault, Japan, (1927 Tango earthquake) are illustrated in Figures 1b and 3. The spatial scale of the strain release is a measure of the fault depth and can be understood in terms of the elastic models (Kasahara 1957, Chinnery & Petrak 1967).

Consider a very long (two-dimensional) vertical strike-slip fault in a half space with uniform slip D extending from the free surface to a depth W. The horizontal displacement $U(y)$ at the free surface is constructed from (2) using a buried screw dislocation and an image:

$$U(y) = \frac{D}{\pi} \tan^{-1}(y/W) \mp D/2. \tag{5}$$

(The sign $\mp$ is chosen: $-$ for $y > 0$ and $+$ for $y < 0$.)

The strain ε is the derivative $\delta U/\delta y$:

$$\varepsilon = \frac{D}{\pi W} \frac{1}{1 + (y/W)^2}. \tag{6}$$

The maximum surface strain and displacement occur at the fault trace, $y = 0$. The falloff of strain and displacement is scaled by the depth W, as illustrated in Figure 4a. Both U and ε drop to half their trace values at a distance $y = \pm W$, which gives a convenient surface measure of the depth

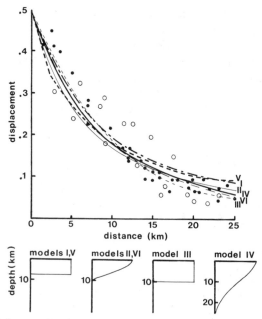

Figure 3   (*Top*) Computed and observed horizontal surface displacements associated with 1927 Tango earthquake. Closed circles—southwest side of the fault; open circles—northeast side (after Chinnery & Petrak 1967). Displacement is normalized by the trace offset; distance is perpendicular to the fault. (*Bottom*) Slip vs depth for six different fault models (see text).

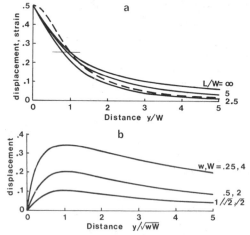

Figure 4   Falloff of surface displacement and strain with distance perpendicular to vertical strike-slip fault models. (*a*) Solid curves show displacement, normalized by the maximum trace offset, for rectangular faults with depth W and length L along the strike. Dashed curve shows shear strain for the two-dimensional case ($L/W = \infty$), arbitrarily normalized by twice the strain at the trace. (*b*) Surface displacement for three buried two-dimensional faults, all having the same mean depth, $\sqrt{wW} = 1$ and unit slip.

for this model. For faults with finite length L along strike, the displacement drops to half its trace value at a distance less than the depth.

Displacements computed for a variety of strike-slip models are compared with the observations from the 1927 Tango earthquake in Figure 3. In the figure, model I is the two-dimensional uniform slip model in Equation (5); model II is a two-dimensional model with slip smoothly tapering to zero with depth (Mahrer & Nur 1979); model III assumes uniform slip over a finite length rectangular fault (Chinnery & Petrak 1967); model IV assumes a rectangular fault with slip smoothly tapering toward zero near the edges (Chinnery & Petrak 1967). Clearly, a variety of uniform half-space models fits the data (the three-dimensional models fit a bit better than the two-dimensional ones) even though the scatter in the data is small. All of the models have a similar slip distribution of about 3 m in the uppermost 7 km, comparable to that predicted by the simplest two-dimensional model. Below 7 km, the models are quite different. This illustrates the general result that geodetic data can define the approximate depth range of greatest slip, but cannot constrain the details of slip (Weertman 1965, Chinnery & Petrak 1967). Details on a spatial scale d cannot be resolved at distances larger than d. Also the sensitivity of surface deformation to small amounts of slip decreases with depth (Thatcher 1978, Savage 1978), so that the estimated depth of faulting can be uncertain by a factor of two or more.

A more precise statement can be made about the models in Figure 3 by defining the geodetic depth as the depth of maximum slip gradient or the depth at which the slip falls to half the maximum value. In this sense the geodetic depths of the two-dimensional models I and II are within 2 km of each other and the depths of the three-dimensional models III and IV are within 4 km.

The simplest model for slip not breaking the surface is also two dimensional. The surface displacement and strain for a fault with uniform slip between depths w and W, not reaching the surface, is modeled with two dislocations and two images (Chinnery 1970):

$$U(y) = \frac{D}{\pi}[\tan^{-1}(y/W) - \tan^{-1}(y/w)], \tag{7}$$

$$\varepsilon = \frac{D}{\pi}\left[\frac{1}{W}\frac{1}{1+(y/W)^2} - \frac{1}{w}\frac{1}{1+(y/w)^2}\right]. \tag{8}$$

The displacement is shown in Figure 4b for several values of w and W. The position of maximum surface displacement, $y = \sqrt{wW}$ indicates the geometric mean of the upper and lower depths. The strain at the trace

Table 1 Geodetic depth and stress drop for strike-slip faults[a]

Earthquake	Magnitude	Depth (km)	Slip (m)	Reference	Stress drop (bars)
San Francisco (1906)	8.3	3.2	4.0	Knopoff (1958)	188
		6	5.0	Kasahara (1958)	125
		5	5.0	Chinnery (1961)	96
		12	4.9	Petrak (1965)	122
Tango (1927)	7.5	15	3.0	Kasahara (1958)	30
		15	3.4	Chinnery (1961)	37
		10	3.4	Chinnery (1961)	39
		25	3.4	Petrak (1965)	27
		20	3.4	Petrak (1965)	55
North Izu (1930)	7.0	8	4.0	Kasahara (1958)	75
		12	3.8	Chinnery (1961)	51
		26	3.8	Petrak (1965)	46
Imperial Valley (1940)	7.1	8	4.2	Kasahara (1958)	79
		6	4.2	Chinnery (1961)	69
		13	4.2	Petrak (1965)	96
		12	4.2	Petrak (1965)	106

[a] From a more complete table by Chinnery (1967).

and the far field displacement ($y \gg \sqrt{wW}, w - W$) are both proportional to the product $D(w - W)$. The depth range $(W - w)$ can therefore be determined only if D is found independently. In practice, the fault area is often found independently from aftershock locations and slip is determined from the moment $\mu D(w - W)$.

Depths of faulting determined geodetically are shown for several strike-slip earthquakes in Table 1 (from a longer list by Chinnery 1967). The range of depths for each event results primarily from the range of models used to fit the data. An important result is that most strike-slip earthquakes are shallower than 10–20 km (Chinnery 1967, Eaton et al 1970a). The lack of deeper earthquakes has been attributed to a transition from stick-slip to stable sliding in the fault zone as the temperature increases with depth (Brace & Byerlee 1970) or to a general increase in ductility of the crust with depth (Lachenbruch & Sass 1973). Thrust faults show a larger scatter in rupture depths but are generally much deeper, particularly at subduction zones.

A complication in determining geodetic depth results from heterogeneity in crustal stiffness, which can distort surface strain fields. Rybicki & Kasahara (1977) and Mahrer (1978), for example, have found from theoretical studies that a relatively soft fault zone embedded in a stiff half

semimajor and semiminor axes L and W

$$M_0 = \frac{3}{2} CW \iint \Delta\sigma \left(1 - \frac{x^2}{L^2} - \frac{y^2}{W^2}\right)^{1/2} dS,$$

$$= CWS\langle\Delta\sigma\rangle \tag{15}$$

where C is now a dimensionless constant that depends on the direction of slip and the ellipticity, $\varepsilon = W/L$, but not on the distribution of slip. The estimated stress drop $\langle\Delta\sigma\rangle$ is an average of the stress drop weighted with a function that emphasizes the stress near the center of the fault:

$$\langle\Delta\sigma\rangle = \frac{3}{2S} \iint \Delta\sigma(x, y) \left(1 - \frac{x^2}{L^2} - \frac{y^2}{W^2}\right)^{1/2} dS. \tag{16}$$

[Mavko & Nur (1979) independently derived the two-dimensional equivalent of (16) for the analogous problem of crack-opening under a heterogeneous pressure distribution.]

For heterogeneous stress drops $\langle\Delta\sigma\rangle$ will usually differ from the simple areal average $\overline{\Delta\sigma}$, though not by much (Madariaga 1979). Both $\langle\Delta\sigma\rangle$ and $\overline{\Delta\sigma}$, however, might be quite different from the actual stress drop. Madariaga considers the example of stress drop $\Delta\sigma_a$ at asperities covering a portion S_a of the total source area S. Assuming negligible stress drop in the rest of the plane the average stress drop is only a fraction of $\Delta\sigma_a$,

$$\langle\Delta\sigma\rangle \simeq \overline{\Delta\sigma} = \Delta\sigma_a S_a/S. \tag{17}$$

The localized or maximum stress drop at complex heterogeneous sources is usually underestimated by the average stress drop. Consider the stress drop on a two-dimensional fault expanded as a polynomial (Mavko & Nur 1978),

$$\Delta\sigma = \sum_{i=0}^{N} a_i T_i(x), \tag{18}$$

where T_i are Chebychev polynomials of the first kind (Abramowitz & Stegun 1964). The weighted stress drop $\langle\Delta\sigma\rangle$ obtained from the seismic moment is obtained from (18) substituted into (14). Because of the orthogonality of the polynomials only T_0 and T_2 contribute to the moment:

$$\langle\Delta\sigma\rangle = a_0 - a_2/2. \tag{19}$$

Wildly fluctuating stresses expressed in the form of T_i, $i \neq 0, 2$, contribute nothing to the moment, regardless of their amplitude. Simple examples are shown in Figure 2. The solid and dashed stress-drop curves correspond to the functions $\Delta\sigma = T_2$ and $\Delta\sigma = T_2 - T_8 + T_{12}/2$ respectively. Both have the same moment and the same average stress $\langle\Delta\sigma\rangle$ even though the maximum stress drop is 50% greater and the maximum stress increase is 100% greater for the dashed function.

POSTSEISMIC, INTERSEISMIC, AND PRESEISMIC DEFORMATION

Observations of Transient Deformation

In a strictly elastic earth, complete elastic rebound would take place in a few seconds, with the characteristic time of strain release determined by the earthquake source rise time, fault dimensions, and rupture velocity. The only slow deformation would be the accumulation of tectonic strain. It appears, however, that an earthquake is often just a fraction of a larger episode of strain release. Pre- and postseismic transients are observed, which indicate a broad relaxation spectrum. For example, during the three years following the 1966 Parkfield, California, earthquake ($M = 5.5$; right-lateral strike-slip) as much as 25 cm of fault creep occurred at a decaying rate, although little or no surface breakage occurred during the main event (Smith & Wyss 1968, Scholz et al 1969). In addition, road damage occurring within several years before, and en-echelon cracks formed within a month before the earthquake (Allen & Smith 1966), suggest a preseismic transient. Rapid surface fault slip of more than 10 cm has occurred within several months following both the August 6, 1979, Coyote Lake, California, and October 15, 1979, Imperial Valley, California, earthquakes (J. Savage, personal communication, USGS 1980).

Even the great 1906 San Francisco earthquake, which led H. F. Reid to propose the elastic rebound mechanism, was followed by transient deformation. Thatcher (1975) suggests that substantial postseismic crustal strains, continuing for at least 30 years following the earthquake, can be inferred from geodetic surveys since 1906. These strains can be explained (though not uniquely) by ~4 m of aseismic fault slip from 10 to 30 km depth, without additional surface slip. Thatcher (1975) also suggests anomalously rapid strain accumulation during the 50 years prior to 1906, although the evidence is weak (Savage 1978, Thatcher 1978).

Perhaps the most spectacular example of postseismic deformation was observed following the 1946 Nankaido, Japan, earthquake ($M = 8.2$; thrust type) where upheavals of as much as 2 m occurred over a 1 to 3 year period. Figure 5a (Matuzawa 1964, Kanamori 1973) shows the rather complicated nature, in space and time, of the vertical displacement. Similar deformations occurred during the 10 years following the 1964 Alaskan ($M = 8.4$; thrust type) earthquake (Brown et al 1977, Prescott & Lisowski 1977, 1980).

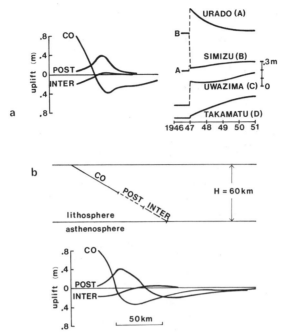

Figure 5 Observed and computed coseismic, postseismic, and interseismic vertical deformation associated with the 1946 Nankaido, Japan, thrust earthquake (after Fitch & Scholz 1971, Thatcher & Rundle 1979). (*a*) Observed profiles perpendicular to the Nankai Trough (*left*) and smoothed tide gage records (*right*) at locations labeled *A–D* in Figure 1*b*. (*b*) Two-dimensional model (see text) and computed profiles, both drawn to the same scale as the profiles, in (*a*).

Particularly short-lived transients have also been observed. Rapid fault slip lasting only several hours was recorded after a Matsushiro shock on September 6, 1966 (Nakamura & Tsuneishi 1967, Scholz 1972). A precursory aseismic slip with time constant of 300 to 600 s, starting about 1000 s before the main shock of the 1960 Chilean earthquake, has been inferred from long-period surface waves and body waves (Kanamori & Cipar 1974) and from free oscillations (Kanamori & Anderson 1975). Ando (1975), Sacks et al (1977), and Pfluke (1978) give evidence of earthquakes with geodetic moments several times that determined from seismic methods. Because large tsunamis were generated, the duration of the aseismic component of slip is apparently minutes to hours, long compared to the response band of seismographs but short compared to the response of the sea. Further examples of transient deformation are reviewed by Scholz (1972), Kanamori (1973), Dunbar (1977), Pfluke (1978), and Thatcher (1979).

Relaxation Mechanisms

An earthquake rupture superimposes a stress perturbation onto the pre-seismic state—decreasing the stress over much of the rupture area and increasing it elsewhere. As discussed earlier, these coseismic fields are now well understood, and numerous studies have shown that the associated abrupt displacements can be explained in terms of fault slip in an elastic medium. Postseismic observations suggest a subsequent viscoelastic response. A material is viscoelastic when its initial response to abrupt changes of stress or strain is elastic, while its longer term response is a viscous relaxation or flow (Fung 1965, Christensen 1971). Most rocks flow to relax shear stresses, as though the rigidity gradually decreases with time. Bulk relaxation is much less common.

What are the viscous elements? A simple mechanical model for the earth's crust and upper mantle, suggested by plate tectonics, consists of a relatively elastic, brittle lithosphere overlying a ductile asthenosphere. Within this framework we can distinguish geometrically three general sources of relaxation.

RELAXATION IN THE ASTHENOSPHERE   The asthenosphere is character-ized by high temperature relaxation mechanisms (Ashby & Verrall 1977, Weertman 1978, Tullis 1979). Solid mineral grains can flow plastically by atomic diffusion and the motion of lattice dislocations (Gordon 1965, Weertman & Weertman 1975, Heard 1976, Carter 1976). This makes the polycrystalline composite fluidlike over long time scales and can account for the large-scale, finite deformation implied by plate motion and the low strength implied by isostatic equilibrium. In addition, enhanced deforma-tion at grain boundaries can occur resulting from dislocation motion and diffusion (Ke 1947, Zener 1948, Anderson 1967) or the viscous flow of melt (Walsh 1969, Mavko & Nur 1975, O'Connell & Budiansky 1977). Other loss mechanisms which are relevant at seismic frequencies include thermoelasticity, dislocation damping, point defect diffusion, and grain boundary effects (Anderson 1967, Jackson & Anderson 1970).

RELAXATION IN THE LITHOSPHERE   The lithosphere, by definition, is a relatively strong, rigid layer that can resist permanent deformation or plastic flow for long periods of time, whereas the asthenosphere cannot (Le Pichon et al 1973). This is consistent with analyses of glacial rebound and lithospheric flexure (McConnell 1968, Walcott 1973, Forsyth 1979), as well as our concept of continental drift.

The important question becomes: How thick is the lithosphere? Or, at least, if we are to construct simple mechanical models for an earthquake cycle, what thickness is appropriate for the elastic layer?

Many investigators agree on a stratified model in which the effective mechanical thickness of the lithosphere depends on time, temperature, strain rate, and deviatoric stress (Melosh 1978b, Forsyth 1979). The upper lithosphere, above approximately the 450° $\pm$ 150°C isotherm (Watts 1978), remains essentially elastic and can support loads for 10^8 to 10^9 years; the lower part is elastic-plastic or viscoelastic and relaxes under stresses with durations a few million years. The mechanisms of relaxation in the lower part are similar to those discussed for the asthenosphere (Kirby 1977) but relaxation times are longer for the lithosphere because of lower temperatures. The effective viscosity at the base of the elastic part of the lithosphere is about 10^{26} Poise and in the asthenosphere 10^{21} Poise or less (Melosh 1977, 1978b). At an ocean trench, for example, the long term flexural thickness of the lithosphere may be only 20–40 km (Hanks 1971, Watts & Talwani 1974) because the strain rate associated with the steady component of subduction is low enough and the temperature below 40 km is high enough for the deviatoric stress to stay relaxed. In contrast, at the same trench the nonsteady strain accumulation and release during a rebound cycle lasting tens or hundreds of years occurs in a lithosphere effectively 70 km thick, which is approximately the seismically determined thickness (Kanamori & Press 1970, Le Pichon et al 1973). Similarly, in continental lithosphere the plate thickness for rebound might be the seismic thickness of 110–130 km. Anderson (1971) and Hadley & Kanamori (1977) suggest, however, that in parts of southern California the shallow crust is mechanically decoupled from the lower crust, so that the moving surface plate is much thinner than is commonly inferred from surface waves. Lachenbruch & Sass (1973) suggest a similar decoupling between the shallow crust (15–20 km) around the San Andreas Fault and the more ductile material below in order to explain a low broad heat flow anomaly. However, in this case, the crustal plate is also undergoing permanent shear flow, generating heat. This uncertainty in plate thickness can affect interpretation of surface strain.

Aside from large scale fluidlike flow, which distinguishes the asthenosphere from the lithosphere, a limited viscoelastic relaxation to changes in the stress field can occur within even the shallow lithosphere. In the shallow lithosphere the relaxed configuration is also essentially elastic, distinguished from the unrelaxed state only by a smaller effective rigidity. Hence, a viscoelastic lithosphere exhibiting transient relaxation times on the order of several years would look elastic at seismic frequencies as well as over the longer periods of flexure and isostatic rebound.

A number of relaxation mechanisms can be considered to account for the viscoelastic response. Concentrated plastic flow at grain boundaries is reasonable in much of the lithosphere (below, say, 20–30 km) where the ratio of absolute temperature T to the melting temperature T_m is greater than one half ($T/T_m > 1/2$). Presumably, motion at grain boundaries could occur while the grains themselves remained essentially elastic, giving to the polycrystalline composite a long term finite strength, yet a short term viscoelastic strain. Pressure solution, a low temperature form of grain boundary diffusion enhanced by water (Tullis 1979), can also relax stresses.

In the shallow crust stress-induced viscous shearing and local squirt of pore fluids (Mavko & Nur 1975, 1979, O'Connell & Budiansky 1977) as well as large scale, regional diffusion (Biot 1941, Nur & Booker 1972) can give a time-dependent deformation qualitatively similar to a viscoelastic response. The regional diffusion might also be enhanced by dilatancy (Nur 1973, Scholz et al 1973).

FAULT CREEP   In addition to direct observations of surface fault creep, aseismic fault slip has been invoked at depth in the lithosphere to explain pre- and postseismic surface deformation (Fitch & Scholz 1971, Smith 1974, Thatcher 1975, Brown et al 1977, Thatcher & Rundle 1979). However, very little is known about the detailed stress-strain behavior of the fault zone at any depth. Nason & Weertman (1973) conclude little more than the existence of an upper yield point phenomenon from observations of shallow creep events. In the laboratory transient stable sliding sometimes precedes stick slip on frictional surfaces (Scholz et al 1969, Dieterich 1979a,b) at conditions corresponding to several kilometers depth. At higher temperatures and pressures Stesky (1974) observes a nonlinear stress-strain rate sliding law similar to that expected for solid-state creep. Laboratory measurements on fault gouge and clay have also been made (Engelder et al 1975, Logan & Shimamoto 1976, Summers & Byerlee 1977). The main problem lies in determining what kind of material is representative of a fault zone at depth.

In addition to creep on the primary fault, creep on nearby faults can have an effect on relaxation. Even though the bulk of the crustal material is elastic, slip on secondary faults and fractures makes the crust effectively more compliant. If the slip is creep-like, the change in compliance is gradual, and the overall effect may not be distinguishable from viscoelastic relaxation.

Models

Many features of observed aseismic deformation can be explained by purely elastic models, much like the coseismic models, in which both steady and episodic aseismic slip occur around edges of the rupture surface (Savage & Burford 1970, Thatcher 1975, Shimazaki 1974). In contrast, a number of authors have attributed the deformation to viscoelastic adjustments, primarily in the asthenosphere (Nur & Mavko 1974, Smith 1974, Rundle & Jackson 1977, Spence & Turcotte 1979, Savage & Prescott 1978a, Thatcher & Rundle 1979). It now appears that the largest post-seismic and interseismic strains are dominated by a combination of these two mechanisms although their relative contributions are difficult to resolve and probably vary from region to region. Other mechanisms, for example the diffusion of pore fluids (Nur & Booker 1972), probably affect deformation much less.

STRIKE-SLIP EARTHQUAKES   A commonly accepted model for a major earthquake cycle on a strike-slip fault like the San Andreas Fault in California is shown in Figure 6a. Two elastic lithospheric plates with thickness H slide past each other with their relative motion occurring across a narrow vertical fault zone. Seismic and geodetic data indicate that seismic slip seldom occurs deeper than ~ 15 km. Therefore, if the concept of strong plates significantly thicker than 15 km is correct, there

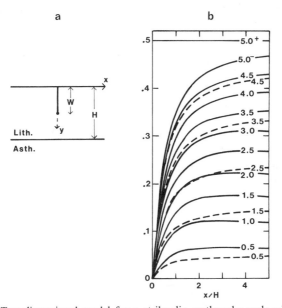

Figure 6 (*a*) Two-dimensional model for a strike-slip earthquake cycle with an elastic lithosphere over a (Maxwell) viscoelastic asthenosphere. (*b*) Surface displacement (solid curves) as a function of distance from the fault, x, for one cycle of periodically occurring earthquake sequences, with $W/H = 0.5$ (after Savage & Prescott 1978a). Displacement is normalized by the seismic slip and is shown relative to the configuration immediately following an earthquake. Curves are labeled with time in increments of $2\eta/\mu$. Dashed curves show the response of an elastic half space to the same earthquake cycle.

can be little doubt that a large amount of aseismic slip at depth is required to accomodate the relative plate offset.

Some of the earliest studies of strain accumulation considered models of deep aseismic slip in an elastic half space (Thatcher 1975, Chinnery 1970, Scholz 1972). Thatcher (1975), for example, explained a rapid episode of postseismic strain ($\sim 1.2 \ 10^{-6} \ yr^{-1}$) during the 30 years following the 1906 San Francisco earthquake with 3–4 m of slip between depths of 10–30 km. In effect, the rupture, which extended coseismically from the surface to ~ 10 km, gradually deepened by a factor of 2 or 3 during the postseismic period. The resulting postseismic displacement fields would have the form shown in Figure 4b, with maximum displacement occurring at a distance of $(10 \cdot 30)^{1/2} = 17.3$ km from the fault, and the far field displacement going to zero. The additional steady component of strain accumulation is simulated with the half space model by uniform slip extending downward to infinity (Savage & Prescott 1978a, Savage et al 1979).

A second series of models attempted to include the effect of a weak fluidlike asthenosphere by considering a plate model with stress-free upper and lower boundaries (Turcotte & Spence 1975, Savage 1975, Spence & Turcotte 1976, Turcotte 1977, Mavko 1977). These models have been criticized (Savage & Prescott 1978a, Spence & Turcotte 1979) for explicitly ignoring the viscous asthenospheric tractions at the base of the plate. It has generally not been recognized, however, that strain rates predicted by plate models with arbitrary nonzero basal tractions under steady motion (if steady motion ever occurs) are exactly the same as for the free plate model (Mavko 1977).

Later studies (Nur & Mavko 1974, Mavko 1977, Rundle & Jackson 1977, Savage & Prescott 1978a, Spence & Turcotte 1979) have included the complete viscoelastic response of the asthenosphere (assuming a Maxwell solid). An example, from Savage & Prescott (1978a), is illustrated in Figure 6b, comparing viscoelastic calculations with half-space results. The viscoelastic model assumes steady uniform slip below a depth $W = 0.5 \ H$ (where H is the plate thickness) at a rate v equal to the far field plate velocity. In the half-space model the same uniform slip rate extends infinitely deep. Shallower than $W = 0.5 \ H$ the fault is usually locked but slips abruptly a uniform amount vT at equally spaced time intervals T. In the example, $T = 5\tau_0$ where $\tau_0 = 2\eta/\mu$, η is the asthenospheric viscosity, and μ is the elastic rigidity of the lithosphere and asthenosphere. The unique feature of the viscoelastic model is the rapid postseismic relaxation that causes the displacement rate at a distance $y/H \simeq 2$ to exceed the far field rate early in the cycle. The viscoelasticity tends to concentrate strain accumulation closer to the fault than it is in the half-space model.

Although it is commonly accepted that both slip and asthenospheric effects are important, there is little consensus on their relative contributions. One reason is the uncertainty in plate thickness. Nur & Mavko (1974) and Thatcher (1975) suggested that postseismic viscoelastic effects were not important for earthquakes on the San Andreas Fault, based on an assumed lithospheric thickness of 75–100 km compared to the 15 km cosiesmic depth. On the other hand, if the thickness of the elastic layer is as small as 20 km (Anderson 1971, Hadley & Kanamori 1977, Lachenbruch & Sass 1973) then the effect may be quite large. A basic problem is our inability to resolve from geodetic observations the difference between deep aseismic slip and distributed viscoelastic relaxation in a layer or half space (Rundle & Jackson 1977, Barker 1976, Savage & Prescott 1978a). In fact, for two-dimensional problems in horizontally layered media, the viscoelastic solution can always be solved using the method of images; that is, a distribution of slip in a uniform half-space model can always be found that precisely duplicates the surface displacement produced by viscoelastic relaxation in one of the buried layers (Savage & Prescott 1978a).

DIP-SLIP EARTHQUAKES Major thrust-type earthquakes at subducting plate margins often rupture through a substantial fraction of the lithosphere, so it is reasonable to expect a large asthenospheric viscoelastic response.

One of the earliest quantitative models for postseismic relaxation in the asthenosphere was the stress guide model introduced by Elsasser (1969, 1971). Patterned after plate tectonics, the model consists of a strong elastic lithospheric plate over a linear viscous fluid asthenosphere. Horizontal displacements, U, in the lithosphere, resulting from long wavelength perturbations in stress, take the form of the diffusion equation,

$$\frac{\delta U}{\delta t} = \frac{h_1 h_2 E}{\eta} \nabla^2 U. \tag{20}$$

Here h_1 and E are the thickness and Young's modulus of the lithosphere and h_2 and η are the thickness and linear viscosity of the asthenosphere, and it is assumed that the scale of lateral variations is large compared to h_1 and h_2. The obvious interpretation (Bott & Dean 1973, Anderson 1975, Savage & Prescott 1978a, Spence & Turcotte 1979) is that localized disturbances (stress drops) associated with earthquakes will diffuse away, qualitatively explaining the transient deformation following some large earthquakes. Anderson (1975) has speculated that this diffusion results in the migration of earthquakes along plate margins. The most important

result is that stress suddenly released at a plate boundary cannot instantaneously affect the whole plate. Disturbances with period T are
damped to e^{-1} of their maximum value at a penetration distance δ (skin
depth) from the fault:

$$\delta = \left(\frac{Eh_1h_2T}{\pi\eta}\right)^{1/2}. \tag{21}$$

For example, disturbances with period $T = 100$ years (the approximate
recurrence time for great earthquakes) are restricted to within a few
hundred kilometers of the plate margin, while the interior of a plate is
affected only by stresses persisting for a million years or more. This
result applies to both strike-slip and dip-slip earthquakes.

The stress guide model has been modified by Melosh (1976) to include a
nonlinear fluid asthenosphere, appropriate for long term steady plate
motion (Weertman & Weertman 1975, Post & Griggs 1973). The non-
linearity introduces a damped, yet somewhat wavelike, propagation of
disturbances which Melosh argues resembles the migration of aftershocks
of the 1965 Rat Island, Alaska, earthquake, therefore proving that the
asthenosphere is nonlinear. While the asthenosphere is generally considered to be nonlinear, Savage & Prescott (1978b) show that Melosh's
(1976) model does not prove it. The short term response of the stress guide
model, either linear or nonlinear, is that of an elastic layer over an infinitely
rigid half space. Initially, strains are confined to within a layer thickness
or so of the fault, and these propagate outward only when the half space
begins to relax. A more realistic model, incorporating the initial elastic
response of the asthenosphere, results in a larger scale coseismic strain
field that subsequently relaxes with less pronounced wavelike propagation.
On the other hand. Melosh (1978a) emphasizes that even if the instantaneous elasticity is included, migration effects may still be significant if the
asthenosphere is nonlinear. It appears that nonlinear effects on an earthquake cycle have yet to be resolved.

The most successful models of postseismic and interseismic deformation
are actually extensions of the purely elastic models developed for coseismic
studies. For example, using the solution for a rectangular fault in an
elastic half space, Equation (1), Fitch & Scholz (1971) modeled the postseismic deformation following the 1946 Nankaido, Japan, earthquake
(Figure 5) with additional forward slip on the down-dip extension of the
fault plane and backslip on portions of the coseismic fault plane. A similar
model incorporating forward and backslip in an elastic medium was
suggested by Scholz & Kato (1978) for deformation following the 1923
Kanto, Japan, earthquake. Brown et al (1977) and Prescott & Lisowski
(1977, 1980) used elastic postseismic slip to model deformation following

the 1964 Alaskan earthquake. The requirement of backslip following the Nankaido earthquake was criticized by Nur & Mavko (1974), who suggested instead that the postseismic deformation was dominated by viscoelastic relaxation in the asthenosphere without additional slip. Their model of coseismic slip in an elastic lithosphere (the free surface only crudely approximated) over a linear viscoelastic asthenosphere was the first relaxation model to include both the near field effects of fault dip angle and the initial elastic response of the asthenosphere. Again, the calculation was based on faulting in an elastic medium, with the subsequent viscoelastic response obtained using the correspondence principle (Fung 1965). A similar model, incorporating both slip and asthenospheric relaxation, was developed independently by Smith (1974, 1980).

Recently, a number of numerical models of movements landward of subduction zones, principally in Japan (Bischke 1974, Thatcher & Rundle 1979, Thatcher et al 1980, Smith 1980) and Alaska (Brown et al 1977, Prescott & Lisowski 1977, 1980), have revealed a fairly consistent pattern: rapid episodic slip, both down-dip and up-dip of the coseismic rupture, during the several year postseismic interval and subsidence due to asthenosphere relaxation during the longer interseismic phase. These are illustrated in Figure 5, patterned after the work of Thatcher & Rundle (1979) using an elastic lithosphere over a linear viscoelastic (Maxwell) half space. The curve labeled CO shows the vertical coseismic displacement, the elastic half-space response to abrupt slip in the upper portion of the lithosphere. This shallow stress release transfers shear stress to the deeper part of the plate and the asthenosphere. Subsequent aseismic slip down-dip of the rupture and the beginnings of viscoelastic response to both the coseismic and postseismic slip cause the postseismic deformation labeled $POST.$ Finally the rapid postseismic deformation merges with the more steady interseismic deformation composed of approximately steady aseismic slip near the bottom of the plate plus viscoelastic subsidence in the asthenosphere, somewhat equivalent to the downward gravitational pull of the slab. Reasonable model parameters for deformation in Japan are a 60 km thick lithosphere and an asthenospheric viscosity of $10^{20}-10^{21}$ P (Thatcher & Rundle 1979, Thatcher et al 1980).

Disagreements in models are usually in detail only, reflecting our inability to resolve fine details of relaxation. While Thatcher & Rundle (1979) model the asthenosphere under Japan with a viscoelastic half space, Smith (1980) chooses a layered upper mantle with a low viscosity asthenosphere with finite thickness. Thatcher & Rundle prefer postseismic and interseismic slip on a discrete fault plane down-dip of the coseismic rupture; Smith chooses instead distributed viscoelastic relaxation in a low viscosity (10^{19} P) pocket down-dip of the fault plane. The least-understood effects on deformation are the downward gravitational pull of the slab, buoyancy, and horizontal convergence of the plates during the inter-seismic period (Thatcher 1979).

HETEROGENEITY IN THE FAULT ZONE

While the emphasis of this review has been on simple quasi-static models, it is important to at least point out the possible role of heterogeneity in stress, material properties, and fault geometry on fault mechanics. Fault models having simple geometries and uniform material properties, like those already discussed, are valuable for understanding large scale low frequency deformation fields associated with an earthquake cycle. How-ever, some heterogeneity is necessary to explain the following fundamental observations: multiple seismic events; high frequency near field ground accelerations; the frequency-magnitude distribution of earthquakes; the termination of rupture (Nur 1978, Andrews 1980, Segall & Pollard 1980).

As discussed by Andrews (1980), rupture termination requires that the difference between the initial shear stress and sliding friction stress vary on the length scale of the rupture, allowing the stress to decrease on much of the slip patch and increase around the borders to stop the rupture (Burridge & Halliday 1971, Andrews 1975). For the same rupture patch, heterogeneity is required on length scales smaller than the rupture length to explain the high frequency ground motion and subsequent smaller earthquakes (Andrews 1978, Nur 1978, Aki 1979). A problem with these frictional models is that the difference between stress and sliding friction becomes smoother with each event until eventually all earthquakes rupture the entire fault. A mechanism is needed to maintain the hetero-geneity between stress and friction.

One of the most obvious sources of heterogeneity is fault geometry. The mapped trace of a fault is never a straight cut or break, but often a collec-tion of bent, offset, and sometimes braided strands. Wallace (1973), for example, found that the longest individual fault strands along active portions of the San Andreas fault are about 10 to 18 km long, comparable to the depth of deepest earthquakes. A frequency count of segments by length suggested a distribution of the form $\log N = a + bL$ where N is the number of strands, L is the length, and a and b are constants. Irregular and discontinuous fault traces occur on all scales in nature, for both strike-slip and dip-slip faults, in a variety of rock types and tectonic settings (for a review see Segall & Pollard 1980).

While certain features of the mapped trace geometry may develop as slip propagates upward through unconsolidated sediments, there is some evidence that faults are discontinuous at appreciable depths (Segall & Pollard 1980). For example, normal faults observed in a South African gold mine are composed of en-echelon segments centimeters to meters in length (McGarr et al 1979). In addition, seismicity patterns often correlate with the surface trace. Aftershocks, at depths of 3–15 km, following the 1966 Parkfield, California, earthquake reflect a 1 km offset in the mapped surface fault trace (Eaton et al 1970b). Bakun et al (1980) and Bakun (1980) report good correlation between the fault trace geometry and epicenter locations (depths 5–8 km), rupture directivity, and aftershock locations on both the San Andreas and Calaveras faults in central California. Hill (1977) and Segall & Pollard (1980) find that earthquake swarms are sometimes localized within fault offsets.

Segall & Pollard (1980) have studied the mechanics of pairs of interacting en-echelon cracks in considerable detail. They suggest that left-stepping offsets on a right lateral fault are sites of increased normal compressive stress that inhibits slip, while right-stepping offsets on right lateral faults have decreased compressive stress, which facilitates slip. Areas of inhibited slip (right lateral, left step) might be sites of strain accumulation and large damaging earthquakes, while areas of enhanced slip (right lateral, right step) might have high seismicity. On a larger scale Mavko (1980) has modeled the interaction of four major faults near Hollister, California, each composed of many individual short segments. Complications in geometry, like large bends, seem to be capable of locking or unlocking sections of the fault which may be important for initiating instability.

SUMMARY

Nearly all fault models are consistent with the concepts of plate tectonics and elastic rebound. Through a combination of remotely applied forces the elastic plates move relative to each other. Whether or not strain accumulates and the way it is released depends on the slip at the common plate boundaries. In terms of the data, the best constrained portion of an earthquake rebound cycle is the rapid coseismic part. Although inelastic deformation in the upper mantle is necessary for long term plate motion and strain accumulation between earthquakes, the short term response of the crust and mantle due to rapid fault slip is essentially elastic. The area, orientation, average slip, and stress drop of the earthquake source can be determined from these coseismic elastic fields using dislocation theory.

A more difficult problem is resolving the sources of aseismic strain. The largest postseismic and interseismic strains appear to be dominated by a combination of aseismic fault slip and viscoelastic adjustments, primarily in the asthenosphere, while crustal effects, like the diffusion of pore fluids, contribute to a lesser extent.

One of the most promising lines of current research concerns the role of heterogeneity. Although much of our understanding of faulting has resulted from the success of greatly simplified models, heterogeneity in stress, material properties, and geometry is ubiquitous in nature. To some extent, these are a source of noise. For example, variations in crustal stiffness distort strain fields and complicate their interpretation. However, heterogeneity offers perhaps the only explanation for the following fundamental observations: the frequency-magnitude distribution of earthquakes; the termination of rupture; high frequency near field ground accelerations; multiple seismic events.

ACKNOWLEDGMENTS

Frequent discussions with Wayne Thatcher and Jim Savage during the preparation of this paper were extremely helpful. Barbara Mavko, John Langbein, Bill Stuart, and Wayne Thatcher provided useful comments on the manuscript.

Literature Cited

Abramowitz, M., Stegun, I. A. 1964. *Handbook of Mathematical Functions*. Washington, DC: Natl. Bur. Stand. 1046 pp.
Aki, K. 1966. Generation and propagation of G waves from the Niigata earthquake of June 16, 1964, 2, estimation of earthquake moment, released energy, and stress-strain drop from G-waves spectrum. *Bull. Earthquake Res. Inst. Tokyo Univ.* 44:73–88
Aki, K. 1979. Characterization of barriers on an earthquake fault. *J. Geophys. Res.* 84:6140–48
Allen, C. R., Smith, S. W. 1966. Parkfield earthquakes of June 27–29, Monterey and San Luis Obispo Counties, California. Pre-earthquake and post-earthquake surficial displacements. *Bull. Seismol. Soc. Am.* 56:966–67
Anderson, D. L. 1967. The anelasticity of the mantle. *Geophys. J.R. Astron. Soc.* 14:135–64
Anderson, D. L. 1971. The San Andreas fault. *Sci. Am.* 225:52–66
Anderson, D. L. 1975. Accelerated plate tectonics. *Science* 187:1077–79
Ando, M. 1975. Source mechanisms and tectonic significance of historical earthquakes along the Nankai Trough, Japan. *Tectonophysics* 27:119–40

Andrews, D. J. 1975. From antimoment to moment: plain strain models of earthquakes that stop. *Bull. Seismol. Soc. Am.* 65:163–82
Andrews, D. J. 1978. Coupling of energy between tectonic processes and earthquakes. *J. Geophys. Res.* 83:2259–64
Andrews, D. J. 1980. A stochastic fault model—I. static case. *J. Geophys. Res.* 85:3867–77
Ashby, M. F., Verrall, R. A. 1977. Micromechanisms of flow and fracture, and their relevance to the rheology of the upper mantle. *Philos. Trans. R. Soc. London Ser. A* 288:59–95
Bakun, W. H. 1980. Seismic activity (1969 to August 1979) on the southern part of the Calaveras fault in central California. *Bull. Seismol. Soc. Am.* 70:1181–98
Bakun, W. H., Stewart, R. M., Bufe, C. G., Marks, S. M. 1980. Implication of seismicity for failure of a portion of the San Andreas fault. *Bull. Seismol. Soc. Am.* 70:185–202
Barker, T. 1976. Quasi-static motions near the San Andreas fault zone. *Geophys. J.R. Astron. Soc.* 45:689–706
Bilby, B. A., Eshelby, J. D. 1968. Dislocations and the theory of fracture. In *Fracture, An Advanced Treatise*, ed. H.

Liebowitz, pp. 99–182. New York: Academic. 597 pp.

Biot, M. A. 1941. General theory of three dimensional consolidation. *J. Appl. Phys.* 12:155–64

Bischke, R. E. 1974. A model of convergent plate margins based on the recent tectonics of Shikoku, Japan. *J. Geophys. Res.* 79:4845–58

Bott, M. H. P., Dean, D. S. 1973. Stress diffusion from plate boundaries. *Nature* 243:339–41

Brace, W., Byerlee, J. 1970. California earthquakes: why only shallow focus? *Science* 168:1573–75

Bracewell, R. 1965. *The Fourier Transform and its Applications.* New York: McGraw-Hill. 381 pp.

Brown, L. D., Reilinger, R. E., Holdahl, S. R., Balazs, E. I. 1977. Post seismic crustal uplift near Anchorage Alaska. *J. Geophys. Res.* 82:3369–78

Brune, J. N. 1968. Seismic moment, seismicity, and rate of slip along major fault zones. *J. Geophys. Res.* 73:777–84

Burridge, R., Halliday, G. S. 1971. Dynamic shear cracks with friction as models for shallow focus earthquakes. *Geophys. J.* 25:261–83

Burridge, R., Knopoff, L. 1964. Body force equivalents for seismic dislocations. *Bull. Seismol. Soc. Am.* 54:1875–88

Canales, L. 1975. *Inversion of realistic fault models.* PhD thesis. Stanford Univ., Stanford, Calif.

Carter, N. L. 1976. Steady state flow of rocks. *Rev. Geophys. Space Phys.* 14:301–60

Chinnery, M. A. 1961. The deformation of the ground around surface faults. *Bull. Seismol. Soc. Am.* 51:355–72

Chinnery, M. A. 1963. The stress changes that accompany strike slip faulting. *Bull. Seismol. Soc. Am.* 53:921–32

Chinnery, M. A. 1964. The strength of the earth's crust under horizontal shear stress. *J. Geophys. Res.* 69:2085–89

Chinnery, M. A. 1965. The vertical displacements associated with transcurrent faulting. *J. Geophys. Res.* 70:4627–32

Chinnery, M. A. 1967. Theoretical fault models. In *A Symposium on Processes in the Focal Region*, ed. K. Kasahara, A. E. Stevens, pp. 211–23. Ottawa: Dominion Astrophys. Obs.

Chinnery, M. A. 1970. Earthquake displacement fields. In *Earthquake Displacement Fields and the Rotation of the Earth*, ed. L. Mansinha et al, pp. 17–38. Dordrecht: Reidel. 308 pp.

Chinnery, M. A., Jovanovich, D. B. 1972. Effect of earth layering on earthquake displacement fields. *Bull. Seismol. Soc. Am.* 62:1629–39

Chinnery, M. A., Petrak, J. A. 1967. The dislocation fault model with a variable discontinuity. *Tectonophysics* 5:513–29

Christensen, R. M. 1971. *Theory of Viscoelasticity.* New York: Academic. 245 pp.

Claerbout, J. F. 1976. *Fundamentals of Geophysical Data Processing.* New York: McGraw-Hill. 274 pp.

Dieterich, J. H. 1974. Earthquake mechanisms and modeling. *Ann. Rev. Earth Planet. Sci.* 2:275–301

Dieterich, J. H. 1979a. Modeling of rock friction, 1, experimental results and constitutive equations. *J. Geophys. Res.* 84:2161–68

Dieterich, J. H. 1979b. Modeling of rock friction, 2, simulation of preseismic slip. *J. Geophys. Res.* 84:2169–76

Dunbar, W. S. 1977. *The determination of fault models from geodetic data.* PhD thesis. Stanford Univ., Stanford, Calif.

Eaton, J. P., Lee, W. H. K., Pakiser, L. C. 1970a. Use of microearthquakes in the study of the mechanics of earthquake generation along the San Andreas fault in central California. *Tectonophysics* 9:259–82

Eaton, J. P., O'Neill, M. E., Murdock, J. N. 1970b. Aftershocks of the 1966 Parkfield-Cholame, California, earthquake: a detailed study. *Bull. Seismol. Soc. Am.* 60:1151–97

Elsasser, W. M. 1969. Convection and stress propagation in the upper mantle. In *The Application of Modern Physics to the Earth and Planetary Interiors*, ed. S. K. Runcorn, pp. 223–45. New York: Wiley

Elsasser, W. M. 1971. Two-layer model of upper-mantle circulation. *J. Geophys. Res.* 76:4744–53

Engelder, J. T., Logan, J. M., Handin, J. 1975. The sliding characteristics of sandstone on quartz fault-gouge. *Pure Appl. Geophys.* 113:69–86

Erdelyi, A., Magnus, W., Oberhettinger, F., Tricomi, F. G. 1954. *Tables of Integral Transforms, vol. 2.* New York: McGraw-Hill. 451 pp.

Eshelby, J. D. 1957. The determination of the elastic field of an ellipsoidal inclusion and related problems. *Proc. R. Soc. London Ser. A* 241:376–96

Fitch, T., Scholz, C. H. 1971. Mechanism of underthrusting in southwest Japan: a model of convergent plate interactions. *J. Geophys. Res.* 80:1444–47

Forsyth, D. W. 1979. Lithospheric flexure. *Rev. Geophys. Space Phys.* 17:1109–14

Freund, L. B. 1979. The mechanics of dynamic shear crack propagation. *J. Geophys. Res.* 84:2199–2209

Fung, Y. C. 1965. *Foundations of Solid Mechanics.* Englewood Cliffs, NJ: Prentice-Hall. 525 pp.

Gordon, R. B. 1965. Diffusion creep in the earth's mantle. *J. Geophys. Res.* 70:2413–18

Goulty, N. R., Gilman, R. 1978. Repeated creep events on the San Andreas fault near Parkfield, California, recorded by a strainmeter array. *J. Geophys. Res.* 83:5415–19

Hadley, D., Kanamori, H. 1977. Seismic structures of the Transverse Ranges, California. *GSA Bull.* 88:1469–78

Hanks, T. C. 1971. The Kuril trench-Hokkaido rise system: Large shallow earthquakes and simple models of deformation. *Geophys. J. R. Astron. Soc.* 23:173–89

Heard, H. C. 1976. Comparison of the flow properties of rocks at crustal condition. *Philos. Trans. R. Soc. London Ser. A* 283:173–89

Hill, D. P. 1977. A model for earthquake swarms. *J. Geophys. Res.* 82:1347–52

Jackson, D. D., Anderson, D. L. 1970. Physical mechanisms of seismic wave attenuation. *Rev. Geophys. Space Phys.* 8:1–63

Kanamori, H. 1973. Mode of strain release associated with major earthquakes in Japan. *Ann. Rev. Earth Planet. Sci.* 1:213–39

Kanamori, H., Anderson, D. L. 1975. Amplitude of the earth's free oscillations and long period characteristics of the earthquake source. *J. Geophys. Res.* 80:1075–78

Kanamori, H., Cipar, J. 1974. Focal processes of the great Chilean earthquake. *Phys. Earth Planet. Inter.* 9:128–36

Kanamori, H., Press, F. 1970. How thick is the lithosphere? *Nature* 226:330–31

Kasahara, K. 1957. The nature of seismic origins as inferred from seismological and geodetic observations (1). *Bull. Earthquake Res. Inst. Tokyo Univ.* 35:473–532

Kasahara, K. 1958. Physical conditions of earthquake faults as deduced from geodetic data. *Bull. Earthquake Res. Inst. Tokyo Univ.* 36:455–64

Ke, T. S. 1947. Experimental evidence of the viscous behavior of grain boundaries in metals. *Phys. Rev.* 71:533

King, C.-Y., Nason, R. D., Tocher, D. 1973. Kinematics of fault creep. *Philos. Trans. R. Soc. London Ser. A* 274:355–60

King, N. E., Savage, J. C., Lisowski, M., Prescott, W. H. 1980. Preseismic and coseismic deformation associated with

the Coyote Lake, California, earthquake. *J. Geophys. Res.* In press

Kirby, S. H. 1977. State of stress in the lithosphere: inferences from the flow laws of olivine. *Pure Appl. Geophys.* 115:245–58

Knopoff, L. 1958. The energy release in earthquakes. *Geophys. J.* 1:44–52

Lachenbruch, A. H., Sass, J. H. 1973. Thermo-mechanical aspects of the San Andreas Fault system. In *Proc. Conf. Tectonic Problems of the San Andreas Fault System*, ed. R. L. Kovach, A. Nur, pp. 192–205. Stanford, Calif: Stanford Univ. Publications

Langbein, J. O. 1980. An interpretation of episodic slip on the Calaveras fault near Hollister, California. *J. Geophys. Res.* In press

Lensen, G. 1970. Elastic and non-elastic surface deformation in New Zealand. *Bull. N.Z. Soc. Earthquake Eng.* 3:131–43

Le Pichon, X., Francheteau, J., Bonnin, J. 1973. *Plate Tectonics.* New York: Elsevier. 300 pp.

Logan, J. M. 1979. Brittle phenomena. *Rev. Geophys. Space Phys.* 17:1121–31

Logan, J. M., Shimamoto, T. 1976. The influence of calcite gouge on the frictional sliding of Tennessee sandstone (abstract). *EOS, Trans. Am. Geophys. Union* 57:1011

Love, A. E. H. 1944. *A Treatise on the Mathematical Theory of Elasticity.* New York: Dover. 643 pp.

Madariaga, R. 1979. On the relation between seismic moment and stress drop in the presence of stress and strength heterogeneity. *J. Geophys. Res.* 84:2243–50

Mahrer, K. 1978. *Strike slip faulting, models for deformation in a nonuniform crust.* PhD thesis. Stanford Univ., Stanford, Calif. 190 pp.

Mahrer, K. D., Nur, A. 1979. Strike slip faulting in a downward varying crust. *J. Geophys. Res.* 84:2296–2302

Maruyama, T. 1964. Statical elastic dislocations in an infinite and semi-infinite medium. *Bull. Earthquake Res. Inst. Tokyo Univ.* 42:289–368

Matuzawa, T. 1964. *Study of Earthquakes.* Tokyo: Uno Shoten

Mavko, G. 1977. *Time dependent fault mechanics and wave propagation in rocks.* PhD thesis. Stanford Univ., Stanford, Calif.

Mavko, G. 1978. Large scale quasi-static fault models. In *Proc. Conf. III Fault Mechanics and Its Relation to Earthquake Prediction*, ed. J. F. Evernden, pp. 339–412. Menlo Park, Calif: US Geol. Surv.

Mavko, G. M. 1980. The influence of

local moderate earthquakes on creep rate near Hollister, California, *Earthquake Notes* 50:71

Mavko, G., Nur, A. 1975. Melt squirt in the asthenosphere. *J. Geophys. Res.* 80:1444–47

Mavko, G., Nur, A. 1978. The effect of non-elliptical cracks on the compressibility of rocks. *J. Geophys. Res.* 83:4459–68

Mavko, G. M., Nur, A. 1979. Wave attenuation in partially saturated rocks. *Geophysics* 44:161–78

McConnell, R. K. 1968. Viscosity of the mantle from relaxation time spectra of isostatic adjustment. *J. Geophys. Res.* 73:7089–7105

McGarr, A., Pollard, D., Gay, N. C., Ortlepp, W. D. 1979. Observations and analysis of structures in exhumed mines. In *Proc. Conf. VIII Analysis of Actual Fault Zones in Bedrock*, ed. J. F. Evernden, pp. 101–20. Menlo Park, Calif.: US Geol. Surv. 594 pp.

Melosh, H. J. 1976. Nonlinear stress propagation in the earth's upper mantle. *J. Geophys. Res.* 81:5621–32

Melosh, H. J. 1977. Shear stress on the base of a lithospheric plate. *Pure Appl. Geophys.* 115:429–39

Melosh, H. J. 1978a. Reply. *J. Geophys. Res.* 83:5009–10

Melosh, H. J. 1978b. Dynamic support of the outer rise. *Geophys. Res. Lett.* 5:321–24

Mescherikov, J. A. 1968. Recent crustal movements in seismic regions: geodetic and geomorphic data. *Tectonophysics* 6:29–39

Mindlin, R. D., Cheng, D. H. 1950. Nuclei of strain in the semi-infinite solid. *J. Applied Phys.* 21:926–30

Muskhelishvili, N. I. 1953. *Some Basic Problems of the Mathematical Theory of Elasticity*. Groningen, Holland: Noordhoff. 704 pp.

Nakamura, K., Tsuneishi, Y. 1967. Ground cracks at Matsushiro probably of strike-slip fault origin. *Bull. Earthquake Res. Inst. Univ. Tokyo.* 45:417–72

Nason, R. D. 1973. Fault creep and earthquakes on the San Andreas Fault. In *Proc. Conf. Tectonic Problems of the San Andreas Fault System*, ed. R. L. Kovach, A. Nur, pp. 275–3895. Stanford, Calif.: Stanford Univ. Publications

Nason, R., Weertman, J. 1973. A dislocation theory analysis of fault creep events. *J. Geophys. Res.* 78:7745–51

Nur, A. 1973. Role of pore fluids in faulting. *Philos. Trans. R. Soc. London Ser. A.* 274:297–304

Nur, A. 1978. Nonuniform friction as a basis for earthquake mechanics. *Pure Appl. Geophys.* 116:964–91

Nur, A., Booker, J. R. 1972. Aftershocks caused by pore fluid flow? *Science* 175:885–87

Nur, A., Mavko, G. 1974. Postseismic viscoelastic rebound. *Science* 183:204–6

O'Connell, R. J., Budiansky, B. 1977. Viscoelastic properties of fluid-saturated cracked solids. *J. Geophys. Res.* 82:5719–35

Petrak, J. A. 1965. *Some theoretical implications of strike-slip faulting.* M.A. thesis. Univ. British Columbia, Vancouver, B.C.

Pfluke, J. H. 1978. Slow earthquakes and very slow earthquakes. In *Proc. Conf. III Fault Mechanics and Its Relation to Earthquake Prediction*, ed. J. F. Evernden, pp. 447–468. Menlo Park, Calif: US Geol. Surv.

Post, R., Griggs, D. 1973. The earth's mantle: evidence of non-Newtonian flow. *Science* 181:1242–44

Prescott, W. H., Lisowski, M. 1977. Deformation at Middleton Island, Alaska, during the decade after the Alaska earthquake of 1964. *Bull. Seismol. Soc. Am.* 66:1013–16

Prescott, W. J., Lisowski, M. 1980. Vertical deformation at Middleton Island *Earthquake Notes* 50:72 (Abstr.)

Press, F. 1965. Displacements, strains, and tilts at teleseismic distances. *J. Geophys. Res.* 70:2395–2412

Reid, H. F. 1910. The mechanics of the earthquake. In *The California Earthquake of April 18, 1906. Rep. State Earthquake Invest. Comm.* Washington, DC: Carnegie Inst.

Rudnicki, J. W. 1980. Fracture mechanics applied to the earth's crust. *Ann. Rev. Earth Planet. Sci.* 8:489–525

Rundle, J. B., Jackson, D. D. 1977. A viscoelastic relaxation model for post-seismic deformation from the San Francisco earthquake of 1960. *Pure Appl. Geophys.* 115:401–11

Rybicki, K. 1971. The elastic residual field of a very long strike-slip fault in the presence of a discontinuity. *Bull. Seismol. Soc. Am.* 61:79–92

Rybicki, K., Kasahara, K. 1977. A strike-slip fault in a laterally inhomogeneous medium. *Tectonophysics* 42:127–38

Sacks, I. S., Suyehiro, S., Linde, A. T., Snoke, J. A. 1977. The existence of slow earthquakes and the redistribution of stress in seismically active regions. *EOS, Trans. Am. Geophys. Union.* 58:437

Savage, J. C. 1975. Comment on "An analysis of strain accumulation on a strike slip fault" by D. L. Turcotte and D. A. Spence. *J. Geophys. Res.* 80:4111–14

Savage, J. C. 1978. Comment on "Strain accumulation and release mechanism of the 1906 San Francisco earthquake" by W. Thatcher. *J. Geophys. Res.* 83:5487–89

Savage, J. C., Burford, R. O. 1970. Accumulation of tectonic strain in California. *Bull. Seismol. Soc. Am.* 60:1877–96

Savage, J. C., Hastie, L. M. 1966. Surface deformation associated with dip-slip faulting. *J. Geophys. Res.* 71:4897–4904

Savage, J. C., Prescott, W. H. 1978a. Asthenospheric readjustment and the earthquake cycle. *J. Geophys. Res.* 83:3369–76

Savage, J. C., Prescott, W. H. 1978b. Comment on "Nonlinear stress propagation in the earth's upper mantle" by H. J. Melosh. *J. Geophys. Res.* 83:5005–8

Savage, J. C., Prescott, W. H., Lisowski, M., King, N. 1979. Geodolite measurements of deformation near Hollister, California, 1971–1978. *J. Geophys. Res.* 84:7599–7615

Scholz, C. H. 1972. Crustal movements in tectonic areas. *Tectonophysics* 14:201–17

Scholz, C. H., Kato, T. 1978. The behavior of a convergent plate boundary: crustal deformation in the South Kanto district, Japan. *J. Geophys. Res.* 83:783–97

Scholz, C. H., Sykes, L. R., Aggarwal, Y. P. 1973. Earthquake prediction: a physical basis. *Science* 181:803–10

Scholz, C. H., Wyss, M., Smith, S. W. 1969. Seismic and aseismic slip on the San Andreas fault. *J. Geophys. Res.* 74:2049–69

Segall, P., Pollard, D. D. 1980. Mechanics of discontinuous faults. *J. Geophys. Res.* 85:4337–50

Shimazaki, K. 1974. Preseismic crustal deformation caused by an underthrusting oceanic plate, in eastern Hokkaido, Japan. *Phys. Earth Planet. Inter.* 8:148–57

Smith, A. T. 1974. Time-dependent strain accumulation and release at island arcs. *EOS, Trans. Am. Geophys. Union* 55:427

Smith, A. T. 1980. *Final Technical Report: Earthquake risk analysis using numerical and stochastic models of time-dependent strain fields.* Santa Cruz, Calif: Univ. Calif.

Smith, S. W., Wyss, M. 1968. Displacement on the San Andreas fault initiated by the 1966 Parkfield earthquake. *Bull. Seismol. Soc. Am.* 58:1955–74

Sneddon, I. N., Lowengrub, M. 1969. *Crack Problems in the Classical Theory of Elasticity.* New York: Wiley. 221 pp.

Spence, D. A., Turcotte, D. L. 1976. An elastostatic model of stress accumulation on the San Andreas fault. *Proc. R. Soc. London Ser. A* 349:319–41

Spence, D. A., Turcotte, D. L. 1979. Viscoelastic relaxation of cyclic displacements on the San Andreas fault. *Proc. R. Soc. London Ser. A* 365:121–44

Starr, A. T. 1928. Slip in a crystal and rupture in a solid due to shear. *Proc. Camb. Philos. Soc.* 24:489–500

Steketee, J. A. 1958a. On Volterra's dislocations in a semi-infinite medium. *Can. J. Phys.* 36:192–204

Steketee, J. A. 1958b. Some geophysical applications of the elasticity theory of dislocations. *Can. J. Phys.* 36:1168–97

Stesky, R. M. 1974. Steady-state creep law for frictional sliding at high temperature and pressure. *EOS, Trans. Am. Geophys. Union* 55:428

Stuart, W. D. 1978. Review of theories for earthquake instabilities. In *Proc. Conf. III Fault Mechanics and Its Relation to Earthquake Prediction,* ed. J. F. Evernden, pp. 541–88. Menlo Park, Calif: US Geol. Surv.

Stuart, W. D. 1979. Quasi-static earthquake mechanics. *Rev. Geophys. Space Phys.* 17:1115–20

Stuart, W. D., Mavko, G. M. 1979. Earthquake instability on a strike-slip fault. *J. Geophys. Res.* 84:2153–60

Summers, R., Byerlee, J. D. 1977. A note on the effect of fault gouge composition on the stability of frictional sliding. *Int. J. Rock Mech. Min. Sci.* 14:155–60

Thatcher, W. 1975. Strain accumulation and release mechanism of the 1906 San Francisco earthquake. *J. Geophys. Res.* 80:4862–72

Thatcher, W. 1978. Reply. *J. Geophys. Res.* 83:5490–92

Thatcher, W. 1979. Crustal movements and earthquake-related deformation. *Rev. Geophys. Space Phys.* 17:1403–10

Thatcher, W., Matsuda, T., Kato, T., Rundle, J. B. 1980. Lithospheric loading by the 1896 Riku-U earthquake, northern Japan: implications for plate flexure and asthenospheric rheology. *J. Geophys. Res.* In press

Thatcher, W., Rundle, J. B. 1979. A model for the earthquake cycle in the underthrust zones. *J. Geophys. Res.* 84:5540–56

Tullis, J. A. 1979. High temperature deformation of rocks and minerals. *Rev. Geophys. Space Phys.* 17:1137–54

Turcotte, D. L. 1977. Stress accumulation and release on the San Andreas fault. *Pure Appl. Geophys.* 115:413–25

Turcotte, D. L., Spence, D. A. 1975. Reply to comments by J. C. Savage, *J. Geophys. Res.* 80:4115

US Geological Survey. 1980. *Professional Paper, The Imperial Valley Earthquake of October 15, 1979.* Menlo Park, Calif: US Geol. Surv.

Walcott, R. I. 1973. Structure of the earth from glacio-isostatic rebound. *Ann. Rev. Earth Planet. Sci.* 1:15–37

Wallace, R. E. 1973. Surface fracture patterns along the San Andreas fault. In *Proc. Conf. Tectonic Problems of the San Andreas Fault System*, ed. R. L. Kovach, A. Nur. Stanford Univ. Publications

Walsh, J. B. 1969. A new analysis of attenuation in partially melted rock. *J. Geophys. Res.* 74:4333

Watts, A. B. 1978. An analysis of isostasy in the world's oceans, 1, Hawaiian-Emperor seamount chain. *J. Geophys. Res.* 83: 5989–6004

Watts, A. B., Talwani, M. 1974. Gravity anomalies seaward of deep sea trenches and their tectonic implications. *Geophys. J.R. Astron. Soc.* 36:57–90

Weertman, J. 1964. Continuum distribution of dislocations on faults with finite friction. *Bull. Seismol. Soc. Am.* 54:1035–58

Weertman, J. 1965. Relationship between displacements on a free surface and the stress on a fault. *Bull. Seismol. Soc. Am.* 55:945–53

Weertman, J. 1978. Creep laws for the mantle of the earth. *Philos. Trans. R. Soc. London Ser. A* 288:9–26

Weertman, J., Weertman, J. R. 1964. *Elementary Dislocation Theory.* London: Macmillan. 213 pp.

Weertman, J., Weertman, J. R. 1975. High temperature creep of rock and mantle viscosity. *Ann. Rev. Earth Planet. Sci.* 3:293–315

Wyss, M., Brune, J. N. 1968. Seismic moment, stress, and source dimensions in the California-Nevada region. *J. Geophys. Res.* 73:4681–94

Zener, C. 1948. *Elasticity and Anelasticity of Metals.* Chicago: Univ. of Chicago Press. 170 pp.

Ann Rev. Earth Planet. Sci. 1981. 9:113–45

ROTATION AND PRECESSION ✖ 10146
OF COMETARY NUCLEI

Z. Sekanina

Harvard-Smithsonian Center for Astrophysics, Cambridge, Massachusetts 02138*

INTRODUCTION

The assumption of isotropic outgassing, still nearly universally employed in physical studies of comets, illustrates the current degree of ignorance in the understanding of the nature of the cometary nucleus. This regrettable situation is caused by the lack of information in two major areas: (*a*) morphology and thermophysical properties of the surface layer of the nucleus; (*b*) bulk, especially rotational, characteristics of the nucleus. This review is concerned with a recent breakthrough in the investigation of polar-axis orientations and spin rates, and with the implications for the surface structure of cometary nuclei. The rotation data for twelve comets are summarized in Table 1.

The progress achieved in the past few years has demonstrated that seemingly complex relations among observed physical and dynamical properties of comets can consistently be interpreted in terms of Whipple's (1950, 1951) model of a rotating icy-conglomerate nucleus. The nucleus, probably a few kilometers in diameter for an average comet, is believed to lose mass more or less continuously, at higher rates near the sun, slowly farther away. The lost mass consists of volatiles and nonvolatiles. The volatile fraction is the ices or frozen gases that sublimate from the surface (or from beneath the surface) of the nucleus in response to the heating by sunlight. The outgassing is accompanied by a momentum transferred to the nucleus in the direction opposite that of the escaping gas. The nonvolatile fraction is the dust, embedded in, or otherwise physically connected with, the ices in the nucleus and released and accelerated away from the nucleus due to a force exerted by momentum transfer from the expanding gas. The situation is complicated by the fact that some of the volatile matter may

* Present address: Jet Propulsion Laboratory, California Institute of Technology, Pasadena, California 91109

0084-6597/81/0515-0113$01.00

Table 1 List of rotation periods and obliquities of the orbit plane to the equator for twelve comets

Comet[a]	Designation or apparition(s)	Rotation period (hr)	Obliquity (deg)	Reference
Bennett	1970 II	10.8	—	Whipple (1981)
		33–36	—	Larson & Minton (1972)
P/Borrelly	1911–1932	—	98–130	Sekanina (1979a and unpublished)
Coggia	1874 III	9.2	—	Whipple (1981)
Daniel	1907 IV	16	91	Horn (1908)
P/d'Arrest	1870, 1963	8.9	—	Whipple (1981)
	1976	5.2	—	Fay & Wisniewski (1978)
	1976	—	~90	Cowan & A'Hearn (1979)
Donati	1858 VI	4.6	—	Whipple (1978a)
P/Encke	1786	—	95	Whipple & Sekanina (1979)
	1855	—	103	" " " "
	1901	6.5	99	" " " "
	1947	—	91	" " " "
	1977	—	84	" " " "
P/Halley	1835, 1910	10.3	—	Whipple (1980b)
P/Schwassmann-Wachmann 1	1927–1977	119	65	Whipple (1980a)
P/Schwassmann-Wachmann 3	1930	—	35–50	Sekanina (1979a)
P/Swift-Tuttle	1862	66.5	80	Sekanina (1981)
P/Tempel 2	1920–1967	—	65–75	Sekanina (1979a and unpublished)

[a] The symbol P/ indicates that the comet is a short-period one, that is having a revolution period about the sun of less than 200 years.

leave the nucleus in the form of icy grains, which, like dust particles, exert no momentum on the comet. Yet the grains may sublimate rapidly enough to add appreciably to the sublimation rate as measured at larger distances from the nucleus. A further complication arises from the fact that some of the expanding gas may be diverted back to the nucleus surface due to molecular collisions, recondense, and subsequently resublimate, in which case the nucleus is subjected to more than one impulse per molecule.

If the sublimation process were characterized by perfect symmetry with respect to the center of mass of the nucleus, momentum contributions from individual molecules would cancel out when integrated over the surface and there would be no net dynamical effect on the nucleus. Since for a number of reasons this is not so, the momentum is generally manifest in three fashions:

1. It affects the spin rate, if the momentum vector does not pass through the rotation axis. Since the thrust is directed normal to the surface of sublimation, the magnitude of this effect is related to the structure of the emission source in the nucleus. In the case of surface sublimation the degree of macroscopic surface roughness is the factor.

2. An off-center component of the momentum exerts a torque on the spin axis, forcing its precession. The forced precession is generally super-imposed on a free precession. However, if cometary nuclei are poorly cemented, their rigidity should be very low. Because of internal energy dissipation, any original deviation of the spin axis from the direction of the axis of maximum moment of inertia should have been damped out and the two axes aligned on an astronomically short time scale.

3. A component of the momentum exerts an acceleration on the center of mass of the comet, which shows up as a perturbation of the comet's gravitational motion about the sun. Such orbital perturbations are known as nongravitational or jet effects.

It has been shown that in the presence of dust the initial gas flow is subsonic (Probstein 1969), except when highly volatile species erupt from subsurface pressurized pockets, in which case it is believed to be perceptibly supersonic (Huebner 1976). Hence, with the possible exception of violent outbursts the initial gas velocity may be approximated by the speed of sound.

THE OUTGASSING ASYMMETRY

Circumstantial evidence based on physical observations of comets mainly at small geocentric distances points to complex surface morphology of cometary nuclei in general and to the apparent existence of isolated active areas in particular. Unfortunately, distinct structures in cometary atmospheres can only rarely be resolved on a linear scale smaller than 10^4 km and exceptionally on a scale smaller than 10^3 km. Yet, a resolution better than 10^2 km would probably be needed to narrow down ambiguities in the interpretation of the emission pattern.

To illustrate this point, I refer to the well-known asymmetry of the coma (atmosphere) of many comets, especially those with revolution periods about the sun shorter than 10 years. Appearing as a fanlike extension emanating from the central condensation (inner atmosphere) of the comet, the asymmetric coma points generally, but not exactly, in the sunward direction. The characteristic extent of the fan, in which virtually all the light is concentrated, is on the order of 10^4 km and its documented existence is restricted to a period of several weeks to several months around the time of perihelion passage, varying from comet to comet. As an example, the photographic appearance of the fan of Periodic Comet Pons-Winnecke, observed under extraordinarily favorable conditions in

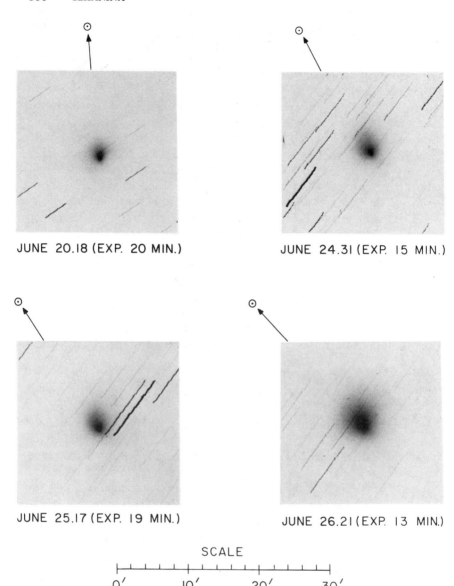

JUNE 20.18 (EXP. 20 MIN.)

JUNE 24.31 (EXP. 15 MIN.)

JUNE 25.17 (EXP. 19 MIN.)

JUNE 26.21 (EXP. 13 MIN.)

SCALE

0'   10'   20'   30'

Figure 1 The fan-shaped coma of Periodic Comet Pons-Winnecke in its 1927 apparition, photographed by G. Van Biesbroeck with the 61-cm reflector of the Yerkes Observatory. North is up, east to the left. The circled dots show the directions to the sun. The nucleus is located near the south-southwest tip of the densest part of the images. Note the deviations of fan orientation from the sunward direction. The linear scale is 2882 km per arcmin on June 20, 1926 km on the 24th, 1808 km on the 25th, and 1727 km on the 26th. (Original photographs courtesy of Yerkes Observatory.)

1927, is shown in Figure 1. The last of the four photographs had been taken shortly before the comet reached the point of closest approach to the earth, being only 5.9 million km away. This circumstance gave observers an exceptional opportunity to detect limited details in the coma on a scale of only hundreds of km from the nucleus. On a densitometer scan of the comet on June 26, shown in Figure 2, the isophotes exhibit two noteworthy features:

1. The position angle of the extension varies from 30° (i.e. pointing 30° east of north) at distances less than 3000 km from the nucleus to 10° at

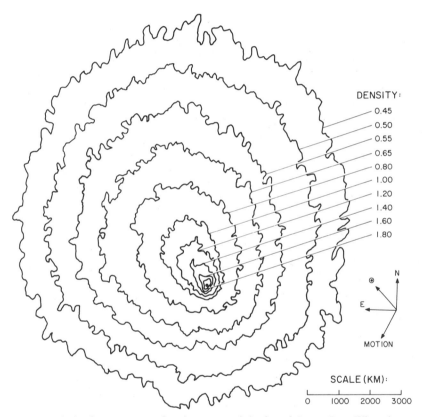

DENSITY:

0.45
0.50
0.55
0.65
0.80
1.00
1.20
1.40
1.60
1.80

N
E
MOTION

SCALE (KM):

0   1000   2000   3000

Figure 2 A densitometer scan of an inner part of the fan of Comet Pons-Winnecke on June 26, 1927 (cf Figure 1). The comet was less than 6 million km distant from the earth and its entire visible coma was more than 120,000 km in diameter. Its motion in the plane normal to the line of sight amounted to some 14,000 km during the 13-minute exposure. Note the conspicuous sunward jet near the nucleus, as described in the text. The circled dot shows again the direction to the sun. The scanning aperture was a square with a side of 50 microns, corresponding to 125 km at the comet's distance.

8000 km. Apparently an effect of kinetic properties of the expanding matter, a similar curvature was also reported for the emission fans of other comets, for example by Stobbe (1938) for Periodic Comet Encke.

2. Both the asymmetry of the contours relative to the nucleus and their elongation increase progressively with decreasing distance from the nucleus. The fan, appearing visually as a sector of $\sim 50°$ opening in Figure 1 on a characteristic scale of 10^4 km, is a narrow expanding jet in Figure 2 on a scale of 10^2 km! The pointedness of the jet would unquestionably show up still more prominently, if the densitometer contours could be corrected for inevitable imperfect guiding of the comet. The apparent elongation of the contours, especially the innermost ones, in the projected direction of motion—here nearly perpendicular to the projected direction of the jet—is the effect of imperfect guiding.

As will be seen from the following, the outgassing asymmetry appears to reflect both the structural heterogeneity of the surface layer and the uneven distribution of the insolation over the nucleus. The deviations of fan orientation from the sunward direction imply that the rate of outgassing is not strictly a measure of the instantaneous local insolation and that the observed pattern of fan orientation should contain information on the position of the spin axis and on the comet's morphological and thermophysical properties at and beneath the nucleus surface.

DIRECTED EJECTION, SPIN-AXIS ORIENTATION, AND A COMET'S "PERSONALITY"

To formulate the geometry of the directed ejection quantitatively, consider a small volume of volatile material emitted radially away from its site on the spherical nucleus (Figure 3). The angular distance θ of the area from the subsolar meridian, i.e. the "hour angle," at the time of ejection is counted positively in the direction of rotation, i.e. $\theta < 0$ in the cometary "morning" and $\theta > 0$ in the "afternoon." The angular distance of the area from the equator is measured by latitude ϕ, positive for the northern hemisphere and negative for the southern.

The two angles that determine the orientation of the polar axis of rotation are the obliquity I of the orbit plane to the equator (the angle subtended by the axis and by the normal to the orbit plane), and the argument Φ of the subsolar meridian at perihelion. The obliquity, reckoned from the equator, determines the sense of rotation, which is prograde, i.e. consistent with the sense of the comet's orbital motion about the sun, when $0 \leqq I < 90°$, or retrograde, when $90° < I \leqq 180°$. Argument Φ, measured along the orbit from its ascending node on the equator (i.e.

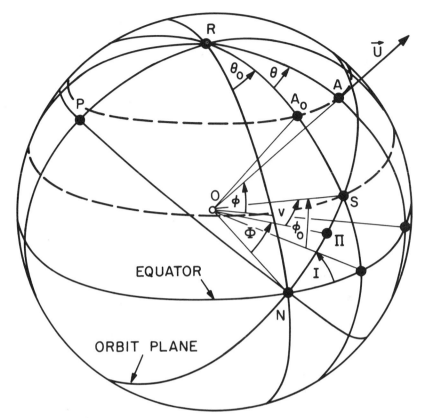

Figure 3   Orientation on a spherical rotating nucleus. O is its center, P the northern orbital pole, R the northern pole of rotation, N the ascending node of the orbit plane on the equator (i.e. the comet's vernal equinox), and Π the subsolar point at perihelion. Angle I is the obliquity of the orbit plane to the equator and Φ the argument of the subsolar meridian at perihelion. At time t the subsolar point is at S, so that v is the true anomaly, θ_0 the longitude of the subsolar point from the vernal equinox, and ϕ_0 its latitude. The position of an active area is given by point A_0 at the time of its pass through the subsolar meridian and by A at the time of (peak) emission U. Then ϕ is the area's latitude and, at the emission time, θ its "hour angle" and $\theta + \theta_0$ its longitude from the vernal equinox.

"vernal equinox") in the sense of the orbital motion, determines which pole faces the sun at perihelion: the sunlit pole is the northern one when $0 < \Phi < 180°$, the southern one when $180° < \Phi < 360°$. When the angles I and Φ are known, the unit vector ω along the spin axis can be calculated from

$$\omega = -\mathbf{P}\sin I \sin \Phi - \mathbf{Q}\sin I \cos \Phi + \mathbf{R}\cos I, \tag{1}$$

where **P**, **Q**, and **R** are the unit vectors directed, respectively, to the perihelion point, to the point in the orbit plane at true anomaly (i.e. angular distance from the direction to perihelion) of $+90°$, and to the northern pole of the orbit. Through ω the ecliptical components of **P**, **Q**, and **R** determine the longitude L and latitude B of the northern pole of rotation.

A unit vector **U** in the direction of the ejection has, at the position in orbit defined by true anomaly v, the following components in the direction of the prolonged radius vector, U_r; in the direction perpendicular to the sunward direction in the orbit plane in the sense of the comet's orbital motion, U_t; and in the direction of the northern orbital pole, from which the comet is seen to orbit the sun counterclockwise, U_n (Sekanina, unpublished):

$$
\begin{pmatrix} U_r \\ U_t \\ U_n \end{pmatrix} = - \begin{pmatrix} \cos(\Phi + v) & \cos I \sin(\Phi + v) & \sin I \sin(\Phi + v) \\ -\sin(\Phi + v) & \cos I \cos(\Phi + v) & \sin I \cos(\Phi + v) \\ 0 & \sin I & -\cos I \end{pmatrix}
$$
$$
\times \begin{pmatrix} \cos\phi\cos(\theta + \theta_0) \\ \cos\phi\sin(\theta + \theta_0) \\ \sin\phi \end{pmatrix}. \tag{2}
$$

The longitude of the subsolar meridian from the ascending node of the orbit plane on the equator, θ_0, and the latitude of the subsolar point, ϕ_0, are related to Φ, I, and true anomaly v by

$$
\begin{aligned}
\sin\phi_0 &= \sin I \sin(\Phi + v), \\
\cos\phi_0 \sin\theta_0 &= \cos I \sin(\Phi + v), \\
\cos\phi_0 \cos\theta_0 &= \cos(\Phi + v).
\end{aligned} \tag{3}
$$

A standard transformation of **U** to the geocentric (or topocentric) equatorial system of coordinates provides the relation between the location of the emission area and the orientation of the polar axis, on the one hand, and the position angle of the projected direction of the ejected material, on the other. In applications to emission fans, one seeks a trial-and-error solution to minimize the position-angle residuals between predicted and observed fan orientation. Since one observes only the projection of the fan direction onto the plane of the sky, and also because various circumstances restrict the precision of observations, a number of fan-orientation measures under diverse geometric conditions are needed to eliminate ambiguous solutions. Only one ambiguity cannot be removed by this approach: if Equations (2) and (3) are satisfied by a particular set of values I, Φ, ϕ_0, θ_0, ϕ, and θ, they are automatically also satisfied by the set

$180° - I$, $\Phi \pm 180°$, $- \phi_0$, $180° - \theta_0$, $- \phi$, and $- \theta$. In short, the sense of rotation remains indeterminate. Fortunately, the sense of rotation of a periodic comet observed at a few apparitions can usually be established dynamically (see the section on the nongravitational effects). As seen from the following, comparison of the two equivalent solutions on a limited sample of comets shows an overwhelming preference for the version with $\theta > 0$, i.e. for the afternoon activity. Angle θ thence may appropriately be termed a lag angle, being a function of both the thermophysical properties of the emission area and the comet's spin rate.

The practical application of the described formalism for directed outgassing turns out to be model dependent, requiring assumptions on ϕ and θ of the ejection. Published results for four short-period comets (Sekanina 1979a) are based on the premise that the activity is controlled entirely by the insolation, with allowance made for a potentially significant, variable lag between maximum insolation (i.e. the subsolar point) and emission peak (i.e. the axis of the emission fan). In this scenario, $\phi \equiv \phi_0$ by definition [cf (3)] and I, Φ, and the time variable lag angle $\theta > 0$ are the unknowns. The insolation-control model can be shown from (2) to provide the following expressions for the components of $\mathbf{U}$:

$$U_r = -[\cos \theta + (1 - \cos \theta) \sin^2 I \sin^2 (\Phi + v)],$$
$$U_t = -[\sin \theta \cos I + (1 - \cos \theta) \sin^2 I \sin (\Phi + v) \cos (\Phi + v)], \quad (4)$$
$$U_n = -[\sin \theta \sin I \cos (\Phi + v) - (1 - \cos \theta) \sin I \cos I \sin (\Phi + v)].$$

As an alternative to the insolation-controlled activity, one may assume that the emission fan is the product of a single, localized area on the spinning nucleus, which is activated for only a relatively small fraction of the rotation period by its exposure to solar radiation. This active-area model is supported by the remarkable success that Whipple (1978a, 1980a,b) has recently achieved in showing that the activity of some comets is indeed spin modulated (see the section on the rotation period). Since I, Φ, ϕ, and θ in (2) are now all independent unknowns, a practical approach is virtually restricted to $\theta = $ const, although this apparently is only a very crude approximation.

To illustrate the sensitivity and accuracy of the devised techniques, the results of comparison of the insolation-control and active-area models are presented in Table 2 for Comet Tempel 2. Based on 20 observations of fan orientation from 55 days before perihelion to 117 days after perihelion in seven apparitions between 1920 and 1967, the two solutions gave rotation-pole positions that differ by 36°. The option of a variable lag was essential for the application of the insolation-control model: the lag

Table 2 Comparison of two models for the rotating nucleus of Comet Tempel 2

| Model of activity | Mean residual in position angle | North pole of rotation[a] | | Obliquity[a] I | Argument[a] Φ | Hour (lag) angle of ejection peak θ | Cometocentric latitude of area ϕ |
		L	B				
Insolation controlled	$\pm 6°\!.4$	$110°$	$+23°$	$65°$	$105°$	variable[b]	$\equiv$ subsolar point[c]
Active-area dominated	± 10.2	76	$+7$	75	142	$+37°$	$+32°$

[a] L, B, I, and Φ are averages for the period 1920–1967; equinox 1950.0.

[b] The variations in θ with time are plotted in Figure 4.

[c] The variations in the latitude of the subsolar point are given by the first of Equations (3).

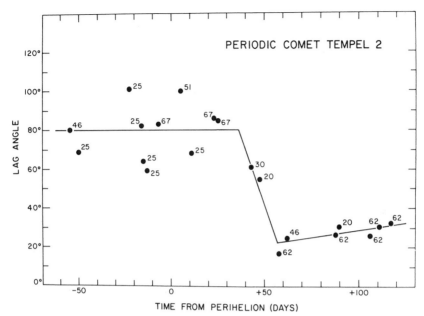

Figure 4   The lag angle θ versus time for the model of insolation-controlled activity of Comet Tempel 2. The number at each entry gives the last two digits of the year of perihelion passage. The solid line is the adopted fit, giving the mean position-angle residual in Table 2. From Sekanina (1979a).

angle θ, plotted versus time in Figure 4, came out to be about 80° before and around perihelion, dropping rapidly to nearly 20° less than two months after perihelion. In the framework of the model, such a large variation must be interpreted in terms of a major morphological discontinuity. The rapid drop in θ occurred between latitudes +35° to +40°.

The fit of fan orientation based on the assumption of a fixed active area was also satisfactory. Significant results of this solution were the area's cometocentric latitude of +32°, close to the zone of the morphological discontinuity indicated by the insolation-control model; again a significant lag angle, corresponding to a time delay of just about 1/10 the rotation period; and a strong correlation between the overall activity of the comet and the calculated altitude of the sun above the active area's horizon. In particular: (*a*) the light curve of the comet, i.e. its integrated-brightness variation with time, was highly correlated with the sun's altitude at "local noon" (Figure 5), including a virtual coincidence, about one or two weeks after perihelion, of the brightness peak with the pass of the sun through the zenith of the active area; (*b*) the first reported appearance of the

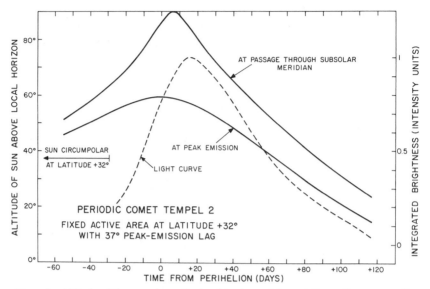

Figure 5   Altitude of the sun above the local horizon of the rotating active area on the nucleus of Comet Tempel 2 at the time of passage through the subsolar meridian and at the time of peak emission. The comet's mean light curve (in relative intensity units) is plotted for comparison.

emission fan coincided with the sun's altitude of about 50° above the horizon at local noon and about 45° at peak emission; and (*c*) the effective termination of the fan occurred at the sun's altitude of somewhat more than 20° at noon and about 15° at peak emission.

This and similar investigations of three more short-period comets (Sekanina 1979a and unpublished) suggest that both models need improvement and that the best first step on the way of achieving it might be their appropriate combination. In the case of Tempel 2 this need was documented by the fact that neglect of the surface heterogeneity in the model of insolation-controlled activity and of the insolation pattern in the model of fixed-area-dominated activity led to clearly recognizable effects involving the neglected factor (lag-angle variation in the first case, the sun's altitude in the latter). In any event, it appears that the "personality" of a comet cannot be understood without the knowledge of the position of its spin axis. This is especially true because available evidence suggests that at least among the short-period comets the obliquity near 90° occurs more frequently than it should statistically and that there also is a tendency for the spin axis to point toward the sun at perihelion. This configuration entails the most extreme insolation regime possible.

The personality of a comet is frequently reflected in its light curve. As illustrated by the memorable case of Comet Kohoutek 1973 XII, the light curve is unpredictable. It varies greatly not only from one comet to another, but sometimes also from apparition to apparition of the same periodic comet. Furthermore, as shown for Comet Tempel 2 in Figure 5, even the mean (i.e. averaged over a number of apparitions) light curve may exhibit a strong asymmetry relative to perihelion passage, an effect that no hypothesis based solely of the average surface exposure to solar radiation can explain.

Taking Comet Kohoutek as an example, Whipple (1978b) discussed potential effects of the spin-axis orientation on the light curve of a "new" (i.e. first time near the sun) comet in qualitative terms. He speculated that the rapid depletion of a super-volatile surface layer, termed "frosting," combined with the insolation variations over the nucleus surface invoked by an obliquity of nearly $90°$ could explain the sharp rise of the comet's intrinsic brightness near discovery long before perihelion, the sudden flare-up near perihelion, and the disappointing performance of the comet after perihelion.

Whipple (1978b) also showed that evolved short-period comets possess light curves that statistically differ from those of the "new" comets such as Kohoutek. A model of a compositionally homogeneous nucleus was applied by Cowan & A'Hearn (1979) with moderate success to Comet d'Arrest, known to exhibit one of the most peculiar light curves among the short-period comets. According to Bortle (1977), this comet is very faint prior to perihelion, but brightens up rapidly near the sun. The maximum is reached about 3 weeks after perihelion, at which point the brightness stabilizes for almost 6 weeks. A decline that follows is so slow that four months past perihelion the comet is still intrinsically as bright as it was at perihelion. Cowan & A'Hearn were able to reproduce the light-curve variations (as well as the variations in the measured production rate of C_2) around and after (but not before) perihelion by the theoretical vaporization curve of an H_2O-dominated rotating nucleus[1], whose spin axis was parallel to the sun-comet line 80 days after perihelion (corresponding in our notation to $\Phi = 23°$, $I = 90°$).

---

[1] From comparison of the measured abundances of OH and C_2, A'Hearn et al (1979) esti- mated the mass production ratio of H_2O/C_2 for the H_2O dominated nucleus of d'Arrest (i.e. assuming that all OH came from H_2O) at ~ 100. This ratio, compatible with the results de- rived for other comets, is believed not to be critically dependent on heliocentric distance. Since d'Arrest is known to have the visual region of the spectrum dominated by the Swan bands (Roemer et al 1970), especially by the $\Delta v = 0$ transition, its visual light curve is closely related to the production of C_2.

The large lag angles listed in Table 1 rule out free surface sublimation from the nucleus of Tempel 2 and suggest instead the presence of an insulating crust of nonvolatile material. As a result, the activity is quenched over most of the surface and curtailed over the rest of it. For Periodic Comet Encke Delsemme & Rud (1973) concluded from entirely independent evidence that not more than 10 percent of the surface is active.

The existence of an insulating layer of dust was predicted by Whipple (1950, 1978b), first to explain nongravitational perturbations of the revolution period of short-period comets, and more recently to elucidate the cometary brightness variation with heliocentric distance. He envisages that the insulating crust, purged near perihelion but rebuilt on the outgoing branch of orbit, is distributed in a spotty fashion over the nucleus surface and that it generally grows with comet age. The basic correctness of this scenario was quantitatively demonstrated by Brin & Mendis (1979), although their model does not include rotation effects.

THE PERIOD OF ROTATION

An obvious prerequisite for a rotation-period determination is observation of a suitable event, preferably near the nucleus, which recurs at demonstrably regular intervals. The choice of brightness variations, applied with considerable success in the rotation studies of minor planets, is less appropriate for comets because of the practical impossibility of eliminating contamination by the coma and because of unpredictable short-term variations in comet activity that are not necessarily associated with rotation. The most ambitious project of this kind was undertaken by Fay & Wisniewski (1978), who used a photoelectric photometer with a 16-arcsec diaphragm attached to the 154-cm Catalina reflector to measure wide- and narrow-band magnitudes of the central condensation of Periodic Comet d'Arrest on three consecutive nights in early August 1976. Calculating a power spectrum for the range of rotation periods P between 0.07 and ~ 5 days, they found only one peak with a confidence level greater than 99.9 percent, required for real solutions by the size of their data set. The peak was centered sharply on $P = 5.17$ hours. The resulting light curve shows two humps, each of an amplitude of 0.15 magnitude (corresponding to a brightness ratio of 1.15), separated by about 1.3 hour, and possibly two or three smaller humps, per rotation period.

Larson & Minton (1972) applied a different approach to Comet Bennett 1970 II. On plates taken with the 154-cm Catalina reflector on several days between March 26 and April 6, 1970, the central condensation of this comet was surrounded by a system of spiral jets, composed apparently of dust particles. Assuming a radial particle velocity of 0.6 km s^{-1} and using two different techniques of reduction, Larson & Minton derived a rotation period of 33 to 36 hours from the separation characteristics of the jets on March 28 and April 4. More recently, Larson (1978) reiterated that dust ejections from discrete areas on a rotating nucleus could explain the observed spiral structure, but admitted uncertainties in the calculated rotation period because of the unknown size distribution and ejection velocity of particles in the jets. One may add that under the given conditions of projection, the "lawn-sprinkler" analogy, employed for the jet structure by Larson & Minton, might be too crude an approximation. Indeed, the two authors did point out that on only two dates could the jet structure be interpreted as a two-dimensional feature.

Two tail phenomena briefly considered but dismissed as potential products of the nucleus rotation include a "wagging" plasma tail of Comet Burnham 1960 II (Malaise 1963) and a system of dust streamers in the tail of Comet West 1976 VI (Sekanina & Farrell 1978).

Whipple (1978a) has recently developed the most productive method for determining the spin rate. The idea is to calculate the zero dates, or initiation times, of concentric halos in the coma from their separation and rate of expansion, and to search for recurrent patterns in the spacing of zero dates. If the halos are produced by a single fixed active area on the nucleus, the time intervals between two consecutive zero dates give the rotation period. Ironically, the best data for this new method were provided by visual micrometric measurements of Comet Donati 1858 VI. Although the halo-formation periodicity of approximately 4 to 5 hours was recognized soon after observation (Schmidt 1863), its significance as a measure of the nucleus rotation was not appreciated until more than a century later!

Most comets do not display the distinct successive halos shown by Comet Donati (Figure 6). Whipple finds that for comets without halos the rotation period still can be determined, if the measured coma diameters are spaced at intervals that are multiples of the period, thus indicating that one active area predominates (Whipple & Sekanina 1979). The rather commonly reported phenomena of a "stellar nucleus" and "central condensation" are logically explained in this framework as manifestations of the halo initiation and early halo development, respectively. The expansion velocity can be calculated from an empirical formula that is based on Bobrovnikoff's (1954) measurements of 57 halos of various comets at heliocentric distances between 0.6 and 7 AU. Unfortunately, coma-diameter measurements spaced at intervals many times the rotation period are unsuitable as they allow multiple solutions. Hence, analysis of rapidly spinning comets is generally more difficult. Also, the toleration of long time intervals increases the danger of severe deformation of the recurrence

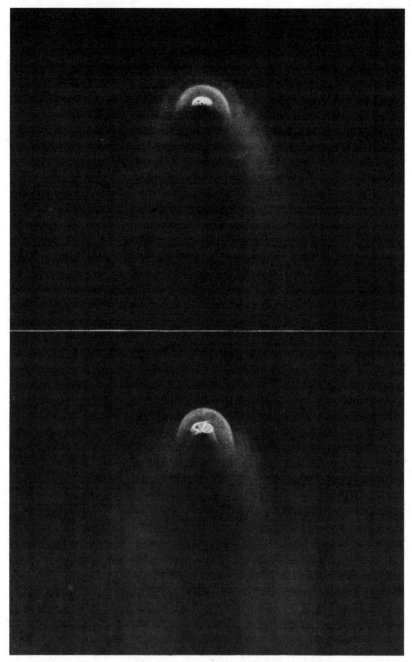

Figure 6 Distinct concentric halos in the coma of Comet Donati 1858 VI, as observed visually by Bond (1862) on 1858 October 5.0 (top) and 6.0 UT. (Courtesy of Harvard College Observatory.)

pattern by perceptible migration of the active area over the nucleus surface.

In practice, Whipple (Whipple & Sekanina 1979) finds that for any arbitrarily chosen value of period P it is possible to find a solution with $P/\sigma \simeq 4$, where σ is the standard deviation of a single zero date. He then introduces a "confidence" factor that measures the significance of the individual solutions by the excess of P/σ over the "noise" level. An additional check on the multiple solutions is available in the case of a short-period comet, for which the rotation period at different apparitions should come out to be virtually the same. Considering the involved uncertainties, the determination of systematic variations in the rotation period with time is very difficult (Whipple & Sekanina 1979). A spin-up is expected for nuclei with rough surfaces (Sekanina 1967, Whipple 1978a, Whipple & Sekanina 1979), but the magnitude of the spin-up is in question.

THE ROLE OF ROTATION IN COMETARY OUTBURSTS AND SPLITTING

Periodic Comet Schwassmann-Wachmann 1, moving in a nearly circular orbit between those of Jupiter and Saturn, is best known for its enormous outbursts that occur at irregular intervals at an average rate of more than one per year. At times, a rapid sequence of outbursts takes place. During an outburst the brightness of the comet increases rapidly to a maximum of up to several hundred times the level in the quiescent phase and its appearance changes radically from a faint, diffuse speck of light to a sharp, completely stellar image, which expands first into a disk-like object and later into an asymmetric shell or fan, with a more complex structure sometimes indicated (Figure 7). As the outburst subsides, the surface brightness gradually decreases, until the shell vanishes completely and the comet returns to its quiescent phase.

Because of the comet's large heliocentric distance the angle sun-comet-earth never exceeds $10.5°$. Hence, the heliocentric distribution of coma orientation can be studied directly from the observed position angles at hardly any loss of accuracy. Recently, Whipple (1980a) undertook a study of this comet, based on 169 directions of asymmetric emission, 433 coma diameters, and a few double halos, reported between discovery in 1927 and 1977. Using different intensity distribution laws for a moderately asymmetric coma and for an elongated fan or jet, he plotted polar diagrams of the mean asymmetric ejection for standard intervals of heliocentric longitude λ of the comet. The result, presented in Figure 8, shows that at λ between $80°$ and $90°$ the extension in the generally southeast direction

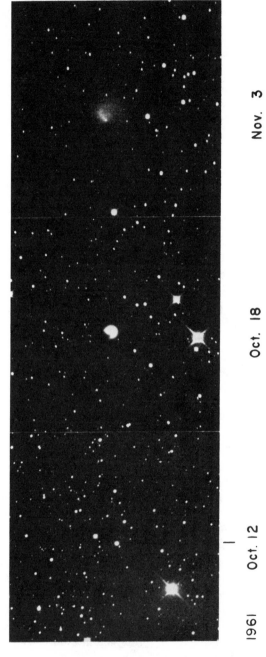

Figure 7 Development of the asymmetric coma of Periodic Comet Schwassmann-Wachmann 1 in October-November 1961, as photographed by Roemer (1962). North is up, east to the left. (Official U.S. Navy photograph).

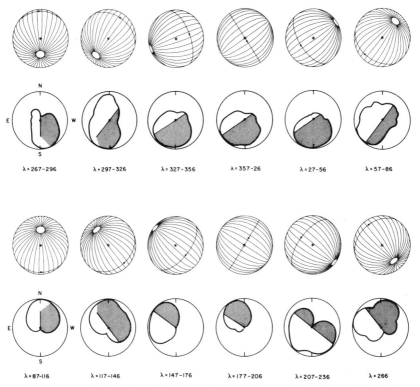

Figure 8 The derived position of the spin axis (upper row) and the observed distribution of asymmetric emission of Periodic Comet Schwassmann-Wachmann 1 versus heliocentric longitude λ. From Whipple (1980a).

suddenly disappears and that at λ between 110° and 120° it reappears in the generally northeast direction. From this and other properties of the orientation pattern Whipple derived the most probable position of the rotation pole, which in our notation is given by $\Phi = 304.5°$, $I = 65°$. The prograde sense of rotation, established from the expected preponderance of halo initiations in the afternoon hemisphere, was confirmed by a sequence of photographs of the halo development in October and November 1961, obtained by Roemer (1962) and reproduced in Figure 7.

Whipple found that the most probable rotation period of Comet Schwassmann-Wachmann 1 is $P = 119 \pm 2$ hours, i.e. practically 5 days. This is a weighted mean of individual solutions from 29 oppositions, for which P was consistently found to lie between 4 and 6.2 days and the ratio P/σ, determining the confidence factor, exceeded 5.

Whipple's interpretation of the activity is based on the spin-modulated interaction between the comet's amorphous-ice reservoir and an insulating crust on its large (~ 10 km) nucleus. The interpretation also includes a fallback effect to explain the secondary outbursts and a slow migration of active areas over the nucleus surface, lagging in the latitude a few tens of degrees behind that of the subsolar point.

Whipple (1961, 1978a) suggested that the spin-up may sometimes cause the nucleus to split into two or more fragments. A critical rotation period, given by the balance between the centrifugal force and the gravitational attraction, is at the equator,

$$P_{crit} = \sqrt{\frac{3\pi}{\rho G}} = \frac{3.30}{\sqrt{\rho}} \text{ (hours)}, \tag{5}$$

where G is the universal gravitation constant and ρ the nucleus density (in g cm^{-3}). The condition for the equatorial rotation velocity to exceed the velocity of escape from the nucleus surface leads to a more stringent limit on the rotation period, $\sqrt{2}$ times shorter than is P_{crit} from (5).

The kinetics of splitting is always complicated by the gravitational interaction between the fragments in the early post-split phase (Sekanina 1979b). As a result, the time of breakup derived from the observed motions of the fragments (Sekanina 1977, 1978, 1979c) refers in fact to the time of their "dynamical separation," i.e. termination of their appreciable attraction. The time interval between the breakup and the dynamical separation depends primarily on the initial velocity of recession. The dynamical separation should virtually coincide with the breakup for a comet whose observed time of splitting is known to be timed perfectly with other observed characteristics of the event, such as a flare-up or a dust outburst (Sekanina & Farrell 1978, Sekanina 1979c). If the splitting is spin induced, the required rotation period must be substantially shorter than is P_{crit} from (5).

THE NONGRAVITATIONAL EFFECTS

The 19th century investigations of the motion of Comet Encke were highlighted by three truly remarkable findings. The first was the determination of the comet's 3.3-year period of revolution about the sun, the shortest ever known among the comets. The second was the detection of the nongravitational effects, i.e. systematic deviations from the gravitational law, indicated by a secular acceleration in the orbital motion. And the last was the discovery of a decrease in the magnitude of the acceleration with time.

Later it turned out that the nongravitational effects can take the form of a secular acceleration in some comets, a deceleration in others, or they can even change slowly from acceleration to deceleration or vice versa. The generally accepted interpretation is due to Whipple (1950), who attributed the nongravitational effects to the momentum transferred to the cometary nucleus by the thrust of subliming ices. Although much earlier Bessel (1836a,b) had argued that ejections must have a dynamical effect on the nucleus, Whipple was the first to formulate a quantitative model, which is consistent with key observations and is physically and dynamically sound. The outstanding features of this model include the predictions of the nucleus rotation; of the formation of an insulating crust of meteoric material on the nucleus surface; and of a total sublimation rate that is a few orders of magnitude higher than calculations based on the species observed in the visual spectrum previously indicated. Whipple concluded that on a rotating nucleus the time lag in sublimation, invoked by the presence of the insulating layer, must generate a systematic tangential component of the nongravitational force, which affects the comet's revolution period. For short-period comets observed at several apparitions, this change in the revolution period is much easier to detect than the very slight reduction in the gravitational pull of the sun on the comet, produced by the (usually) larger radial component of the nongravitational force. For a comet whose spin axis is tilted appreciably to the orbit plane, Whipple formulated the classical relationship between the sense of rotation and the resulting change in the revolution period: the prograde rotation (i.e. in the sense of the comet's orbital motion) causes a secular deceleration (i.e. a systematic increase in the revolution period); the retrograde rotation results in a secular acceleration[2].

Recently, this relationship was confirmed brilliantly by Periodic Comet Schwassmann-Wachmann 3. From the measures of fan orientation in 1930, the comet's only pre-1979 apparition, I established that its nucleus had a prograde rotation and moderate obliquity (Sekanina 1979a); and following Whipple's theory, I predicted a secular deceleration. Not observed at eight intervening returns to the sun, this comet of a 5.4-year revolution period was due for a very favorable return in 1979. Its expected path in the sky was calculated from the 1930 orbital elements, based on the

---

[2] It should be remarked that, following Bessel's ideas, Dubiago (1948) also accounted for nongravitational changes in the revolution period by tangential ejection of matter from the nucleus. However, he failed to provide an explanation for the systematic deviation of the ejection from the direction to the sun in general, and for the acceleration in some comets and deceleration in others in particular. Also, he argued that the imparted momentum comes from the expulsion of solid particles and not from the sublimation. In truth, of course, particle ejection is an energy sink, not source.

assumption of a purely gravitational motion with the planetary perturbations fully accounted for. The uncertainty in the time of perihelion passage was estimated at not more than several days. Discovered in August 1979 as a new comet, it was located almost 25° from the predicted position! An updated orbit, calculated by Marsden (1979) from a dozen 1979 observations, showed that the comet passed through perihelion 37.5 days later than had been expected. This tardiness corresponds to an average deceleration rate of more than 0.8 day per revolution per revolution, by far the highest rate known among comets with periods less than ~ 10 years.

The knowledge of the character of the nongravitational motions of comets has been enlarged immensely since Marsden's (1969) introduction of a new method of orbit determination, in which the nongravitational terms are included directly in the equations of motion. Marsden has shown that for comets with revolution periods less than ~ 10 years, two neighboring apparitions can always be linked on the assumption of a purely gravitational motion. Although the situation is slightly more complicated for other comets, especially for some with nearly parabolic orbits, one may state that, as a rule, at least three apparitions are needed to determine the nongravitational effects with some confidence.

Marsden's method offers options to solve for parameters related to any of the three nongravitational-acceleration components: the radial, Z_1, directed away from the sun; the transverse, Z_2, perpendicular to Z_1 in the orbit plane and in the sense of the comet's motion; and the normal, Z_3, directed to the north orbital pole. The employed law $g(r)$ of variation of the nongravitational acceleration with heliocentric distance, first purely empirical, was later modified to comply with the vaporization curve of H_2O snow. In the absence of data on the nucleus rotation, this law is believed to be the best first approximation from both the physical and empirical standpoints (Delsemme 1974). Normalizing the law so that $g(1 \text{ AU}) = 1$, the components of the nongravitational acceleration are

$$Z_i(r) = A_i g(r), \qquad i = 1, 2, 3, \tag{6}$$

where the parameters A_1, A_2, and A_3 are the three acceleration components at a heliocentric distance of 1 AU.

The calculations by Marsden (1969, 1970, 1972), by Marsden & Sekanina (1971, 1974), by Marsden et al (1973), and by Yeomans (1972a,b, 1974, 1975, 1977) provide a wealth of data on the nongravitational motions of comets. The data on more than two dozen short-period comets confirm a conclusion I reached on the basis of a smaller sample, namely that most of these comets tend to discriminate into two groups (Sekanina 1972).

Group I includes comets whose radial nongravitational acceleration A_1 comes out from the calculations to be always positive and well determined, and whose transverse acceleration A_2 is very small compared to the A_1. Typical examples are comets Borrelly, Faye, Giacobini-Zinner, Halley, Kopff, and Schwassmann-Wachmann 2.

Group II consists of comets whose A_1, calculated to be sometimes positive, sometimes negative, is poorly determined, and whose A_2 may be comparable with the A_1 in magnitude. This group includes, among others, comets d'Arrest, Encke, Grigg-Skjellerup, Pons-Winnecke, and Tempel 2.

The two groups have roughly the same number of known members. It appears that there is no difference between them in the degree of determinacy of A_2, which is usually good to excellent, and in the degree of determinacy of A_3, which so far has always been poor. The interpretation of the tendency to group is complicated by circumstances inherent in the nature of the problem, such as the smallness of the nongravitational effects; the imbalance in the sensitivity of the method to the radial, transverse, and normal components; the uncertainty in the adopted law for the force; the errors introduced by a possible displacement of the comet's center of mass from its apparent center of light; etc. The likelihood that the quality of an orbital fit based on law (6) is affected significantly by the nucleus rotation is illustrated by the consideration of the insolation-control model of cometary activity [Equations (4)]. Law (6) should then be replaced by the following expressions:

$$
\begin{aligned}
Z_1 &= Z_0 g(r) a_1 \left[1 - p_1 \cos 2(\Phi + v) \right], \\
Z_2 &= Z_0 g(r) a_2 \left[1 + p_2 \sin 2(\Phi + v) \right], \\
Z_3 &= Z_0 g(r) a_3 p_3 \cos (H + \Phi + v),
\end{aligned}
\tag{7}
$$

where Z_0 is the magnitude of the acceleration at 1 AU from the sun,

$$
\begin{aligned}
a_1 &= \cos 2H \cos^2 K, \\
a_2 &= \sin 2H \cos^2 K, \\
a_1 p_1 &= a_2 p_2 = \sin^2 K, \\
a_3 p_3 &= \sin 2K
\end{aligned}
\tag{8}
$$

and

$$
\begin{aligned}
\tan H &= \tan \tfrac{1}{2}\theta \cos I, \\
\sin K &= \sin \tfrac{1}{2}\theta \sin I.
\end{aligned}
\tag{9}
$$

The spin-dependent nongravitational law (7) can be reduced to the forced law (6) only if the amplitudes of all three periodic terms are zero, i.e. if obliquity $I = 0$ or $180°$, or if lag angle $\theta = 0$, or both. Otherwise, the nongravitational parameters A_i are subject to error (Sekanina 1972). Since the normal component consists of a periodic term only, A_3 is always indeterminate, as ascertained empirically. The amplitudes p_i of the periodic terms in expressions (7) for the radial and transverse components of the force are plotted as functions of the lag angle and obliquity in Figure 9. Analysis of relations (8) and (9) shows that the determinacy of A_1 depends chiefly on the lag angle, which ought to be fairly small for Group I comets and large for Group II comets. A small lag angle also guarantees that $A_2 \ll A_1$, which is characteristic of Group I comets. These conclusions are individually confirmed by the three short-period comets for which independent data exist on both the lag angle (Sekanina 1979a) and the nongravitational parameters (Marsden & Sekanina 1971, 1974, Yeomans 1972b): an average lag angle was found to be 31° for Borrelly, a Group I comet; and 45° and 58°, respectively, for Encke and Tempel 2, two Group II comets. The determinacy of A_2 should be good to excellent except when the spin axis lies nearly in the orbit plane. For $I \simeq 90°$ Figure 9 shows that $|p_2| \gtrsim 1$, so that the positive A_2 (secular deceleration) does not

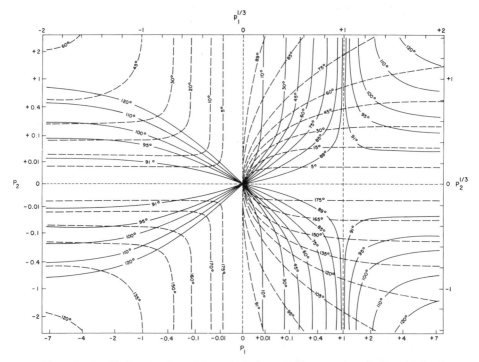

Figure 9   Amplitudes p_1 and p_2 of the periodic terms in the expressions for the radial and transverse components of the nongravitational acceleration as functions of the lag angle θ (solid curves) and obliquity I (broken curves) for the insolation-control model. Dotted lines show singular solutions. The linear scales refer to the cube roots of p_1 and p_2. From Sekanina (1972).

necessarily imply the direct sense of rotation, and vice versa (Sekanina 1972). This is confirmed by Whipple & Sekanina's (1979) detailed calculations for Comet Encke.

The derived magnitudes of the nongravitational effects were used in the past to estimate the relative loss of mass from comets. From the early data Whipple (1950, 1951) and Hamid & Whipple (1953) obtained typical loss rates on the order of magnitude of a few tenths of a percent to a few percent of the comet's total mass per revolution. The more recent results by Marsden et al give much the same numbers, with greatest weight attached to values near the lower limit. About twenty years after having been justified dynamically, the high mass-loss rates were fully confirmed by satellite- and rocket-borne observations of huge hydrogen halos about several comets, detected in Lyman alpha emission in the far ultraviolet [see Keller (1976) for an extensive review].

A perplexing property of an ever increasing number of short-period comets is the occasional occurrence of what appear to be "sudden" discontinuities in their nongravitational motions. The term "sudden" as used is of course subject to qualification in view of the long time intervals needed in the nongravitational runs of orbit determination. Nevertheless, Marsden et al isolated a number of discontinuities by determining the immediately preceding and following passages through perihelion. Marsden (1969, 1970) was the first to notice that some comets tend to behave "erratically" after close approaches to Jupiter, although there exist erratic comets without an approach and "normal" comets with an approach. Since the discrete "jumps" appear to take place generally at large heliocentric distances, their interpretation as products of sublimation is very unattractive. For lack of a better mechanism, an ad hoc hypothesis was advanced (Marsden & Sekanina 1971, Sekanina 1973), which attributed the jumps to possible collisions of the comets with interplanetary boulders, meter-sized objects first postulated by Harwit (1967) to explain other phenomena. Although a preferable explanation for many of these irregular nongravitational effects has recently been proposed (Whipple & Sekanina 1979; see the section on the nucleus precession), some of them are still difficult to account for.

Similarly, it is unclear to what extent the variations in A_2 are related to secular variations in the sublimation rate. Although a systematic change in the magnitude of A_2 with time—whether it amounts to an increase or decrease—can in principle be interpreted as a mass-loss related effect (Sekanina 1971), the discovery of comets whose nongravitational motions have changed from acceleration to deceleration, or vice versa, has eliminated interpretations based on the loss-rate variation alone. Comparing the slightly accelerated motion of Comet Pons-Winnecke in the 19th century with its slightly decelerated motion in the 20th century, Marsden (1970) recognized that, in terms of the Whipple model, the nucleus obliquity must have passed through the value 90°. He suggested next that obliquities of some comets could experience periodic variations about 90° with an amplitude proportional to the lag angle (Marsden 1972). Following a parallel line of reasoning, I formulated (and applied to Comet Faye) two simple precession models (Sekanina 1972).

At the present time, three comets are known to have undergone changes from a deceleration to an acceleration; two from an acceleration to a deceleration; and one from a deceleration to a slight but definite acceleration and back to a deceleration. The secular variations in the motions of these comets are exhibited in Figure 10. To illustrate the variety in the character of the nongravitational forces still further, Table 3 correlates A_2 with the average perihelion distance $\langle q \rangle$ and average orbit inclination $\langle i \rangle$ to the ecliptic for selected time intervals of four short-period comets: (a) Comet Wolf shows a strong correlation of A_2 with $\langle q \rangle$, indicative of an effect of varying sublimation; (b) Comet Comas Solá exhibits large variations in A_2, although both $\langle q \rangle$ and $\langle i \rangle$ are practically constant—a probable sign of precession, especially because A_2 passes through zero;

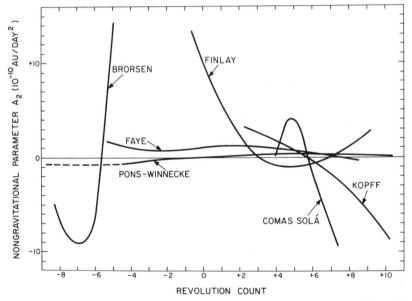

Figure 10   Nongravitational parameter A_2 versus revolution count for six short-period comets. Positive A_2 indicates a secular deceleration; negative A_2, an acceleration. The revolutions are counted from the perihelion passage nearest the year 1900.

Table 3 Nongravitational parameter A_2 for four short-period comets

Comet	Apparitions	$\langle q \rangle$ (AU)	$\langle i \rangle$	A_2 (10^{-12} AU day^{-2})
Wolf	1884–1918	1.59	25°2	$+81 \pm 1$
	1925–1967	2.47	27.3	0^a
Comas Solá	1927–1944	1.77	13.7	$+15 \pm 13$
	1935–1952	1.77	13.6	$+402 \pm 12$
	1944–1961	1.77	13.5	-173 ± 11
	1952–1970	1.77	13.5	-737 ± 11
Kopff	1906–1932	1.70	8.7	$+267 \pm 3$
	1932–1951	1.59	8.0	-216 ± 6
	1945–1964	1.51	6.0	-455 ± 5
	1958–1970	1.53	4.7	-804 ± 19
d'Arrest	1870–1897	1.31	15.7	$+960 \pm 6$
	1890–1910	1.31	15.7	$+937 \pm 8$
	1910–1943	1.34	17.3	$+957 \pm 11$
	1943–1963	1.38	18.0	$+961 \pm 2$

[a] No nongravitational term necessary.

(c) Comet Kopff indicates a definite correlation between A_2 and $\langle i \rangle$, which can qualitatively be understood as an effect of variable obliquity due to the orbit-plane instability; and (d) Comet d'Arrest has a constant A_2 in spite of variations in both $\langle q \rangle$ and $\langle i \rangle$.

NUCLEUS PRECESSION

The application of a simple, quantitative model to comets with well-determined nongravitational parameters was the next logical step toward gaining greater insight into the character of the forced precession of cometary nuclei. Periodic Comet Encke was best suited for such an undertaking because of a large amount of information on its nongravitational motion over a time interval of nearly 200 years; because of its relatively stable orbit, restraining uncertainities in difficult requisite assumptions, such as the choice of the sublimation law; and because it is one of only a few comets for which two essential quantities, the spin-axis position and the effective sublimation lag angle, have been determined (Sekanina 1979a). In approaching the problem, Whipple & Sekanina (1979) assumed that the nucleus is an oblate spheroid rotating about its minor axis. Except at the equator and poles, the thrust of the subliming ices subjects the axis to a torque, which is directed normal to the plane defined by the thrust vector and the axis. Since the resulting precession entails slight changes in the components of the momentum transferred to the nucleus, it can be "monitored" by the dynamically determined variations in A_2. Whipple & Sekanina's solution is highly satisfactory in terms of both the fit to A_2 and the representation of fan orientation during the nearly 200 years of observation. Also, their determination of the spin-axis position and the lag angle is in excellent agreement with Sekanina's (1979a) earlier, entirely independent, result. Extrapolating back to 1000 B.C. and forward to A.D. 3000, Whipple & Sekanina show that the motion of the spin axis is strongly time dependent (Figure 11). At present it maintains a systematic rate of about 1° per revolution, but attained a maximum of 2.7° per revolution in the mid-19th century and should reach another peak of 2.6° per revolution around the year 2070. The spin axis appears to have been "locked" in a quasi-stable configuration, practically in the orbit plane, for hundreds of revolutions before the 18th century. The obliquity slowly grew to about 103° by the mid-19th century, but dropped back to 90° by the 1950s and amounts to 83° at present. The comet is still subjected to a slight secular acceleration, which should change to a deceleration after 1990 as the spin axis tends to "straighten up" and ultimately to get practically aligned with the normal to the orbit plane. Whipple & Sekanina note, however, that this evolutionary trend must sooner or later be upset by the gradual transformation of the shape of the nucleus from the oblate spheroid to a dynamically unstable prolate figure, as a consequence of the excessive sublimation from the equatorial zone under the conditions of a low obliquity.

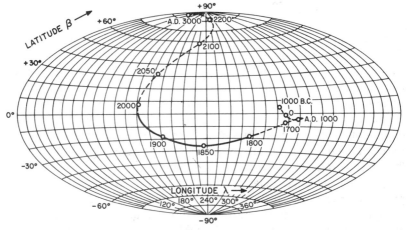

Figure 11 The calculated motion of the spin axis of Periodic Comet Encke between 1000 B.C. and A.D. 3000 in the ecliptic coordinates. The heavy portion of the curve is the observed time interval, the broken portions are extrapolations. From Whipple & Sekanina (1979).

COMET ROTATION AND PRECESSION 141

During the observed time interval of almost 200 years, the spin axis described an arc of more than 100°, a change that could have affected the evolution of the nucleus surface. Comet Encke's activity is known to be strongly asymmetric with respect to perihelion: at 1 AU from the sun the comet is about 10 times brighter before than after perihelion. The derived position of the spin axis can qualitatively explain this peculiarity, if the comet's major ice reservoir is located on the polar hemisphere that is turned toward the sun before perihelion (Sekanina 1979a). A surprising result of Whipple & Sekanina's calculations is the fact that the currently more active hemisphere is the one that was exposed to the sun near perihelion for hundreds of revolutions. Hence, the "dormant" hemisphere is either more rocky per se, or became insulated by impacts of massive meteoroidal debris, ejected on ballistic trajectories from the active hemisphere. Whipple & Sekanina prefer the latter interpretation.

The two authors also point out that in addition to pursuing its systematic precessional motion, the spin axis wobbles during each revolution with an amplitude that is sometimes greater than is the net shift in the pole position from aphelion to next aphelion. The wobbling is limited essentially to heliocentric distances less than 1 AU and its properties are correlated with the obliquity.

The geometry on an oblate spheroid can be described by the equations used previously for a spherical nucleus, except that ϕ is now a cometographic latitude related to the cometocentric latitude ϕ' by[3]

$$\tan \phi = (1 - \alpha)^{-2} \tan \phi', \tag{10}$$

where α is the oblateness defined as a difference between the equatorial and the polar radii of the spheroid in units of the equatorial radius. Thus $0 < \alpha < 1$ for a spheroid, $\alpha = 0$ for a sphere. In the case of the insolation-control model, used by Whipple & Sekanina, the instantaneous precession rate $\dot{\chi}$ at time t, given generally by a more complicated expression, can for a small oblateness be approximated by

$$\dot{\chi}(t) \doteq \text{const} \, \frac{\alpha P}{RM} \, \dot{M}(t) V(t) \sin 2\phi_0(t), \tag{11}$$

where M and R are the mass and the equatorial radius of the nucleus, $\dot{M}$ the sublimation rate, and V the initial expansion velocity of the gas.

---

[3] Analogous to the definition of geographic and geocentric latitudes, the cometographic latitude is an angle between the equator and the normal to the surface, the cometocentric latitude is the angular distance from the equator. For a sphere the two latitudes are of course identical.

The constant includes a momentum-transfer coefficient. The other quantities have been defined before. Note that $\dot{\chi} = 0$ for $\phi_0 = +90°$, $0°$, and $-90°$. Expression (11) can be used to assess the oblateness of the nucleus. Assuming that the nucleus has a depth-independent bulk density of 1 g cm^{-3}, estimating its radius at 1.0 ± 0.3 km, and deriving the rotation period of 6.5 hours, Whipple & Sekanina find that Comet Encke has an oblateness of only 3 ± 1 percent. They also show that if the bulk density increases with depth, the same precession rate will be achieved at a smaller oblateness still.

The experience with Comet Encke suggests that slowly rotating comets, small comets, and comets with highly oblate nuclei should have precession rates that could exceed that of Encke considerably. Hence, if the comets in Figure 10 possess some or all of these properties, the rapid changes from a secular acceleration to deceleration, or vice versa, can qualitatively be understood as by-products of the nucleus precession.

Finally, Whipple & Sekanina noticed definite discontinuities in the calculated nongravitational parameters of Comet Encke in most, but not all, instances when the orbit-plane inclination changed near aphelion due to Jupiter's perturbations. Minor, but perceptible, discontinuities in A_2 were sometimes caused by inclination changes as small as 0.05°! Marsden's (1969, 1970) conclusion on occasional jumps in the nongravitational parameters of comets following close approaches to Jupiter (which often result in major changes in the inclination) is under these circumstances readily understood.

CONCLUSIONS

Recent work has shown that Whipple's model of a rotating, icy-conglomerate nucleus, the heterogeneous surface of which is assumed to be covered in part with an insulating crust of dust, is generally consistent with observations of the sunward fanlike coma and its orientation pattern, with personality signatures of individual comets, and with the character and special features of the nongravitational motions of comets. It is argued that discrete emission areas, activated for only a fraction of the rotation period, must exist on the nuclei of many, especially short-period comets. The activity, reflecting both the existence of the emission centers and the distribution of insolation over the nucleus surface, tends to peak in the cometary afternoon, clearly an effect of the insulating crust. The lag increases with comet age as the crust is growing and spreading over the surface. At least for the short-period comets, the spin axis has a tendency to lie near the orbit plane and to be roughly parallel to the line of apsides, thus implying the most extreme surface insolation regime possible. Derived mostly from expanding halo measurements on the assumption of a single, predominant active area, the rotation periods of comets come out to be in the general range from several hours to several days, with no definite evidence for a spin-up. Besides the measures of emission fans and expanding halos, data on the nongravitational motions of comets provide the major source of information on the nucleus rotation. Brought about by the momentum transferred to the nucleus by the thrust of subliming ices, the nongravitational effects were shown to indicate essentially the sense of rotation and to provide an estimate for the sublimation lag angle. The introduction of the model of an oblate nucleus made it possible to study the relation between time variations in the nongravitational effects and the nucleus precession, caused by a torque that the thrust applies to the spin axis. Two intriguing properties of the nongravitational motions, namely gradual changes from a secular acceleration to a deceleration (or vice versa) and occasional discontinuities, appear to be products of the nucleus rotation and precession also. The change of sign of the secular variation in the revolution period can be commonplace among comets with small, slowly rotating, or highly oblate nuclei, whose polar axis lies near the orbit plane. The discontinuities are mostly associated with "sudden" changes in the obliquity caused by changes in the orbit-plane inclination following close approaches to Jupiter. The effect might be augmented in the case of a rapidly precessing comet with a near 90° obliquity.

While a breakthrough has truly been achieved in the study of the rotation and precession of comets, an enormous amount of work still remains to be completed. Numerous sources of data, such as published and unpublished reports of the appearance of many comets by visual observers, a number of existing comet plate collections, etc, should be searched and inspected for information on emission fans, expanding halos, and possibly other features indicative of cometary rotation; the data should be systematically analyzed and the rotation characteristics determined; and, last but not least, the results should be carefully interpreted and improved models of the nucleus structure formulated. Although the physical descriptions of comets from groundbased observations provide data that can by no means compete with qualitatively superior information from future rendezvous-type comet missions, their very availability and, for a number of short-period comets, also their coverage for long time intervals are offering us a valuable basis for the investigation of some of the most fundamental properties of the cometary nucleus. The same is of course also true about the data acquired from the sophisticated orbital calculations.

I thank Drs. E. Roemer, University of Arizona, and F. L. Whipple, Center for Astrophysics, for permission to reproduce Figures 7 and 8,

144 SEKANINA

respectively; and Dr. L. M. Hobbs, Yerkes Observatory, for permission to use the original photographs assembled in Figure 1. This work was supported by the Planetary Atmospheres program of the National Aeronautics and Space Administration under Grant NGR 09-015-159.

Literature Cited

A'Hearn, M. F., Millis, R. L., Birch, P. V. 1979. Gas and dust in some recent periodic comets. *Astron. J.* 84:570–79

Bessel, F. W. 1836a. Beobachtungen über die physische Beschaffenheit des Halleyschen Kometen und dadurch veranlasste Bemerkungen. *Astron. Nachr.* 13:185–232

Bessel, F. W. 1836b. Bemerkungen über mögliche Unzulänglichkeit der die Anziehungen allein berücksichtigenden Theorie der Kometen. *Astron. Nachr.* 13:345–50

Bobrovnikoff, N. T. 1954. Physical properties of comets. *Astron. J.* 59:357–58

Bond, G. P. 1862. Account of the Great Comet of 1858. *Ann. Astron. Obs. Harvard Coll.* 3:1–372

Bortle, J. E. 1977. The 1976 apparition of Periodic Comet d'Arrest. *Sky & Telesc.* 53:152–57

Brin, G. D., Mendis, D. A. 1979. Dust release and mantle development in comets. *Astrophys. J.* 229:402–8

Cowan, J. J., A'Hearn, M. F. 1979. Vaporization of comet nuclei: Light curves and lifetimes. *Moon & Planets* 21:155–71

Delsemme, A. H. 1974. The vaporization of the volatile fraction in comets. In *Asteroids, Comets, Meteoric Matter*, ed. C. Cristescu, W. J. Klepczynski, B. Milet, pp. 313–22. Bucharest: Academy. 333 pp.

Delsemme, A. H., Rud, D. A. 1973. Albedos and cross-sections for the nuclei of comets 1969 IX, 1970 II, and 1971 II. *Astron. Astrophys.* 28:1–6

Dubiago, A. D. 1948. On the secular acceleration of motions of the periodic comets. *Astron. J. USSR* 25:361–68 (In Russian)

Fay, T. D., Jr., Wisniewski, W. 1978. The light curve of the nucleus of Comet d'Arrest. *Icarus* 34:1–9

Hamid, S. E., Whipple, F. L. 1953. On the motions of 64 long-period comets. *Astron. J.* 58:100–8

Harwit, M. 1967. The cloud of interplanetary boulders. In *The Zodiacal Light and the Interplanetary Medium*, ed. J. L. Weinberg, pp. 307–13. *NASA SP-150*. Washington, DC. 430 pp.

Horn, G. 1908. Struttura e rotazione della Cometa Daniel (1907d). *Mem. Soc. Spettroscop. Ital.* 37:65–75

Huebner, W. F. 1976. The nucleus: Panel discussion. In *The Study of Comets*, ed.

B. Donn, M. Mumma, W. Jackson, M. A'Hearn, R. Harrington, pp. 597–605. *NASA SP-393*. Washington, DC. 1083 pp.

Keller, H. U. 1976. The interpretations of ultraviolet observations of comets. *Space Sci. Rev.* 18:641–84

Larson, S. 1978. A rotation model for the spiral structure in the coma of Comet Bennett (1970 II). *Bull. Am. Astron. Soc.* 10:589

Larson, S. M., Minton, R. B. 1972. Photographic observations of Comet Bennett, 1970 II. In *Comets: Scientific Data and Missions*, ed. G. P. Kuiper, E. Roemer, pp. 183–208. Tucson: Univ. Ariz. Press. 222 pp.

Malaise, D. 1963. Photographic observations of the tail activity of Comet Burnham 1960 II. *Astron. J.* 68:561–65

Marsden, B. G. 1969. Comets and nongravitational forces. II. *Astron. J.* 74:720–34

Marsden, B. G. 1970. Comets and nongravitational forces. III. *Astron. J.* 75:75–84

Marsden, B. G. 1972. Nongravitational effects on comets: The current status. In *The Motion, Evolution of Orbits, and Origin of Comets*, ed. G. A. Chebotarev, E. I. Kazimirchak-Polonskaya, B. G. Marsden, pp. 135–43. Dordrecht: Reidel. 521 pp.

Marsden, B. G. 1979. Periodic Comet Schwassmann-Wachmann 3. *Minor Planet Circ./Minor Planets & Comets No. 5031*

Marsden, B. G., Sekanina, Z. 1971. Comets and nongravitational forces. IV. *Astron. J.* 76:1135–51

Marsden, B. G., Sekanina, Z. 1974. Comets and nongravitational forces. VI. Periodic Comet Encke 1786–1971. *Astron. J.* 79:413–19

Marsden, B. G., Sekanina, Z., Yeomans, D. K. 1973. Comets and nongravitational forces. V. *Astron. J.* 78:211–25

Probstein, R. F. 1969. The dusty gasdynamics of comet heads. In *Problems of Hydrodynamics and Continuum Mechanics*, pp. 568–83. Philadelphia: Soc. Industr. Appl. Math. 815 pp.

Roemer, E. 1962. Activity in comets at large heliocentric distance. *Publ. Astron. Soc. Pac.* 74:351–65

Roemer, E., Owen, T. C., White, R. E. 1970. Periodic Comet d'Arrest (1970d). *IAU Circ. No. 2267*

Schmidt, J. F. J. 1863. Ueber Donati's Cometen. *Astron. Nachr.* 59:97–108

Sekanina, Z. 1967. Non-gravitational effects in comet motions and a model of an arbitrarily rotating comet nucleus. V. General rotation of comet nuclei. *Bull. Astron. Inst. Czech.* 18:347–55

Sekanina, Z. 1971. A core-mantle model for cometary nuclei and asteroids of possible cometary origin. In *Physical Studies of Minor Planets*, ed. T. Gehrels, pp. 423–28. *NASA SP- 267*. Washington, DC. 687 pp.

Sekanina, Z. 1972. Rotation effects in the nongravitational parameters of comets. See Marsden 1972, pp. 294–300

Sekanina, Z. 1973. New evidence for interplanetary boulders? In *Evolutionary and Physical Properties of Meteoroids*, ed. A. F. Cook, C. L. Hemenway, P. M. Millman, pp. 199–208. *NASA SP-319*. Washington DC. 378 pp.

Sekanina, Z. 1977. Relative motions of the split comets. I. A new approach. *Icarus* 30:574–94

Sekanina, Z. 1978. Relative motions of the split comets. II. Separation velocities and differential decelerations for extensively observed comets. *Icarus* 33:173–85

Sekanina, Z. 1979a. Fan-shaped coma, orientation of rotation axis, and surface structure of a cometary nucleus. I. Test of a model on four comets. *Icarus* 37:420–42

Sekanina, Z. 1979b. The split comets: Gravitational interaction between the fragments. In *Dynamics of the Solar System*, ed. R. L. Duncombe, pp. 311–14. Dordrecht & Boston:Reidel. 330 pp.

Sekanina, Z. 1979c. Relative motions of the split comets. III. A test of splitting and comets with suspected multiple nuclei. *Icarus* 38:300–16

Sekanina, Z. 1981. Large-scale nucleus surface topography and outgassing pattern analysis of Periodic Comet Swift-Tuttle. In preparation

Sekanina, Z., Farrell, J. A. 1978. Comet

West 1976 VI: Discrete bursts of dust, split nucleus, flare-ups, and particle evaporation. *Astron. J.* 83:1675–80

Stobbe, J. 1938. Photographische Beobachtungen des Kometen 1937f (Finsler) und des Enckeschen Kometen 1937h. *Astron. Nachr.* 265:321–30

Whipple, F. L. 1950. A comet model. I. The acceleration of Comet Encke. *Astrophys. J.* 111:375–94

Whipple, F. L. 1951. A comet model. II. Physical relations for comets and meteors. *Astrophys. J.* 113:464–74

Whipple, F. L. 1961. Problems of the cometary nucleus. *Astron. J.* 66:375–80

Whipple, F. L. 1978a. Rotation period of Comet Donati. *Nature* 273:134–35

Whipple, F. L. 1978b. Cometary brightness variation and nucleus structure. *Moon & Planets* 18:343–59

Whipple, F. L. 1980a. Rotation and outbursts of Comet P/Schwassmann-Wachmann 1. *Astron. J.* 85:305–13

Whipple, F. L. 1980b. Periodic Comet Halley. *IAU Circ. No. 3459*

Whipple, F. L. 1981. On observing comets for nuclear rotation. In *Modern Observational Techniques for Comets*, ed. J. C. Brandt, B. Donn, J. Rahe. In press

Whipple, F. L., Sekanina, Z. 1979. Comet Encke: Precession of the spin axis, nongravitational motion, and sublimation. *Astron. J.* 84:1894–1909

Yeomans, D. K. 1972a. Nongravitational forces and Periodic Comet Giacobini-Zinner. See Marsden 1972, pp. 181–86

Yeomans, D. K. 1972b. A nongravitational orbit for Periodic Comet Borrelly. See Marsden 1972, pp. 187–89

Yeomans, D. K. 1974. The nongravitational motion of comet Kopff. *Publ. Astron. Soc. Pac.* 86:125–27

Yeomans, D. K. 1975. The orbital motion of comets Wolf and Barnard 3. *Publ. Astron. Soc. Pac.* 87:635–37

Yeomans, D. K. 1977. Comet Halley—the orbital motion. *Astron. J.* 82:435–40

Ann. Rev. Earth Planet. Sci. 1981. 9:147–74

THE SIGNIFICANCE OF TERRESTRIAL ELECTRICAL CONDUCTIVITY VARIATIONS

�># 10147

G. D. Garland

Department of Physics, University of Toronto, Toronto, Canada

INTRODUCTION

The degree to which different materials of the earth can conduct electricity varies enormously, from the insulating qualities of cold, quartz-rich crystalline rock to the metallic behavior of the liquid core. A well-recognized physical property, the specific resistance or resistivity expresses quantitatively the resistance to current flow, and the fact that it can be sensed through measurements on the earth's surface makes it an important geophysical parameter.

The history of the investigation of resistivity is similar to that of other geophysical quantities, such as seismic velocity. When it was recognized that the earth as a whole was in some degree a conductor, the initial aim was to determine a global average value of resistivity; this was followed by the search for a function describing the radial variation. While the latter effort is by no means completed, a great deal of recent work has been concerned with departures from the average behavior and that aspect forms the basis of this paper. Lateral variations in electrical resistivity may provide information on the composition and water content of the crust and on the temperature and degree of partial melting in zones of the mantle. The success of any investigation depends on the adequacy of electromagnetic measurements at the earth's surface, the ability to mathematically infer a distribution of resistivities on the basis of induction theory, and, by no means least, the knowledge, through laboratory measurements, of electrical properties of earth materials under different physical conditions. Recent progress in each of these aspects is considered.

First, a word as to units. Resistivity is conveniently expressed in ohmmeters (Ωm), the resistance of a meter cube of the material. As an indication

147

of the variation to be expected, the resistivity of cold granite may exceed 10^4 Ωm, that of sea water is about 0.25 Ωm, while the resistivity inferred for the earth's core is 10^{-6} Ωm. When it is desired to emphasize that a material is a very good conductor, the reciprocal of the resistivity, or conductivity, may be quoted. Conductivity is measured in reciprocal Ωm, now usually written as Siemens per meter (Sm^{-1})

INDUCTION PROCESSES

While it is not the purpose of this paper to discuss the techniques of measurement and analysis in great detail, some appreciation of these is essential for the critical evaluation of the conductivity anomalies that have been located. The geophysicist is limited to observations on, or very close to, the earth's surface or seafloor, and for the inference of electrical conductivity within the earth there are three basic requirements. First, there must be an energy source on or above the earth's surface. This may be a natural or man-made alternating magnetic field, of amplitude and frequency content sufficient to induce electric currents in the region under investigation, or it may be a source of electric current applied conductively to the earth. Second, a combination of magnetic and electric field components must be measured at stations over the region. The choice of quantity measured, and of whether the measurements are made simultaneously at many stations, or sequentially, has led to the naming of specific techniques, such as array studies, geomagnetic depth sounding, and magnetotelluric sounding. Third, there must be a mathematical procedure for inverting the observed field quantities into a model of underground conductivity distribution. All such procedures will be based upon the fundamental laws of induction and conduction, as expressed by Maxwell's equations.

Let us look first at induction processes. Because of the skin effect, or decrease in amplitude of a time-varying magnetic field within a conductor, signals useful for investigating the lower crust or upper mantle must have frequencies less than about 10 Hz. The skin effect may be visualized by considering the depth of penetration, or depth below the boundary at which a field is diminished to $1/e$ of its surface value. For a rock of resistivity 10^4 Ωm, this depth is about 50 km for a frequency of 1 cps, and would be about 400 km for a frequency of 1 cycle per minute, if the resistivity remained constant to that depth. Since, with portable stations, it is difficult to obtain observations of frequency components with periods greater than a few hours, the useful period range lies between about 0.1 s and 10^5 s. Within this range, typical natural magnetic field variations are produced by electric current systems external to the earth, including the sources of geomagnetic substorms (Matsushita 1975) which are fluctuations in the strength of ionospheric electrojets flowing east-west above the auroral zone.

The skin effect is a consequence of the induction of electric currents near the bounding surface of a conductor, a fact of great geophysical importance. For example, in the case of a conductive half-space, the secondary magnetic field of the induced currents is the same as that of an image of the primary source in the boundary. As Figure 1 indicates, this immediately leads to the reduction, at the earth's surface, of the vertical component magnetic field of an external source, and to the strengthening of the horizontal component. For this reason, induction by vertical and horizontal fields has often been treated separately, although this distinction may obscure the complete physical process. Nevertheless, it is a useful rule that the vertical component of time-varying fields over uniform

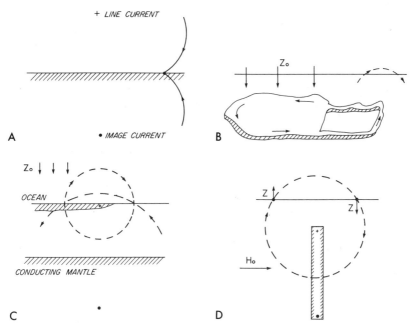

Figure 1 Schematic view of induction situations in the earth, with magnetic flux shown by arrows and continuous lines: (A) reduction of the vertical component and doubling of the horizontal component at the surface of a conductor, by the image of an overhead source; (B) current channeling in which current induced by the vertical field Z_0 over a large region leaks into a conductive channel; (C) lines of force around an ocean edge, produced by the edge current in the ocean, and the image of this current in the conducting mantle; (D) reversal in the sign of the secondary field Z across a tabular body subject to induction by the primary horizontal field H_0.

regions of the earth is small. Wherever the near-surface conductivity varies laterally, as at the edge of an ocean, the vertical component is enhanced and may show abrupt changes in sign.

One of the difficulties of analysis of inductive effects is that it is impossible to treat a limited volume of the earth in isolation. Electric currents are continuously induced over the whole outer region of the earth, and locally these will flow preferentially through zones of higher conductivity. It would be erroneous to consider only local induction in the zone. This phenomenon of current concentration is known as "channeling," although it is simply a part of the complete induction process.

All measuring systems have as their aim the separation of the secondary effects due to induced currents from the primary inducing fields, to provide a field quantity from which the internal conductivity may be inferred.

THE RADIAL VARIATION OF CONDUCTIVITY WITHIN THE EARTH

As is the case with many other geophysical parameters, lateral variations in electrical conductivity are best studied as departures from the globally averaged radial variation. For this reason, it is important to consider the results of global studies, although the methods of analysis are not described here in detail. Recent reviews include the works of Rikitake (1966) and Hutton (1976).

To probe the mantle to depths of hundreds of kilometers requires external magnetic field sources with periods from one day to many days, a fact that normally limits the data available to magnetic observatory records. Indeed, the first recognition that the interior of the earth is a remarkably good electrical conductor came from the analysis, on a global scale, of observatory records of the daily variation. Just as the static geomagnetic field may be analyzed in terms of spherical harmonics to separate internal and external contributions, by the method introduced by Gauss, so may instantaneous values of any time variation. Analyses showed (Lahiri & Price 1939) that approximately 30% of the total daily variation is of internal origin, arising from currents induced in the earth by the primary external sources. The problem then reduces to determining a radial variation of conductivity which will provide the observed internal portion.

The great advance in recent years has come partly from the successful isolation of longer-period variations, and partly from the application of inverse theory. For the former, Banks (1969) utilized frequencies of one, two, and three cycles of the 27-day storm recurrence period. The application of inverse theory to the problem is due to Bailey (1970) and Parker (1971), with the results discussed in Bailey (1973).

For our purpose, we may consider the estimates shown in Figure 2, both based upon data assembled by Banks (1969), who established his curve by forward modeling. Parker's data, shown as individual points with uncertainty bars, result from inversion. It is to be noted that there is a discrepancy at shallower depths, due partly to the different techniques of analysis and partly to data uncertainty at the higher frequencies (Bailey 1973), and it is also most important that no estimate is available at depths greater than 1500 km. The major jump in conductivity at a depth of about 400 km is believed to be related to the phase change in olivine to spinel structure. At shallower depths neither the highly resistive continental crust nor the conducting aesthenosphere are resolved. Thus, while any observed large departure from the radial curves has become known as "conductivity anomaly," some of these may in fact be normal features of the earth.

The conductivity of the deeper mantle remains very uncertain, largely because external variations of sufficiently long periods are not available. Estimates of the maximum possible conductivity have been made, based on the assumption that variations of a few years period, seen at magnetic observatories, are produced in the core and propagate through the entire

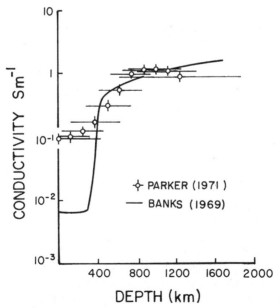

Figure 2 Estimates of the radial variation in conductivity.

mantle. On this basis, the maximum conductivity for the lower mantle is usually taken to be about 10^2 Sm^{-1}, although Alldredge (1977) has attributed the same variations to external causes, and suggested that lower mantle conductivity could be as high as 10^5 Sm^{-1}, within a factor of 10 of the assumed conductivity of the core. Recently, Ducruix et al (1980) have re-examined the evidence for secular variation impulses, and have concluded that effects of internal origin with characteristic times as short as four years are observed at the earth's surface, with the consequence that the very high value of conductivity proposed by Alldredge (1977) is not possible.

REGIONAL STUDIES

Arrays

There is an obvious advantage, in many situations, to the simultaneous measurement of magnetic field variations over a large grid or array of temporary magnetic observatories. This procedure became possible only with the development of an inexpensive, automatically recording magnetometer by Gough & Reitzel (1967), but arrays have by now been deployed over many parts of the earth's surface (Gough 1973, Lilley 1976, de Beer et al 1976). The quantities measured are simultaneous values of three magnetic components for specific magnetic events. Analysis of the records by Fourier spectral techniques permits the display of amplitude and phase variation over the region, for given periods [Figure 3, after Woods (1979)].

Qualitatively, a separation of internal and external effects may be made if distinctive variations in amplitude or phase, particularly of the vertical component, are shown over the survey area for different magnetic events. The quantitative separation depends upon the fact that, for all periods of geophysical interest, displacement currents are negligible, and the magnetic field outside the earth is derivable from a potential. Simultaneous values in the time domain may be subjected to standard procedures derived for the analysis of gravity or static magnetic fields. In particular, the internally produced field component at any station of the array is then given by a surface integral, over the complete horizontal plane, of a function involving the observed field components at all other stations. In principle, therefore, a complete separation into external and internal contributions is possible. There are, however, practical limitations to this procedure. The most serious is that the array may not provide sufficient coverage of an anomalous effect, so that the integration over the complete horizontal plane is not possible. Second, the separation resulting from

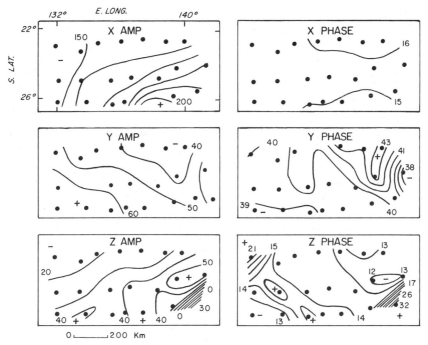

Figure 3 Example of array results from central Australia. Solid circles represent the locations of simultaneously operated three-component magnetometers. Contours show Fourier amplitudes (in nT-min, that is, gammas per cycle per minute) and phase lag (in min) for a period of 46 min. Note the remarkable uniformity except at the southeast corner. After Woods (1979).

any one analysis is for a single time; if the spectral content of the internal field is desired, repeated analyses for a series of times is required.

An alternative approach is to utilize the ability of the array to define gradients of field components (Schmucker 1970). Outside of the earth, the magnetic field is derivable from a scalar potential U which satisfies Leplace's equation

$$\frac{\partial^2 U}{\partial x^2} + \frac{\partial^2 U}{\partial y^2} + \frac{\partial^2 U}{\partial z^2} = 0.$$

Since $H_x = \partial U / \partial x$ and $H_y = \partial U / \partial y$, the measurable combination of gradients $(\partial H_x)/\partial x + (\partial H_y)/\partial y$ is thus equivalent to $-\partial H_z / \partial z$. Furthermore, for a plane-layered earth, the vertical variation of the vertical component H_z is determined by a function $f(z)$ which in any layer satisfies

$$\frac{d^2 f(z)}{dz^2} - (k^2 + 4\pi i \omega \sigma) f(z) = 0,$$

with k and ω the wave number and angular frequency of the primary disturbance and σ the conductivity of the layer. The result is that if one computes, for a given frequency, the ratio at the earth's surface

$$\frac{H_z}{\frac{\partial H_x}{\partial x} + \frac{\partial H_y}{\partial y}} = -\frac{H_z}{\frac{\partial H_z}{\partial z}}$$

a quantity is determined that is a function of the layered conductivity structure. The interpretation of the frequency variation of this quantity, as obtained from array observations, has usually been carried out by direct comparison with computed results for assumed models consisting of horizontal layers (Schmucker 1970, Woods 1979, Lilley et al 1980).

Geomagnetic Depth Sounding

If an array of magnetometers is not available, information on conductivity structure may still be determined, from statistical relationships between field components observed at individual stations. The method is an outcome of the study of magnetic effects at coastlines by Parkinson (1959). For a station exposed only to a random distribution of sources, there should, for a sufficient length of record, be no correlation between components. But if either the source is repetitive, as for example an auroral electrojet lying always to the north of a station, or if current is preferentially induced or channeled in a structure near the station, magnetic variations will tend to lie in well-defined planes. In other words, the vertical field variations at frequency f, $Z(f)$, will be a linear function of the two horizontal components, $X(f)$ and $Y(f)$.

$$Z(f) = A(f)X(f) + B(f)Y(f)$$

where $A(f)$ and $B(f)$ are, in general, complex because of phase differences between components. The transfer functions $A(f)$ and $B(f)$ may be represented graphically, by plotting, for both real and imaginary parts, the magnitude M and azimuth θ, where

$$M(f) = |A(f)^2 + B(f)^2|^{1/2},$$

$$\theta(f) = \tan^{-1}\frac{B(f)}{A(f)}.$$

The resulting quantities are known as induction arrows; it is usual, however, to reverse the sense of the real arrows, so that they will tend to point toward internal current concentrations (otherwise, because by convention Z is positive downward, arrows tend to point away from internal

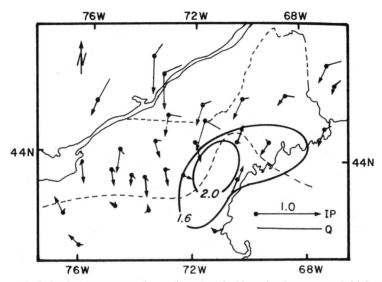

Figure 4 Induction arrows over the northern Appalachians; in-phase arrows (with heads) tend to point toward current concentrations, one of which is indicated by the southerly broken line. Solid contours show heat flow in heat-flow units (10^{-6} cal. cm^{-2} s^{-1}) in the vicinity of the White Mountains. After Bailey et al (1978).

sources). The magnitude of the plotted arrows, on a scale of 0 to 1 (Figure 4) is a measure of the ratio of vertical to horizontal magnetic field variation.

Ideally, induction arrows should be statistical properties of a station, in the sense that the Bouguer gravity anomaly is such a property. However, as mentioned above, there may be a systematic bias introduced by a source effect (Beamish 1979). In the absence of source effects, and over a horizontally layered earth, the induction arrows are vanishingly small. Where local structure is present, the arrows, for a range of periods, will tend to point toward internal current concentrations, reversing direction immediately over the current axis (Figure 4). Quantitative interpretation is usually carried out by assuming model conductivity distributions, and computing the transfer functions for them.

Magnetotelluric Surveys

In favorable circumstances, a sounding of the conductivity structure beneath a single station may be made, provided both the electric and magnetic fields are measured. The magnetotelluric method, as it is known, is due to Tikhonov (1950) and Cagniard (1953). On the assumption that an electromagnetic disturbance, of period T, at the earth's surface arises from a plane-polarized wave traveling downward from ionospheric sources, there is a simple physical interpretation to the ratio of the two fields. If we let the wave be polarized with its electric vector in the x-direction and magnetic vector parallel to the y-axis, and assume a uniform of conductivity σ, the electromagnetic wave equations require that

$$\frac{1}{2T}\left(\frac{H_y}{E_x}\right)^2 = \sigma,$$

provided consistent units are used throughout. On a real earth of varying conductivity, the application of the equation to the observations leads to an apparent conductivity, which will be some average value for the region sampled by the wave. Because of the skin effect, the depth of sampling increases with the period T of the disturbance analyzed, so that a curve of σ apparent against T is in effect a sounding. Further control is possible if the phase difference between H_y and E_x is also measured. The interpretation is straightforward if conductivity is a function of depth only, but is much more complex when there are lateral changes. Futhermore, it is recognized that the plane-wave assumption is rarely completely satisfied, but the effect of this on interpretation is difficult to evaluate and is frequently ignored.

The great attraction of the magnetotelluric method is the relative simplicity of the field apparatus: a single recording magnetometer for H and a potentiometer, electrode pair, and cable for E, together with the fact that no reference to another station is required. Recording is now usually performed digitally, and the records of E and H are spectrally analyzed for the direct computation of apparent conductivity. For probing to the lower crust or upper mantle periods up to 10^4 s are required; a sufficient length of record for analysis is usually obtained after three to five days recording at a station, depending upon the level of magnetic activity.

Controlled-Source Experiments

The methods described above depend upon natural geomagnetic fluctuations. They share the disadvantages that the level of activity varies with the solar cycle, and that the geometry of the external sources is never completely known. In controlled-source techniques, the geophysicist produces his own signal and applies it to the earth either inductively or conductively. The approach is therefore similar to methods of geophysical exploration, but on a much larger scale and with signals of lower frequency. For inductive coupling, a large loop on the surface of the earth is a convenient source. Nekut et al (1977) have described a sounding in the Adirondacks with a loop of 1.5 km diameter. Alternatively, a section of straight power line, grounded at each end, may be employed to apply current to the earth. Experiments have been conducted with very large current transients (Keller 1971), and also with currents of a few amperes (Duncan et al 1980), using a distinctive frequency content to permit isolation of very small signals from background noise. With both the loop and power line sources of this type, the quantity actually recorded at distant stations is a magnetic field component. The influence of currents induced at depth on the measured field is most easily extracted when the source-receiver distance approximates the depth involved. For example, in the experiment described by Duncan et al (1980), designed to probe to the lower crust, magnetic fields were recorded at distances up to 85 km from a power-line source 20.5 km long. The frequency components of the magnetic fields, normalized to the values expected from the same source, over a uniform earth, may be inverted (Anderson 1977) to give a conductivity distribution, for laterally uniform situations.

Power lines may also be used in a method analoguous to the geophysical exploration technique known as Schlumberger sounding. In this technique, current is applied to the earth through two electrodes at various spacings, while the potential drop is measured across two other electrodes. In an experiment described by Blohm et al (1977), current-electrode spacings up to 1250 km were employed, using a high voltage power line. The spacing of the potential electrodes was held at 100 m, this being sufficient because of the very large currents employed (up to 2950 amperes), and the fact that several cycles of on-off switching could be stacked.

Measurements at Sea

Apart from measurements on oceanic islands, results of which will be discussed below, induction surveys at sea are limited to the recording of the time variations of magnetic and electric field components on the seafloor with special instrumentation (Cox et al 1971). The cost of the instrumentation has precluded array studies, but results from magnetotelluric type soundings have been reported. Because of the small depth of penetration in sea water (about 0.25 km for a frequency of 1 cps), short-period oscillations are not recorded on the sea floor, but sufficient longer-period information is obtained for the sounding of the oceanic crust. The electric field may be recorded with special electrodes (Filloux 1977), or, if measurements on both the sea surface and seafloor are available, the electric field may be computed. From Maxwell's equations

curl $\mathbf{H} = \sigma \mathbf{E}$

where σ is the local conductivity, or, in component form

$$\frac{\partial H_z}{\partial y} - \frac{\partial H_y}{\partial z} = \sigma E_x.$$

The first term on the left vanishes under the assumed magnetotelluric conditions, and the second is obtained from the surface and bottom measurements. Thus, a component of the seafloor electric field is determined, given the conductivity σ of seawater. Law & Greenhouse (1980) have described soundings over the Juan de Fuca ridge in which, as a further simplification, measurement of the local surface magnetic field was replaced by observations taken at the nearest coastal observatory.

Application of Linear Inverse Theory

The methods described above employ various sources and depend upon the measurement of different physical quantities, but all attempt to determine the distribution of conductivity beneath the region under investigation. A valid question is the degree of uncertainty in the estimates of depths and conductivities, as provided by the different methods. Edwards et al (1980) have adapted the now widely used generalized linear inverse theory (Wiggins 1972) to provide information on this. Application of the theory requires the adoption of a class of models described by a family of parameters and the ascribing of realistic errors to the measured quantities. Edwards et al (1980) chose the horizontally layered section shown in Figure 5. The model is in fact one proposed by Blohm et al (1977), but may be typical of many areas. The results that would be obtained through probing by Schlumberger, controlled-source grounded power-

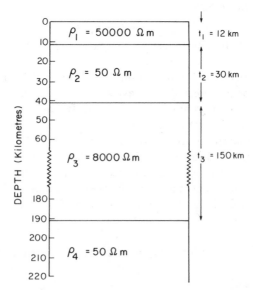

Figure 5   Conductivity section in South Africa, determined by Blohm et al (1977).

Table 1 Significant parameters, listed in order of decreasing certainty, of the model shown in Figure 5, as determined by four methods of sounding. Numbers in brackets are fractional standard errors.

Schlumberger	Magnetotelluric	Controlled bipole source[a]	Geomagnetic depth sounding[b]
t_1 (0.01)	t_1 (0.03)	t_1 (0.02)	$\sigma_2 t_2$ (0.09)
$\sigma_2 t_2$ (0.05)	$\sigma_2 t_2$ (0.03)	ρ_2 (0.04)	t_3 (0.12)
ρ_1 (0.19)	ρ_1 (0.12)	t_2 (0.12)	t_1 (0.28)
$\rho_3 t_3$ (0.52)	t_3 (0.15)	ρ_1 (0.12)	ρ_3 (0.66)
	$\rho_2 t_2$ (0.21)		
	$(\sigma_2 = 1/\rho_2)$		

[a] Vertical magnetic field measured at 70 km, bipole length $\ll$ 70 km.
[b] Period 1.2 min, for which the conductive layer 2 is exactly one skin-depth thick.

line, magnetotelluric, and geomagnetic depth sounding were then investigated. For the latter case, the model had to be converted into a two-dimensional situation by the termination of layer 2 at a vertical plane beneath the region, because over a horizontally layered earth, the transfer function is everywhere zero. The results are summarized in Table 1, which shows, for each method, the parameters determined in order of decreasing certainty. Noteworthy is the fact that combinations of parameters, such as the conductivity-thickness product of a conductive layer, rather than separated parameters, are often determined. Second, the depth to the upper conductor, which lies at lower-crustal depths, is determined with surprisingly small error by most methods. Information on the lower conductor, which lies in the upper mantle, is much less certain. Geomagnetic depth sounding gives relatively poor depth determination of even the upper conductor, but a fact not shown in the table is that the horizontal position of the edge of the upper conductor is well determined. These results are, as mentioned above, for a specific model, but it is believed to be a model representative of the crust and upper mantle in many areas, and the findings will be referred to again when the results of surveys over specific areas are considered.

MINERAL AND ROCK CONDUCTIVITIES

As mentioned in the introduction, conduction in the earth may take place by two distinct families of mechanisms: ionic conduction in fluids occupying fractures and pore spaces, and by ionic or electronic conduction through the mineral fabric itself. These two classes of conduction exhibit very different behaviors with varying pressure and temperature, and it is necessary to consider them separately. Table 2 indicates the extreme range of values of resistivity that may be expected in earth materials. Fluid conduction is the dominant process in near-surface rocks, particularly unconsolidated sedimentary rocks, and probably also in the seawater-saturated basalts of the ocean floor (Drury 1979). Laboratory measurements confirm that pore conductivities may be predicted by Archie's Law (Dvorak 1975), which gives the rock conductivity σ as

$$\sigma = \sigma_f P^n$$

where σ_f is the conductivity of the pore fluids, P is the porosity, and n is an exponent varying between 1 and 2. The effect of increasing pressure is immediately seen to be that of decreasing porosity and therefore conductivity, and it was formerly thought that conduction through pores could only be significant at very shallow depths. However, there are two factors that tend to increase the effectiveness of pore conduction at greater depths. First, the effect of increasing temperature is usually to permit the fluids to take into solution additional material from the pore walls, increasing the fluid conductivity. Second, the pressure that reduces pore volume is the effective pressure, that is, the lithostatic pressure reduced by the pore pressure. In a rock regime including pores that are interconnected

Table 2   Ranges or orders of magnitude of the resistivity of some earth materials.

Material	Typical resistivity (Ωm)
Sea water	0.25–0.33
Wet, poorly consolidated, sedimentary rocks	2–10
Granite, room temperature	10^4–10^5
Seafloor basalt, saturated, room temperature, atmospheric pressure	0.67 (Drury 1979)
Graphite	3×10^{-4}–4×10^{-3}
Olivine (2–8 kbar pressure)	
650°C	10^5 (Duba et al 1974)
900°C	10^4 (Duba et al 1974)
1250°C	10^3 (Duba et al 1974)
Tholeitic basalt melt	
1400°C	1.1 (Waff 1974)
Mantle, 1500 km depth	1

but isolated from shallower depths, the effective pressure may be very much less than the lithostatic pressure.

Conduction mechanisms involving the mineral lattice were systematically investigated by Tozer (1959), and the parameters have been subjected to many laboratory investigations in more recent years (Duba et al 1974, Shankland 1975, Volarovich & Parkhomenko 1976). Conduction may be by ions of the lattice, impurity ions, or electron-hole pairs, with the effect of temperature expressed through an activation equation:

$$\sigma_i = \sigma_{0i}e^{-E_i/kT}$$

or

$$\sigma_e = \sigma_{0e}e^{-E_e/2kT}$$

where the subscripts i and e refer to ionic and electronic conduction, and k is Boltzmann's constant. However, both the initial conductivities σ_0 and the activation energies E are pressure dependent. Furthermore, in the case of olivine, they are dependent upon the crystal class or phase of the mineral. Akimoto & Fujisawa (1965) observed in the laboratory the decrease in activation energy E, and consequent increase in conductivity, upon the phase change of olivine to spinel structure. At the temperatures corresponding to lower crustal and upper mantle depths, conduction by electrons and holes is probably dominant in basalt and olivine. At greater depths, as Tozer (1959) predicted, it may be matched by ionic conduction, although increasing pressure inhibits the latter mechanism.

The construction of prediction curves for conductivity in the outer part of the earth requires the assumption of pressure and temperature variations with depth, of mineral composition, and of the parameters required to express both fluid and mineral conduction. Examples of these predictions, which are extremely useful as standards against which to compare the results of sounding experiments, are shown in Figure 6. The earlier estimate of Brace (1971) for eastern United States predicts lower conductivities at lower crustal depths than does the "wet" continental model of Dvorak (1975), chiefly because of a more conservative estimate of the contribution of pore fluids. As is discussed below, the interpretation of recent observations requires conductivities at least as high as those predicted by Dvorak.

Within the upper mantle, there is an additional effect to be considered, that of partial melting at the depth of the aesthenosphere. It is well-established that melting greatly increases the conductivity of a rock (Waff & Weill 1975, Shankland & Waff 1977), but the difficulty of modeling a zone of partial melting lies in ascribing the proportion of molten material

RESISTIVITY Ωm

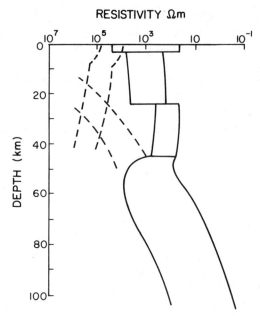

Figure 6   Estimates of bounds on crustal and upper mantle conductivity, based upon laboratory measurements and assumed conditions. Broken lines indicate the prediction by Brace (1971) for the eastern United States, showing pore (upper) and mineral (lower) conduction separately. Solid lines give the results by Dvorak (1975) for combined conduction in saturated rocks.

and the manner in which molten portions are interconnected. Waff (1974) has carried out calculations for both cubic and spherical elements with their boundaries "wetted" by the conductive melt. At a given temperature, the conductivity of molten basalt exceeds that of solid olivine by over three orders of magnitude. For example, at 1300°C, the conductivity of Red Sea olivine is 10^{-3} Siemens m^{-1}, that of tholeitic basalt magma, almost 10 Siemens m^{-1}. However, while the temperature dependence in the solid phase depends upon the precise composition and oxygen fugacity (Duba et al 1974), the conductivity of melts is remarkably independent of these factors. Even water content does not appear to influence the conductivity directly, although it presumably determines the proportion of molten material at a given temperature. The bulk conductivity is therefore almost completely determined by melt fraction and connectivity. A major conclusion, therefore, is that conductivity anomalies are to be expected if the aesthenosphere is a zone of partial melting. Conversely, the estimate of temperature from the conductivity, without independent control on the degree of melting and texture, would be impossible.

EXAMPLES OF CONDUCTIVITY ANOMALIES

While the electromagnetic sounding results over any area are influenced by conductivities over a range of depths, and the precise location of any anomaly may not be distinguishable, it is convenient to make a preliminary separation into these anomalies whose cause is almost certainly crustal, and those which very probably involve mantle material. Crustal anomalies may themselves be separated into those made up of extended, linear zones of enhanced conductivity, and those which seem to arise from high conductivity layers within the crust.

Linear Crustal Anomalies

A narrow belt of conductive rocks within a more resistive crust is usually located with a high degree of certainty by either an array survey or by transfer function measurements at stations on either side. Alabi, Camfield & Gough (1975) have described the detection, by array studies, and the significance, of a linear region of the enhanced conductivity extending from Southern Wyoming to Northern Saskatchewan. Complementary geophysical evidence suggests that the region lies along a zone of fracturing or shearing in the Precambrian basement, within which more conductive minerals, such as graphite, have developed. On a smaller scale, it is well known from exploration geophysics that shear zones in basement rocks often carry conductive graphite. In the Carpathians, Jankowski et al (1977) traced a very narrow conductive zone for 300 km along strike, by means of transfer functions. The zone follows the Klippen belt, which marks the boundary between the outer and central Carpathians. Following a separation of anomalous fields into external and internal contributions, a process facilitated by the two-dimensional geometry, the authors produced a numerical model of the conductive zone. The final model consists of a 16 km wide belt of rocks, with resistivity as low as 1 Ωm, lying at a depth between 22 and 45 km. It is probable that the development of metamorphic minerals is here also the chief source of the conductivity, although locally enhanced temperature may be a factor.

   Linear anomalies of the above types may be treated quantitatively, but it is most important that the effects of current channeling be included. By their very nature, these extended, narrow conductive zones provide channels for current flow between large volumes of the earth's crust, within which large potential differences are developed by regional inductive processes. The effects of channeling will, of course, be automatically included in the electromagnetic calculations, but only if a sufficiently large region is considered, and if the anomalous body is treated three-dimensionally.

Rift valleys provide an important example of linear crustal features with associated conductivity effects. Both the East African Rift in Kenya (Banks & Ottey 1974, Rooney & Hutton 1977) and the Rhine Graben (Winter 1973, Reitmayr 1975) have been investigated by transfer function and magnetotelluric methods. Higher than normal conductivity is indicated, but, in spite of the two-dimensional geometry, the location in depth of the conductive material has proven to be very difficult. The rifts are usually infilled with material that is itself conductive, and resolution of sources beneath this surface conductor is not easy. A deeper conductor is required by the observations, but it appears to straddle the crust-mantle interface. Both percolation of fluids along rift-related fractures, and enhancement of temperature, have been proposed to explain the higher conductivity.

In all of the above cases, a correlation between the linear conductor and some known feature of the crust is apparent. For the transfer-function study of the northern Appalachians, already shown in Figure 4 (Bailey et al 1978), a narrow zone of high conductivity indicated by reversal of the arrows has no surface expression. The axis of the feature strikes east-west, almost normal to the trend of known tectonic structures. As mentioned in the discussion of uncertainties, induction arrow or transfer-function analyses do not yield depths with precision, although the horizontal position of a conductor axis is well determined. In this case, the authors found that a linear conductor, extending in depth between approximately 15 and 40 km, could produce the effect. There is a possibility that the zone marks the northern edge of an extensive area of higher conductivity in the lower crust beneath the southern, as opposed to the northern, Appalachians. The fact that the current axis is apparently warped by the White Mountains heat flow anomaly (Figure 4) suggests a thermal relationship, but the lack of more obvious correlation with surface tectonic features is puzzling.

The Lower Continental Crust

In very many parts of the world, a layer of remarkably conductive rock is indicated, at depths corresponding to the lower crust. Many of the earlier determinations were based on magnetotelluric observations alone, a fact that led to some skepticism based on the thought that limitations of the method, or oversimplification of the models, were leading to erroneously high conductivities. But the phenomenon has been found by Schlumberger resistivity and controlled-source surveys also, and while the concept of uniform horizontal layering is an oversimplification, there is now good evidence for the existence of a widespread, but locally discontinuous, conductor beneath the continents, even outside of tectonic zones. As noted above, all three methods indicate the depth to the top of a conducting layer with remarkable precision, so that there is little reason to doubt the lower crustal location. On the other hand, they give in the first instance the conductivity-thickness product of the layer, rather than the properties individually. Nevertheless, the indicated values are such that, in the areas where the conductor has been found, most of the lower crust must be composed of material considerably more conductive than was formerly though probable.

Turning to specific examples, the section obtained by Blohm et al (1977) by means of Schlumberger soundings in South Africa has already been shown in Figure 5. Beneath the Midland Valley of southern Scotland, Jones & Hutton (1979a,b) determined, on the basis of magnetotelluric sounding, a resistivity of 35–60 Ωm for a depth range of 12–67 km. They showed also the discontinuous nature of the conductor, for beneath the Southern Uplands to the south the crust above 42 km was found to have much higher resistivities, up to 6000 Ωm. In a controlled-source power-line sounding conducted near Timmins, Ontario, on the Canadian Precambrian shield, Ducan et al (1980) found the resistivity to decrease from 15,000 Ωm to 270 Ωm, but at depths varying between 17 and 29 km, within a horizontal distance of 35 km. The lower resistivities quoted are typical of the values found beneath virtually every continent (Fournier 1978). Figure 6 suggests that the explanation probably lies in a considerable efficiency of pore conduction possibly supplemented by unusual composition rather than in the influence of temperature on normal lower crustal rocks. Among the possible compositional effects that could increase conductivity are the presence of graphite, some oxides such as magnetite. and hydrated minerals including serpentine.

Oceanic Crust and Uppermost Mantle

For soundings in normal oceanic areas, away from ridges and volcanic islands, the ocean-floor magnetometer technique described above is suitable. Poehls & Von Herzen (1976), from such measurements made in the Atlantic Ocean south of Bermuda, infer a resistivity of 10 Ωm extending from the seafloor to a depth of 100 km, so that both the oceanic crust and the uppermost mantle are highly conducting. Filloux (1977) has described a seafloor sounding experiment at a station approximately 800 km northeast of Hawaii, to the results of which a six-layer conductivity model was fitted. The upper 60 km were found to have the rather high resistivity of 10^2 Ωm, but within the depth range 140 to 220 km, a decrease toward 10 Ωm was found, possibly related to the aesthenosphere. Measurements made on oceanic islands are complicated by what is known as the "island effect," the direct electromagnetic effect of a relatively insulating island in a conducting ocean. Nevertheless, a series of independent observations made in Iceland on the Mid-Atlantic Ridge (Hermance & Grillot 1970, Beblo 1978) confirms that the island is underlain by highly conducting rock, the conductivity at shallow depths being consistent with that of basalt at a temperature of about 1000°C. Hawaii, which may be located over a mantle plume, has been studied by Larsen (1975) using a long series of electric field measurements and observatory magnetograms from the island of Oahu. The electromagnetic response was examined for frequencies in the range 0.1 to 6.0 cycles per day, and the island effect isolated by assuming initially a mantle conductivity profile given by global studies. This showed that the island effect was frequency independent for frequencies less than 6 cycles per day, a fact that permitted the inversion of the impedance tensor to give a layered

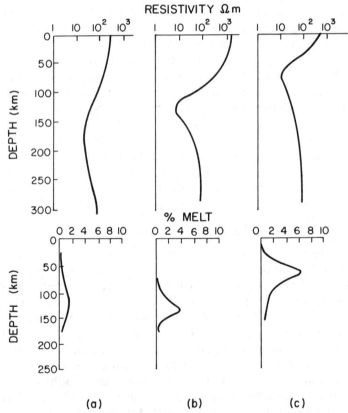

Figure 7 Possible models of resistivity (upper) and inferred proportion of melt (lower) beneath three ages of ocean floor in the Pacific: (a) 72 my, (b) 30 my, (c) 1my. After Oldenburg (1980).

mantle solution. Resistivities of approximately 10 Ωm were found in the uppermost mantle, with a decrease to 1 Ωm found, not at the depth usually assumed for the aesthenosphere, nor at that found by Filloux (1977) to the northeast, but at approximately 350 km. At greater depths, the profile is similar to that given by global studies. Larsen suggests that the conducting zone at 350 km may indicate a source of volcanism related to a mantle plume.

Oldenburg (1980) has performed a consistent inversion of three sounding experiments carried out over ocean floor of very different ages. The oldest (72 my) was studied in the work of Filloux (1977) described above. Observations made over oceanic crust of 30 my age were reported by Filloux (1980), while the work of Law & Greenhouse (1980) near the Juan de Fuca ridge, where the age is 1 my, has also been noted above. Oldenburg's analysis shows that, whereas many layered models will fit the data in each case, it is possible to pick models of minimum complexity. All of these (Figure 7 upper) show resistivity minima, which Oldenburg has interpreted, on the basis of the known behavior of partially molten material described earlier, in terms of minimum required percentage of melt (Figure 7 lower). The increase in depth of the partially molten zone, and the decrease in its severity, with increasing age of the ocean floor, are striking.

Other Mantle Anomalies

Although the definition of conductivity anomalies from sources within the mantle presents difficulties as compared to those arising from crustal conditions, much of the earliest work in the field was carried out in Japan, where the mantle is almost certainly involved. Large differences in response between stations were observed, even at diurnal periods, and it was shown (Rikitake 1969, 1975) that these could not be due entirely to the island or coastline effect. The original interpretation was that the depth to the conducting region of the mantle, where the resistivity first decreases to the order of 10 Ωm, is variable, with an insulating wedge beneath a portion of Japan. This insulating wedge could correspond with the upper part of the subducted lithospheric slab, and, if this is the case, it means that the Japanese electromagnetic studies provided the first evidence of a cold, descending slab. More recent work (Honkura 1978) has provided much more detail on mantle conductivities beneath Japan and the surrounding seas, but has not changed the basic interpretation. For example, the profiles shown in Figure 8, relating to central Japan, show that both the Japan Sea and the Pacific Ocean are underlain by relatively conducting mantle, as contrasted to Japan itself. Farther north, it appears that the mantle beneath the Pacific Ocean east of Japan

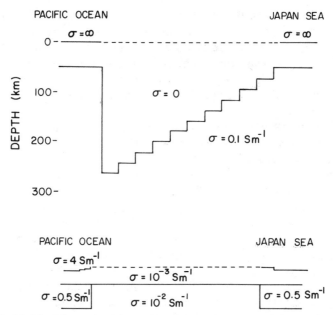

Figure 8   Models of the conductivity structure beneath Japan, proposed by Rikitake (1975) (upper) and Honkura (1978) (lower). Horizontal scale is equal to vertical. The characteristic of both models is the low mantle conductivity beneath Japan itself (broken line).

is much less conducting (resistivity 1000 Ωm) and therefore colder than that beneath the inland Japan Sea, where resistivities as low as 2 Ωm may exist within the uppermost mantle. The implication is that the rising of molten material above the descending slab has greatly enhanced the conductivity.

The earlier discussion on rock properties indicated that if the aesthenosphere, over some range of its depth, contains partially molten material there should be a dramatic increase in conductivity. The depth, except possibly beneath regions of extremely thick lithosphere, would be much less than that for the sharp increase in radially averaged conductivity shown by global studies (approximately 400 km). While there are good indications of high conductivity at upper mantle depths from a number of continents, it must be admitted that the electromagnetic probing of the aesthenosphere is in a very early stage. One reason is that discussed earlier: the definition of a lower conductor beneath a shallower, crustal conductor is less certain by any of the usual techniques. A corollary is that areas in which the crustal conductor is absent or at least very uniform, provide windows to the aesthenosphere, an example of which is the sounding by Woods (1979) in central Australia, shown in Figure 9.

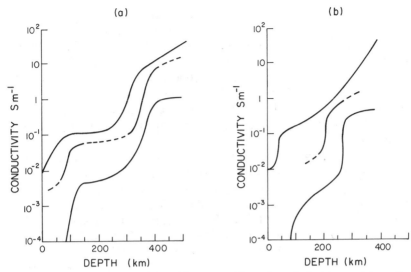

Figure 9   Estimates of conductivity, with upper and lower bounds, for (*a*) central Australia and (*b*) southeastern Australia by Woods (1979).

The determination of conductivity was by the array gradient method, the array providing good evidence of near-surface uniformity in the area for which the sounding is shown. The aesthenospheric effect is presumably the rise in conductivity seen at a depth of about 100 km beneath central Australia. It is possible that the conductivity then decreases before experiencing the sharp rise at a depth of about 400 km. The latter increase appears to be the same effect noted in global studies (Figure 2) and may correspond to a phase change in the mantle (Akimoto & Fujisawa 1965). This result is one of the best indications, from a regional survey, of the variation found globally. The explanation for the very different behavior beneath southeastern Australia, with an apparent increase in conductivity at the intermediate depth of 200 km, is not yet clear.

In western North American (Camfield & Gough 1975), there is a widespread conductor whose depth is difficult to resolve, but whose conductivity-thickness product appears to require mantle material. The conductor appears to be thickest in areas of high heat flow and severe P-wave velocity minima. Measurements in other parts of North America, including the Canadian Precambrian Shield, have not provided unequivocal evidence of an aesthenospheric conductor. The conductor may well be absent, but, as noted earlier, the Shield is characterized by discontinuous crustal conductors and by linear features, which make deeper sounding difficult. Beneath the Andes in Peru and Bolivia (Aldrich et al 1975), there is evidence of enhanced conductivity, but again the uncertainty in depth resolution, and the fact that the crust is thick beneath the mountains, make the assignment to mantle or crust somewhat difficult. South Africa has a prominent linear conductor associated with the Cape fold belt (Gough et al 1973) and a second, apparently separate, conductor (de Beer et al 1976) which may represent an extension of the Rift Valley zone. Of these, the first is perhaps the best candidate for a mantle source.

As might be expected, the variety of tectonic regimes within the USSR has led to many different types of resistivity-depth profiles. Berdichevski et al (1978) propose that, in view of the lack of depth resolution in many cases, the electrical nature of the aesthenosphere be divided into simply three categories on the basis of the conductivity-thickness product S: "poor" for $S < 1000$ Siemens, "moderate" $S \doteq 5000$ Siemens, and "well-developed", $S > 10,000$ Siemens. To put these figures in perspective, if we consider the seismic low velocity zone of the aesthenosphere to be 50 km thick, the resistivities in the zone would be $> 50 \,\Omega m$, $\doteq 10 \,\Omega m$, and $< 5 \,\Omega m$ respectively. According to Berdichevski et al (1978), well-developed aesthenosphere is found beneath the Pacific coast, the South-Caspian depression, and the Carpathians, while the East European and Siberian platforms are characterized by "poor" aesthenosphere. The Lake Baikal area, one of great tectonic interest, is subject to the complications mentioned earlier in connection with other rifts, that of near-surface high conductivities. However, long-period (1–2 hour) magneto-telluric observations suggest that the aesthenosphere beneath is "poor."

CONCLUSIONS AND FUTURE NEEDS

The chief contribution of regional conductivity studies to the knowledge of the earth can be expected in the areas of the structure and composition of the crust, especially the lower crust, and in the degree of partial melting, and possibly temperature, of the aesthenosphere. Recent advances have been notable in instrumentation, in interpretation theory, including the application of inverse theory to the estimate of uncertainties, and in laboratory measurements of rocks under conditions of high temperature, high pressure, and partial melting. But, as has been indicated above, a number of fundamental problems remain.

While the depth to the first conducting layer can be well established, it is the conductivity-product of the layer itself, rather than these parameters individually, that is best established. Sounding through a crustal conductor to a deeper zone produces uncertainties in the depth and properties of the latter. There are many continental regions characterized by a lower crust of very high conductivity, so high that one must invoke extreme contributions of pore fluids or compositional effects including hydrated minerals and oxides, to explain it. Laboratory measurements on rocks at lower crustal conditions should include critical evaluation of these current carriers. Those areas in which the crustal conductor is absent provide windows for the sounding to aesthenospheric depths, and should be exploited. However, it should be accepted that the information on the aesthenosphere will probably be limited to general classifications of the type of Berdichevski et al (1978), which may reflect the degree of partial melting. The estimate of temperature in a partially molten zone will be extremely difficult, because of the dependence of melting temperature upon fluid content.

While there is a valid reason, on electromagnetic grounds, for the deeper conductor to be seen most clearly where the lower crustal conductor is absent, we should be alert for a causal relationship. It is just possible that these conductors *are* developed beneath different regions, the crustal effect being produced where fluids have migrated upward, leaving an impoverished upper mantle. There is already some evidence from the British Isles (Jones & Hutton 1979a,b) and eastern Canada (Kurtz & Garland 1976) that conductive crust is associated with resistive uppermost mantle, and vice versa. The relationship in distribution between the two conducting zones will be an important avenue for future research.

Many features of global tectonics, for example subduction zones and the ridges away from islands, remain to be investigated. Because of the coastline effect, it is extremely desirable that measurements be extended in the uniform environment of the ocean floor. The field operations are expensive, but the behavior of conductivity with temperature under true mantle conditions would become much more certain if the signature of a downgoing slab, known from other evidence to be cold, could be obtained. Still more elusive problems for the future include the detection of supposed mantle plumes, and of relict subduction zones. Their investigation will require the simultaneous application of several sounding techniques, and the joint inversion of the observations, using full three-dimensional geometry.

Finally, the estimate of mantle temperatures beneath the aesthenosphere, one of the original aims of global induction studies, remains fraught with many difficulties. It has been pointed out that investigations of mantle conductivity by means of external sources are limited to the upper 1500 km; at greater depths only an upper bound can be placed by the analysis of fields of internal origin. Furthermore, the application of inverse theory shows that only an average conductivity over a range of depths can be expected, so that the conductivity-depth gradient at any depth is poorly known. It is noteworthy that none of the global investigations has yet included the presence of a possible "spike" in conductivity, correlated with the aesthenosphere. If regional studies show that the aesthenospheric conductor is widespread, the effect on the estimates of deeper conductivity, when the spike is ignored, should be investigated. In addition to these uncertainties in the electrical properties themselves, there is the problem of inferring temperature from the conductivity. The parameters σ_0 and E for the activation equation are required, not only as functions of pressure, but as functions applicable to the three presumed phases of mantle material, separated by depths near 400 and 600 km. Ducruix et al (1980), in estimating that the accepted lower mantle conductivity of 100 Siemens m^{-1} corresponds to a temperature of 3400 K, make the statement "it may still seem quite presumptuous to use conductivity estimates to infer temperatures." It is difficult to disagree with this expression of caution.

Literature Cited

Note: The literature on all aspects of electromagnetic induction in the earth has become very large, and the literature cited has intentionally been kept within modest limits, with the omission of many papers dealing with techniques and with surveys of specific continental areas. Some of the references (e.g. Hutton 1976) are themselves detailed reviews.

The International Association of Geomagnetism and Aeronomy organizes Workshops on Electromagnetic Induction every two years. Papers from these meetings may be found in the following special volumes:

Edinburg, Scotland (1972): *Phys. Earth Planet. Inter.* 7, 1973
Ottawa, Canada (1974): *Phys. Earth Planet. Inter.* 10, 1975
Sopron, Hungary (1976): *Acta Geodaet. Geophys. Montanist., Acad. Sci. Hungary* 12, 1977
Murnau, F. R. G. (1978): *Geophys. Surv.* 4(1 & 2), 1980

Akimoto, S., Fujisawa, H. 1965. Demonstration of the electrical conductivity jump produced by the olivine-spinel transition. *J. Geophys. Res.* 70:443–49

Alabi, A. O., Camfield, P. A., Gough, D. I. 1975. The North American central plains conductivity anomaly. *Geophys. J. R. Astron. Soc.* 43:815–33

Aldrich, L. T., Bannister, J. R., del Pozo, S., Salguiro, R., Beach, L. 1975. Electrical conductivity studies in South America: Chile-Bolivia. *Carnegie Inst. Washington Yearbook*, p. 292

Alldredge, L. R. 1977. Deep mantle conductivity. *J. Geophys. Res.* 82:5427–31

Anderson, W. L. 1977. Marquhardt inversion of the vertical magnetic field from a grounded wire source. *Rep. GD.-77-003.* USGS, Natl. Technical Information Service, Springfield, Va.

Bailey, R. C. 1970. Inversion of the geomagnetic induction problem. *Proc. R. Soc. London Ser. A* 315:185–94

Bailey, R. C. 1973. Global magnetic sounding-methods and results. *Phys. Earth Planet. Inter.* 7:234–44

Bailey, R. C., Edwards, R. N., Garland, G. D., Greenhouse, J. P. 1978. Geomagnetic sounding of eastern North America and the White Mountain heat flow anomaly. *Geophys. J. R. Astron. Soc.* 55:499–502

Banks, R. J. 1969. Geomagnetic variations and the electrical conductivity of the upper mantle. *Geophys. J. R. Astron. Soc.* 17:457–87

Banks, R. J., Ottey, P. 1974. Geomagnetic deep sounding in and around the Kenya rift valley. *Geophys. J. R. Astron. Soc.* 36:321–35

Beamish, D. 1979. Source field effects on transfer functions at midlatitudes. *Geophys. J. R. Astron. Soc.* 58:117–34

Beblo, M. 1978. Electrical structure of the lower crust and upper mantle beneath Iceland. Fourth Workshop on Electromagnetic Induction in the Earth and Moon. Murnau, F.R.G.

Berdichevski, M. N., Vanyan, L. L., Kuznetsov, V. A., Nechaeva, G. P., Okulesski, B. A., Shilovski, P. P. 1978. The geoelectrical model of the Baikal region and the USSR asthenosphere zone. Fourth Workshop on Electromagnetic Induction on the Earth and Moon. Murnau F.R.G. (Abstr.)

Blohm, E. K., Worzyk, P., Scriba, H. 1977. Geoelectrical deep soundings in Southern Africa using the Cabora Bassa power line. *J. Geophys.* 43:665–79

Brace, W. F. 1971. Resistivity of saturated crustal rocks to 40 km based on laboratory measurements. In *The Structure and Physical Properties of the Earth's Crust*, ed. J. G. Heacock. *Monogr. 14*, pp. 243–55. Washington DC: Am. Geophys. Union

Cagniard, L. 1953. Basic theory of the magneto-telluric method of geophysical prospecting. *Geophysics* 18:605–35

Camfield, P. A., Gough, D. I. 1975. Anomalies in daily variation magnetic fields and structure under north-western United States and southwestern Canada. *Geophys. J. R. Astron. Soc.* 41:193–218

Cox, C. S., Filloux, J. H., Larsen, J. C. 1971. Electromagnetic studies of ocean currents and electrical conductivity below the ocean floor. In *The Sea*, 4(Pt. 1):637–93. New York: Wiley

de Beer, J. H., van Zihl, J. S. V., Hyssen, R. M. J., Hugo, P. L. V., Joubert, S. J., Meyer, R. 1976. A magnetometer array study in South-West Africa, Botswana and Rhodesia. *Geophys. J. R. Astron. Soc.* 45:1–17

Drury, M. J. 1979. Electrical resistivity models of the oceanic crust based on laboratory measurements on basalts and gabbros. *Geophys. J. R. Astron. Soc.* 56:241–53

Duba, A., Heard, H. C., Schock, R. N. 1974. Electrical conductivity of olivine at high pressure and under controlled oxygen fugacity. *J. Geophys. Res.* 79:1667–73

Ducruix, J., Courtillot, V., Le Mouël, J. L. 1980. The late 1960's secular variation impulse, the eleven-year magnetic variation and the electrical conductivity of the deep mantle. *Geophys. J. R. Astron. Soc.* 61:75–94

Duncan, P. M., Hwang, A., Edwards, R. N., Bailey, R. C., Garland, G. D. 1980. The development and applications of a wideband electromagnetic sounding system using a pseudo-noise source. *Geophysics* 45:1276–96

Dvorak, Z. 1975. Electrical conductivity models of the crust. *Can. J. Earth Sci.* 12:962–70

Edwards, R. N., Bailey, R. C., Garland, G. D. 1980. Conductivity anomalies: lower crust or aesthenosphere? *Phys. Earth Planet. Inter.* In press

Filloux, J. H. 1977. Ocean-floor magnetotelluric sounding over north-central Pacific. *Nature* 269:297–301

Filloux, J. H. 1980. Magnetotelluric soundings over the northeast Pacific may reveal spatial dependence of depth and conductance of the asthenosphere. *Earth Planet. Sci. Lett.* 46:244–52

Fournier, H. 1978. Magnetotelluric and geomagnetic investigations of crustal and uppermost mantle conductivity. Fourth Workshop on Electromagnetic Induction in the Earth and Moon. Murnau, F.R.G.

Gough, D. I. 1973. The interpretation of magnetometer array studies. *Geophys. J. R. Astron. Soc.* 35:83–98

Gough, D. I., Reitzel, J. S. 1967. A portable three-component magnetic variometer. *J. Geomag. Geoelectr.* 19:203–15

Gough, D. I., de Beer, J. H., van Zihl, J. S. V. 1973. A magnetometer array study in Southern Africa. *Geophys. J. R. Astron. Soc.* 34:421–433

Hermance, J. F., Grillot, L. R. 1970. Correlation of magnetotelluric, seismic and temperature data from Southwest Iceland. *J. Geophys. Res.* 75:6582–91

Honkura, Y. 1978. Electrical conductivity anomalies in the earth. *Geophys. Surv.* 3:225–53

Hutton, V. R. S. 1976. The electrical conductivity of the Earth and planets. *Rep. Prog. Phys.* 39:487–572

Jankowski, J., Szymanski, A., Pěc, K., Pěcora, J., Petr, V., Praus, O. 1977. Electromagnetic studies of the Carpathian conduction anomaly. *Acta Geodaet. Geophys. Montanist., Acad. Sci. Hungary.* 12:99–109

Jones, A. G., Hutton, R. 1979a. A multistation magnetotelluric study in southern Scotland: I. Fieldwork, data analysis and results. *Geophys. J. R. Astron. Soc.* 56:329–49

Jones, A. G., Hutton, R. 1979b. A multistation magnetotelluric study in southern Scotland: II. Monte Carlo inversion of the data and its geophysical and tectonic implications. *Geophys. J. R. Astron. Soc.* 56:351–68

Keller, G. V. 1971. Electrical studies of the crust and upper mantle. In *The Structure and Physical Properties of the Earth's Crust*, ed. J. G. Heacock. *Monogr. 14*, pp. 107–25. Washington DC: Am. Geophys. Union

Kurtz, R. D., Garland, G. D. 1976. Magnetotelluric measurements in eastern Canada. *Geophys. J. R. Astron. Soc.* 45:321–47

Lahiri, B. N., Price, A. T. 1939. Electromagnetic induction in non-uniform conductors and the determination of the conductivity of the earth from terrestrial magnetic variations. *Philos. Trans. R. Soc. London Ser. A* 237:509–40

Larsen, J. C. 1975. Low frequency (0.1–6.0 cpd) electromagnetic study of deep mantle electrical conductivity beneath the Hawaiian Islands. *Geophys. J. R. Astron. Soc.* 43:17–46

Law, L. K., Greenhouse, J. P. 1980. Geomagnetic variation sounding of the asthenosphere beneath the Juan de Fuca ridge. *Geophys. J. R. Astron. Soc.* In press

Lilley, F. E. M. 1976. A magnetometer array study across southern Victoria and the Bass Strait area, Australia. *Geophys. J. R. Astron. Soc.* 46:165–84

Lilley, F. E. M., Woods, D. V., Sloane, M. N. 1980. Electrical conductivity from Australian magnetometer arrays using spatial gradient data. *Phys. Earth Planet. Inter.* In press

Matsushita, S. 1975. Morphology of slowly-varying geomagnetic external fields—a review. *Phys. Earth Planet. Inter.* 10:299–312

Nekut, A., Connerney, J. E. P., Kuckes, A. F. 1977. Deep crustal electrical conductivity: evidence for water in the lower crust. *Geophys. Res. Lett.* 4:239–42

Oldenburg, D. W. 1980. Conductivity structure of oceanic upper mantle beneath the Pacific Ocean. *Geophys. J. R. Astron. Soc.* In press

Parker, R. L. 1971. The inverse problem of electrical conductivity in the mantle. *Geophys. J. R. Astron. Soc.* 22:121–38

Parkinson, W. D. 1959. Directions of rapid geomagnetic fluctuations. *Geophys. J. R. Astron. Soc.* 2:1–14

Poehls, K. A., Von Herzen, R. P. 1976. Electrical resistivity structure beneath the north-west Atlantic Ocean. *Geophys. J. R. Astron. Soc.* 47:331–46

Reitmayr, G. 1975. An anomaly of the upper mantle below the Rhine graben, studied by the inductive response of natural electromagnetic fields. *J. Geophys.* 41:651–58

Rikitake, T. 1966. *Electromagnetism and the Earth's Interior.* Amsterdam: Elsevier

Rikitake, T. 1969. The undulations of an electrically conductive layer beneath the islands of Japan. *Tectonophysics* 7:257–64

Rikitake, T. 1975. A model of the geoelectric structure beneath Japan. *J. Geomag. Geoelect.* 27:223–44

Rooney, D., Hutton, V. R. S. 1977. A magnetotelluric and magnetovariational study of the Gregory rift valley, Kenya. *Geophys. J. R. Astron. Soc.* 51:91–119

Schmucker, U. 1970. Anomalies of geomagnetic variations in the southwestern United States. *Bull. Scripps Inst. Oceanogr.* 13. 165 pp.

Shankland, T. J. 1975. Electrical conduction in rocks and minerals: parameters for interpretation. *Phys. Earth Planet. Inter.* 10:209–19

Shankland, T. J., Waff, H. S. 1977. Partial melting and electrical conductivity anomalies in the upper mantle. *J. Geophys. Res.* 82:5409–17

Tikhonov, A. N. 1950. Determination of the electrical characteristic of the deep strata of the Earth's crust. *Dokl. Akad. Nauk. SSSR* 73:295–97

Tozer, D. C. 1959. The electrical properties of the earth's interior. *Phys. Chem. Earth* 3:414–36

Volarovich, M. P., Parkhomenko, E. I. 1976. Electrical properties of rocks at high temperatures and pressures. In *Geoelectric and Geothermal Studies*, ed. A. Adam. KAPG Geophys. Monogr., Budapest.

Waff, H. S. 1974. Theoretical considerations of electrical conductivity in a partially molten mantle and implications for geothermometry. *J. Geophys. Res.* 79:4003–10

Waff, H. S., Weill, D. F. 1975. Electrical conductivity of magmatic liquids: effects of temperature, oxygen fugacity and composition. *Earth Planet. Sci. Lett.* 28:254–60

Wiggins, R. A. 1972. The general linear inverse problem: implication of surface waves and free oscillations for earth structure. *Rev. Geophys. Space Phys.* 10:251–85

Winter, R. 1973. *Der Oberrheingraben als Anomalie der Elektrischen Leitfähigkeit Untersucht mit Methoden der erdmagnetischen Tiefensondierung.* Dissertation. Univ. Göttingen, Göttingen

Woods, D. V. 1979. *Geomagnetic depth sounding studies in central Australia.* PhD thesis. Aust. Natl. Univ., Canberra

Ann. Rev. Earth Planet. Sci. 1981. 9:175–98

METAMORPHOSED LAYERED IGNEOUS COMPLEXES IN ARCHEAN GRANULITE-GNEISS BELTS

✖ 10148

Brian F. Windley

Department of Geology, University of Leicester, Leicester, England LEI 7RH

Finley C. Bishop

Department of Geological Sciences, Northwestern University, Evanston, Illinois 60201

Joseph V. Smith

Department of Geophysical Sciences, University of Chicago, Chicago, Illinois 60637

INTRODUCTION

Archean granulite-gneiss belts are deeply eroded segments of crust formed before 2.7 Ga B.P. (1 Ga = 1000 m.y.). They contain remnants of intrusive layere digneous complexes that were highly deformed and metamorphosed to high amphibolite and granulite grades. The petrography and mineral chemistry of the complexes can provide some understanding of the igneous stratigraphy despite the metamorphic overprint, and selected geochemical features provide important constraints on the many tectonic models for evolution of Archean crust (summarized by Windley 1977).

Two contrasting types of layered igneous complexes are recognized. Anorthosite-leucogabbro complexes occur in many high-grade Archean regions (Table 1). Ultramafic-gabbro complexes in the Scourian belt of northwest Scotland and the Limpopo belt of southern Africa (Table 1) are in different sub-belts than anorthosite-leucogabbro complexes (F. B.

175

0084-6597/81/0515-0175$01.00

Table 1   Occurrences of Archean layered igneous complexes

Locality	Complex	References
Anorthosite-Leucogabbro complexes		
West Greenland	Fiskenaesset complex	Windley 1969, Windley et al 1973, Myers 1976a
Malagasy	Sakeny	Boulanger 1959
Limpopo, southern Africa	Messina layered intrusion	Hor et al 1975, Barton et al 1977a, 1979
Tamil Nadu and Karnataka, southern India	Sittampundi complex	Subramaniam 1956, Windley & Selvan 1975, Ramadurai et al 1975, Janardhanan & Leake 1975, Ramakrishnan et al 1978
Outer Hebrides, Scotland	Ness and Rodil	Watson 1969, Wittey 1975
Kola and Aldan shields, USSR	Belomorides	Moshkin & Dagelaiskaja 1972, Bogatikov 1974
Labrador	Nain Province	Collerson et al 1976
Sierra Leone	—	Wilson 1965, Andrew-Jones 1966, Williams & Williams 1976
Ontario	Shawmere complex	Simmons et al 1980
Ultramafic-Gabbro complexes		
Scourian mainland, NW Scotland	—	Bowes et al 1964, Davies 1974, J. D. Sills et al, in preparation
Limpopo, southern Africa	Neshuro complex and Chingwa-Ma-Karorro complex	Robertson 1973, 1974

Davis et al, unpublished data). This review concentrates on the anortho-site-leucogabbro complexes, which have been studied more thoroughly than the ultramafic-gabbro complexes.

GEOLOGICAL SETTING

Quartzo-feldspathic orthogneiss of tonalitic to granodioritic composition typically makes up 80–85% of the surface of Archean granulite-gneiss belts. Older and younger gneisses, separated chronologically by amphibo-lite dike swarms, occur in West Greenland (McGregor 1973), Labrador (Bridgwater et al 1975), and the Limpopo belt (Barton et al 1977b). Metamorphic ages are: older gneisses, West Greenland 3.6 to 3.8 Ga B.P., Labrador 3.6, Limpopo 3.86 ± 0.12; younger gneisses, West Green-land 2.8 to 3.0, Labrador 2.8 ± 0.2, Limpopo 3.13 ± 0.08 Ga B.P. Igneous tonalitic-granodioritic precursors can be seen in zones of low strain. The gneisses have calc-alkaline type of trace-element chemistry (Tarney 1976). Low initial strontium isotope ratios (~ 0.701) suggest derivation from a low-Rb source, probably in the mantle, and a short time interval between emplacement and metamorphism (Moorbath 1975). Within the gneisses are conformable layers up to 1 km thick of metamorphosed sedimentary (mica schist, marble, quartzite) and volcanic rocks with rare relict pillows. Rare, but locally thick, banded iron formations occur.

Deformed fragments of layered igneous complexes form lenses (1 m) to layers (up to 1 km thick and tens of kilometers wide) in the gneisses, and are commonly bordered by metamorphosed volcanic and sedimentary rocks (Figure 1). The fragmentation of the igneous complexes was caused partly by thinning during deformation and partly by intrusion of the tonalitic precursors of the gneisses. The complexes are usually conformable with their wall rocks but rarely, as at Fiskenaesset, West Greenland, discordant contacts demonstrate intrusion into supracrustal rocks (Escher & Myers 1975).

The western part of the Fiskenaesset complex (Figure 1) was intruded into pillow-bearing basaltic lavas which are chemically akin to modern oceanic basalts (Rivalenti 1976, B. L. Weaver et al, unpublished data). Marble and schist units occur sporadically along the top. The complex was isoclinally folded together with bordering basaltic rocks in an early episode (Figure 2) associated with thrusting; this probably occurred in a nappe-thrust regime (Bridgwater et al 1974). Two later phases of tight to isoclinal folding produced complex interference patterns (Figure 2). Granulite-facies metamorphism of the region has a Pb-Pb whole-rock isochron age of 2.81 ± 0.10 Ga B.P. (Black et al 1973) and a U-Pb zircon

178   WINDLEY ET AL

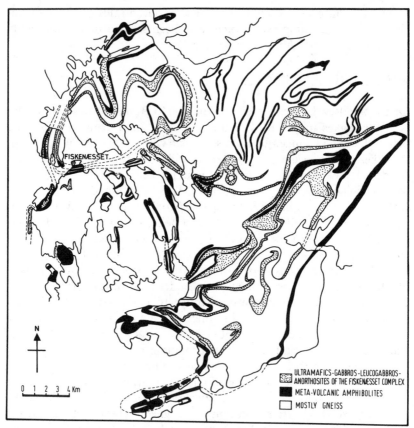

Figure 1   Geological map of western part of Fiskenaesset region, West Greenland. Approximately 50°30′ W, 63° N. After Kalsbeek & Myers (1973).

age of 2.80 ± 0.01 Ga B.P. (Pidgeon & Kalsbeek 1978). Amphibolite-facies metamorphism has zircon ages not older than 2.66 ± 0.02 Ga B.P.

In the Limpopo belt of southern Africa the Messina layered intrusion was emplaced at 3.16 ± 0.5 Ga B.P. (Rb-Sr isochron of Barton et al 1979) into near-surface rocks, now metamorphosed to interbedded iron formation, quartzite, pelitic gneiss, and hornblendite. The Sand River Gneisses of 3.86 Ga B.P. age form an elongated strip close to and partly bordering the intrusion (Figure 3). The exact details of the subsequent deformation are uncertain because of different structural interpretations (Hor et al 1975, Coward et al 1976, Barton et al 1979); however, there is agreement on an early phase of nappe-like isoclinal folding and several phases of refolding. Repetition of individual layers of the complex suggests the

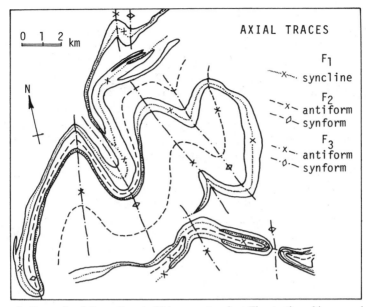

Figure 2 Successive deformations in Fiskenaesset region. The stratigraphic succession of the Fiskenaesset complex is duplicated by the F_1 isoclinal folding.

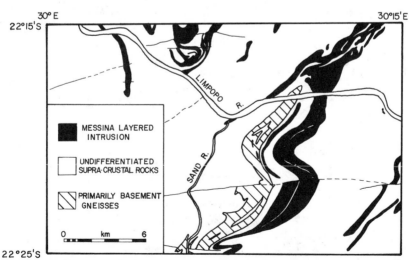

Figure 3 Simplified geological map of the Messina Layered Intrusion, southern Africa. After Barton et al 1979.

presence of early isoclines like those in the Fiskenaesset complex (Barton et al 1977a, 1979).

Layered complexes of leucogabbro and anorthosite are common in the high-grade gneisses of Tamil Nadu (Windley & Selvan 1975) and Karnataka (Ramakrishnan et al 1978) in southern India. They occur as concordant sheets, close to but not usually bordered by metamorphosed sedimentary and volcanic rocks. The Sittampundi complex is known best (Subramaniam 1956, Ramadurai et al 1975, Janardhanan & Leake 1975). The layered complexes lie in those parts of the Archean gneisses that have undergone only an amphibolite grade of metamorphism.

In NW Scotland lithological units in the Scourian mobile belt have a NE trend (F. B. Davies et al, unpublished data). On the mainland metamorphosed ultramafic-gabbro complexes contain only minor anorthosite (see, for example, Bowes et al 1964, Davies 1974), whereas in the Outer Hebrides there are several anorthosite-leucogabbro complexes including Ness (Watson 1969) and Rodil (Dearnley 1963, Wittey 1975).

The ultramafic-gabbro complexes have been so deformed that they occur in the gneisses as relict conformable lenses up to 100 m thick and 1 km long; they are preserved best in the Scourie region (Figure 4). In

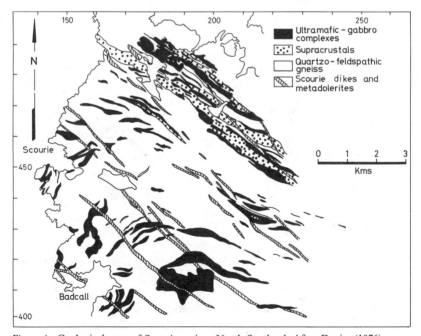

Figure 4 Geological map of Scourie region, North Scotland. After Davies (1976).

places they are bordered by mica schists of presumed sedimentary origin. Early isoclinal folding caused repetition of igneous stratigraphy and symmetrical distribution of schists.

In the northern marginal zone of the Limpopo mobile belt of southern Africa there are poorly exposed remnants of metamorphosed and deformed ultramafic-gabbro complexes. The Neshuro complex comprises serpentinite, tremolite-orthopyroxene rocks, pyroxene granulite, chromitite, and metanorite (Robertson 1974). The Chingwa-Ma-Karoro complex, 4 km long and 600 m thick, consists of layers of hornblende-two pyroxene-plagioclase, hornblende-orthopyroxene-plagioclase, and tremolite-orthopyroxene. The ultramafic layers contain chromitite seams up to 30 m thick and there is a 100 m thick layer of anorthosite (Robertson 1973). The complexes were recrystallized in the granulite facies and locally retrogressed. There are no modern mineralogical and petrochemical data on these Limpopo rocks.

STRATIGRAPHY

Anorthosite-Leucogabbro Complexes

The stratigraphic successions of the Fiskenaesset complex, West Greenland at Majorqap qâva and 30 km west on Qeqertarssuatsiaq Island are given in Table 2. Ultramafic rocks range from dunite and peridotite to pyroxenite and hornblendite. Thin ultramafic interlayers, 30–100 cm thick and up to 30 m long, occur between and within many of the above zones. Chromitite forms major seams up to 20 m thick in anorthosite, and magnetite is a major cumulus mineral in both plagioclase and ultramafic layers (Myers 1976a). Abundant relict igneous textures prove that the zones formed by differential precipitation of cumulus olivine, orthopyroxene, clinopyroxene, and plagioclase; the origin of hornblende is uncertain (see below). Mineral layering and megacrysts (usually plagioclase) which preserve the original cumulate textures are common. Trough layering and channel deposits demonstrate the action of currents on the floor of the magma chamber (Myers 1976b).

The rocks responded variably to the successive superimposed tectonic strains depending on position relative to limb thinning, degree of mineral layering, and viscosity contrast with respect to surrounding or adjacent rocks. Deformation tended to accentuate igneous layering.

In the Messina layered intrusion, anorthosite and gabbroic anorthosite make up over 90% of the succession (Hor et al 1975, Barton et al 1977a, 1979). Anorthositic gabbro, gabbro, melagabbro, and ultramafic rocks including serpentinites and pyroxenites form layers $10-10^3$ cm thick.

Table 2 Stratigraphies of selected layered igneous complexes

	Fiskenaesset		Messina[c]		Sittampundi[d]	
Majorqap qâva[a]	Qeqertarssuatsiaq[b]					
General deformed thickness (m)		Maximum thickness (m)		Thickness of one section (m)		Maximum thickness (m)
Upper gabbro 50	Upper anorthosite	75	Garnet anorthosite	285	Clinozoisite anorthosite	150
Anorthosite 250	Chromitite	20	Hornblende anorthosite	200	Hornblende anorthosite	75
Upper leucogabbro 50	Lower anorthosite	130	Gabbroic anorthosite	30	Gabbro (with 100 m long)	750
Middle gabbro 40	Upper leucogabbro	250	Pyroxenite	15	inclusions of pyroxenite)	
Lower leucogabbro 50	Gabbro	60				
Ultramafic rocks 40	Lower leucogabbro	100				
Lower gabbro 50	Ultramafic rocks	100				

[a] Myers 1976a
[b] Windley et al 1973
[c] Hor et al 1975
[d] Ramadurai et al 1975

Layers of chromitite up to 60 cm thick and magnetitite also occur. The stratigraphic succession depends on the structural interpretation: that for one section (Hor et al 1975) is given in Table 2. Cumulate plagioclase megacrysts, size-graded bedding, and mafic to felsic grading in some pyroxenitic and mafic layers were used to deduce younging directions (Barton et al 1977a, 1979).

In southern India the layered complexes consist predominantly of anorthosite and leucogabbro with minor gabbro and ultramafic rocks; relict cumulate textures and graded mafic and felsic layers occur (Ramakrishnan et al 1978). The stratigraphic succession of the Sittampundi complex is given in Table 2.

The Rodil complex in Scotland has a stratigraphic succession similar to that of the Fiskenaesset complex except for absence of the lower ultramafics and the chromitites (Wittey 1975).

Ultramafic-Gabbro Complexes

The Scourie bodies (J. D. Sills et al, unpublished data) consist of ultramafic units, composed mainly of amphibole pyroxenites, and of gabbroic units composed of garnet, two pyroxenes, plagioclase, amphibole, and spinel. Locally there is a cyclic repetition of ultramafic and gabbroic units with an intervening horizon of sulfide minerals. Some of the Scourie bodies contain minor anorthosite (Davies 1974). The ultramafic units have pronounced mineralogical layering, and serpentinized bands up to 5 m thick may have been dunites. Increase of plagioclase in gabbro layers up to 50 m thick provides a criterion for determination of younging direction. An early interpretation that the gabbroic units were a metasomatic corona between ultramafic and gneissic rocks has been discarded in favor of a consanguinous igneous origin of gabbroic and ultramafic units (Bowes et al 1964).

MINERALOGY

Although there is local preservation of igneous textures, almost all the igneous minerals have recrystallized in response to granulite, amphibolite, or greenschist grades of metamorphism. At least some amphibole replaced pyroxene, thereby requiring influx of hydrogen, but most chemical trends appear to have been preserved. Mineral compositions must have changed during metamorphism, e.g. Mg/Fe exchange between silicates and oxides, and NaSi/CaAl exchange between plagioclase and amphibole. Because

Table 3   Mineralogy of the Fiskenaesset complex at Qeqertarssuatsiaq

Zone	Mineralogy[a]
Upper anorthosite	Hb + Pl ± Gt ± Bi ± Il ± Mt ± Po ± Py
Chromitite	Hb + Pl ± Cpx ± Op ± Bi ± Mu + AC + Rt ± Il + Py ± Cp ± Mi ± Pn
Lower anorthosite	Hb + Pl ± Cpx ± Sph ± Qt ± Ep ± Mt ± Il ± Py ± Cp
Upper leucogabbro	Hb + Pl ± Op ± Bi ± Qt ± Ep ± Mt ± Rt ± Py ± Cp
Gabbro	Hb + Pl ± Op ± Bi ± Kf ± Ol ± Cpx ± Qt ± Sph ± Zr ± Mt ± AC ± Pn ± Po ± Cp ± Py
Lower leucogabbro	Hb + Pl ± Cpx ± Op ± Bi ± Mt
Ultramafic	Hb + Ol ± Cpx ± Op ± Tr ± Bi ± Qt ± Se ± Ac ± At + Il + Cm ± Mt ± Sp ± AC ± Cm + Pn ± Po ± Cp ± Py ± Mi ± Hz ± Cp ± Pd ± Gd ± Vi ± Cb ± Di

[a] Abbreviations: Ac actinolite, AC Al-chromite, At anthophyllite, Bi biotite, Cb cubanite, Cm chromite, CM Cr-magnetite, Cp chalcopyrite, CP Co-pentlandite, Cpx clinopyroxene, Di digenite, Ep epidote, Gt garnet, Hb hornblende, Hz heazlewoodite, Il ilmenite, Kf K-feldspar, Mi millerite, Mt magnetite, Mu muscovite, Ol olivine, Op orthopyroxene, Pl plagioclase, Pd polydymite, Pn pentlandite, Po pyrrhotite, Py pyrite, Qt quartz, Rt rutile, Se serpentine, Sp spinel, Sph sphene, Tr tremolite, Vi violarite, Zr zircon.

of poorer exposure, the mineralogical variations of the Messina layered intrusion and Sittampundi complex cannot be correlated with stratigraphic position as reliably as in the Fiskenaesset complex.

Anorthosite-Leucogabbro Complexes

Table 3 shows the principal minerals as a function of stratigraphic position in the Fiskenaesset complex. Plagioclase and amphibole are dominant, but about 40 minerals have been described.

SILICATES Amphibole occurs throughout the Fiskenaesset complex, usually as mosaic aggregates, rarely as inclusions in plagioclase megacrysts (Windley et al 1973), sometimes surrounding diopside and hypersthene relicts, and also as rims adjacent to peridotite layers (Myers & Platt 1977). Most amphiboles from the ultramafic and gabbro zones are magnesio-hornblendes. In plagioclase-free rocks, the *mg* ratio [= Mg/(Mg + Fe) mol] varies between 0.95 and 0.79. In plagioclase-bearing rocks, including the chromitites, the *mg* ratio tends to be lower 0.82–0.48 (Steele et al 1977), and locally displays a systematic trend with stratigraphic position (Myers & Platt 1977). The TiO_2 and Cr_2O_3 contents of hornblende are lower (mostly < 1 wt%) in the ultramafic and gabbro zones than in the chromitite zone (~ 1.2 wt% TiO_2 ; ~ 1.5 wt% Cr_2O_3), and the Al_2O_3 content tends to increase from 2–11 wt% in the ultramafic zone to 10–16 wt% in the chromitites and upper anorthosites. Coexisting tremolite and hornblende occur in the ultramafic zone.

Calcic plagioclase is the most abundant mineral in the Fiskenaesset complex, occurring as rare megacrysts partly recrystallized along rims and fractures (Myers & Platt 1977), but mainly as polygonal mosaics. Compositions range from An_{95} to An_{70} (Steele et al 1977) with considerable scatter in each zone, and only a weak tendency for the An content to decrease with increasing stratigraphic height. Although secondary plagioclase grains have similar composition to primary megacrysts (Myers & Platt 1977), a weak tendency for the Na content of plagioclase to increase with the amphibole/plagioclase ratio suggests transfer of Na from amphibole to plagioclase during metamorphism (J. V. Smith et al, unpublished data).

Olivine has mosaic textures and ranges from Fo_{87} to Fo_{72} with no obvious correlation with stratigraphic position (Steele et al 1977). Olivine also occurs in cores of coronas in the gabbro zone (Myers & Platt 1977). Orthopyroxene (En_{86-62}, CaO <0.8 wt%) and clinopyroxene high in Ca are common in the ultramafic rocks, and occur in a few gabbros and anorthosites. Garnet ($Alm_{46-50}And_4Gross_{3-8}Pyr_{36-46}Sp_{1-2}$) up to 2 cm in diameter occurs in the upper anorthosite. Phlogopite, tremolite, and biotite occur sporadically.

In the Messina layered intrusion (Hor et al 1975, Barton et al 1979), the silicate minerals have similar compositions to those in the Fiskenaesset complex with a tendency for higher Fe/Mg in amphibole and garnet, and higher Na/Ca in plagioclase. Irregular zoning and mechanical twinning of some plagioclase indicate lack of chemical equilibrium, and Na metasomatism may have occurred. Plagioclase coexisting with hornblende tends to be more sodic (50–70% An) than other plagioclase, suggesting that the hornblende grew in part at the expense of the anorthite component (Barton et al 1979). Some megacrysts of clinopyroxene and orthopyroxene contain magnetite dust, as seen in a few Fiskenaesset specimens.

In the Sittampundi complex, tschermakitic hornblende forms triple junctions with plagioclase (mostly An_{80}) in anorthositic rocks (Subramaniam 1956, Janardhanan & Leake 1975) and is joined by anthophyllite and chlorite in the chromitites (J. V. Smith et al, unpublished data). Vermicular sphene and networks of clinozoisite up to 20 cm across occur in anorthositic rocks, and the latter is the basis for stratigraphic distinction of clinozoisite anorthosite (Table 2). Garnet occurs as an accessory in anorthositic rocks, and as a major phase in garnet granulite. Remarkable plagioclase-pyroxene-hornblende-magnetite symplectites are developed around garnet in some granulites, and the minerals have a wide range of compositions, with most plagioclase being more sodic than in the anorthosites. Sulfate-rich scapolite, cordierite, sillimanite, and biotite occur rarely.

OXIDES   In the Fiskenaesset complex, spinel and Cr-magnetite occur in the ultramafic zone and interlayers. Ilmenite and magnetite are common in the ultramafic zone, and ilmenite occurs rarely in plagioclase-rich rocks. Coexisting magnetite and spinel neatly outline a miscibility gap in the $(Mg,Fe)(Fe,Al,Cr)_2O_4$ system (Steele et al 1977) between $(Mg,Fe)Fe_2O_4$ and $(Fe,Mg)Al_2O_4$ with a limit on the solvus near $(Mg,Fe)(Fe_{0.6}Al_{0.6}Cr_{0.8})O_4$.

Chromite-rich layers alternate with anorthosite bands to form a unit 10 cm–20 m thick between the upper and lower anorthosite zones (Ghisler & Windley 1967, Windley et al 1973). Some Al-chromite grains contain zones of rutile and silicate inclusions (Ghisler 1976, Steele et al 1977) while other grains are clean. Al-chromite occurs rarely in the ultramafic and plagioclase-rich zones. Ilmenite and magnetite occasionally coexist in the latter but rutile is rare. The composition range of the Al-chromite [mg 0–0.3, $Cr/(Cr+Al)$ 0.8–0.3, $Fe^{3+}/(Fe^{3+}+Al+Cr)$ 0–0.4] is distinct from that of chromite from spinel peridotites and alpine-type peridotites in ophiolites (Figure 5). The higher Al-content than for chromites from dry igneous layered complexes (e.g. Bushveld) may indicate a more aluminous magma in the Fiskenaesset complex.

Al-chromite is also a distinctive mineral in the Messina layered intrusion (Hor et al 1975, Barton et al 1979) and at Sittampundi (Subramaniam 1956, Ramadurai et al 1975). At Messina, chromitite layers up to 60 cm thick in anorthosite occur discontinuously over 14 km (van Zyl 1950), and a magnetitite layer in anorthosite occurs along strike for 1.5 km (Barton et al 1979). The compositions of Messina and Sittampundi chromite are generally similar to those at Fiskenaesset (Figure 5). At Sittampundi, chromite-hornblende-anthophyllite-chlorite layers in anorthosite reach 5 m width (Ramadurai et al 1975). Ruby corundum occurs sparsely in chromitite and adjacent anorthosite at Sittampundi in crystals up to 15 cm across.

SULFIDES   Sulfides in the Cu-Fe-Ni-Co-S system occur as accessory minerals throughout the entire Fiskenaesset complex (Bishop et al 1980), especially in the ultramafic zone and interlayers. Local occurrences of molybdenite may be metasomatic (Myers 1974).

Pentlandite, pyrrhotite, and subordinate chalcopyrite occur with one or more oxides as interstitial globules in ultramafic rocks. Fine-scale exsolution is prominent: pentlandite-pyrite, chalcopyrite-cubanite, chal-

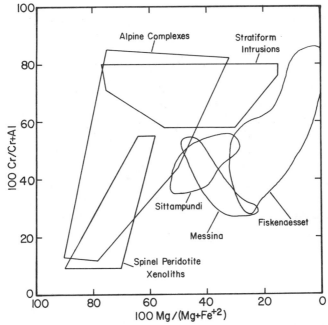

Figure 5 Comparison of spinel compositions. Data for Archean anorthosite-leucogabbro complexes derived from Ghilser (1976) and Steele et al (1977) for Fiskenaesset, Hor et al (1975) for Messina, Subramaniam (1956) and J. V. Smith et al (unpublished data) for Sittampundi.

copyrite-pyrite, pentlandite-heazlewoodite, and pentlandite-god levskite-polydymite. Millerite, Co-pentlandite, violarite, and digenite also occur. Pyrrhotite and pentlandite still occur in the gabbro zone, but become very rare in overlying leucocratic rocks. Euhedral pyrite with blebs of chalcopyrite and millerite becomes more common in leucocratic rocks, and pyrite inclusions in chromite occur in chromitite.

Sulfide minerals are disseminated throughout the Sittampundi complex (J. V. Smith et al, unpublished data). Pyrrhotite, pentlandite, and chalcopyrite form interstitial globules in pyroxenite and garnet-granulite layers within gabbro. Primary pyrite occurs, especially in pentlandite-free specimens. Secondary pyrite with textures similar to those in oxidized zones of supergene deposits (Ramdohr 1969) indicates weathering has affected primary sulfides. The thiospinel siegenite also occurs as a weathering product. Unlike Fiskenaesset sulfides, the Sittampundi ones do not commonly coexist with primary oxide; instead, goethite is found. Pentlandite, chalcopyrite, and pyrite occur interstitially in chromitites and as inclusions in chromite. Pyrite and chalcopyrite occur in anorthosite. As at Fiskenaesset, pyrrhotite and pentlandite tend to be replaced by pyrite and chalcopyrite with increasing stratigraphic height and decreasing color index of the host rock.

Ultramafic-Gabbro Complexes

In the Scourie bodies (J. D. Sills et al, unpublished data), the ultramafic rocks contain orthopyroxene, olivine (Fo_{86-82}, NiO 0.5 wt%), clino-pyroxene, pargasitic amphibole, spinel and magnetite, and the gabbroic rocks contain clinopyroxene, orthopyroxene, plagioclase, garnet, par-gasite, magnetite, spinel, ilmenite, and Fe,Ni,Cu sulfides. In general, the mineral compositions are typical for the granulite grade of metamorphism: pyroxenes have a wide composition gap and the orthopyroxenes contain 2–4 wt% Al_2O_3; almandine-pyrope garnets hold 14–19% grossular; par-gasitic amphiboles have high Ti and alkalis; spinel and magnetite hold up to 7 and 4 wt% Cr_2O_3, respectively. The Scourie ultramafic-gabbro bodies show a stratigraphic trend of mineral compositions in spite of the regional metamorphism. With increasing height in the gabbro, *mg* changes from 0.69 to 0.61 in orthopyroxene, 0.77 to 0.72 in clinopyroxene, 0.45 to 0.35 in garnet, and 0.66 to 0.61 in amphibole. Concomitantly in the amphibole, TiO_2 changes from 1.4 to 4 wt% and K_2O from 0.4 to 2.0 wt%. The Ab content of plagioclase increases from $\sim An_{90}$ to $\sim An_{60}$ as the Na_2O content of amphibole decreases from 2.7 to 1.7 wt%.

Pleonaste spinel occurs as single grains or composites with magnetite in the ultramafic rocks, and green hercynite replaces garnet in magnesian gabbros. Magnetite and ilmenite occur as composite grains in gabbros.

GEOTHERMOMETRY AND BAROMETRY OF ANORTHOSITE-LEUCOGABBRO COMPLEXES

In the Fiskenaesset complex, the following blocking temperatures for min-eral reactions were deduced: two-pyroxene (Wood & Banno 1973 model) 830–890°C; ilmenite-mafic silicate (Bishop 1976, 1980, Anderson & Lindsley 1979 models) $\sim 200°C$ lower; magnetite-ilmenite (Buddington & Lindsley 1964 model) $\sim 500°C$ lower than pyroxene temperature; pent-landite-pyrite (Misra & Fleet 1973) $< 230°C$. This range of $> 600°C$ was attributed (Steele et al 1977, Bishop et al 1980) to increasing diffusion rate from pyroxene to sulfides, with consequent lowering of the blocking tem-perature during retrograde metamorphism. Pressure estimates are more difficult, but symplectites around olivine cores suggest 6–9 kb (Myers & Platt 1977), which is consistent with the composition of the sapphirine (Bishop & Newton 1975) in metasomatised sedimentary rocks at the mar-gin of the Fiskenaesset complex.

For the Messina layered intrusion, evidence for intrusion as a sill suggested emplacement at <12 km (Barton et al 1979) and the sapphirine-kornerupine-corundum assemblage indicates metamorphism of adjacent sediments of at least 4 kb and 700°C (Schreyer & Abraham 1976).

For the Sittampundi complex, microprobe analyses of symplectites in a metamorphosed gabbro allowed recognition of a high-pressure granulite event producing garnet-clinopyroxene-plagioclase (An_{34-37}) at ~800°C and 12 kb and a retrogressive event producing higher-Fe garnet + higher-Ca plagioclase + orthopyroxene + clinopyroxene + hornblende at 650–700°C and 7–8 kb (J. V. Smith et al, unpublished data). A few samples of gabbro contain scattered olivine, indicating <5 kb before retrogression ceased. Such conditions are consistent with the occurrence of sillimanite, kyanite, sapphirine, and staurolite in the adjacent aluminous rocks.

For the Rodil complex the appearance of garnet in olivine-normative rocks and the absence of garnet in quartz-normative rocks, the presence of kyanite, and Fe/Mg exchange between garnet and clinopyroxene indicate 800–860°C and 10–13 kb (Wood 1975).

Mineral compositions in the Scourie ultramafic-gabbro complexes are consistent with granulite metamorphism. O'Hara (1977) has suggested a complex history for the surrounding gneisses, indicating that the rocks were uplifted from the base of the crust about 50 km deep (15 kb, 1250°C) to about 30 km deep (7–11 kb, 450°C) in the late Archean.

PETROGENESIS AND WHOLE-ROCK CHEMISTRY

It is difficult to provide an interpretation of the petrogenesis and whole-rock chemistry of the Archean layered complexes because of incomplete exposure, possible loss of layers and definite change of thickness during deformation, change of mineral compositions during metamorphism, change of rock composition by metasomatism, and uncertainty of mineral-liquid partition coefficients. Particularly troublesome is the likelihood that the complexes consist largely of crystals precipitated in a chamber open to gain of new liquid from the mantle and to loss of liquid at the top. Indeed such escaping liquids might be represented by adjacent amphibolites. In this section, an attempt is made to select those features of the petrogenesis and whole-rock chemistry that are least dependent on models.

Anorthosite-Leucogabbro Complexes

In the Fiskenaesset complex (Steele et al 1977) the petrographical and mineralogical evidence leads to the reliable conclusion that the following minerals formed cumulates: Ti-Cr-Al-magnetite in certain layers of ultramafic rocks, especially prominent in the lower zones; Al-chromite in oxide-rich zones of the anorthosites and especially in the chromitites; Ca-plagioclase in the anorthosites and at least some leucogabbros; and anhydrous Fe,Mg-silicates in the gabbros. Certainly some amphibole must be secondary as evidenced by the textural relationships to pyroxenes. A critical question is whether some hornblende is a pre-metasomatic phase, and if so, whether the hornblende is a cumulus or an intercumulus phase or both.

Irrespective of the origin of the hornblende, the other cumulus minerals clearly point to a distinction from the mineralogy of the dry basaltic layered intrusions such as the Skaergaard and Stillwater complexes. It is deduced from the presence of magnetite that the Fiskenaesset magma was more oxidized than the dry tholeiitic basalts of the above complexes. Early precipitation of magnetite was proposed (Windley et al 1973) as a mechanism for reducing the fractionation of Mg in the Fiskenaesset complex, in contrast to the strong fractionation producing Fe-enrichment in dry tholeiitic complexes, but it could be argued that continuous feeding into a magma chamber and simultaneous bleeding from the top would be a viable alternative. In plagioclase-rich rocks and chromitites, the plagioclase remains Ca-rich throughout the Fiskenaesset complex in contrast to the strong Na enrichment in the Skaergaard complex. Prolonged crystallization of Na-rich hornblende was proposed (Steele et al 1977) as a mechanism for retaining high Ca/Na in the liquid in spite of abundant precipitation of plagioclase; however, a feed-and-bleed model might again obviate this need. Although the role of hornblende is still uncertain, the presence of primary magnetite and hornblende would be consistent with a magma wetter than the dry tholeiitic basalts in the Skaergaard, Bushveld, and Stillwater complexes. Abundant crystallization of Al-rich plagioclase and Al-rich chromite in the upper zones suggests that the Fiskenaesset magma was Al-rich (Windley et al 1973).

Because the Messina layered intrusion has less magnetitite and chromitite than at Fiskenaesset, lower oxygen fugacity is indicated (Barton et al 1979). Metasomatic growth of hornblende from Fe,Mg-silicates, the anorthite component of plagioclase, and from incoming hydrogen is indicated by the positive correlation between albite content of plagioclase and volume of hornblende; this contrasts with a weak and erratic correlation at Fiskenaesset. Trends in whole-rock chemistry indicate that olivine and/or orthopyroxene are cumulate phases in mafic rocks, and that plagioclase is a cumulate phase throughout most of the complex.

For the Sittampundi complex, Al-chromite and plagioclase are definitely cumulate phases, and the evidence on hornblende is equivocal but mostly favors a metamorphic origin (Janardhanan & Leake 1975).

Indication of the types of magma responsible for the complexes can be obtained by interpretation of whole-rock chemistry. Table 4 lists some

Table 4 Selected whole-rock analyses from Archaen layered igneous complexes

	Fiskenaesset complex		Messina layered intrusion		Sittampundi complex		Scourian, Scotland	
	Anorthosite[a]	Gabbro[b]	Anorthosite[c]	Gabbro[d]	Anorthosite[e]	Garnet-Granulite[f]	Ultramafic[g]	Gabbro[h]
SiO_2	45.38	46.62	53.57	47.72	45.45	43.91	46.73	47.31
Al_2O_3	33.91	18.09	26.63	15.00	32.04	17.88	7.37	14.13
TiO_2	0.04	0.11	0.13	1.37	0.15	0.08	0.35	0.86
Fe_2O_3	0.60[i]	7.31[i]	2.62[i]	14.90[i]	0.82	0.98	3.38	2.42
FeO	—	—	—	—	1.44	7.08	8.91	10.82
MnO	0.01	0.12	0.04	0.18	0.03	0.14	0.21	0.25
MgO	0.46	13.68	1.47	8.17	1.60	15.48	22.64	9.16
CaO	17.64	12.68	12.58	10.47	16.06	12.29	7.09	11.66
Na_2O	1.09	1.27	2.45	1.18	1.31	0.66	0.67	1.87
K_2O	0.06	0.08	0.51	0.89	—	—	0.34	0.27
P_2O_5	0.001	0.004	0.01	0.12	0.05	0.04	0.03	0.09
Total	99.19	99.96	100.01	100.00	99.81	99.19	99.61	100.24
Rb	1	2	6.17	9.35	1.5	0.3	7	1
Sr	70	27	20.3	144	189	16	51	130
Cr	7	618	—	—	40	461	2991	461
Ni	15	434	—	—	22	296	986	167
Cu	—	—	—	—	21	8.5	19	182
Zn	1	27	—	—	—	—	118	162
Ga	24	14	—	—	25	10.5	—	—

[a] 132016 (Weaver et al, unpublished data)
[b] 132006 (Weaver et al, unpublished data)
[c] B-75-33 (Barton et al 1979)
[d] B-76-127 (Barton et al 1979)
[e] JS 155 (Janardhanan & Leake 1975)
[f] JS 173 (Janardhanan & Leake 1975)
[g] Average Ultramafic, Camas nam Buth, 1.89 L.O.I. (J. D. Sills et al, unpublished data)
[h] Average Mafic Gabbro, Camas nam Buth, 1.40 L.O.I. (J. D. Sills et al, unpublished data)
[i] Total Fe as Fe_2O_3

representative analyses of major rock types. Interpretation of parental liquids depends on the assumed fractions of cumulus and intercumulus minerals, especially hornblende. As an illustration of the problems, Henderson et al (1976) concluded that the Fiskenaesset magma had a strong REE (rare-earth element) fractionation (normalized La/Lu 24), whereas Morgan et al (1976) postulated a magma unfractionated with respect to REE and volatiles.

For the Sittampundi complex, the AFM $[(K_2O + Na_2O)\text{-}(FeO + Fe_2O_3)\text{-}MgO]$ diagram shows a calc-alkaline trend toward A with no Fe enrichment for the granulites and meta-anorthosites (Janardhanan & Leake 1975). For the Messina layered intrusion, there is extensive scatter on an AFM diagram because of the cumulate nature of the rocks, and the parent magma was interpreted to be a low-K, quartz tholeiite with low Sr (< 150 ppm) resembling tholeiitic rocks in some Archean greenstone belts (Barton et al 1979). Therefrom it was suggested that the sill-like complex formed at moderate depth from a chamber in the feeder system for volcanic rocks in a greenstone belt removed by erosion. A calc-alkaline trend had been deduced earlier by Hor et al (1975). For the Shawmere complex, Simmons et al (1980) suggest a basaltic magma with mildly fractionated REE pattern.

Preliminary whole-rock analyses by Windley et al (1973) for the Fiskenaesset complex are being superseded by comprehensive analyses by B. L. Weaver et al (unpublished data). Just as for the Messina layered intrusion, interpretation is complicated by the cumulate-rich nature of the rocks, and the large amount of magma that must have escaped; escape of magma is supported by the chemical consanguinity between the layered complex and adjacent amphibolites, and by the presence of rafts of roof amphibolite within the complex (Windley et al 1973). Two correlated groups of elements (Fe-Mg-Ni-Zn-Cr) and (Al-Ga-Sr-Na-Si-Ca) were attributed to cumulus ferromagnesian and plagioclase components, and a third group (Ti-Zr-Y-K-Ba-P) to a different source or sources. Although models utilizing only accumulations of olivine, orthopyroxene, clinopyroxene, plagioclase, and Al-chromite are sufficient to explain the chemistry of most of the cumulate rocks, petrographic evidence requires that Ti-Cr-Al-magnetite must also have been a cumulate for magnetite-rich rocks. Hornblende is not needed as a cumulus phase, but the presence of intercumulus hornblende seems likely because hornblende is concentrated in the intercumulus areas between cumulus plagioclase grains and, indeed, the alkalis in hornblende may be difficult to explain unless it is an intercumulus phase. Detailed calculations lead to the following estimates for the parent liquid: Zr ~ 24 ppm, Sr ~ 90 ppm, Ba ~ 30 ppm, Y ~ 8 ppm,

P_2O_5 0.02%, TiO_2 ~0.5%. For the REE, an essentially unfractionated pattern is deduced. The slight variation of Fe/Mg throughout the complex is attributed to tholeiitic nature of the magma.

Ultramafic-Gabbro Complexes

Few data are available at present, and this account is based solely on the studies of the Scourie bodies by J. D. Sills et al (unpublished data).

The granulite metamorphism of Scourie bodies was apparently iso-chemical except for possible loss of Rb, U, Th, and Cs from the gabbros. Chemically, the mafic gabbros are iron-rich, alkali-poor olivine tholeiites which resemble Tertiary basalts from Skye, and the ultramafic rocks resemble picritic basalts with high Ni (up to 2500 ppm) and Cr (up to 5000 ppm). An AFM diagram shows an alkali-poor tholeiitic trend with mod-erate Fe enrichment. Progressive decrease of Ni and Cr abundances through the stratigraphic sequence is consistent with petrographic evi-dence for early crystallization of olivine and pyroxene. Delayed crystal-lization of magnetite and ilmenite is indicated by increase of Ti and V as Mg decreases. This contrasts with early crystallization of magnetite in the Fiskenaesset complex, and indicates a lower oxygen fugacity for the Scourie bodies. Depletion of Ni by early olivine was suggested to explain the low Ni content of the sulfides, and influx of new magma apparently was associated with sulfide precipitation. Because the REE patterns are only slightly fractionated, and the Ni and Cr contents are so high, a high degree of melting of the mantle was deduced; furthermore, the smallness of Eu anomalies indicates a minor role for plagioclase, in contrast to the anorthosite-leucogabbro complexes.

TECTONIC ENVIRONMENT

The following features are important for understanding the tectonic envi-ronment: (a) the layered complexes abut a wide variety of rocks: meta-morphosed pillow basalts with oceanic-type chemistry (Fiskenaesset), meta-sediments with affinity to modern continental margins (Messina), older gneisses (Messina), and younger gneisses (Fiskenaesset; Sittam-pundi), (b) the complexes were engulfed in calc-alkaline material, which was deformed and recrystallized later into tonalitic gneisses of amphibolite to granulite grade under an overburden of 20–40 km, and (c) the complexes were deformed contemporaneously with intrusion of tonalitic sheets by thrusting and isoclinal folding in a tectonic regime characterized by sub-horizontal movements (West Greenland, Bridgwater et al 1974; Messina, Barton et al 1979; Sittampundi, Ramadurai et al (1975).

Complications arise because in spite of considerable similarities in stratigraphy, structure, and mineralogy, it is not certain that the complexes formed in the same tectonic environment. Moreover, it is not certain to what extent the thrust-nappe movements obliterated stratigraphic relations between the complexes and earlier rocks, and whether these rocks were continental or oceanic in origin. For example, Barton et al (1979) suggested emplacement of the Messina layered intrusion between supracrustal rocks and underlying older Sand River Gneisses, while Chadwick & Nutman (1979) suggested that major thrusting in the Buksefjorden region of W. Greenland made the original unconformity between Amîtsoq gneisses and overlying Malene supracrustals into which anorthosites were intruded appear conformable. McGregor (1973) and Sutton & Windley (1974) suggested that sedimentary and volcanic rocks in W. Greenland formed in an oceanic environment and were then interthrusted with older continental gneisses. Wells (1979) implied that volcanic rocks were part of the simatic crust injected later by tonalitic magmas in a Cordilleran-type continental margin.

Can the Archean layered complexes be interpreted in plate-tectonic terms? The complexes are comparable in stratigraphy and general mineralogy with complexes within modern Cordilleran batholiths (Windley & Smith 1976). In particular, the complexes of the Peninsular Range batholith of southern California (Nishimori 1974) are bordered by quartzite, calc-silicate rock, and micaceous schist and gneiss units (Walawender 1976), and are engulfed by tonalite. However, the surface rocks of southern California would need to be buried 20–40 km to match the metamorphic grade of the Archean complexes. If there was, at least locally, preexisting Archean continental basement, Central Peru provides a modern analogy where submarine basaltic-andesitic lavas of island-arc tholeiitic association were erupted onto continental crust, and were then intruded by the predominantly tonalitic Coastal batholith (Thorpe et al 1980). Pillow-bearing basaltic lavas have basic and ultrabasic intrusions along the strike in Ecuador. In conformity with this interpretation, Barton et al (1979) attributed the Messina layered intrusion to a continental environment.

Alternatively, the volcanic rocks and associated igneous complexes might have formed in an Archean oceanic environment. Consistent with this interpretation would be lack of any older continental rocks at Fiskenaesset, the Sr isotopic data indicating that tonalitic rocks have not reacted with older continental material (Moorbath 1975), and the pillow structure and bulk tholeiitic chemistry of the volcanic rock. The trace element patterns of the Fiskenaesset complex and its bordering metavolcanics indicate a similar source for both. Anorthosite-gabbro-ultramafic complexes occur in modern oceanic crust (Hedge et al 1979), though their mineral chemistry is not identical to that of Archean complexes (Windley & Smith 1976). Wet basaltic material at 20 km depth should melt partially to yield calc-alkaline magma (Wyllie 1979), and sinking of an oceanic slab carrying Archean layered complexes and volcanics over a basaltic base should have generated voluminous tonalitic material for engulfment of the layered complexes in an evolving arc or Cordilleran belt.

At present there are insufficient data to differentiate between these two models, and indeed the greater instability of Archean crust than for modern crust may preclude simple analogy.

CONCLUSIONS

Although the present studies of the mineralogy, petrochemistry, and tectonic environment of layered igneous complexes in Archean granulite-gneiss belts are only at a preliminary stage, it is obvious that a detailed understanding is needed to place constraints on the establishment of the Earth's crust. The present data demonstrate that the complexes underwent strong horizontal deformation as well as loading to a depth of 20 to 40 km. Better understanding of the exchange reactions between minerals during the deep-seated metamorphism should lead to better reconstruction of the original igneous stratigraphy. Therefrom it should be possible to test petrochemical models for the whole-rock chemistry. Currently it seems best to be rather cautious in trying to assign a continental or an oceanic character to the magmas responsible for the layered complexes. In spite of these uncertainties, it is gratifying to have reached the stage in Archean geology in which enough data exist to allow the posing of sensible questions.

ACKNOWLEDGMENTS

We thank NSF for grants EAR 77-02711 (JVS) and EAR 7815475 (FCB), NASA for 14-001-171 (JVS), and NERC for GR3/3198 (BFW). We thank J. D. Sills, D. Savage, and J. V. Watson for providing unpublished data on the Scourie bodies, and B. L. Weaver for providing unpublished data on the chemistry of the Fiskenaesset complex. A. S. Janardhanan, G. R. McCormick, and R. C. Newton also provided unpublished data. We thank the Geological Survey of Greenland for field support in Greenland.

Literature Cited

Anderson, D. J., Lindsley, D. H. 1979. The olivine-ilmenite thermometer. *Proc. Lunar Planet. Sci. Conf., 10th*, pp. 493–507

Andrew-Jones, D. A. 1966. Geology and mineral resources of the northern Kambui schist belt and adjacent granulites. *Geol. Surv. Sierra Leone Bull. 6.* 100 pp.

Barton, J. M. Jr., Fripp, R. E. P., Horrocks, P., McLean, N. 1977a. The geology, age and tectonic setting of the Messina Layered Intrusion, Limpopo mobile belt. *Proc. Seminar Pertaining to the Limpopo Mobile Belt*, eds. I. F. Ermanovics, G. McEwen, pp. 75–82. *Geol. Surv. Botswana Bull. 12*

Barton, J. M. Jr., Fripp, R. E. P., Ryan, B. 1977b. Rb/Sr ages and geological setting of ancient dykes in the Sand River area, Limpopo mobile belt, South Africa. *Nature* 267:487–90

Barton, J. M. Jr., Fripp, R. E. P., Horrocks, P., McLean, N. 1979. The geology, age and tectonic setting of the Messina Layered Intrusion, Limpopo mobile belt, southern Africa. *Am. J. Sci.* 279:1108–34

Bishop, F. C. 1976. *Partitioning of Fe^{2+} and Mg between ilmenite and some ferromagnesian silicates*. PhD thesis. Univ. Chicago, Chicago, Ill. 137 pp.

Bishop, F. C. 1980. The distribution of Fe^{2+} and Mg between coexisting ilmenite and pyroxene with applications to geothermometry. *Am. J. Sci.* 280:46–77

Bishop, F. C., Newton, R. C. 1975. The composition of low-pressure synthetic sapphirine. *J. Geol.* 83:511–17

Bishop, F. C., Smith, J. V., Windley, B. F. 1980. The Fiskenaesset complex, West Greenland, Part 4. Chemistry of sulphide minerals. *Grønlands Geol. Unders. Bull. 137.* 35 pp.

Black, L. P., Moorbath, S., Pankhurst, R. J., Windley, B. F. 1973. $^{207}Pb/^{206}Pb$ whole rock age of the Archaean granulite facies metamorphic event in West Greenland. *Nature Phys. Sci.* 244:50–53

Bogatikov, O. A. 1974. *Anorthosites of the USSR*. Moscow: Nauka 122 pp. (In Russian)

Boulanger, J. 1959. Les anorthosites de Madagascar. *Ann. Geol. Madagascar* 26:1–71 (In French)

Bowes, D. R., Wright, A. E., Park, R. G. 1964. Layered intrusive rocks in the Lewisian of the north-west Highlands of Scotland. *Q. J. Geol. Soc. London* 120:153–91

Bridgwater, D., McGregor, V. R., Myers, J. S. 1974. A horizontal tectonic regime in the Archaean of Greenland and its implications for early crustal thickening. *Precamb. Res.* 1:179–97

Bridgwater, D., Collerson, K. D., Hurst, R. W., Jesseau, C. W. 1975. Field characters of the early Precambrian rocks from Saglek, coast of Labrador. *Geol. Surv. Can. Pap. 75–1*:pt. A, pp. 287–96

Buddington, A. F., Lindsley, D. H. 1964. Iron-titanium oxide minerals and synthetic equivalents. *J. Petrol.* 5:310–57

Chadwick, B., Nutman, A. P. 1979. Archaean structural evolution in the northwest of the Buksefjorden region, southern West Greenland. *Precamb. Res.* 9:199–226

Collerson, K. D., Jesseau, C. W., Bridgwater, D. 1976. Crustal development of the Archaean gneiss complex, eastern Labrador. In *The Early History of the Earth*, ed. B. F. Windley, pp. 237–53. London: Wiley. 619 pp.

Coward, M. P., James, P. R., Wright, L., 1976. Northern margin of the Limpopo mobile belt, southern Africa. *Geol. Soc. Am. Bull.* 87:601–11

Davies, F. B. 1974. A layered basic complex in the Lewisian south of Loch Laxford, Sutherland. *J. Geol. Soc. London* 130:279–84

Davies, F. B. 1976. Early Scourian structures in the Scourie-Laxford region and their bearing on the evolution of the Laxford Front. *Q. J. Geol. Soc. London* 132:543–54

Dearnley, R. 1963. The Lewisian complex of South Harris with some observations on the metamorphosed basic intrusions of the Outer Hebrides, Scotland. *Q. J. Geol. Soc. London* 119:243–312

Escher, J. C., Myers, J. S. 1975. New evidence concerning the original relationship of early Precambrian volcanics and anorthosite in the Fiskenaesset region, southern West Greenland. *Grønlands Geol. Unders. Rapp.* 75:72–76

Ghisler, M. 1976. The geology, mineralogy and geochemistry of the preorogenic Archaean stratiform chromite deposits at Fiskenaesset, West Greenland. *Monogr. Series Miner. Deposits 14.* 156 pp.

Ghisler, M., Windley, B. F. 1967. The chromite deposits of the Fiskenaesset region, west Greenland. *Grønlands Geol. Unders. Rapp.* 12:1–39

Hedge, C. E., Futa, K., Engel, C. G., Fisher, R. L. 1979. Rare earth abundances and Rb-Sr systematics of basalts, gabbro, anorthosite and minor granitic rocks from the Indian Ocean Ridge System, western Indian Ocean. *Contrib. Mineral. Petrol.* 68:373–76

Henderson, P., Fishlock, S. J., Laul, J. C., Cooper, T. D., Conard, R. L., Boynton, W. V., Schmitt, R. A. 1976. Rare earth element abundances in rocks and minerals from the Fiskenaesset complex, West Greenland. *Earth Planet. Sci. Lett.* 30:37–49

Hor, A. K., Hutt, D. K., Smith, J. V., Wakefield, J., Windley, B. F. 1975. Petrochemistry and mineralogy of early Precambrian anorthositic rocks of the Limpopo belt, southern Africa. *Lithos* 8:297–310

Janardhanan, A. S., Leake, B. E. 1975. The origin of the meta-anorthositic gabbros and garnetiferous granulites of the Sittampundi complex, Madras, India. *J. Geol. Soc. India* 16:391–408

Kalsbeek, F., Myers, J. S. 1973. The geology of the Fiskenaesset region. *Grønlands Geol. Unders. Rapp.* 51:5–18

McGregor, V. R. 1973. The early Precambrian gneisses of the Godthaab district, West Greenland. *Philos. Trans. R. Soc. London Ser. A.* 273:343–58

Misra, K., Fleet, M. E. 1973. The chemical compositions of synthetic and natural pentlandite assemblages. *Econ. Geol.* 68:518–39

Moorbath, S. 1975. The geological significance of early Precambrian rocks. *Proc. Geol. Assoc.* 86:259–79

Morgan, J. W., Ganapathy, R., Higuchi, H., Krahenbuhls, U. 1976. Volatile and siderophile trace elements in anorthositic rocks from Fiskenaesset, West Greenland: comparison with lunar and meteorite analogues. *Geochim. Cosmochim. Acta* 40:861–87

Moshkin, V. N., Dagelaiskaja, I. N. 1972. The Precambrian anorthosites of the USSR. *24th Int. Geol. Congr.*, Montreal, Sect. 12:329–33

Myers, J. S. 1974. Molybdenite in the Fiskenaesset anorthosite complex, southern West Greenland. *Grønlands Geol. Unders. Rapp.* 65:63–64

Myers, J. S. 1976a. Stratigraphy of the Fiskenaesset anorthosite complex, southern West Greenland, and comparison with the Bushveld and Stillwater complexes. *Grønlands Geol. Unders. Rapp.* 80:87–92

Myers, J. S. 1976b. Channel deposits of peridotite, gabbro and chromitite from turbidity currents in the stratiform Fiskenaesset anorthosite complex, Southwest Greenland. *Lithos* 9:281–91

Myers, J. S., Platt, R. G. 1977. Mineral chemistry of layered Archaean anorthosite at Majorqap qâva, near Fiskenaesset, southwest Greenland. *Lithos* 10:59–72

Nishimori, R. K. 1974. Cumulate anorthositic gabbros and peridotites and their relation to the origin of the calc-alkaline trend of the Peninsular Ranges batholith. *Geol. Soc. Am. Abstr. with Program* 6:229–30 (Abstr.)

O'Hara, M. J. 1977. Thermal history of excavation of Archaean gneisses from the base of the continental crust. *Q. J. Geol. Soc. London* 134:185–200

Pidgeon, R. T., Kalsbeek, F. 1978. Dating of igneous and metamorphic events in the Fiskenaesset region of southern West Greenland. *Can. J. Earth Sci.* 15:2021–25

Ramadurai, S., Sankaran, M., Selvan, T. A., Windley, B. F. 1975. The stratigraphy and structure of the Sittampundi complex, Tamil Nadu, India. *J. Geol. Soc. India* 16:409–14

Ramakrishnan, M., Viswanatha, M. N., Chayapathi, N., Narayanan Kutty, T. R. 1978. Geology and geochemistry of anorthosites of Karnataka craton and their tectonic significance. *J. Geol. Soc. India* 19:115–34

Ramdohr, P. 1969. *The Ore Minerals and their Intergrowths.* Oxford: Pergamon. 1174. pp.

Rivalenti, G. 1976. Geochemistry of metavolcanic amphibolites from southern West Greenland. See Collerson et al 1976, pp. 213–23

Robertson, I. D. M. 1973. The geology of the country around Mount Towla, Gwanda district. *Rhod. Geol. Surv. Bull. 68.* 168 pp.

Robertson, I. D. M. 1974. Explanation of the geological map of the country south of Chibi. *Rhod. Geol. Surv. Rept. 41*

Schreyer, W., Abraham, K. 1976. Natural boron-free kornerupine and its breakdown products in a sapphirine rock of the Limpopo belt, southern Africa. *Contrib. Mineral. Petrol.* 54:109–26

Simmons, E. C., Hanson, G. N., Lumbers, S. B. 1980. Geochemistry of the Shawmere Anorthosite Complex, Kapuskasing structural zone, Ontario. *Precamb. Res.* 11:43–71

Steele, I. M., Bishop, F. C., Smith, J. V., Windley, B. F. 1977. The Fiskenaesset complex, West Greenland, Part 3. Chemistry of silicate and oxide minerals from oxide-bearing rocks, mostly from Qeqertarssuatsiaq. *Grønlands Geol. Unders. Bull.* 124:1–38

Subramaniam, A. P. 1956. Mineralogy and petrology of the Sittampundi complex, Salem district, Madras State, India. *Geol. Soc. Am. Bull.* 67:317–79

Sutton, J., Windley, B. F. 1974. The Precambrian. *Sci. Prog.* 61:401–20

Tarney, J. 1976. Geochemistry of Archaean high-grade gneisses, with implications as to the origin and evolution of the Precambrian crust. See Collerson et al 1976, pp. 405–17

Thorpe, R. S., Francis, P. W., Harmon, R. S., 1980. Andean andesites and crustal growth. *Philos. Trans. R. Soc. London.* In press

van Zyl, J. S. 1950. Aspects of the geology of the northern Soutpansberg area. *Univ. Stellenbosch Ann.* 26:1–95

Walawender, M. J. 1976. Petrology and emplacement of the Loss Pintos pluton, southern California. *Can. J. Earth Sci.* 13:1288–1300

Watson, J. 1969. The Precambrian gneiss complex of Ness, Lewis, in relation to the effects of Laxfordian regeneration. *Scott. J. Geol.* 5:269–85

Wells, P. R. A. 1979. Chemical and thermal evolution of Archaean static crust, southern West Greenland. *J. Petrol.* 20:187–226

Williams, H. R., Williams, R. A. 1976. The Kasila Group, Sierra Leone, an interpretation of new data. *Precamb. Res.* 3:505–8

Wilson, N. W. 1965. Geology and mineral resources of part of the Gola forest, southeastern Sierra Leone. *Geol. Surv. Sierra Leone Bull. 4.* 102 pp.

Windley, B. F. 1969. Anorthosites of southern West Greenland. *North Atlantic— Geology and Continental Drift*, ed. M.

Kay, pp. 899–915. *Am. Assoc. Pet. Geol. Mem. 12*

Windley, B. F. 1977. *The Evolving Continents.* London: Wiley. 385 pp.

Windley, B. F., Selvan, T. A. 1975. Anorthosites and associated rocks of Tamil Nadu, southern India. *J. Geol. Soc. India* 16:209–15

Windley, B. F., Smith, J. V. 1976. Archaean high-grade complexes and modern continental margins. *Nature* 260:671–75

Windley, B. F., Herd, R. K., Bowden, A. A. 1973. The Fiskenaesset complex, West Greenland, Part 1. A preliminary study of the stratigraphy, petrology and whole rock chemistry from Qeqertarssuatsiaq. *Grønlands Geol. Unders. Bull.* 106:1–80

Wittey, G. J. 1975. *The geochemistry of the Roneval anorthosite, South Harris, Scothland.* PhD thesis. Univ. London, London, England

Wood, B. J. 1975. The influence of pressure, temperature and bulk composition on the appearance of garnet in orthogneiss— an example from South Harris, Scotland. *Earth Planet. Sci. Lett.* 26:299–311

Wood, B. J., Banno, S. 1973. Garnet-orthopyroxene and orthopyroxene-clino-pyroxene relationships in simple and complex systems. *Contrib. Mineral. Petrol.* 42:109–24

Wyllie, P. J. 1979. Magmas and volatile components. *Am. Mineral.* 64:469–500

Ann. Rev. Earth Planet. Sci. 1981. 9:199–225

ICE AGE GEODYNAMICS   ✖ 10149

W. R. Peltier

Department of Physics, University of Toronto, Toronto, Ontario, Canada, M5S 1A7

INTRODUCTION

The study of glacial isostasy is concerned with the interpretation of that set of geophysical observables unambiguously associated with the response of the Earth to the massive continental deglaciations that marked the end of the last glacial epoch of the current ice age. These globally synchronous deglaciations, which began ca 18 KBP (thousands of years before present), resulted in the addition to the oceans of approximately 10^{19} kgm of water (see, for example, Peltier & Andrews 1976), which effected a mean global rise of sea level on the order of 80 meters. The major ice sheets that served as the source of this meltwater were the Laurentide-Innuitian complex of North America and the Fennoscandian complex of North Western Europe. On the basis of time series of the relative abundances of the isotopes 0^{16} and 0^{18}, obtained from deep sea sedimentary cores (see, for example, Broecker & van Donk 1970), from which one may infer the extent of northern hemisphere continental ice coverage, we know that the major ice sheets had existed on the surface for roughly 10^5 years prior to their disintegration, a time scale over which they were slowly, if not monotonically, approaching their maximum extents (Andrews & Barry 1978). Because of the enormous scale of these glacial complexes they exerted considerable mechanical stress on the underlying planet by virtue of their mutual gravitational attraction. It is fortunate for the subject of geodynamics that the planet has conspired to "remember" its response to this applied stress and it is for this reason that Reginald Daley (1934) was led to refer to the deglaciation event itself as "nature's great experiment."

The importance of this experiment is that, through interpretation of the data it produced, we may directly investigate the nature of the rheological law governing the long timescale response of mantle material to an applied stress. Such information is a crucial ingredient required in the construction of thermal convection models for plate tectonics (Peltier

199

Figure 1 Photograph of a flight of raised beaches in the Richmond Gulf of Hudson Bay near the center of rebound. Photo courtesy of Professor C. Hilaire-Marcel.

1980b). A second type of information that may also be extracted through interpretation of the results of this natural experiment concerns the detailed history of the melting event itself and this is important in understanding the mechanism of climatic change.

The main observational data that we are interested in explaining consist of histories of relative sea level obtained by the C^{14} dating of marine shells or other in situ carbonaceous material from relict beaches found either above or below present-day sea level. The photograph shown in Figure 1 is of a flight of raised beaches found in the Richmond Gulf of Hudson Bay, near what was the center of the ancient Laurentide ice sheet. When the height above present-day sea level of each of the beach horizons is plotted as a function of isotopic age (corrected to give proper sidereal age) then one obtains the relaxation curve shown in Figure 2 in which the oldest beach is found at the greatest height above m.s.l. (mean sea level). These data show that the surface of the solid earth at this site has been continuously elevated above the geoid since the removal of the ice load and that the rate of this uplift has been an exponentially decreasing function of time.

Complementing such pointwise relaxation data and providing (as we shall see) an equally important key to the rheological puzzle, are maps of

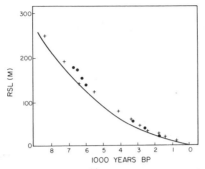

Figure 2   Relaxation curve based upon C^{14} dates of the beaches in Richmond Gulf compared with a theoretical prediction using the ICE-3 deglaciation history (to be discussed in a later section) and an Earth model with constant mantle viscosity. Figure is from Peltier (1980).

the present-day free-air gravity anomaly over the once ice-covered regions. Free-air anomaly data for Fennoscandia are discussed in Balling (1980), for example, and reveal a maximum negative anomaly over the once ice-covered region of between 15 and 20 milligals. An equivalent map for the Laurentide region was compiled by Walcott (1970) and is discussed in some detail later in this article. Here the maximum observed negative anomaly is on the order of 35 milligals.

That the sea level and gravity data do contain important rheological information should be immediately clear. Since the melting of the ice sheets was certainly complete by 5 KBP and most probably by 6.5 KBP (Prest 1969), relaxation data such as those shown in Figure 2 conclusively establish that the Earth cannot be described as a Hookes' Law solid. If it could be so described then the quasi-static deformation forced by the shifting surface load would have ceased with the cessation of melting. Even though the Hookean rheology is universally agreed to provide a reasonable description of short timescale seismic processes, it is completely inadequate insofar as the understanding of postglacial rebound is concerned. Even in the seismic regime, however, anelastic processes are important and produce observable effects, such as the dispersion of body wave velocities (Liu et al 1976), and endow the elastic gravitational free oscillations with finite Q (see, for example, Buland & Gilbert 1978). Such high frequency anelasticity is apparently well described by a linear viscoelastic absorbtion-band model (Anderson & Minster 1979, Minster 1980), which is capable of delivering the weak dependence of intrinsic Q upon frequency observed in the seismic regime. Such a model, however, does not allow steady-state creep so that if it were a complete representation of mantle rheology then thermal convection would be impossible and continental drift an indefensible hypothesis. That this is in fact the case is a view still held by a minority of Earth scientists (e.g. Jeffreys 1973).

More likely, however, is the possibility that the transient creep of the absorbtion-band model is followed by steady-state deformation and this is in accord with laboratory observations of the creep of Olivine single crystals (Kohlstedt & Goëtze 1974). It is presumably such steady-state creep that governs both postglacial rebound and mantle convection (Peltier et al 1980a). A linear viscoelastic approximation to the complete rheology must then consist of an absorbtion-band model (i.e. a standard linear solid with a continuous spectrum of relaxation times) in series with a Maxwell element to give a modified Burgers body solid (Peltier et al 1980b). In the long time limit such a solid behaves as a Newtonian viscous fluid, which is characterized by a linear relation between stress and strain rate. Laboratory data suggest that this might only be an approximation since they show, in the high creep rate regime where the experiments must be performed, that the stress-strain relation is nonlinear. It is still unclear whether or not at relevant mantle strain rates the deformation mechanism might be Newtonian but recent analyses (Twiss 1976) suggestive of structural superplasticity conclude that this could well be the case.

In total (and perhaps blessed!) ignorance of these microphysical complexities, early attempts to interpret the relaxation data from glacial isostasy were based almost exclusively upon the assumption that the Earth behaved entirely as a Newtonian viscous fluid insofar as the long timescale adjustment phenomenon was concerned. A rather complete account of the major achievements of interpretation up to and including the year 1973 will be found in the review by Walcott (1973). In Walcott's view the preferred model for the mantle viscosity profile based upon this work consisted of "a 110 km thick lithosphere which, although requiring a viscosity of 10^{25} P[oise] to explain some long-term behavior, behaves elastically on time scales of a few thousand years; a thin, low viscosity channel some $100-500$ km thick with a viscosity dependent upon thickness of 10^{19} to 10^{21} P[oise]; and a lower mantle viscosity greater than 10^{24} P[oise]." In Walcott's opinion the totality of the evidence then available tended to support the previously stated conclusions of McConnell (1968) and others to the effect that the viscosity of the lower mantle was significantly in excess (by at least two orders of magnitude) of the viscosity of the upper mantle. He considered the lower mantle value of 10^{24} Poise to be a *lower* bound upon the actual value, which he believed could be very much higher.

The question of the viscosity of the lower mantle relative to that of the upper mantle is an extremely important one insofar as models of mantle convection are concerned. The simple geometric argument given in Peltier

(1973), for example, suggests that for a mantle of uniform chemical composition an increase of viscosity by more than two orders of magnitude across the 670 km seismic discontinuity would be required to focus a distinct circulation into the upper mantle region. Otherwise, whole mantle convection would be preferred (Peltier 1973, 1980b, Sharpe & Peltier 1978, 1979) with a single cellular motion in the radial direction having a vertical extent comparable to the mantle thickness. The view expressed by Walcott in 1973 was necessarily conditioned by extant models of the rebound process. These models were deficient in several important respects and since that time a more complete and internally self-consistent theory of postglacial rebound has become available, in terms of which a more accurate description of the phenomenon is possible. In the following sections, the main ingredients of this new theory are reviewed and the revision of our notion of the mantle viscosity profile, which the new analyses require, is assessed.

THE THEORY OF GLACIAL ISOSTASY

The physical basis of the new model of isostatic adjustment is embodied in a mechanical constitutive relation that provides a macroscopic connection between stress and strain. Insofar as the long timescale rebound process is concerned, we shall assume that the mantle may be described as a linear viscoelastic Maxwell solid for which the stress tensor τ_{ij} and the strain tensor e_{ij} are related as (Peltier 1974)

$$\dot{\tau}_{ij} + \frac{\mu}{\nu}(\tau_{ij} - \tfrac{1}{3}\tau_{kk}\delta_{ij}) = 2\mu\dot{e}_{ij} + \lambda\dot{e}_{kk}\delta_{ij}, \tag{1}$$

where the dot denotes time differentiation, μ and λ are the conventional Lamé parameters of Hookean elasticity, and ν is the molecular viscosity. In the Laplace transform domain of the imaginary frequency s, (1) takes the form

$$\tilde{\tau}_{ij} = \lambda(s)\tilde{e}_{kk}\delta_{ij} + 2\mu(s)\tilde{e}_{ij} \tag{2}$$

where the tilde indicates implicit dependence upon s, and where $\lambda(s)$ and $\mu(s)$ are the compliances

$$\lambda(s) = \frac{\lambda s + \mu K/\nu}{s + \mu/\nu}; \qquad \mu(s) = \frac{\mu s}{s + \mu/\nu}, \tag{3}$$

where $K = \lambda + 2\mu/3$ is the elastic bulk modulus. When it is subject to an applied stress, the Maxwell solid exhibits an initial response which is

Hookean elastic. As time proceeds, however, the behavior tends asymptotically to that of a Newtonian fluid for which the deviatoric stress
depends linearly upon the strain rate. Inspection of the Laplace transform
domain compliances (3) shows that the time scale over which this transition in behavior is achieved is the Maxwell time $T_m = v/\mu$. In the upper
mantle, where the viscosity is on the order of 10^{22} Poise, the elastic shear
modulus is such that $T_m \simeq 200$ years. One of the ideas we hope to test by
fitting the Maxwell model to the postglacial rebound data is whether the
radial profile $v(r)$ that we obtain is acceptable from the point of view of
models of the mantle convection process. It is important to note that,
although the Maxwell model is in itself an incomplete description of mantle rheology, the missing short timescale anelasticity required to understand seismic Q does not affect the inference of $v(r)$ (Peltier et al 1980b).

Because the strains produced by ice sheet and ocean loading are sufficiently small, the deformation history may be described in terms of the
Laplace transformed and linearized field equations for the conservation
of momentum and for the gravitational field. These are respectively

$$\nabla \cdot \tilde{\tau} - \nabla(\rho g \mathbf{u} \cdot \mathbf{e}_r) - \rho \nabla \phi + g \nabla \cdot (\rho \tilde{\mathbf{u}}) \mathbf{e}_r = 0, \tag{4a}$$

$$\nabla^2 \tilde{\Phi} = -4\pi G \nabla \cdot (\rho \tilde{\mathbf{u}}), \tag{4b}$$

where $\rho = \rho(r)$ is the density field in the background (undeformed)
hydrostatic and spherical equilibrium configuration, $g = g(r)$ is the corresponding gravitational acceleration, $\mathbf{u}$ is the displacement field, $\mathbf{e}_r$ is a
unit vector in the radial direction, ϕ is the perturbation of the ambient
gravitational potential associated with the deformation, and G is the
gravitational constant. The inertial force in (4a) is suppressed because of
the long time scale of the adjustment process.

The essential purpose of the theory is to employ (4) to determine the
response of the Earth to the space- and time-dependent ice and water
loads applied on its surface. We proceed first to calculate the response of
the planet to impulsive loading by a point mass and thus to construct a
Green's function for the gravitational interaction problem. We may then
invoke the principle of superposition to describe the response to a realistic
loading history.

If in its unperturbed state the Earth may be well approximated as a
radially stratified sphere, then the response to a point mass load on its
surface will be a function only of radius r, angular distance from the applied
load θ, and time t. The solution three vector $(\mathbf{u}, \phi)$ to equations (4) may then
be expanded in vector spherical harmonics as

$$\tilde{\mathbf{u}} = \sum_{n=0}^{\infty} \left(U_n(r, s) P_n(\cos \theta) \mathbf{e}_r + V_n(r, s) \frac{\partial P_n}{\partial \theta}(\cos \theta) \mathbf{e}_\theta \right), \tag{5a}$$

$$\tilde{\phi} = \sum_{n=0}^{\infty} \Phi_n(r,s)P_n(\cos\theta), \tag{5b}$$

where U_n, V_n, Φ_n are spectral amplitudes for the harmonic disturbances of degree n in radial and tangential displacement and in the perturbation to the ambient gravitational potential respectively. When (5) are substituted in (4) they reduce the field equations to a set of six coupled ordinary differential equations of the form

$$\frac{d\mathbf{Y}}{dr} = \mathscr{A}\mathbf{Y} \tag{6}$$

where $\mathbf{Y} = (U_n, V_n, T_{rn}, T_{\theta n}, \Phi_n, Q_n)$ and $\mathscr{A}$ is a complicated 6×6 matrix of coupling coefficients given explicitly in Peltier (1974). T_{rn} and $T_{\theta n}$ are spectral amplitudes in the expansions for radial and tangential stress and Q_n are those in the expansion for the auxilliary variable $q = \partial\phi/\partial r + (n+1)\phi/r + 4\pi G\rho u_r$. To determine the response of the Earth to the point mass load we solve (6) subject to the following boundary conditions, which obtain at the Earth's surface $r = a$:

$$T_{rn} = -g(2n+1)/4\pi a^2, \tag{7a}$$

$$T_{\theta n} = 0, \tag{7b}$$

$$Q_n = -4\pi G(2n+1)/4\pi a^2, \tag{7c}$$

which follow respectively (Peltier 1974) from the condition that the normal stress balance the applied load, that the tangential stress vanish, and that the radial derivative of the potential possess the correct discontinuity to account for the surface density of the load.

Solutions of (6) subject to (7) are most economically expressed in terms of the triplet of dimensionless scalars called surface load Love numbers which are denoted by (h_n, l_n, k_n) and defined through

$$\begin{bmatrix} U_n(r,s) \\ V_n(r,s) \\ \Phi_n^1(r,s) \end{bmatrix} = \Phi_n^2(r) \begin{bmatrix} h_n(r,s)/g \\ l_n(r,s)/g \\ k_n(r,s) \end{bmatrix} \tag{8}$$

where the expansion $\Phi_n = \Phi_n^1 + \Phi_n^2$ has been employed, in which Φ_n^2 are the harmonic coefficients in the expansion for the field of force which produces U_n, V_n, Φ_n^1 in response. Φ_n^2 are independent of the Laplace transform variable because the surface point mass load is applied impulsively (i.e. as a Dirac delta function in the time domain).

In Figure 3 we show an example of the surface $h_n^V(a,s)$ for a realistic viscoelastic model of the Earth, which has an elastic lithosphere of thickness $T = 120$ km, a constant viscosity mantle with $v = 10^{22}$ Poise, and an

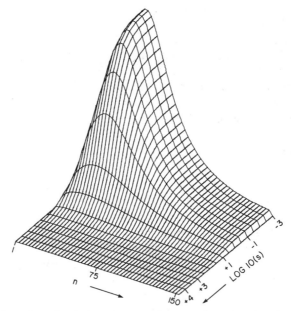

Figure 3 Three-dimensional view of the Laplace transform domain relaxation surface $h_n^V(s)$. Note that as $n \to \infty$, $h_n^V \to 0$ and the viscous relaxation is completely suppressed. Figure is from Peltier et al (1978).

inviscid core. The elastic structure of the model $[\rho(r), \lambda(r), \mu(r)]$ is that obtained from the inversion of elastic gravitational free oscillation data (Gilbert & Dziewonski 1975). The viscous part of the surface load Love number is just $h_n^V(a, s) = h_n(a, s) - h_n^E(a)$ where h_n^E is the Love number previously calculated by Farrell (1972) and others for the Hookean elastic rheology. Figure 3 shows that for short deformation wavelengths (large n), the viscous part of the response to the applied point mass load tends to zero, an affect that is due to the presence of the elastic lithosphere. Also evident from this figure is the existence of spectral asymptotes at both large and small values of s, which correspond via the Tauberian theorems to small and large values of time t respectively.

As shown in Peltier (1976), each spectrum $h_n(a, s)$ has an *exact* normal mode expansion of the form

$$h_n(a, s) = \sum_j \frac{r_j^n}{s + s_j^n} + h_n^E, \qquad (9)$$

where the s_j^n are a set of poles on the negative real s-axis of the complex s-plane and where the r_j^n are the residues at these poles which measure the extent to which each of the associated modes is excited by the point forcing. The relaxation times $\tau_j^n = 1/s_j^n$ associated with each of these modes may be computed by solving the associated homogeneous eigenvalue problem. Figure 4 shows such normal mode relaxation spectra for several different viscoelastic models of the planet. These models have been constructed to illustrate the individual components of the spectrum for a realistic Earth model, which is shown in plate *d*. The spectrum in plate *a* is for a homogeneous viscoelastic sphere, which has only one mode for each wavenumber *n*. This modal branch has relaxation time increasing with *n* such that for *n* large we have $s \simeq g\rho/2vk_H$ which is the relaxation spectrum for a viscous halfspace (k_H is the horizontal wavenumber). Plate *b* shows the effect of adding a lithosphere to the otherwise homogeneous model and this is seen to introduce a second mode for each *n* *and* to force the relaxation time of the fundamental mode to *decrease* with increasing *n* for $n \gtrsim 30$. It was on the basis of his observation of this effect in the Fennoscandia data that McConnell (1968) was originally led to infer the existence of a lithosphere with a thickness of about 110 km. In plate *c* we

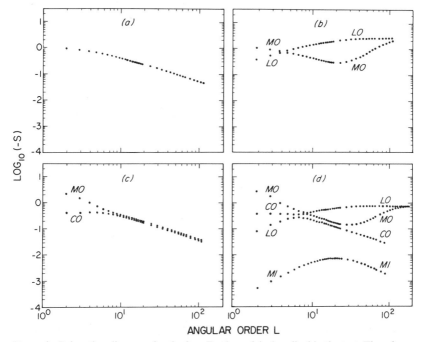

Figure 4   Relaxation diagrams for the four Earth models described in the text. The relaxation times $\tau_n = s_n^{-1}$ are plotted on a logarithmic scale and measured in units of 10^3 years. Figure is from Peltier (1980).

illustrate the effect of adding a high density inviscid core to the homogeneous model and note that this feature also introduces a second relaxation time for each n and that for this extra mode, as for the fundamental, relaxation time increases with n. Plate d is the spectrum for a complete Earth model which is seen to contain all of the individual features mentioned above plus additional modes which have particularly long relaxation times and which are due to the sharp increases of density in the transition zone associated with the Olivine-Spinel and the Spinel–post Spinel phase changes.

Since the Love number spectra have simple normal mode decompositions of the form (9) they possess trivial time domain representations

$$h_n(a, t) = \sum_j r_j^n e^{-s_j^n t} + h_n^E \, \delta(t), \tag{10}$$

which are the time domain spectral amplitudes for the impulsive point forcing. The r_j^n in (10) and (9) are determined by employing the known s_j^n as pivots in the collocation method discussed in Peltier (1976). If the point mass were allowed to remain on the surface for $t \geq 0$, then the coefficients for the resulting deformation may be obtained simply from (10) by convolution with a Heaviside step function. This operation yields (suppressing a)

$$h_n^H(t) = \sum_j \frac{r_j^n}{s_j^n} (1 - e^{-s_j^n t}) + h_n^E \tag{11}$$

$$= h_n^{H,V}(t) + h_n^E.$$

Note that (9) and (11) are related by

$$\lim_{t \to \infty} h_n^H(t) = \sum_j \frac{r_j^n}{s_j^n} + h_n^E = \lim_{s \to 0} h_n(s), \tag{12}$$

which clearly reveals the isostatic compensation mechanism since the number $\sum_j r_j^n / s_j^n$ for each n is just the viscous contribution to the final isostatically adjusted amplitude and this is a unique number for all Earth models which differ from one another only through their mantle viscosity profiles (Peltier 1976, Wu & Peltier 1981a). Examples of $h_n^{H,V}(t)$ spectra are shown in Peltier et al (1978) for the realistic Earth model with constant mantle viscosity. For $n \gtrsim 150$ the viscous response vanishes and such short wavelength loads are supported elastically.

In order to describe the response of the planet to an arbitrary surface load that is variable in both space and time, we construct Green's functions for the inhomogeneous boundary value problem and invoke the principle of superposition. These Green's functions may be computed for various signatures of the response in terms of the Love numbers h_n, l_n, k_n. Here we shall concern ourselves only with those for radial displacement, gravity anomaly, and the perturbation to the potential field, which have the respective representations (Peltier 1974):

$$u_r^H(\theta, t) = \frac{a}{m_e} \sum_{n=0}^{\infty} h_n^H(t) P_n(\cos \theta), \tag{13a}$$

$$\alpha^H(\theta, t) = \frac{g}{m_e} \sum_{n=0}^{\infty} [n - 2h_n^H(t) - (n + 1)k_n^H(t)] P_n(\cos \theta), \tag{13b}$$

$$\phi^H(\theta, t) = \frac{ag}{m_e} \sum_{n=0}^{\infty} [1 + k_n^H(t) - h_n^H(t)] P_n(\cos \theta). \tag{13c}$$

Depending upon the nature of the Earth model, the above infinite series may be slowly convergent and acceleration techniques such as the Euler transformation may be required to sum them (Peltier 1974). In Figure 5 the $u_r^{H,V}(\theta, t)$ Green's function is shown for the realistic Earth model with constant mantle viscosity. For plotting purposes the function has been normalized by multiplication with "$a\theta$" to remove the geometric singularity at $\theta = 0$. Note that the θ scale is logarithmic. As may be seen by inspection of (11), the viscous part of the response vanishes at $t = 0$, then

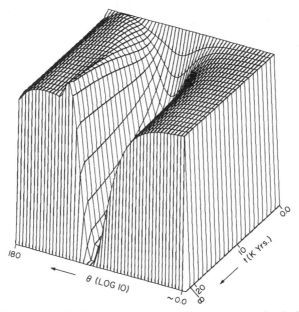

Figure 5   Viscous part of the Green's function for radial displacement for the Earth model whose relaxation spectrum is shown in plate (*d*) of Figure 4. The response has been normalized by multiplication with "$a\theta$" to remove the effect of the geometric singularity at $\theta = 0$ for plotting purposes. Figure is from Peltier et al (1978).

increases with time as the surface sags under the applied load, eventually reaching the deformed equilibrium state in which isostatic (gravitational) equilibrium prevails. Note that in the region peripheral to the load the planetary radius increases above its equilibrium value to create a "peripheral bulge" surrounding the point mass. It is the collapse of this bulge, following removal of a compensated load, that explains the observed submergence in the immediately peripheral regions (Peltier 1974).

Given these response functions, we may next consider how they are to be employed to predict observations of the type described in the introduction. In order to make a prediction we require a priori information concerning the deglaciation history and this may be obtained with reasonable accuracy in the manner described in Peltier & Andrews (1976). Inspection of relative sea level variations (drowned beaches) in the far field of the ice sheets provides an estimate of the amount of water that was added to the ocean basins as a function of time during melting. Knowing the location of the ice sheet margins as a function of time from end moraine data, the water load is simply partitioned back to the ice sheets in the ratio of their instantaneous areas. Ice mechanical arguments (Patterson 1972) allow us to obtain an approximation to the instantaneous topography. On the basis of such information, Peltier & Andrews (1976) produced approximate disintegration histories for the major ice sheets. In Figure 6a–f we show a sequence of time slices through their disintegration histories for both the Laurentide-Innuitian complex and for Fennoscandia (this model will be referred to subsequently as ICE-1). It should be emphasized that these reconstructions are to be viewed as first approximations, which will undoubtedly require modification to correct certain characteristic misfits between observation and theoretical prediction. It is a characteristic feature of the problem of glacial isostasy that its solution involves the determination of two unknown functionals of the theoretical model. Neither the deglaciation history nor the viscosity profile are known perfectly in advance and so the inverse problem for either is intrinsically nonlinear.

If there were a set of geophysical data that provided an accurate memory of the change in local radius of the planet produced by ice sheet disintegration, then we could make a theoretical prediction to compare with the observation simply by convolving the radial displacement Green's function (13a) with the surface load history to give

$$\Delta R(\theta, \phi, t) = \iint d\Omega' \int u_r(\theta|\theta', \lambda|\lambda', t|t')M(\theta', \lambda', t') \, dt', \qquad (14)$$

where M is the (assumed known) load history. Not withstanding the fact that there is *no* geophysical data that provides a direct memory of

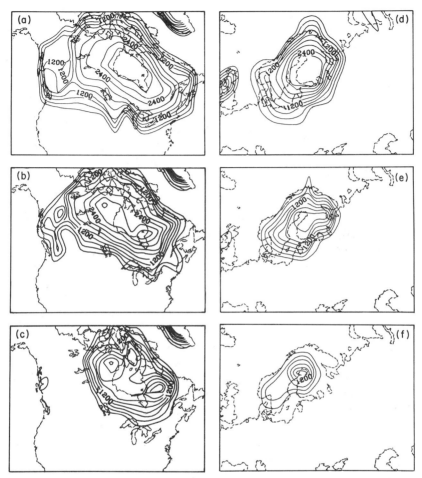

Figure 6   Plates (*a*), (*b*), and (*c*) show three time slices through the ICE-1 melting history at 18 KBP, 12 KBP, 8 KBP respectively for the North American ice complex. Plates (*d*), (*e*), (*f*) show the same sequence for Fennoscandia.

$\Delta R(\theta, \lambda, t)$, there is a severe problem involved in implementing (14) which has to do with M. This surface mass load functional may be expanded as

$$M = \rho_I L_I(\theta, \lambda, t) + \rho_w L_w(\theta, \lambda, t), \tag{15}$$

where ρ_I and ρ_w are the densities of ice and water respectively and where L_I and L_w are the corresponding thicknesses. The point to note here is that the functional M consists not only of the detailed ice history which we may estimate directly, but also of the water load in the ocean basins produced by the melting ice. Since we may safely assume the hydrological cycle to be closed, mass conservation demands

$$\int d\Omega \, \rho_w L_w = M_w = -\int d\Omega \, \rho_I L_I = M_I \qquad (16)$$

which is an integral constraint upon the load history. Given $M_I(t)$, the time-dependent mass flux to the oceans (assumed negative for melting), we can convert the corresponding M_w to a uniform equivalent time-dependent rise of *mean* sea level which we call

$$S_{EUS}(t) = \frac{M_w(t)}{\rho_w A_0} \qquad (17)$$

where A_0 is the surface area of the Earth's oceans. $S_{EUS}(t)$, the so-called eustatic water rise, for the Peltier & Andrews history, agrees with the eustatic curve suggested by Shepard (1963) based upon the submergence curve for the Gulf of Mexico. This curve is clearly supposed to represent the average increase in water thickness over the ocean basins. If the actual rise of sea level locally were everywhere equal to this average, then L_w in (15) would be known from L_I and the convolution integral (14) could be evaluated. In fact, L_w is not everywhere constant and this poses the severe problem mentioned above. That L_w cannot increase uniformly over the ocean basins follows from the fact that if it did then the surface of the ocean would not remain equipotential during and after the deglaciation event. However, any deviation of the surface away from equipotential would set up currents in the ocean, which would redistribute water in such a way that this surface condition was restored. In order to predict relative sea level variations accurately, we must determine *where* in the oceans the meltwater actually goes. It turns out to be possible to answer this question rather accurately when a slightly more abstract view of sea level history is adopted, and this subject will be discussed in the next section.

RELATIVE SEA LEVELS

In order to satisfy the constraint that the surface of the oceans remain equipotential, we are obliged to consider the perturbations of potential associated with an arbitrary history of loading. These perturbations may clearly be calculated by convolving the load with the appropriate Green's function, which in this case is the third of those listed in (13). Let us suppose that the major ice sheets all melted at a single instant $t = 0$ so that the change of potential at $r = a$ may be computed as

$$\Delta\phi(\theta, \lambda, t) = \rho_I \phi^H \underset{I}{*} L_I + \rho_w \phi^H \underset{w}{*} S \qquad (18)$$

where $(\overset{*}{\text{I}})$ and $(\overset{*}{\text{w}})$ represent convolutions over the ice and water respectively and where we have replaced L_w by S for later convenience. Now the change in potential $\Delta\phi$ will itself cause a change in sea level S, which is such as to ensure that the surface of the new ocean remains equipotential. This adjustment of sea level is in the amount (Farrell & Clark 1976, Peltier et al 1978)

$$S = \frac{\Delta\phi(\theta, \lambda, t)}{g} + C \qquad (19)$$

where the constant C is fixed by the requirement of mass conservation. Equation (19) is a result from first-order perturbation theory and is therefore valid only for sufficiently small variations of bathymetry. From the manner in which it has been constructed, S is the local variation of sea level with respect to the solid surface of the planet and is therefore precisely the sea level history recorded in raised beaches such as the sequence shown in Figure 1. Substituting (19) into (18) we obtain

$$S = \rho_I \frac{\phi^H}{g} \overset{*}{\underset{\text{I}}{}} L_I + \rho_w \frac{\phi^H}{g} \overset{*}{\underset{\text{w}}{}} S + C. \qquad (20)$$

An explicit expression for C is obtained by multiplying (20) by ρ_w and integrating over the surface area of the oceans; solving for C then gives

$$C = \frac{-M_I(t)}{\rho_w A_o} - \frac{1}{A_o} \left\langle \rho_I \frac{\phi^H}{g} \overset{*}{\underset{\text{I}}{}} L_I + \rho_w \frac{\phi^H}{g} \overset{*}{\underset{\text{w}}{}} S \right\rangle_o, \qquad (21)$$

since $-M_I(t) = \langle \rho_w S \rangle_o$ is the known mass loss history for the ice sheets in which $\langle \ \rangle_o$ indicates integration over the oceans.

With C given by (21), Equation (20) now constitutes an integral equation for S which we have called the sea level equation. Since it is an integral equation, with the unknown $S(\theta, \lambda, t)$ both on the left-hand side and under the convolution integral on the right, it must be solved using a matrix method. What we do is to divide the "active" area of the surface of the Earth (ice plus ocean) into a number of finite elements. The system is also discretized in time and to date we have elected to sample the S-history at equispaced intervals of 10^3 years, which is the same sampling interval employed to describe the input deglaciation model L_I. We assume that the mass load upon the finite element with centroid at $\mathbf{r}'$ may be described by

$$L(\mathbf{r}',t) = \sum_{l=1}^{P} L_l(r')H(t - t_l) \qquad (22)$$

where the $t_l (l = 1, P)$ are a series of times that span the deformation history and the L_l are the loads applied or removed at the discrete times t_l. A complete description of the numerical methods employed for the solution of the general form of (20) will be found in Peltier et al (1978) and we will not repeat the detailed discussion here. Experience with the gravitationally self-consistent model embodied in (20) has shown that in the near field of the ice sheets the differences between its predictions and those based upon the assumption that the relative sea level history is essentially a measure of the local change in radius corrected by the eustatic sea level rise (e.g. Peltier & Andrews 1976) are small. The reason why these errors might be small is provided in Peltier (1980a) although it took explicit calculation to establish the fact.

Inspection of a sample solution to (20) shows that the rise of sea level in the ocean basins at points distant from the ice is quite nonuniform, with the greatest variation away from the 76.6 meter present-day average of ICE-1 being confined to the region nearest the ice. This may be taken as rather strong evidence to the effect that the notion of a eustatic rise of sea level should be used cautiously. By piercing such theoretically predicted global histories at longitude-latitude coordinates for which C^{14} data on relative sea level history exists, we may extract a specific prediction to compare with observations. Such predictions, as mentioned above, are dependent upon two unknown functionals of the model, namely, the viscosity profile of the Earth's interior and the assumed disintegration history. It is our ability to perform an initial linearization of the problem using a priori knowledge of the melting history that makes it possible to proceed. We first fix the ice history and determine a best $v(r)$. We then fix $v(r)$ and refine $L_l(\theta, \phi, t)$, continuing the iterative process until convergence is achieved. This procedure is precisely analogous to the use of free oscillation data to simultaneously constrain the interior elastic structure and the seismic moment tensor (Gilbert & Dziewonski 1975).

The first step along the iterative path was made by Peltier & Andrews (1976) who calculated sea levels using the approximate formula (14) in which ocean loading effects were neglected entirely, although a single calculation was shown, which suggested that these effects were weak in the near field of the ice, as mentioned above. Their conclusion from these initial calculations was that a model with uniform mantle viscosity was preferred. This conclusion was based almost entirely upon the fact that the model with high viscosity in the lower mantle predicted present-day rates of emergence in the central region that were much higher than observed. Very little data from the peripheral region had been compiled at that time and the model fit the far field data poorly because of the neglect of ocean loading. Under the ice sheets the fit to the data was quite good when the uniform mantle viscosity model was employed.

A second step in this analysis was made in the papers by Clark et al (1978) and Peltier et al (1978) in which the gravitationally self-consistent model was first introduced. These calculations focused upon the ability of this general model to make accurate predictions at sites arbitrarily distant from the ice sheets. Use of the Peltier & Andrews load history showed that the predictions for the far field sites were time shifted from the observations such as to imply that melting should be delayed by about 2×10^3 years from the timing assumed by Peltier & Andrews. This time shift was therefore introduced and the far field data were fit rather accurately in consequence. This forward shift in timing of the disintegration, of course, had a disastrous effect upon the fits to the data at sites near the ice sheets; in particular, misfits to the emergence-submergence curves exceded 300% at sites along the eastern seaboard of North America and at locations that were once under the Laurentide ice sheet (Peltier 1980a). We currently refer to the shifted version of the Peltier & Andrews melting history as ICE-2.

Given that the delayed melting characteristic of ICE-2 seemed required to fit the far field emergence data, the third step in the iterative process of refinement was to attempt to improve the agreement between theory and observation in the near field of the Laurentide ice sheet. The obvious way of effecting the desired reconciliation was to reduce the ice sheet thickness (Peltier et al 1978), since only in its vicinity were the misfits significant, and under the ice the errors were such that excessive emergence was predicted. The approach that we adopted (Wu & Peltier 1981b) was simply to reduce the thickness of Laurentide ice by the amount required to give the observed emergence of the 6 KBP beach at sites under the ice. This thickness correction was therefore a strong function of location (though a weak function of the viscosity model) and when it was applied led to an ice sheet with a much more irregular topography than that of the Peltier & Andrews model (ICE-1). This new ice history (to which we shall refer as ICE-3) was found to provide a good fit to the r.s.l. observations not only at sites under the ice sheet (which cannot be considered surprising) but also at sites in the peripheral region along the eastern seaboard of North America where the previous misfits had been enormous (Peltier et al 1978). As described by Wu & Peltier (1981b), with this new melting history the model exhibits some preference for an increase of viscosity with depth, and if a single step increase of viscosity is introduced at the seismic discontinuity at 670 km, then some of the sea level data prefer a lower mantle viscosity of about 10^{23} Poise. Values of lower mantle viscosity that are significantly *greater* than this degrade the fit to the data, and not all data demand such an increase.

The basis of these statements is illustrated in Figure 7, where we show the observed r.s.l. histories at six sites, compared with the corresponding theoretical predictions from three viscosity models, which differ from one another only in the magnitude of the viscosity in the lower mantle below the 670 km discontinuity. The first three sites labelled (*a*), (*b*), (*c*) are, respectively, for Churchill, the Ottawa Islands, and Cape Henrietta Maria, all of which were located under the Laurentide ice sheet around Hudson Bay. Inspection of these emergence curves shows that they are not extremely sensitive to a modest increase of lower mantle viscosity although the model with a lower mantle viscosity of 5×10^{23} Poise is firmly rejected by all three data sets. Both the Churchill and Ottawa Islands sites prefer the uniform viscosity model and this preference is strongest at Ottawa Islands where the 10^{23} Poise lower mantle model predicts both too little net emergence and too high a present-day emergence rate [in accord with the previous conclusions of Peltier & Andrews (1976)]. At Churchill the amount of emergence is best fit by the uniform model, while the present-day emergence rate is best fit by the model with the 10^{23} Poise lower mantle. At Cape Henrietta Maria the model with the 10^{23} Poise lower mantle fits both characteristics best.

Plates (*d*), (*e*), and (*f*) of Figure 7 show the emergence data for Boston, Clinton, Connecticut, and Bermuda, respectively, all of which are located beyond the ice margin and at successively greater distances from the ice sheet center. The non-monotonic sea level history at Boston is characteristic of sites through which the peripheral bulge propagates, an effect which seems to be inhibited by high lower mantle viscosity (Peltier 1974). These data, therefore, prefer the uniform viscosity model. At both Clinton, Connecticut, and Bermuda, which are located at greater distance from the ice margin, the situation is reversed since at these locations the uniform model predicts too much submergence. Again, the model with the 5×10^{23} Poise lower mantle is nowhere preferred and so, to the extent that the ICE-3 load history is reasonable, this number is established as a firm upper bound on the viscosity of the lower mantle insofar as the relative sea level data are concerned.

The ICE-3 load history is, however, deficient from several points of view. Since melting is delayed by 2,000 years compared to the Peltier & Andrews history ICE-1, and since the locations of the terminal moraines in ICE-1 were controlled by data, it follows that the disintegration isochrones for ICE-3 may significantly violate the observational constraints. An equally serious criticism of ICE-3 is that it does not contain sufficient ice to explain the observed submergence in the far field (e.g. the r.s.l. curve

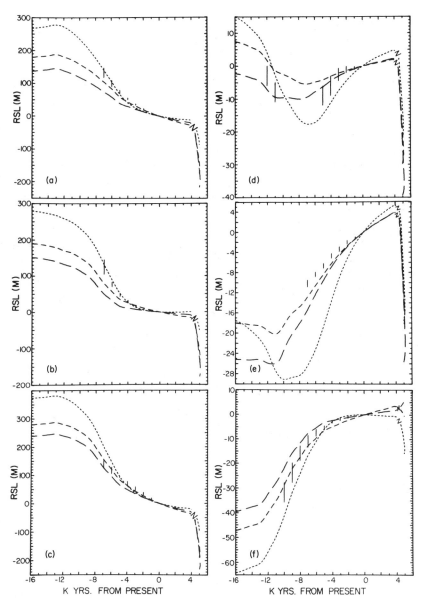

Figure 7   RSL histories at (*a*) Churchill, (*b*) Ottawa Islands, (*c*) Cape Henrietta Maria, (*d*) Boston, (*e*) Clinton, Conneticut, (*f*) Bermuda. The observed data with attached error bars are compared with predictions from three Earth models which differ from others only in the value of the mantle viscosity beneath a depth of 670 km. The short, intermediate, and long dashed curves are, respectively, for the model with lower mantle viscosity of 10^{23} Poise, with constant viscosity of 10^{22} Poise, and with lower mantle viscosity of 5×10^{23} Poise. The final dramatic excursion of each r.s.l. prediction ends at the point corresponding to the total r.s.l. variation that would be achieved in infinite time. The absisca of this point is just the amount of emergence (submergence) remaining. The actual curves are, of course, discontinuous between the last two points since these should be separated by infinite time.

for the Gulf of Mexico). Although it may be possible to remedy this defect by melting an amount of ice equivalent to a eustatic rise of 15–20 meters from West Antarctica, it is not yet clear that this will not produce a significant (and perhaps adverse) effect on the northern hemisphere r.s.l. histories.

This brief discussion of the sea level data suffices to demonstrate that the inference of mantle viscosity from them is conditioned in a non-negligible way by the ice sheet history that one employs to make the predictions. Before we may be confident of the inferred viscosity we are obliged to investigate its sensitivity to plausible variations in the deglaciation model. As it happens [see Wu & Peltier (1981b) for details] the ICE-1 history with constant mantle viscosity also produces quite good fits to the r.s.l. curves at most near field sites and firmly rejects the model in which the lower mantle viscosity is 5×10^{23} Poise. In the next section, we investigate the extent to which free-air gravity data are able to provide confirmation of this inference from the r.s.l. information.

THE FREE-AIR GRAVITY ANOMALY

For a given viscoelastic Earth model and deglaciation history, solution of the sea level equation (20) yields a complete record of the loading of the ocean basins. Convolution of the ocean and ice loads with the Green's function for the free-air anomaly (13b) then provides a prediction of the free-air signal on the Earth's surface as

$$\Delta g(\theta, \lambda, t) = \Delta g'(\theta, \lambda, t) - \Delta g'(\theta, \lambda, \infty) \tag{23}$$

where

$$\Delta g'(\theta, \lambda, t) = \iint d\Omega' \int_0^t dt' \alpha(\theta|\theta', \lambda|\lambda', t|t') M(\theta', \lambda', t').$$

This form ensures that the free-air anomaly vanishes in the isostatic equilibrium configuration that obtains at infinite time. The free-air anomaly is a measure of the extent of isostatic disequilibrium. Clearly the calculation of $\Delta g(\theta, \lambda, \infty)$ requires the infinite-time Green's function $\alpha(\theta, \infty)$, the Heaviside form for which may be obtained from $\lim_{t \to \infty}$ of (13b). In this limit the Love numbers have the forms (12) and so correspond to the small s asymptotes of the spectra shown in Figure 3. It might be thought that one could compute the required $\alpha(\theta, \infty)$ simply by direct calculation of the Love numbers for sufficiently small s, but this turns out to be impossible since as $s \to 0$ the system (6) becomes "stiff" as its order degenerates from sixth to second. As explained in Wu & Peltier (1981a), the required function $\alpha(\theta, \infty)$ must be calculated from Love numbers determined by solution of the degenerate second-order system. Our first attempt to calculate a free-air map (Peltier 1980b) was somewhat in error because of the failure to recognize this problem.

In Figure 8 we show the observed free-air gravity map for the Laurentide region (Walcott 1970). If one compares this anomaly with the ice sheet topography shown in Figure 6a, it is clear that both have the same elliptical form and that the minimum in the free-air map is coincident with the maximum of the ice sheet topography. The minimum value of the observed anomaly is approximately -35 milligals. In Figure 9 are shown six computed present-day free-air anomaly maps, three for each of the deglaciation models ICE-1 and ICE-3. The three calculations for each of the two ice histories are for the same three viscosity models as were employed in the previous discussion of relative sea levels.

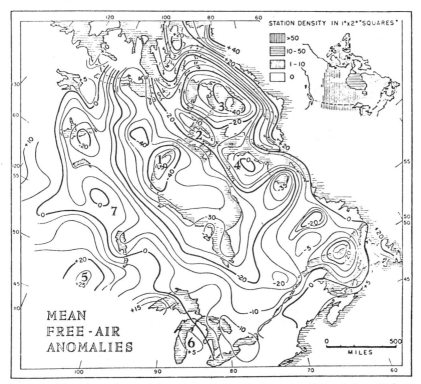

Figure 8   Free-air gravity anomaly map for the Laurentide region. Figure is from Walcott (1970).

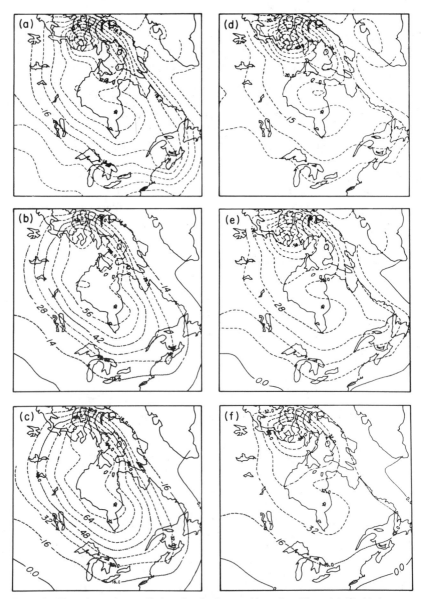

Figure 9   Free-air gravity predictions for the Laurentide region. Plates (*a*), (*b*), (*c*) show results from calculations based upon ICE-1 for the uniform viscosity model, that with a lower mantle viscosity of 10^{23} Poise, and that with a lower mantle viscosity of 5×10^{23} Poise. Plates (*d*), (*e*), and (*f*) show the same sequence of calculations but based upon the ICE-3 disintegration history.

Inspection of plates (a), (b), and (c), which correspond to the Peltier & Andrews history ICE-1, shows that the uniform viscosity model (plate a) gives a gravity anomaly very near that which is observed. Increasing the viscosity in the lower mantle (below the 670 km transition) from 10^{22} Poise to 10^{23} Poise [plate (b)] leads to a maximum anomaly in the central region on the order of 60 milligals, which is so much in excess of the observation that this model must be completely ruled out. The problem is only amplified when the lower mantle viscosity is further increased to 5×10^{23} Poise [plate (c)] since the maximum negative anomaly then increases to approximately 75 milligals. If the ICE-1 model were correct then the uniform viscosity mantle would be strongly preferred by the data as previously argued by Peltier & Andrews (1976) using only the r.s.l. information. However, as pointed out in the last section, ICE-1 cannot fit the far field emergence data, which seem to require that melting be delayed somewhat from the timing in ICE-1. With this delayed timing, the ice sheet thickness must be reduced in order to fit the sea level data and this new history (called ICE-3) was employed in the last section in demonstrating the ability of the model to fit the r.s.l. observations.

Plates (d), (e), and (f) of Figure 9 show the gravity anomaly for ICE-3 for the three mantle viscosity models. Plate (d) for the model with a uniform mantle viscosity of 10^{22} Poise shows that the maximum negative anomaly over Hudson Bay is significantly less than the observed (~ 20 mgals rather than ~ 35 mgals). Increasing the viscosity of the lower mantle to 10^{23} Poise [plate (e)] increases this anomaly to about 33 mgals, which is near the observed magnitude. A further increase of the viscosity of the lower mantle to 5×10^{23} Poise [plate (f)] leaves a present-day anomaly on the order of 38 mgals, which is somewhat larger than observed.

This sequence of comparisons very clearly demonstrates the extent to which the inference of the viscosity of the lower mantle is influenced by the assumed ice sheet melting history. For the thick ice sheet model that melts early (ICE-1), the uniform viscosity mantle is strongly preferred. For the thin ice sheet that melts late (ICE-3), an increase of the viscosity of the lower mantle is required to fit the observed free-air gravity anomaly but the lower mantle value cannot exceed about 10^{23} Poise, otherwise neither the gravity anomaly [plate (f)] nor the r.s.l. data (Figure 7) can be reconciled. The model with a lower mantle viscosity of 5×10^{23} Poise is rejected irrespective of the ice sheet history.

We consider the deglaciation histories ICE-1 and ICE-3 to be bounds upon the actual surface mass load functional. ICE-1, although it contains sufficient ice to explain the net submergence observed at far field sites (e.g. Gulf of Mexico) of about 80 meters, cannot explain the time (~ 6 KBP) at which raised beaches first appear along distant continental coastlines (e.g.

Reciffe, Brazil; New Zealand; Islands in the Indian Ocean) due to the tilting forced by the offshore water load. ICE-3, on the other hand, contains too little ice to explain the net emergence of raised beaches along distant continental coastlines at 6 KBP. It may be possible to remedy the mass defect of ICE-3 by introducing a substantial melting event in West Antarctic, but it remains to be demonstrated that such a melting event will not severely impact upon the northern hemisphere r.s.l. data. ICE-3, however, suffers from the additional defect that its disintegration isochrones do not match the observations (Prest 1969). The correct melting history is probably intermediate between ICE-1 and ICE-3 and therefore the viscosity of the lower mantle is probably intermediate between the values inferred with these two melting chronologies. This argument implies that an *upper* bound upon the viscosity of the mantle beneath the 670 km seismic discontinuity is 10^{23} Poise. A more detailed discussion of these ideas will be found in Wu & Peltier (1981b).

All of the discussion of gravity anomalies given here has been based upon the assumption that the planet and its ice sheets were in isostatic equilibrium at Wisconsin maximum. To the extent that isostatic equilibrium was not complete at that time, this initial disequilibrium will also impact upon the inference of mantle viscosity from the uplift data. Although, as discussed in Wu & Peltier (1981a), this effect should not be too severe, it might lead to the requirement for somewhat higher lower mantle viscosities than those mentioned above.

SUMMARY

The discussion in the previous sections of this article has illustrated several ways in which the planet has conspired to remember its response to the last deglaciation event of the current ice age. Each of these different modes of memory contributes a significant constraint upon the viscosity of the planetary mantle, and, in fact, provides information that is highly complementary rather than redundant. In order to simultaneously fit the relative sea level and free-air gravity data over Wisconsin Laurentia, the viscosity of the lower mantle is strongly restricted to be significantly less than 5×10^{23} Poise. The preferred value is a function of the unloading history assumed. If the Peltier & Andrews (1976) ICE-1 history is employed, then the preferred value of the viscosity of the lower mantle is the same as that of the upper, i.e. $\sim 10^{22}$ Poise. However, if the ICE-3 model is used, then some increase of viscosity in the lower mantle is necessary and a value in the neighborhood of 10^{23} Poise reconciles both data sets nicely. Neither the ICE-1 nor the ICE-3 loading histories are acceptable from all points of view. Melting occurs too early in ICE-1, as is evident from far field emer-

ICE AGE GEODYNAMICS   223

gence curves. On the other hand, melting is so much delayed in ICE-3 that its thickness must be reduced substantially in order to fit the near field r.s.l. data. This leads to the failure of ICE-3 to deliver sufficient water to the oceans to account for the observed net submergence (e.g. in the Gulf of Mexico). ICE-3 also violates the observed isochrones for the disintegration of the Laurentide sheet, which suggest a somewhat earlier retreat. It seems clear, therefore, that the ICE-1 and ICE-3 models are bounds upon the actual disintegration history and that the estimates of lower mantle viscosity which one obtains by employing them in the solution of the direct problem are themselves bounds upon the actual lower mantle viscosity. This leads to the conclusion that a value near 10^{23} Poise is the *upper* bound on the viscosity of the lower mantle beneath the seismic discontinuity at 670 km depth. This conclusion, it will be recalled from the introduction, is quite different from that of Walcott (1973) who believed that this value represented a *lower* bound upon the viscosity in this region. However, the conclusion is also different from that stated in Peltier & Andrews (1976) who, using the ICE-1 history, concluded that the lower mantle viscosity could not be significantly in excess of the upper mantle value of 10^{22} Poise. This conclusion is only valid to the extent that ICE-1 is valid and this model possesses the flaws mentioned above. Likewise, our conclusion contradicts that reached by Cathles (1975, 1980) who employed an ice sheet model for the Laurentide that was most like ICE-3, in that it was based upon the melting chronology of Bryson et al (1969) rather than the better controlled, and therefore more likely reliable, chronology of Prest (1969), which we used as the basis for ICE-1. Cathles' Laurentide model is thinner than ICE-1 and melting is delayed. Our analysis clearly shows that such a melting history will require an increase of lower mantle viscosity in order to explain the observed free-air gravity anomaly. Cathles did not employ these data to constrain his model and this led him to reach an overly strong conclusion regarding the value of the viscosity in the deep mantle.

One further implication of the results of the postglacial rebound analyses discussed here concerns the question of the depth extent of the convective circulation in the mantle which we suppose is responsible for driving the surface plates in the course of their relative motion. It has often been assumed in the past, on the basis of arguments to the effect that the viscosity of the deep mantle is extremely high, that convection must be confined to the upper mantle. The results presented here and discussed in detail in Wu & Peltier (1981a,b) show that this argument is untenable. If convection is confined to the upper mantle, confinement must be accomplished by some mechanism other than through a large increase of viscosity across the 670 km seismic discontinuity. In order to continue to support this hypothesis, one is more or less obliged to invoke the existence of a marked change of mean atomic weight (i.e. chemistry) at this depth. As argued in Peltier (1973, 1980b), Jarvis & Peltier (1980a,b), Yuen & Peltier (1980a,b), Yuen et al (1980), and as assumed in the recent thermal history models of Sharpe & Peltier (1978, 1979), an equally plausible assumption is that the whole mantle is involved in the convective circulation. This model fits the surface observations very well and does so with a minimum of complexity. Its viability is strongly reinforced through interpretation of the observed geodynamic response to the current ice age.

ACKNOWLEDGMENTS

I have enjoyed profitable discussions with many colleagues concerning the subject matter of this paper, most recently with Patrick Wu, David Yuen, and Gary Jarvis, but initially also with Bill Farrell, Jim Clark, and John Andrews. I am indebted to all of them for useful observations and helpful suggestions. My research at the University of Toronto is sponsored by NSERCC grant A-9627. Several of the numerical computations discussed here were done on the CRAY-1 computer at the National Center for Atmospheric Research, which is sponsored by the National Science Foundation, and I am particularly indebted to Patrick Wu for his assistance with this phase of the work.

Literature Cited

Anderson, D. L., Minster, J. B. 1979. The frequency dependence of Q in the Earth and implications for mantle rheology and Chandler wobble. *Geophys. J. R. Astron. Soc.* 58:431–40

Andrews, J. T., Barry, R. G. 1978. Glacial inception and disintegration during the last glaciation. *Ann. Rev. Earth Planet. Sci.* 6:205–28

Balling, N. 1980. The land uplift in Fennoscandia, gravity field anomalies and isostasy. In *Earth Rheology Isostasy and Eustasy*, ed. N. A. Morner, pp. 297–321. Chichester: Wiley. 599 pp.

Broecker, W. S., Van Donk, J. 1970. Insolation changes, ice volumes, and the O^{18} record in deep-sea cores. *Rev. Geophys.* 8:169–98

Bryson, R. A., Wendland, W. M., Ives, J. D., Andrews, J. T. 1969. Radiocarbon isochrones on the disintegration of the Laurentide ice sheet. *Arctic Alp Res.* 1:1–13

Buland, R., Gilbert, F. 1978. Improved resolution of complex eigenfrequencies in analytically continued seismic spectra. *Geophys. J. R. Astron. Soc.* 52:457–70

Cathles, L. M. 1975. *The Viscosity of the Earth's Mantle*. Princeton Univ. Press. 386 pp.

Cathles, L. M. 1980. Interpretation of postglacial isostatic adjustment phenomena in terms of mantle rheology. In *Earth Rheology Isostasy and Eustasy*, ed. N. R. Mormer, pp. 11–43. Chichester: Wiley. 599 pp.

Clark, J. A., Farrell, W. E., Peltier, W. R. 1978. Global changes in postglacial sea level: a numerical calculation. *Quat. Res.* 9:265–87

Daley, R. A. 1934. *The Changing World of the Ice Age*. Yale Univ. Press. 271 pp.

Farrell, W. E. 1972. Deformation of the Earth by surface loads. *Rev. Geophys. Space Phys.* 10:761–97

Farrell, W. E., Clark, J. A. 1976. On postglacial sea level. *Geophys. J. R. Astron. Soc.* 46:647–67

Gilbert, F., Dziewonski, A. M. 1975. An application of normal mode theory to the retrieval of structural parameters and source mechanisms from seismic spectra. *Philos. Trans. R. Soc. London Ser. A* 278:187

Jarvis, G. T., Peltier, W. R. 1980a. Ocean floor bathymetry profiles flattened by radiogenic heating in a convecting mantle. *Nature* 285(5767):649–51

Jarvis, G. T., Peltier, W. R. 1980b. Mantle convection as a boundary layer phenomenon. *Geophys. J. R. Astron. Soc.* Submitted

Jeffreys, H. 1973. Developments in geophysics. *Ann. Rev. Earth Planet. Sci.* 1:1–13

Kohlstedt, D. L., Goetze, C. 1974. Low stress and high temperature creep in Olivine single crystals. *J. Geophys. Res.* 79:2045–51

Liu, H. P., Anderson, D. L., Kanamorie, H. 1976. Velocity dispersion due to anelasticity: Implications for seismology and mantle composition. *Geophys. J. R. Astron. Soc.* 47:41–58

McConnell, R. K. 1968. Viscosity of the mantle from relaxation time spectra of isostatic adjustment. *J. Geophys. Res* 73:7089–7105

Minster, J. B. 1980. Anelasticity and attenuation. In *Physics of the Earth's Interior*, ed. A. M. Dziewonski and E. Boschi. New York: Elsevier. In press

Patterson, W. S. B. 1972. Laurentide ice sheet: Estimated volumes during late Wisconsin. *Rev. Geophys. Space Phys.* 10:885–917

Peltier, W. R. 1973. Penetrative convection in the planetary mantle. *Geophys. Fluid Dynam.* 5:47–88

Peltier, W. R. 1974. The impulse response of a Maxwell Earth. *Rev. Geophys. Space Phys.* 12:649–69

Peltier, W. R. 1976. Glacial isostatic adjustment—II: The inverse problem. *Geophys. J. R. Astron. Soc.* 46:669–706

Peltier, W. R. 1980a. Ice sheets, oceans, and the Earth's shape. In *Earth Rheology Isostasy and Eustasy*, ed. N. A. Morner, pp. 45–63. Chichester: Wiley. 599 pp.

Peltier, W. R. 1980b. Mantle convection and viscosity. In *Physics of the Earth's Interior*, ed. A. M. Dziewonski and E. Boschi. New York: Elsevier

Peltier, W. R. 1980c. Models of glacial isostasy and relative sea level. In *Dynamics of Plate Interiors*, ed. R. I. Walcott. Washington: AGU Publ.

Peltier, W. R., Andrews, J. T. 1976. Glacial isostic adjustment—I: The forward problem. *Geophys. J. R. Astron. Soc.* 46:605–46

Peltier, W. R., Farrell, W. E., Clark, J. A. 1978. Glacial isostasy and relative sea level: a global finite element model. *Tectonophysics* 50:81–110

Peltier, W. R., Yuen, D. A., Wu, P. 1980a. Postglacial rebound and transient rheology. *Geophys. Res. Lett.* 7(10):733–36

Peltier, W. R., Wu, P., Yuen, D. A. 1980b. The viscosities of the Earth's mantle. In *Anelasticity of the Mantle*,

Prest, V. K. 1969. Retreat of Wisconsin and Recent Ice in North America. *Geol. Surv. Can., Map 1257A*

Sharpe, H. N., Peltier, W. R. 1978. Parameterized mantle convection and the Earth's thermal history. *Geophys. Res. Lett.* 5:737–44

Sharpe, H. N., Peltier, W. R. 1979. A thermal history model for the Earth with parameterized convection. *Geophys. J. R. Astron. Soc.* 59:171–203

Shepard, F. P. 1963. 35,000 years of sea level. In *Essays in Marine Geology.* Univ. Southern Calif. Press.

Twiss, R. J. 1976. Structural superplastic creep and linear viscosity in the Earth's mantle. *Earth Planet. Sci. Lett.* 33:86–100

Walcott, R. I. 1970. Isostatic response to loading of the crust in Canada. *Can. J. Earth Sci.* 7:716–27

Walcott, R. I. 1973. Structure of the Earth from glacio-isostatic rebound. *Ann. Rev. Earth Planet. Sci.* 1:15–37

Wu, P., Peltier, W. R. 1981a. Viscous gravitational relaxation. *Geophys. J. R. Astron. Soc.* Submitted

Wu, P., Peltier, W. R. 1981b. The free air gravity anomaly as a constraint upon deep mantle viscosity. *Geophys. J. R. Astron. Soc.* Submitted

Yuen, D., Peltier, W. R. 1980a. Mantle plumes and thermal stability of the D" layer. *Geophys. Res. Lett.* 7(9):625–28

Yuen, D., Peltier, W. R. 1980b. Temperature dependent viscosity and local instabilities in mantle convection. In *Physics of the Earth's Interior*, ed. A. M. Dziewonski and E. Boschi. New York: Elsevier.

Yuen, D., Peltier, W. R., Schubert, G. 1980. On the existence of a second scale of convection in the upper mantle. *Geophys. J. R. Astron. Soc.* In press

Ann. Rev. Earth Planet. Sci. 1981. 9: 227–49
Copyright © 1981 by Annual Reviews Inc. All rights reserved

NUCLEATION AND GROWTH OF STRATOSPHERIC AEROSOLS

x 10150

A. W. Castleman, Jr.

Department of Chemistry and Chemical Physics Laboratory, CIRES, University of Colorado, Boulder, Colorado 80309

Robert G. Keesee

National Aeronautics and Space Administration, Ames Research Center, Moffett Field, California 94035

INTRODUCTION

Despite early evidence of the presence of small particles in regions above the troposphere, it was only about two decades ago that the existence of a globally distributed, temporally persistent, stratospheric aerosol layer was established. Early twilight measurements by Gruner & Kleinert (1927) and Gruner (1942–1961) provided evidence that aerosols exist above the tropopause, but the first careful systematic direct measurements of change in aerosol concentration with altitude were reported by Junge et al (1961). Their *in situ* measurements showed the aerosol concentration to decrease monotonically up to and through the tropopause, a fact both expected and confirmed by other investigators (see Reiter 1971 and Junge 1963). A rather unexpected finding from these original investigations was that a globally distributed persistent stratospheric aerosol having a broad concentration maximum for large particles (radius $\geq 0.15 \ \mu$m) around 20 km in elevation existed. On the basis of twilight measurements, Volz (1970) concluded that the layer persisted to the highest altitudes in the tropical latitudes, and up to an elevation of at least 25 km. However, direct measurements by Bigg et al (1970) indicated that the influence of this layer on the total aerosol content of the stratosphere persists up to approximately 37 km. Using optical radar techniques (Lidar), Schuster (1970) showed the existence of significant quasistable layers between 25 and 40 km, all presumably associated with the same sulfate layer, often termed the Junge

227

0084-6597/81/0515-0227$01.00

layer or stratospheric sulfate aerosol layer. Numerous early investigations of the distribution of this layer were made by Junge and co-workers (Junge et al 1961, Junge & Manson 1961, Junge 1963). Its general distribution with regard to altitude and latitude was further documented by Friend (1966), Bigg et al (1970) and later by Lazrus et al (1971).

Although some investigators consider explosive volcanic eruptions to be the only source with sufficient energy to penetrate the tropopause to any great extent, early attempts (Hogg 1963, Volz 1965, Volz & Goody 1962) to correlate both the physical properties and the concentration of these aerosols with volcanic activity met with limited success. The catastrophic eruption of Mount Agung (8.5°S, 115.5°E) in March 1963 was the first well-documented major perturbation to the total stratospheric sulfate aerosol layer. Subsequent to the eruption on 18 March 1963, a marked change was noted in the nature and concentration of aerosol particles collected in the stratosphere of both the northern and southern hemispheres. Mossop (1964) reported as much as a ten-fold increase in number concentration of particles collected in the southern hemisphere following the eruption, and obtained evidence of some insoluble particles within the droplet-like phase of which the aerosol was mainly composed. Friend (1966) found a three-fold increase in particle number concentration for stratospheric samples collected during the same time periods at middle latitudes in the northern hemisphere.

Other investigators attempted to correlate volcanic activity and other perturbations such as meteoritic influx with optical measurements in both the northern (Meinel & Meinel 1963) and southern hemispheres (Hogg 1963). Volz (1965) summarized a number of optical measurements and, in addition, reported independent measurements of twilight intensity at a fixed angle of elevation prior to and following the Agung eruption (Volz & Goody 1962). Their data have the advantage of utilizing the same observational method both before and after the eruption and show that atmospheric turbidity increased by up to a factor of 20 during the latter part of 1963.

A direct quantitative comparison of data obtained by various direct and optical sampling methods is difficult to make, and the extent to which volcanoes contributed to the stratospheric aerosol layer has been the subject of some controversy. Cronin (1971) reviewed the subject and suggested that disagreement is also due to a failure among investigators to recognize the character and location of volcanic eruptions. He discussed the significance of two major belts of volcanic activity located in the arctic and equatorial latitudes, and noted that in the arctic latitudes where the tropopause is lower, eruptions may be of considerably less magnitude than those in equatorial regions and still be of comparable importance. However, since there is a general poleward stratospheric circulation, this suggestion would have to be modified if the particles arise primarily from relatively slow gas-to-particle reactions. For a compilation of the relative magnitudes and observed trends with the dates and locations of major volcanic eruptions, the interested reader is referred to work by Cronin (1971) and Lamb (1970). Studies by Cronin (1971), and others by Hofmann & Rosen (1977) concerning the Fuego eruption, have provided important evidence concerning the origin of the stratospheric aerosol layer.

During the last decade, the group under D. J. Hofmann and J. M. Rosen and co-workers at the University of Wyoming has undertaken the most consistent studies of the change in concentration of the aerosol layer, both in altitude and in time. A summary of recent trends is given in an article by Hofmann & Rosen (1980). It is now well recognized that the early measurements of Junge and co-workers were made during a period of very low volcanic activity, and the measurements probably were made at a time when the layer was at a "background" concentration level. In an attempt to ascertain contributions to this layer, in the absence of volcanic emissions, Hofmann and Rosen have compared their concentration measurements in recent years with those of Junge in the 1959 to 1960 time period. Based on this comparison, there seems to be as much as a 9% per year increase in aerosol concentration, which Hofmann & Rosen suggested may be due to anthropogenic activities. Even though a careful comparison of concentrations measured by various techniques was made, this conclusion is, of course, speculative at this point. In addition, the comparison by Hofmann & Rosen suggested an increase in the upper troposphere over the same period. No generally accepted mechanisms of origin have been proposed that account for this increase.

Early study of the stratospheric aerosol layer was prompted by curiosity concerning the upper atmosphere surrounding the earth. However, more recently there has developed considerable concern about the impact of this aerosol layer on the earth's environment, especially its contribution to the earth's total albedo. Because of both size range and location, stratospheric aerosols tend to scatter the earth's incident radiation and are believed to have had an impact on the earth's global temperature following certain major volcanic eruptions (Lamb 1970). In this context the aerosol layer can have a direct impact on climate, and since climate has a direct bearing on agriculture as well as on the energy needs of the nation and the world, there has been a considerable growth of interest in the nature, formation, and distribution of the stratospheric aerosol layer. The climatic effects due to volcanic eruptions are reviewed by Toon & Pollack (1980).

In addition to their role in scattering sunlight, atmospheric aerosols might affect climate if they have a substantial impact on certain atmospheric chemical reactions. Particles themselves potentially provide a reservoir whereby gas-phase constituents can be removed from participating in further homogeneous reaction mechanisms. Furthermore, although no definitive evidence currently exists, calculations have suggested that aerosols may play a role as heterogeneous catalytic sites in certain reactions (see section on the role of aerosols in the upper atmosphere). Finally, since ions readily charge exchange with aerosols upon collision, the presence of aerosols has a potential bearing on ion concentration in the upper atmosphere. Aerosols may, therefore, partly govern atmospheric electrical parameters.

FORMATION MECHANISMS

The pioneering studies of Junge established that a substantial component of the stratospheric aerosol layer is sulfate. As a result of the work of Lazrus and co-workers (Cadle et al 1970, Lazrus & Gandrud 1974), these aerosols also are known to contain elements such as Si, Na, Cl, Mn, Br, and Ca.

Early speculations that the aerosol layer might be formed as a result of tropospheric aerosols carried upward into the stratosphere or as the result of meteoritic ablation products are inconsistent with some of the compositional findings. This, and the surprising finding that sulfate represented the most abundant single chemical component of the aerosol, led to extensive work to ascertain the origin and mechanisms of sulfate aerosol formation (see Friend 1966, Shedlovsky & Paisley 1966). Although early measurements indicated that the sulfate ions were chemically combined with ammonium ions, studies by Manson et al (1961) indicated that this conclusion must be viewed with caution. Handling procedures often introduce ammonia as contamination, which sometimes leads to the formation of ammonium sulfate by subsequent reactions *after* collection.

More recent compositional analysis of the stratospheric aerosol by Lazrus & Gandrud (1974) suggests that the quantity of cations (other than protons) is insufficient to chemically balance the sulfate. Consequently, it is generally assumed that sulfuric acid represents the major component of stratospheric aerosol.

Toon & Pollack (1973) concluded that, from a thermodynamic point of view, sulfuric acid droplets should be stable at the temperature and water vapor concentrations of the stratosphere. This conclusion follows from the fact that the partial pressures of both sulfuric acid vapor and water vapor in equilibrium with concentrated sulfuric acid solutions are very low at stratospheric temperatures. The authors estimate that the stratospheric droplets should be either a supercooled liquid or a solid composed of an approximately 75% sulfuric acid solution. Attempts to ascertain chemical composition have been made by Rosen (1971) in which the stratospheric aerosol was evaporated *in situ*, and its temperature of evaporation measured. These findings also indicate that a 75% sulfuric acid solution is the major portion of the aerosol.

In early publications, Junge and his co-workers provided circumstantial evidence that the sulfate aerosols were not being carried directly into the stratosphere but were most likely generated by gas-to-particle conversion reactions of SO_2 or H_2S imported into the stratosphere (Junge et at 1961). Subsequently, Castleman et al (1974) made a detailed study of the sulfur isotopic ratio of the stratospheric aerosol, which led to similar conclusions. The observations of Castleman et al have shown that during periods of large volcanic eruptions sulfate aerosol is generated by an *in situ* oxidation mechanism rather than by simple input of tephra from the original eruption cloud. Studies of the oxidation of SO_2 with OH (Castleman & Tang 1976/77, Moortgat & Junge 1977) have indicated that these reactions may be dominant in the formation of the stratospheric aerosol layer from SO_2 generated by input via the volcanic eruptions.

It is well known that various sources of sulfur compounds have specific ratios of ^{32}S to ^{34}S and that this isotope ratio can be used as an indication of the original source of the sulfur, if some measure of potential reactions is taken into consideration. A systematically varying trend in the isotopic ratio following input of the reactive species may provide further information on subsequent chemical reactions. A clear correlation between sulfur concentration and volcanic activity was found from the data of Castleman et al (1974). The perturbation to the total stratospheric dust burden by the paroxysmal eruption of Mount Agung has been well documented and there was a large change in the sulfate concentration as well. The data show that the sulfate concentration in the southern hemisphere increased by approximately two orders of magnitude within a year after the eruption but took some seven months or more to reach a peak concentration. Tephra are known to settle out relatively rapidly, and it must be concluded that this increase in sulfate concentration is evidence for a gas-to-particle conversion mechanism.

Further evidence for a gas-to-particle conversion mechanism was established by considering the isotopic ratios of sulfur for a similar time period. Characteristic trends are invariably observed following a major volcanic event. The data show an abrupt decline in the ^{34}S enrichment following the eruption of Mount Agung and a much smaller decline in magnitude after the smaller eruption of Fernandina. The trends with time and altitude provide very good evidence for an *in situ* formation process (Castleman et al 1974). After the passage of several years without another major eruption, the isotopic ratios always show a tendency to return to pre-eruption values.

The data have been interpreted by recognizing that for short time intervals following major eruptions the additional sulfate contribution by all other processes is comparatively negligible, and the observed continual change in the sulfur isotopic composition for stratospheric aerosols is the result of fractionation processes. Since fractionation would not be expected on a purely physical basis after a few days following an eruption, these changes are clearly indicative of an *in situ* reaction. As a reaction proceeds to completion, the sulfate product would be expected to exhibit a unidirectional change in isotopic composition over a time span related to the eruption.

Isotopic data for sulfur also suggest a common source of sulfur in the stratosphere in both the northern and southern hemispheres, indicating that in the absence of major volcanic eruptions the sulfate in the two hemispheres probably has a common origin. It is speculated that biogenic sulfur released into tropical upwellings may be this source. Most of the tropospheric sulfur compounds such as SO_2, H_2S, and dimethyl sulfide are believed to be sufficiently reactive as to make it unlikely that these compounds would survive in the troposphere long enough to diffuse in sufficient quantities into the stratosphere (Crutzen 1976). Based on these considerations, Crutzen has suggested that stratospheric aerosol formation in the absence of direct sulfur injection by volcanic eruptions may result from the diffusion of OCS into the stratosphere where it becomes photodissociated, eventually reacts to SO_2 and subsequently converts to sulfate aerosol. CS_2 might also make a minor contribution to the stratospheric aerosol layer. Recent measurements of stratospheric sulfur compounds (Inn & Vedder 1980) seem to bear out this conclusion.

NUCLEATION

An aerosol is formed when a gas-phase species undergoes a phase transformation to condensed-phase particles. This process may proceed via a nucleation mechanism involving (*a*) stepwise clustering of gas-phase molecules or (*b*) interaction between small molecular clusters of the aerosol-forming species. The development of an aerosol and its size distribution can in principle be completely described by a general dynamic equation (Gelbard & Seinfeld 1979), which describes the formation and loss processes for the spectrum of particles ranging from gas molecules through small molecular clusters to large aerosol particles. However, the complete kinetics required to solve this equation are not generally known, especially in the case of the formation and interaction of molecular clusters containing only tens of molecules. Therefore, the formation of an aerosol particle is often treated as a problem separate from the other processes such as condensation and coagulation. Following this pattern, the present section of the review first considers aerosol formation and then aerosol growth processes as applicable to the stratosphere. Many of these general processes for aerosols are treated in several recent texts (Pruppacher & Klett 1978, Friedlander 1977, Twomey 1977).

Nucleation processes may be classified according to the material basically responsible for inducing nucleation. These are (a) homogeneous, (b) heterogeneous, and (c) heteromolecular. In the homogeneous case only those molecular species that constitute the condensing phase are involved in the nucleation process. Heterogeneous nucleation operates when these species condense onto the surface of a foreign body. In the heteromolecular case, a molecular entity acts as the foreign body that induces the nucleation of the condensable components.

A further distinction among the nucleation processes is made to denote the number of condensable substances that participate simultaneously to form new aerosol particles. Thus, nucleation may be unary, binary, ternary, etc. It should be noted that in the literature binary homogeneous nucleation is often referred to as heteromolecular nucleation, and so confusion with the above definition should be avoided.

Classical nucleation theory (see Pruppacher & Klett 1978) assumes that a quasi-steady state population of molecular clusters exists and that the rate of nucleation or aerosol formation is determined by the rate of collision of the monomeric gas molecules with critical clusters. The critical cluster is that cluster which corresponds to a free energy maximum during its growth starting from a free gas-phase molecule to a condensed-phase particle. The nucleation barrier $\Delta G^{\mp}$ is then the difference between this free energy maximum and the free energy minimum corresponding to the most stable subcritical (i.e. smaller) cluster. Thus, the nucleation rate J is expressed in terms of a Boltzmann-type relation, which determines the critical cluster concentration, given by

$$J = K_{exp}(-\Delta G^{\mp}/kT), \tag{1}$$

where k is the Boltzmann constant, T the absolute temperature, and the preexponential factor includes terms that express the collision dynamics of the monomer with the critical cluster and the deviation of the quasi-steady state population of critical clusters from a constrained equilibrium one. This formulation neglects any contribution of forming aggregates larger than the critical size by means of cluster-cluster collision. However, during heterogeneous or heteromolecular nucleation, a stable small cluster population may exist and cluster-cluster interactions should not be ignored. Zurek & Schieve (1980) have shown that even for homogeneous nucleation cluster-cluster interactions can be important under some circumstances.

The demarcation between whether an aggregate of molecules is a gas-phase cluster or a condensed-phase aerosol particle, i.e. the point at which nucleation has occurred and growth commences, is not intuitively well defined. From nucleation theory this point can be thermodynamically defined by the critical cluster size, since further growth by condensation for particles larger than this size is assured. When no nucleation barrier exists, a critical cluster is not defined thermodynamically and nucleation is limited only by the kinetics of the forward growth rates. In this case, no dividing line exists. In practice, the critically sized cluster is not directly observed, and the properties of nucleation are inferred from the aerosol particles that are observed. Therefore, the demarcation may be conveniently defined by a minimum detectable particle size.

When a nucleation barrier exists, the problem of determining the nucleation rate is largely one of calculating the requisite free energy changes. Furthermore, this calculation is necessary to determine under what conditions the barrier disappears. The most broadly applicable approach is the use of the capillarity approximation, often called the Thomson drop model, which is based on the properties of the bulk phase. In this model the free energy of forming a cluster of n molecules from the monomeric gas is due to condensation of n molecules, the formation of a surface, and effects due to a foreign nucleus if any. For the unary homogeneous case, a spherical drop of n molecules, the free energy is expressed by the equation

$$\Delta G(n) = -nkT \ln S + 4\pi r^2 \sigma, \tag{2}$$

where r is the radius (related to n by the density of the condensed medium) of the drop, and σ the surface tension. The saturation ratio S is defined as the partial pressure p of a component divided by its vapor pressure p_0 over the bulk condensed phase. Whenever a system is supersaturated ($S > 1$), the first term of Equation (2) is negative and favors cluster growth. The second term is responsible for the existence of the nucleation barrier and is commonly referred to as the Kelvin effect. In other words, the surface energy results in a higher vapor pressure for a curved drop than for a flat surface of bulk. The Kelvin effect is such that the vapor pressure of a water drop of 0.1 μm radius is about 2% higher than that for a flat surface and about 25% higher for a 0.01 μm radius drop. Evidence in support of this effect and its validity is reviewed by Skinner & Sambles (1972).

A macroscopic interpretation of a surface with a surface tension in nucleation theory, however, is at best questionable when one considers critical clusters of typically a few tens of molecules. However, the Thomson model remains popular because of its simplicity and its relative success in interpreting experimental results (Jaeger et al 1969, Castleman et al 1978). Another approach using macroscopic-type parameters is the Fisher droplet model (Stauffer & Kiang 1977).

Molecular models using statistical mechanics (Hale & Plummer 1973, Hoare et al 1980) to calculate cluster properties have a firmer theoretical footing and qualitatively produce results similar to macroscopic models. However, such models are not as yet applicable to problems of stratospheric nucleation because of their complexity and the lack of the necessary information. Recent computer simulation techniques such as Monte Carlo (Abraham 1974) and molecular dynamics (Briant & Burton 1975) methods circumvent the necessity of directly determining thermodynamic quantities. Using these methods, Zurek & Schieve (1978) have confirmed the concept of a critical cluster. As with molecular models, these methods have been limited to the treatment of only very elementary systems.

NUCLEATION IN THE STRATOSPHERE

Assuming water and sulfuric acid as the primary components, the requirement that a gaseous constituent must be supersaturated with respect to a condensed phase in order to initiate the formation of an aerosol eliminates unary homogeneous nucleation as a viable mechanism for the creation of a stratospheric aerosol. The vapor pressure at stratospheric temperatures (around $-50°C$) of water and sulfuric acid over their pure solutions are well above the measured partial pressures of water and estimated sulfuric acid vapor concentrations (Hamill et al 1977a). More recently, the existence of low sulfuric acid concentrations in the stratosphere has been supported experimentally by Arnold & Fabian (1980). Similarly, the unary heterogeneous or heteromolecular process could only occur if the vapor pressure is reduced by the foreign nucleus as, for instance, a "soluble" particle. However, a particle that may be soluble in the bulk phase may not act as a soluble nucleus as far as nucleation is concerned. A critical saturation must be exceeded in order that the soluble crystal absorb a sufficient number of solvent molecules to dissolve. With water, for instance, the critical saturations are 0.55 (55% relative humidity) for sodium sulfate and 0.76 for sodium chloride (O'Brien 1948). The relative humidity of the stratosphere is typically below 1% so that such salts will behave as insoluble nuclei. Since water is much more abundant than sulfuric acid and the acid itself is hygroscopic, self-nucleation of sulfuric acid onto an acid soluble crystal is improbable. Consequently, any discussion of nucleation in the natural stratosphere must include at least a binary process.

Binary Nucleation

As discussed in the foregoing section, binary nucleation is the most likely mechanism to lead to new particle formation in the stratosphere. When two pure substances are mixed, the vapor pressures of the individual components over that mixture may be lower than those over their pure states. Thus, a supersaturation in a mixed system may develop without the supersaturation of any single component with respect to its pure state.

In the Thomson formulation for the homogeneous case, the free energy of forming a drop of a composition defined by the number of molecules n_i of each component is given by the equation

$$\Delta G(n_1, n_2, \ldots n_l) = -kT \sum_{i=1}^{l} n_i \ln \frac{p_i}{p_{0i}} + 4\pi r^2 \sigma, \tag{3}$$

where p_i is the partial pressure of the ith component, and p_{0i} is its vapor pressure over a bulk solution of the composition determined by $(n_1, \ldots, n_l)$. The determination of the free energy barrier to nucleation in the critical size and composition is complicated by the fact that σ, p_{0i}, and the solution density to which r is related are implicit functions of the variables $n_1, \ldots, n_l$. Yue (1979) concisely described the various numerical and graphical methods that have been applied to determine these properties for a binary system (where l equals 2).

In the stratosphere the gaseous concentration of sulfuric acid molecules is much less than that of water. Therefore, a cluster or aerosol particle is quickly equilibrated with respect to water vapor, and the clustering and growth is determined by the collision rate of sulfuric acid molecules with these clusters and aerosol particles (Mirabel & Katz 1974). Consequently, the ratio p_i/p_{0i} for water may always be assumed to be unity.

Ternary Nucleation

The preceding discussion has considered only the simultaneous nucleation of two components since sulfuric acid and water are thought to be the major condensable species in the stratosphere. However, the possibility that three components promote nucleation under stratospheric conditions had been suggested by Kiang & Hamill (1974), with nitric acid as the third component. Kiang et al (1975a,b) subsequently extended nucleation theory to ternary systems and considered the nitric acid–sulfuric acid–water system in the stratosphere. Unfortunately, vapor pressure data of ternary systems, particularly at stratospheric temperatures, are largely lacking. The suggestion by Kiang & Hamill (1974) that nitric acid might be saturated in the stratosphere with respect to HNO_3-H_2SO_4-H_2O solutions containing a significant amount of nitric acid (about 15% by weight) was based on an admittedly crude extrapolation using data available in the International Critical Tables (Zeisberg 1928) which are from several different sources and are evidently inconsistent when examined carefully. The low vapor pressures deduced by Kiang & Hamill's extrapolation for 15% nitric acid solution appear to arise from this inconsistency.

At present no evidence exists for nitric acid participation in stratospheric aerosol formation (Lazrus & Gandrud 1974). One should note, however, that compositional analysis of aerosols may also be misleading in distinguishing between the possible nucleation processes. The composition of stratospheric aerosols has routinely been determined at room temperature where a component may become relatively more volatile than at stratospheric temperatures. On the other hand, a ternary mixture may develop subsequent to nucleation due to the reaction of aerosol particles with other gaseous species. The possibility of ternary nucleation cannot as yet be discounted until more adequate data such as partial pressures of gaseous species over mixed systems becomes more readily available.

Binary Heterogeneous Nucleation

Hamill et al (1977a) have shown that homogeneous processes are very improbable due to the general availability in the stratosphere of foreign nuclei along with preexisting aerosol particles. The nuclei include particles transported into the stratosphere from tropospheric, extraterrestrial, or volcanic sources (Castleman 1974). Nuclei created *in situ* are generally of molecular dimensions such as ions and, therefore, may create aerosol particles via heteromolecular nucleation, which is discussed in the next subsection.

Usually nuclei of tropospheric, volcanic, or extraterrestrial origin are large in comparison to the critical size of homogeneous nucleation. These nuclei would actually be aerosol particles in their own right. Heterogeneous nucleation as applied to the stratosphere may be considered a conversion of a non-sulfate aerosol into a sulfate-containing aerosol. The heterogeneous process on insoluble nuclei, which act only as sites for the sticking of the nucleating phase, modifies the surface energy required to reach the critical size. For instance, for a completely "wettable" nucleus of radius r_0 the surface free energy of forming a drop of radius r is given by $4\pi(r^2 - r_0^2)\sigma$ since the bare nucleus plus the monomeric gas molecules is the reference state of the system. If r_0 is larger than the critical radius, then no nucleation barrier would exist, and nucleation would be kinetically controlled by the time required to "wet" the nucleus.

For the case of a partially wettable nucleus, the macroscopic concept of a contact angle φ is often employed (see Chapter 12 in Pruppacher & Klett 1978). For a completely wettable nucleus, the contact angle is $0°$, whereas for a completely non-nucleating unwettable nucleus φ would be $180°$. Basically, the contact angle determines the curvature and consequently the surface energy of the nucleating medium on a nucleus of a given shape or radius. A soluble nucleus would affect the vapor pressure in the system and act as a completely wettable nucleus.

Nucleation on completely wettable insoluble nuclei has been calculated to be a very effective mechanism (Hamill et al 1977a). However, the problem of the "wettability" of these nuclei has not been adequately addressed, and Farlow et al (1977) found that undissolved granules are present in only about one third of the stratospheric aerosol particles during periods of low volcanic activity. In addition, the number of such nuclei in the stratosphere is still uncertain. Measurements with condensation nuclei counters (Podzimek et al 1977) cannot uniquely distinguish between particles formed in the stratosphere and those transported to it from other sources.

Heteromolecular Nucleation

IONS   Due to the electrostatic forces whereby neutral molecules, especially polar ones, readily cluster around an ion (see Castleman 1979) ion-induced nucleation is known to happen preferentially to homogeneous nucleation. The continual ionization of the stratosphere (see Meyerott et al 1980) provides a significant source of ions. If the lifetime against neutralization for the clustering ion is sufficiently long compared to the time required to induce nucleation, then this mechanism may be operative. The average lifetime of an ion in the stratosphere is around 10^3 seconds, but the low sulfuric acid concentration implies slow H_2SO_4 clustering rates onto ions. Therefore, clustering kinetics, instead of thermodynamics as assumed in classical nucleation theory, limit ion-induced nucleation in the stratosphere.

Nevertheless, Castleman & Tang (1972) established that small ion clusters actually represent a segment of the overall size distribution of atmospheric species. Chan & Mohnen (1980) estimated a stable ion cluster distribution and relative critical cluster population using the Thomson drop model for the binary sulfuric acid–water system. The Thomson model is a very crude assumption in that specific ion-neutral interactions are not accounted for. Evidence now exists that stratospheric negative ions are mixed clusters of acids such as nitric acid, sulfuric acid, and possibly hydrochloric acid, while positive cluster ions contain water and possibly sodium hydroxide or sodium chloride (see Viggiano et al 1980, Ferguson 1978). The rather specific acid-base type interactions of ions of opposite charge invalidate the use of the simple Thomson theory to predict the type of cluster ions present in the stratosphere.

Since ions form small stable clusters at a concentration of a few thousand per cubic centimeter, an attractive hypothesis is that ion cluster-cluster interactions may be important in producing nuclei upon which condensation can eventually occur. Mohnen (1971) first suggested this possibility. Since clustering stabilizes an ion by effectively reducing the ionization potential of positive ions and increasing the electron affinity of negative ones, E. E. Ferguson (1977, private communication) subsequently pointed out that although the neutralization of oppositely charged ions may be exothermic, clustering may cause the charge neutralization of two ions to be endothermic so that their combination could produce a stable solvated ion pair. One should note that the products of ion-ion collisions and the stability of many of the feasible ion-neutral association clusters in the stratosphere are not known. Consequently, the degree of clustering required for a pair of ions to form a solvated ion pair in contrast to dissociated neutral products can at present only be qualitatively estimated.

Since an ion pair has a large dipole moment, Arnold (1980) has suggested that further clustering of these ion pairs with ions to create multi-ion complexes may effectively promote nucleation by producing small electrolytic droplets as nuclei upon which condensation can occur. In this case, of course, classical nucleation theory is quite inadequate for describing the rate of phase transformation and, therefore, aerosol formation.

RADICALS   The reaction $OH + SO_2 + M \rightarrow HSO_3 + M$ is recognized as an important loss mechanism of sulfur dioxide (Castleman & Tang 1976/77). The formation of aerosols initiated by this reaction in the presence of water from which OH is formed by photolysis is also well established (Davis et al 1979). Niki et al (1980) have spectroscopically shown that the products appear to be liquid sulfuric acid containing varying amounts of water. However, the role of the HSO_3 radical in forming the aerosol is not understood. Friend et al (1973) have demonstrated that the products of the photolysis of H_2O-SO_2-O_2 mixtures rapidly produce aerosol particles when introduced into a condensation nuclei counter in which the count rate (related to the nucleation rate) is dependent on the initial relative humidity of the preceding reaction zone. Friend et al (1980) have suggested that single sulfur molecules such as $H_2S_2O_6$, $H_2S_2O_8$, SO_3, or H_2SO_4 act as nuclei analogous to ions in ion-induced nucleation in the highly supersaturated condensation counter. These sulfur compounds are possible products in reactions or combinations of the free radicals created in the reaction zone. Another possibility is that the production rate of sulfuric acid is sensitive to humidity due to the reaction chemistry of hydrated radicals such as HSO_5, $HSO_5 \cdot H_2O$, and $HSO_5 \cdot 2H_2O$.

Davis et al (1979) have estimated that the dominant initial reaction path for HSO_3 in the stratosphere should be association with O_2. Association of water molecules to these radicals may also occur. The number of associated water molecules will be dependent on the relative humidity analogous to the cases for H_2SO_4 (Heist & Reiss 1974) or ions (Castleman & Tang 1972). In general the prevalence and role of neutral association complexes in atmospheric phenomena has been considered only recently (Carlon 1979, Calo & Narcisi 1980) and deserves quantitative investigation.

In regard to the stratosphere Davis et al (1979) have established that reactions of sulfur radicals with radicals such as NO_2 may also be competitive pathways. Interestingly, species such as $NOHSO_4$ (Farlow et al 1978) and $(NH_4)_2S_2O_8$ (Friend 1966) have been tentatively identified in samples of stratospheric aerosols. Such compounds as $H_2S_2O_6$, $H_2S_2O_8$, and $NOHSO_4$ are also known to decompose or hydrolyze to sulfuric acid in acidic solutions, for example (*Gmelins Handbuch* 1960),

$$H_2S_2O_8 + H_2O \rightarrow H_2SO_4 + H_2SO_5, \tag{4}$$

$$H_2SO_5 + H_2O \rightarrow H_2SO_4 + H_2O_2. \tag{5}$$

GROWTH

Aerosol particles may grow by coagulation, scavenging, condensation, or gas-aerosol reactions; the reverse process, namely, evaporation, must also be considered.

In the stratosphere the coagulative process due to Brownian diffusion is of primary importance. Gravitational coagulation, which results from a larger particle falling at a net rate with respect to a smaller one, is unimportant in the stratosphere (Turco et al 1979). Also, coagulation due to turbulent motion is usually not relevant to stratospheric conditions.

The general treatment of Brownian coagulation for stratospheric aerosol is complicated by particle dynamics. The particle size ranges of interest occur in a transition region where neither free molecular kinetic nor slip flow diffusive kinetic motion are fully applicable (Castleman 1974). Equations have been formulated to interpolate this region so as to approximate the particle dynamics over the entire range from free molecular to diffusive motion (Fuchs & Sutugin 1971).

Whether condensation or evaporation operates on an aerosol particle depends on the sulfuric acid vapor pressure at given conditions. An aerosol particle is generally assumed always to be essentially in equilibrium with the surrounding water vapor. Thus, at a given temperature water vapor concentration determines the concentration of sulfuric acid in the aqueous sulfuric acid aerosol particle. This in turn specifies the sulfuric acid vapor pressure over the particle (Toon & Pollack 1973). Evaporation of the particle occurs if the sulfuric acid partial pressure is less than its corresponding vapor pressure. Condensation occurs in the opposite case. During these processes the water content of the aerosol is adjusted practically instantaneously on the time scale of sulfuric acid addition or loss to maintain its equilibrium. Whereas coagulation decreases the total number of aerosol particles and shifts the size spectrum to larger particles, condensation has no direct effect on the total particle concentration. Many of the details of coagulation and condensation are discussed by Hamill et al (1977b).

The scavenging of molecular clusters by aerosol particles lies between the domain of coagulation and condensation. All three processes are necessary to describe the effect of an existing aerosol on the total size spectrum. The major significance of scavenging is its effect on the steady-state molecular cluster distributions and consequently on heteromolecular and homogeneous nucleation (Gelbard & Seinfeld 1979). The decrease of nucleation rates by a preexisting aerosol represents a feedback mechanism such that a balance is created between the total aerosol surface area and the particle production rate required to maintain that aerosol against losses due to sedimentation of large particles (see McMurry & Friedlander 1979). For the stratosphere, however, the contribution to the aerosol of particles transported from volcanic, tropospheric, and meteoric sources needs to be considered to establish the magnitude of *in situ* particle production. At present only rough estimates of these contributions exist (Turco et al 1979, Hunten et al 1980).

The reaction of gases with an aerosol, i.e. heterogeneous chemistry, may also lead to growth of the particle. For example, sulfur species may be absorbed and directly oxidized to sulfuric acid within the aerosol particle, or gaseous species may react with the sulfuric acid in an aerosol particle to form salt species. The kinetics of such reactions may be controlled by either surface or volume characteristics (see Friedlander 1977). At present the details of gas-aerosol heterogeneous chemistry are poorly understood and only a limited amount of quantitative data exists (Baldwin & Golden 1979).

NUCLEATION AND GROWTH IN MODELS

Since the discovery of the sulfate aerosol layer, its formation and properties have been the object of several modeling efforts. In an attempt to explain observed condensation nuclei distributions, Junge et al (1961) developed a model in which sedimentation, diffusion, and coagulation were considered. The particle source was implied by fixing particle number densities at the top and bottom altitudes of the model, and aerosol formation as well as condensation and evaporation of aerosol particles was neglected. Despite this, the model helped to explain some observations of condensation nuclei distribution. Burgmeier & Blifford (1975) also considered basically only sedimentation, diffusion, and coagulation but included an empirical source term to balance the loss of aerosols by sedimentation.

In a subsequent modeling attempt, Rosen et al (1978) introduced a fixed H_2SO_4 distribution where the mixing ratio was defined by a narrow Gaussian distribution centered at 20 km altitude. This model allowed condensation, but evaporation was ignored as no significant fraction of aerosol particles in the model reached the model's upper boundary level. The important advancement in this model was that the supply of sulfuric acid was explicitly considered.

Most recently, Turco et al (1979) included evaporation and nucleation. In addition, the H_2SO_4 concentration was calculated interactively within the aerosol model. Thus, the concentration of sulfuric acid depended on the rate of production of sulfuric acid (based on a simple chemical scheme to oxidize sulfur species) versus that of its incorporation into the aerosol particles. Nucleation was assumed to proceed by heterogeneous nucleation onto chemically inert nuclei transported from the troposphere. The nucleation rate was taken to involve a fixed induction time (10^6 seconds) required before condensation could occur on these nuclei. Toon et al (1979) demonstrated that an order of magnitude variation in time did not significantly affect the model results. These models have successfully explained many of the observed properties of the stratospheric sulfate aerosol. Although several details are simplified or ignored, they have significantly contributed to understanding the relative importance of various mechanisms in determining the features of the stratospheric aerosol.

ROLE OF AEROSOLS IN THE UPPER ATMOSPHERE

Potentially important chemical and physical effects may be divided into three categories. These include charge exchange between ions and aerosols, surface catalysis, and the production of trace species. In this regard, the aerosol may provide a sink or source for some gas-phase constituents. Rough estimates of the surface area in the stratospheric aerosol layer based on particle size distributions indicate about 10^{-8} cm^2 cm^{-3}. Based on these values and an elementary kinetic theory analysis, a given gas-phase molecule suffers a collision with a stratospheric aerosol surface approximately every 10^4 to 10^5 seconds.

There is one situation where stratospheric aerosols have been thought to play a role in the production of trace species, namely, the production of HCl from the interaction of H_2SO_4 with NaCl (Castleman et al 1975). Calculations show that the reaction is exothermic and the quantity of HCl so produced is likely to be limited by the quantity of NaCl, potentially from sea salt, which may be deposited in the stratosphere following major storms in equatorial regions. However, the kinetics of the reactions are not known under the low temperature conditions prevailing in the stratosphere, and laboratory studies of this reaction are needed.

There are three possible ways in which aerosol surfaces may play a role in affecting the gas-phase concentration of chemically reactive species. These include 1. the destruction of reactive intermediates such as free radicals, which would be normally important in gas-phase reactions, 2. a catalytic influence on reactions between stable gas-phase constituents, or 3. the stabilization of a product molecule in the condensed state, which might otherwise readily dissociate in the gas phase.

Quantitatively, very little is known about catalytic reaction mechanisms or rates that might be important in the upper atmosphere. Therefore, in order to assess which mechanisms warrant attention, model calculations were made where it was assumed that one species is adsorbed and others react with it upon each collision (Castleman et al 1975). By comparing these rates to known homogeneous ones involving the same species, a crude prediction was made regarding what species might have potentially important surface-controlled chemistry. Admittedly this does not provide an unequivocal assessment of the importance, or lack thereof, for the species under consideration, but it gives a way of screening and establishing priorities for laboratory investigation. Table 1 is a list of reactants, possible products, and an estimate of the maximum possible ratio of heterogeneous to homogeneous rates.

Much of the stratospheric aerosol is believed to be composed of H_2SO_4, which is notoriously non-catalytic for most reactions. Based on the listing in Table 1, few of the systems warrant any attention whatsoever, especially if H_2SO_4 is assumed to be distributed over all surfaces to an equal extent (a condition that is suspected but not as yet established with any certainty). Yet, since sulfuric acid aerosols contain H_2O, reaction with N_2O_5 might be important and deserves attention in this context. Olszyna et al (1979) have shown that ozone destruction is not important on sulfuric acid surfaces. Therefore, this marginally important case may be omitted from further consideration unless substantial metallic oxide aerosols are found,

Table 1   Comparison of surface and gas-phase reaction rates[a]

Reactants	Products	Ratio, surface to gas-phase reaction rate
N_2O_5, H_2O	$2 HNO_3$	$>10^3$
NO_2, O_3	NO_3, O_2	~ 10
CH_3O_2, NO	CH_3O, NO_2	~ 1
O_3, O	$2 O_2$	200
NO, O_3	NO_2, O_2	~ 10
$HO_2\cdot, O_3$	$OH\cdot, O_2$	5.3×10^{-2}
$OH\cdot, O_3$	$HO_2\cdot, O_2$	1×10^{-3}
$CH_3O\cdot, O_2$	$CH_2O\cdot, HO_2\cdot$	1.6×10^{-5}
$Cl\cdot, O_3$	$ClO\cdot, O_2$	1×10^{-6}
$O\cdot, O_2$	O_3	2.5×10^{-7}
$CH_3\cdot, O_2$	$CH_3O_2\cdot$	5.8×10^{-8}
$H\cdot, O_2$	$HO_2\cdot$	1.7×10^{-10}

[a] Maximum surface reaction rate based on kinetic theory for collision rates, and assuming unit reaction probability upon collision.

in which case they may play a somewhat important role. [Ferguson (1978) suggests the possible importance of ozone reactions with metal atoms which might be a source of oxide aerosols.]

Three other possibly important surface reactions are the reaction of ClO with H_2SO_4/H_2O to produce HCl (Martin et al 1978). $ClONO_2$ with H_2O giving rise to HOCl (J. W. Birks 1979, private communication), and sulfur dioxide by H_2O_2. There is some evidence, based on laboratory measurements, that the first reaction occurs to a small extent. Estimates suggest that the inclusion of this reaction in models could decrease by a few percent the estimated Cl/HCl ratios existing in the stratosphere (Martin et al 1978). Laboratory studies are needed to assess the importance of surface reactions with $ClONO_2$. There are no laboratory data suitable for assessing the potential importance of SO_2 reaction with H_2O_2 on surfaces (Cadle et al 1975).

A consideration of the potential importance of stratospheric heterogeneous reactions on the behavior of the fluorocarbons has failed to provide any evidence in support of such a mechanism. Ausloos et al (1977) have demonstrated that unexpectedly large surface photochemical effects can result on certain surfaces, notably certain sands, leading to the destruction of some fluorocarbon compounds and N_2O. The mechanisms and significance of these catalytic processes are not well known, but even if they do occur, the quantity of potentially active surface material is far too low to be of importance in the stratosphere.

Ion concentrations are established by the balance between the rates of formation and destruction. The latter is the result of two processes and includes recombination between negative and positive ions, and the process of ion charge exchange with free surfaces provided by aerosol particles. Studies (Keefe et al 1959) show that to a good approximation aerosols maintain a Boltzmann charge distribution. The rate of change of ion concentration n is given by

$$\frac{dn}{dt} = Q - \alpha n^2 - n\left(\beta A\right),\tag{6}$$

where t is time, Q ion production rate, α recombination coefficient for negative and positive ions, β the coefficient of ion attachment to aerosols, and A the aerosol surface area.

The stratosphere contains approximately 10^3 ion pairs/cm^3. Under stratospheric conditions, the recombination coefficient α is 10^{-7} to 10^{-6} cm^3/sec from which it follows that the ratio of the loss of ions due to mutual recombination to that due to charge exchange upon collision with an aerosol particle is

$$\frac{\alpha n}{\beta A} \approx 10.\tag{7}$$

Therefore, the charge exchange process with aerosols are of only secondary importance except when the aerosol concentration is increased following periods of intense volcanic activity.

CONCLUSION

Various sampling studies and numerical models have provided evidence that the *in situ* oxidation of sulfur-bearing gases (Junge & Manson 1961, Turco et al 1978) is responsible for the sulfate mass of the stratospheric aerosol. An extensive study of the temporal and spatial distribution of the sulfur isotope ratio by Castleman et al (1974) has borne this out. These data suggest that there is a common source of sulfur compounds for the stratosphere of both the northern and southern hemispheres. Using elementary modeling calculations and the results of laboratory experiments, Castleman & Tang (1976/77), Davis et al (1979), and Moortgat & Junge (1977) have speculated that the stratospheric aerosol layer originates, at least in part, from SO_2 oxidation via OH. Other candidates for the origin of the sulfur component of the aerosol layer are COS (Crutzen 1976) and CS_2 as a precursor to COS [Sze & Ko 1979]. It is still unclear to what degree the oxidation of the sulfur species is completely homogeneous (creating a supersaturation of H_2SO_4 with subsequent nucleation and condensation) or heterogeneous with the ultimate sulfate production occurring directly on preexisting particles.

Very little is known concerning the origin of the primary small particles that form as a result of processes following the generation of the precursors to the prenucleation embryos. It is almost certain that homogeneous nucleation does not operate in the atmosphere and that the more relevant processes are those termed heteromolecular and heterogeneous nucleation. Furthermore, in the stratosphere, processes that involve the interaction of more than one gaseous species participating in the formation of the condensed phase (particle) are certain to dominate under situations where new particle generation occurs.

The relative contribution of the various mechanisms proposed for the introduction of particles in the stratosphere has not been established. This problem is further complicated because the various nucleation processes and condensation compete for a limited supply of sulfur. Nevertheless, the general characteristics and extent of the stratospheric sulfate aerosol are reasonably well understood in terms of coagulation, condensation, evaporation, and sedimentation when a source for the generation of new particles is assumed.

ACKNOWLEDGMENTS

Support of the Department of Energy under Grant No. DE-AC02-78EV04776, the National Aeronautics and Space Administration under Grant No. NSG 2248, the Atmospheric Sciences Section of the National Science Foundation under Grant No. ATM 79-13801, and the US Army Research Office under Grant No. DAAG 29-79-C-0133 is gratefully acknowledged. This work was conducted while one of us (R. G. K.) was a National Research Council Research Associate at NASA-Ames.

Literature Cited

Abraham, F. F. 1974. Monte Carlo simulation of physical clusters of water molecules. *J. Chem. Phys.* 61:1221–22

Arnold, F. 1980. Multi-ion complexes in the stratosphere—implications for trace gases and aerosol. *Nature* 284:610–11

Arnold, F., Fabian, R. 1980. First measurements of gas phase sulfuric acid in the stratosphere. *Nature* 283:55–57

Ausloos, P., Rebbert, R. E., Glasgow, L. 1977. Photo-decomposition adsorbed on silica surfaces. *J. Res. Natl. Bur. Stand.* 82:1–8

Baldwin, A. C., Golden, D. M. 1979. Heterogeneous atmospheric reactions: sulfuric acid aerosols as tropospheric sinks. *Science* 205:562–63

Bigg, E. K., Ono, A., Thompson, W. J. 1970. Aerosols at altitudes between 20 and 37 km. *Tellus* 22:550–63

Briant, C. L., Burton, J. J. 1975. Molecular dynamics study of the structure and thermodynamic properties of argon microclusters. *J. Chem. Phys.* 63:2045–58

Burgmeier, J. W., Blifford, I. H. 1975. A reinforced coagulation-sedimentation model for stratospheric aerosols. *Water, Air, and Soil Pollut.* 5:133–47

Cadle, R. D., Lazrus, A. L., Pollack, W. H., Shedlovsky, J. P. 1970. Chemical composition of aerosol particles in the tropical stratosphere. *Proc. Symp. Tropical Meteorol.* Section K-IV. 7 pp.

Cadle, R. D., Crutzen, P., Ehhalt, D. 1975. Heterogeneous chemical reactions in the stratosphere. *J. Geophys. Res.* 80:3381–85

Calo, J. M., Narcisi, R. S. 1980. van der Waals molecules—possible roles in the atmosphere. *Geophys. Res. Lett.* 7:289–92

Carlon, H. R. 1979. Variations in emission spectra from warm water fogs: evidence for clusters in the vapor phase. *Infrared Phys.* 19:49–64

Castleman, A. W. Jr. 1974. Nucleation processes and aerosol chemistry. *Space Sci. Rev.* 15:547–89

Castleman, A. W. Jr. 1979. Nucleation and molecular clustering about ions. *Adv. Colloid Interface Sci.* 10:73–128

Castleman, A. W. Jr., Tang, I. N. 1972. Role of small clusters in nucleation about ions. *J. Chem. Phys.* 57:3629–38

Castleman, A. W. Jr., Tang, I. N. 1976/77. Kinetics of the association reaction of SO_2 with the hydroxyl radical. *J. Photochem.* 6:349–54

Castleman, A. W. Jr., Munkelwitz, H. M., Manowitz, B. 1974. Isotopic studies of the sulphate component of the stratospheric aerosol layer. *Tellus* 26:222–34

Castleman, A. W. Jr., Davis, R. E., Tang, I. N., Bell, J. A. 1975. Heterogeneous process and the chemistry of aerosol formation in the upper atmosphere. *Proc. Conf. Climate Impact Assess. Prog., 4th*, pp. 470–77

Castleman, A. W. Jr., Holland, P. M., Keesee, R. G. 1978. The properties of ion clusters and their relationship to heteromolecular nucleation. *J. Chem. Phys.* 68:1760–67

Chan, L. Y., Mohnen, V. A. 1980. The formation of ultrafine ion $H_2O-H_2SO_4$ aerosol particles through ion-induced nucleation process in the stratosphere. *J. Aerosol Sci.* 11:35–45

Cronin, J. F. 1971. Recent volcanism and the stratosphere. *Science* 172:847–49

Crutzen, P. 1976. The possible importance of CSO for the sulfate layer of the stratosphere. *Geophys. Res. Lett.* 3:73–76

Davis, D. D., Ravishankara, A. R., Fisher, S. 1979. SO_2 oxidation via the hydroxyl radical atmospheric fate of HSO_4 radicals. *Geophys. Res. Lett.* 6:113–16

Farlow, N. H., Hayes, D. M., Lem, H. Y. 1977. Stratospheric aerosols: undissolved granules and physical state. *J. Geophys. Res.* 82:4921–29

Farlow, N. H., Snetsinger, K. G., Hayes, D. M., Lem, H. Y., Tooper, B. M. 1978. *J. Geophys. Res.* 83:6207–11

Farlow, N. H., Ferry, G. V., Lem, H. Y., Hayes, D. M. 1979. Latitudinal variations of stratospheric aerosols. *J. Geophys. Res.* 84:733–43

Ferguson, E. E. 1978. Sodium hydroxide ions in the stratosphere. *Geophys. Res. Lett.* 5:1035–38

Friedlander, S. K. 1977. *Smoke, Dust, and Haze*. New York: Wiley. 317 pp.

Friend, J. P. 1966. Properties of the stratospheric aerosol. *Tellus* 18:465–73

Friend, J. P., Leifer, R., Trichon, M. P. 1973. On the formation of stratospheric aerosols. *J. Atmos. Sci.* 30:465–79

Friend, J. P., Barnes, R. A., Vasta, R. M. 1980. Nucleation by free radicals from the photooxidation of sulfur dioxide in air. *J. Phys. Chem.* 84:2423–36

Fuchs, N. A., Sutugin, A. G. 1971. Highly dispersed aerosols. In *Topics in Current Aerosol Research*, ed. G. M. Hidy, J. R. Brock, 2:1-60. New York: Pergamon

Gelbard, F., Seinfeld, J. 1979. The general dynamic equation for aerosols. *J. Colloid Interface Sci.* 68:363–82

Gmelins Handbuch. 1960. Schwefel teil. Berlin: Springer. 825 pp. (In German)

Gruner, P. 1942–61. In *Handbuch der Geophysik*, ed. F. Linke, F. Moller, Vol. 8. Berlin: Borntrager. 432 pp.

Gruner, P., Kleinert, H. 1927. In *Probleme der Kosmischen Physik*, Vol. 10. Hamburg: Henri Grand

Hale, B. N., Plummer, P. L. M. 1973. On nucleation phenomena I: a molecular model. *J. Atmos. Sci.* 31:1615–21

Hamill, P., Kiang, C. S., Cadle, R. D. 1977a. The nucleation of $H_2SO_4-H_2O$ solution aerosol particles in the stratosphere. *J. Atmos. Sci.* 34:150–62

Hamill, P., Toon, O. B., Kiang, C. S. 1977b. Microphysical processes affecting stratospheric aerosol particles. *J. Atmos. Sci.* 34:1104–19

Heist, R. H., Reiss, H. 1974. Hydrates in supersaturated binary sulfuric acid-water vapor. *J. Chem. Phys.* 61:574–81

Hoare, H. R., Pal, P., Wegener, P. P. 1980. Argon clusters and homogeneous nucleation: comparison of experiment and theory. *J. Colloid Interface Sci.* 75:126–37

Hofmann, D. J., Rosen, J. M. 1977. Balloon observations of the time development of the stratospheric aerosol event of 1974–1975, *J. Geophys. Res.* 82:1435–40

Hofmann, D. J., Rosen, J. M. 1980. Stratospheric sulfuric acid layer: evidence for an anthropogenic component. *Science* 208:1368–70

Hogg, A. R. 1963. The Mount Agung eruption and atmospheric turbidity. *Aust. J. Science* 26:119–20

Hunten, D. M., Turco, R. P., Toon, O. B. 1980. Smoke and dust of meteoric origin in the mesosphere and stratosphere. *J. Atmos. Sci.* 37:1342–57

Inn, E. C. Y., Vedder, J. F. 1980. Measurements of stratospheric sulfur constituents. *Geophys. Res. Lett.* In press

Jaeger, H. L., Wilson, E. J., Hill, P. G. Russell, K. C. 1969. Nucleation of supersaturated vapors in nozzles. I. H_2O and NH_3. *J. Chem. Phys.* 51:5380–88

Junge, C. E. 1963. *Air Chemistry and Radioactivity.* New York: Academic. 382 pp.

Junge, C. E., Manson, J. E. 1961. Stratospheric aerosol studies. *J. Geophys. Res.* 66:2163–82

Junge, C. E., Chagnon, C. W., Manson, J. E. 1961. Stratospheric aerosols. *J. Meteorol.* 18:81–108

Keefe, D., Noland, P. J., Rich, T. A. 1959. Charge equilibrium in aerosols according to the Boltzmann law. *Proc. R. Irish Acad. A.* 60:27–45

Kiang, C. S., Hamill, P. 1974. H_2SO_4-HNO_3-H_2O ternary system in the stratosphere. *Nature* 250:401–2

Kiang, C. S., Cadle, R. D., Hamill, P., Mohnen, V. A., Yue, G. K. 1975a. Ternary nucleation applied to gas to particle conversion. *J. Aerosol Sci.* 6:465–74

Kiang, C. S., Cadle, R. D., Yue, G. K. 1975b. H_2SO_4-HNO_3-H_2O ternary aerosol formation mechanism in the stratosphere. *Geophys. Res. Lett.* 2:41–44

Lamb, H. H. 1970. Volcanic dust in the atmosphere, with a chronology and assessment of its meteorological significance. *Proc. R. Soc. London Ser. A.* 226:425–533

Lazrus, A. L., Gandrud, B. W. 1974. Stratospheric sulfate aerosol. *J. Geophys. Res.* 79:3424–31

Lazrus, A. L., Gandrud, B., Cadle, R. D. 1971. Chemical composition of air filtration samples of the stratospheric sulfate layer. *J. Geophys. Res.* 76:8083–88

Manson, J. E., Junge, C. E., Chagnon, C. W. 1961. *Chemical Reactions in the Lower and Upper Atmospheres.* New York:Interscience

Martin, L. R., Wrenn, A. G., Wun, M. 1978. *Chlorine Atom and ClO Wall Reactions Products.* Final report prepared for the National Science Foundation

McMurry, P. H., Friedlander, S. K. 1979. New particle formation in the presence of an aerosol. *Atmos. Environ.* 13:1635–51

Meinel, M. P., Meinel, A. B. 1963. Late twilight glow of the ash stratum from the eruption of Agung volcano. *Science* 142:582–3

Meyerott, R. E., Reagan, J. B., Jainer, R. G. 1980. The mobility and concentration of ions and the ionic conductibity in the lower stratosphere. *J. Geophys. Res.* 85:1273–78

Mirabel, P., Katz, J. L. 1974. Binary homogeneous nucleation as a mechanism for the formation of aerosols. *J. Chem. Phys.* 60:1138–44

Mohnen, V. A. 1971. Discussion of the formation of major positive and negative ions up to the 50 km level. In *Mesospheric Models and Related Experiments,* ed. G. Fiocco, pp. 210–19. Dordrecht, Holland: Reidel

Moortgat, G. K., Junge, C. E. 1977. The role of SO_2 oxidation for the background stratospheric sulfate layer in the light of new reaction rate data. *Pure Appl. Geophys.* 115:759–74

Mossop, S. C. 1964. Volcanic dust collected at an altitude of 20 km. *Nature* 203:824–27

Niki, H., Maker, P. D., Savage, C. M., Breitenbach, L. P. 1980. Fourier transform infrared study of the HO radical initiated oxidation of SO_2. *J. Phys. Chem.* 84:14–16

O'Brien, F. E. M. 1948. The control of humidity by saturated salt solutions. *J. Sci. Instrum.* 25:73–76

Olszyna, K., Cadle, R. D., de Pena, R. G. 1979. Stratospheric heterogeneous decomposition of ozone. *J. Geophys. Res.* 84:1771–75

Podzimek, J., Sedlacek, W. A., Haberl, J. B. 1977. Aitken nuclei measurements in the lower stratosphere. *Tellus* 29:116–27

Pruppacher, H. R., Klett, J. D. 1978. *Microphysics of Clouds and Precipitation.* Dordrecht, Holland: Reidel. 714 pp.

Reiter, E. R. 1971. *Atmospheric Transport Processes Pt2: Chemical Tracers.* AEC Critical Review Series, US Atomic Energy Commission, Division of Technical Information, Washington, DC

Rosen, J. M. 1971. The boiling point of stratospheric aerosol. *J. Appl. Meteorol.* 10:1044–46

Rosen, J. M., Hofmann, D. J., Singh, S. P. 1978. A steady-state stratospheric aerosol model. *J. Atmos. Sci.* 35:1304–13

Schuster, B. G. 1970. Detection of tropospheric and stratospheric aerosol layers by optical radar (LIDAR). *J. Geophys. Res.* 75:3123–32

Shedlovsky, J. P., Paisley, S. 1966. On the meteoritic component of stratospheric aerosols. *Tellus* 18:499–503

Skinner, L. M., Sambles, J. R. 1972. The Kelvin equation—a review. *J. Aerosol Sci.* 3:199–210

Stauffer, D., Kiang, C. S. 1977. Fisher's droplet model and nucleation theory. *Adv. Colloid Interface Sci.* 7:103–30

Sze, N. D., Ko, M. K. W. 1979. Is CS_2 a precursor for atmospheric COS? *Nature* 278:731–32

Toon, O. B., Pollack, J. B. 1973. Physical properties of stratospheric aerosols. *J. Geophys. Res.* 78:7051–59

Toon, O. B., Pollack, J. B. 1980. Atmospheric aerosols and climate. *Am. Sci.* 68:268–78

Toon, O. B., Turco, R. P., Hamill, P., Kiang, C. S., Whitten, R. C. 1979. A one-dimensional model describing aerosol formation and evolution in the stratosphere II. Sensitivity studies and comparison with observations. *J. Atmos. Sci.* 36:718–36

Turco, R. P., Hamill, P., Toon, O. B., Whitten, R. C., Kiang, C. S. 1979. A one-dimensional model describing aerosol formation and evolution in the stratosphere: I. Physical processes and mathematical analogy. *J. Atmos. Sci.* 36:699–717

Twomey, S. 1977. *Atmospheric Aerosols.* New York: Elsevier. 302 pp.

Viggiano, A. A., Perry, R. A., Albritton, D. L., Ferguson, E. E., Fehsenfeld, F. C.

1980. The role of H_2SO_4 in stratospheric negative-ion chemistry. *J. Geophys. Res.* In press

Volz, F. E. 1965. Note on the global variation of stratospheric turbidity since the eruption of Agung Volcano. *Tellus* 17:513–5

Volz, F. E. 1970. On dust in the tropical and mid-latitude stratosphere from recent twilight measurements. *J. Geophys. Res.* 75:1641–46

Volz, F. E., Goody, R. M. 1962. The intensity of the twilight and atmospheric dust. *J. Atmos. Sci.* 19:385–406

Yue, G. K. 1979. A quick method for estimating the equilibrium size and composition of aqueous sulfuric acid droplets. *J. Aerosol Sci.* 10:75–86

Zeisberg, F. C. 1928. Vapor pressures, boiling points and vapor compositions for the system H_2O-H_2SO_4-HNO_3. In *International Critical Tables*, ed. E. W. Washburn, 3:306–8. New York: McGraw-Hill

Zurek, W. H., Schieve, W. C. 1978. Molecular dynamics evidence for vapor-liquid nucleation. *Phys. Lett.* 67A:42–45

Zurek, W. H., Schieve, W. C. 1980. Multistep clustering and nucleation. *J. Phys. Chem.* 84:1479–82

Ann. Rev. Earth Planet. Sci. 1981. 9:251–84

ANCIENT MARINE PHOSPHORITES

✖ 10151

Richard P. Sheldon

National Center MS 953, US Geological Survey, Reston, Virginia 22092

INTRODUCTION

Marine phosphorites have been studied for more than a hundred years, both because of their commercial value as fertilizer and because of their importance to sedimentary petrology. Phosphorus is, of course, one of the basic plant nutrients and is intimately involved in biospheric processes; however, its role in biogenic and chemogenic sedimentation is difficult to study and is not completely understood. Marine phosphorites on the seafloor today occur on continental shelves, rises, and seamounts where they are difficult to observe and collect. In the laboratory, the mineral structure and crystal chemistry of sedimentary apatite has been difficult to analyze because of its complexity and small crystal size, and apatite has not yet been synthesized at low temperatures in solutions that approximate seawater in composition. Physical-chemical equilibrium studies have been difficult because of the slow rates of reaction, the complex composition of apatite, and the formation of complex surface phases and complex phosphorous ions in solution, as well as the complex nature of seawater itself. As a result, progress has been slow in understanding the origin of sedimentary phosphorites, including the source and the processes of concentration of the phosphorus and the processes and the environment of deposition of the phosphorite. Hypotheses and arguments have proliferated; however, in recent years consensus has been reached on many points. Arguments still exist on a number of other points, stimulating continued research in the field.

Sedimentation processes commonly can be studied directly in the modern environment, and the results can be applied by analogy to interpret ancient deposits according to Lyell's principle. In the case of phosphorites, marine geology studies, particularly in the last decade, have produced information that has been extremely interesting and somewhat

251

unexpected. Very little apatite has been found forming in the ocean today, and what recent phosphorite has been found is not on the ocean floor but in organic-rich muds, where it is forming diagenetically. Thus the presently forming apatite has not given crucial evidence for selecting among the earlier ideas on the origin of ancient phosphorites, but has caused the formulation of a new hypothesis. This result has raised the question of whether, in the case of phosphorites, the present is the key to the past, or whether the present time of negligible accumulation of slightly phosphatic mud is atypical of past times when large amounts of phosphorite were deposited.

These problems are the subject of a number of review articles, which have either been published since 1978 or are in press at this writing. Cook & McElhinny (1979) reviewed and reevaluated phosphogenesis in space and time in relation to episodes of glaciation, iron and evaporite sedimentation, volcanism, orogeny, and plate tectonics. Baturin (1978) has published in Russian a comprehensive review of phosphorite on the ocean floor, in which he included data and analyses of the sources of phosphorus, phosphorus in marine waters, biologic processes and sedimentation of phosphorus, diagenetic reactions of phosphorus, and the distribution, and origin of phosphorites. This book has been submitted for publication in English. Burnett (1980b) has reviewed the ideas on oceanic phosphate deposits presented at the Marine Phosphatic Sediments Workshop held in 1979, which covered much of the same scientific ground as Baturin, and represented the input of 30 scientists. Y. Kolodny, who contributed to the Marine Phosphatic Sediments Workshop, has prepared an extensive review of phosphorites to be published in Volume 7 of *The Sea* (1980). His emphasis is on phosphorites found on the marine continental shelves around the world, although not to the exclusion of ancient phosphorites, and he deals with the petrography, mineralogy, major and trace element geochemistry, age of submarine phosphorites, and genesis of marine phosphorites. Finally, P. J. Cook is preparing and commenting on a collection of important past papers on phosphorites for the Benchmark Series.

One can question, and certainly I did, whether one more review article is of value at this time. In deciding in the affirmative I felt that although these recent reviews have been excellent in their depth and scholarship, they have either not addressed the full subject or have not given the perspective of geologists working with ancient phosphorites.

This review draws heavily on these and other earlier reviews, as well as on the conclusions of a number of recent workshops on phosphorite geology. These include two, in particular, (a) the International Geological Correlations Program (IGCP) Project 156 Field Workshop/Seminar held in Queensland, Australia, in 1978 and summarized by Cook & Shergold

(1979), and (*b*) the Fertilizer Raw Materials Resources Workshop sponsored by the East-West Resource Systems Institute of the East-West Center, held in Honolulu, Hawaii, in 1979 (Sheldon & Burnett 1980).

DEVELOPMENT OF HYPOTHESES

The development of hypotheses about the origin of phosphorites went through an early period in which basic disagreements existed between scientists concerning the mode of deposition. As Gulbrandsen (1969) in his excellent historical review pointed out, most hypotheses called on organic activity as essential to phosphate deposition. Suggested modes of deposition ranged through coprolites (Buckland 1829), decomposition of seaweed and animals (Seeley 1866), precipitation from solutions secreted by animal agency (Fisher 1873), animal remains through mass mortalities (Cornet 1886), redeposition of phosphate dissolved from bones and phosphatic matter from dead animals (Keyserling in Penrose 1888), and deposition in swamps in association with peat (Shaler in Penrose 1888). Fisher (1873) also proposed that nodular phosphate beds represented a condensed section. Blackwelder (1916) described the geochemical cycle of phosphorus from its source in igneous rocks, through its path in the biologic cycle, its storage in the ocean, its sedimentation, and its reincorporation in the earth's crust. Grabau (1919) pointed out that phosphate beds occur at minor and major unconformities. Pettijohn (1926) interpreted the zones of phosphatic limestone which are rich in phosphate nodules as a residuum on corrosion surfaces due to submarine erosion. Gulbrandsen (1969) pointed out that by 1888 many modes of formation of phosphate and factors now considered significant in the formation of phosphate deposits were known. It was thought by many, including Blackwelder (1916), that mixing of cold and warm ocean water caused the death of organisms and the deposition of their remains on the ocean floor. Still missing was an hypothesis as to the locale of phosphorus deposition, the oceanographic processes causing its concentration, and, finally, the conditions of deposition of relatively pure phosphorite. Kazakov (1937) supplied this hypothesis and a new one on the physical-chemical precipitation of apatite on the seafloor.

Kazakov (1937) used ideas advanced by earlier workers and added more of his own ideas to formulate a comprehensive model that dealt with the origin and localization of phosphorus, its mode of deposition, paleogeography of depositional basins geologic environment, geometry of beds, and the petrography of the phosphorites. This comprehensive hypothesis had a stimulating effect on phosphorite research. It put order into previously collected data and ideas, focused data gathering, and suggested many problems for additional research. The vitality of Kazakov's hypothesis is shown by the importance that it—or as it now is called, the upwelling model—holds in modern treatments of phosphorite petrology. Although his ideas on the complex chemical precipitation of apatite from seawater have proved too simple, later research has broadened, refined, and corrected the hypothesis while continuing its basic tenets.

McKelvey and others (1953) analyzed the Kazakov hypothesis and adapted it, in a slightly modified form as explaining the Permian phosphorites of the Rocky Mountains. The Kazakov hypothesis pointed out that phosphorus content in ocean water increases with depth from a minimum in the zone of photosynthesis to a maximum at about 500 meters. Phosphate is precipitated on shelving, shallow ocean bottoms where deep, cold, phosphorus-rich waters ascend. In this circulating system, which Kazakov related to conditions in the South Atlantic, the deep waters rising on one side of a basin are depleted of phosphorus by deposition. They then circulate back to the other side of the basin where phosphate deposition does not occur. Physical-chemical changes in ascending water include increasing temperature and decreasing partial pressure of CO_2 which cause phosphate deposition. These changes cause saturation with respect to apatite. Kazakov reasoned from the standpoint of solution chemistry and hydrographic data that below 200 meters the cold water contained enough dissolved CO_2 to be undersaturated with respect to apatite. Above 50 meters, in the zone of photosynthesis, the available phosphorus is assimilated by phytoplankton so no apatite is deposited. He reasoned that ascending water first became saturated with respect to calcium carbonate, which was deposited before the apatite, and thereby separated the two facies.

The geographic environment of phosphorite deposition, according to Kazakov, is the continental shelf at depths of 50–200 meters in areas of ascending deep ocean water. The phosphorite facies are deposited with synchronous shallow-water sediments on the shoreward side and deep-water sediments on the basinward side. Two phosphorite facies are found. First is the shallow-water platform facies, which is distinguished by nodular, pebbly phosphorites, low to moderate P_2O_5 content, and associated glauconitic and arenaceous materials. Second the deeper water geosynclinal facies which is thin bedded, high in P_2O_5, and associated with limestone and shales. According to Kazakov, phosphorites increase in thickness and P_2O_5 content with increasing distance from the shoreline, though thickness and quality are independent. The quality of phosphorites depends on the composition of the enclosing rocks and on the relative sedimentation rates of terrigenous materials and phosphates. Phosphorite sequences typically consist of a basal conglomerate, overlain by reworked, pre-phosphorite materials, a phosphate layer, and other materials derived from nearby terrigenous sources. Phosphorites are generally associated with marine transgressions, so phosphorite facies may migrate across time lines.

In formulating his hypothesis, Kazakov discarded a number of the then current hypotheses. He replaced the biogenic hypotheses with the chemogenic elements of his hypothesis. He reasoned that the deposition of apatite could be accomplished by the physical-chemical processes alone, and although he recognized the abundant evidence of organic remains and organic matter in phosphorites, he apparently did not regard them as necessary to phosphorite formation. He also discarded the hypothesis requiring volcanic activity in the formation of phosphorites, apparently believing that his model was sufficient to explain phosphorite deposition without relying on volcanic sources of phosphorus.

The reaction to Kazakov's ideas was mixed. In Russia his hypothesis was contested vigorously (Strakhov 1962, Bushinski 1966a, Librovich 1966). Bushinsky, who was a colleague of Kazakov and who advocated a different hypothesis, strongly opposed Kazakov's ideas and published his criticism along with an historical summary of the development of hypotheses on the origin of phosphorites (1966a,b). These writings give a good insight into the evolution of ideas in Russia where much active research on phosphorites was and is going on, only a part of which has crossed the language barrier. Outside of Russia, the Kazakov hypothesis was more widely accepted, particularly by McKelvey and his colleagues. At present, many if not most phosphorite petrologists accept the Kazakov hypothesis, modified to some extent and excluding the simple chemogenic concept.

A major development of phosphorite petrology after the Kazakov hypothesis has been in the physical chemistry of apatite depositions. Gulbrandsen (1969) pointed out that the optimum conditions for the formation of large amounts of apatite seem to be the coincidence of a special steady supply of phosphate, originally derived from organic matter, and a decreased capacity of seawater for phosphate. He suggested that these conditions were best met in shallow parts of seas on continents where large amounts of organic matter accumulate in oxygenated waters of higher than normal temperature, pH, and salinity. The mechanism faces the problem pointed out by Martens & Harriss (1970), that the Mg/Ca ratio in seawater is sufficiently high to inhibit apatite precipitation from the ocean, a conclusion later supported and enlarged on by Atlas (1975). Also, Stumm (1972) noted that return of phosphorus to solution by oxidation of organic debris liberates large amounts of CO_2, causing a lowering of the pH. As suggested much earlier by Blackwelder (1916) and elaborated by Kassin (1925) and Bushinski (1966b), anaerobic bacterial action provides an alternative to oxidation for releasing the phosphorus to the solution from the organic matter. Blackwelder stated that anaerobic bacterial action yields ammonia, H_2S, phosphate, hydrocarbons, and CO_2, depending on the supply of oxygen. In the late 1960s and early 1970s intensive marine geologic work confirmed that the process of anaerobic sulfate-reducing bacterial action on organic oozes produces interstitial water of high alkalinity and phosphate content and low magnesium content and that apatite pellets form chemically in this diagenetic environment. The results of this research were well described by Baturin (1978) and summarized by Burnett (1980b) and Kolodny (1980).

This diagenetic process produces phosphatic pellets but not phosphorite, according to Baturin (1971). He proposed an hypothesis that slightly phosphatic muds could be reworked at times of lower sea level, thereby washing away the mud and concentrating the apatite pellets. Most earlier workers had called on primary sedimentation for marine phosphorites, although reworking or winnowing to produce higher-grade deposits has been advocated for the Florida deposits (Altschuler et al 1964) and the Permian deposits of the Rocky Moutains (Sheldon 1957, 1963, Cressman & Swanson 1964). Baturin's hypothesis of "upwelling currents–diagenetic precipitation–low sea level reworking" received much support. Cook (1976) and Kolodny (1980) have accepted it for the general case for ancient phosphorites. Burnett (1977) initially accepted the hypothesis but later expressed reservations regarding the universality of application of the hypothesis (Burnett 1980a). I questioned the hypothesis as a mechanism for producing the major phosphorite deposits (Sheldon 1980) and, following concepts developed by Fischer & Arthur (1977), presented an alternative nonuniformitarianism hypothesis of episodic phosphogenesis.

Not all current research is guided by the Kazakov upwelling hypothesis. Bushinski (1964, 1966a), Pevear (1966), and F. de Keyser (in de Keyser & Cook 1972) favor a river source and shallow-water shelf or estuarine deposition to explain the source and paleogeography of ancient phosphorite deposits with which they were concerned. Some workers propose volcanic emanations as the source of phosphorite (Dzotsenidze 1969, Brodskaya 1974, Riggs 1979b, 1980). Hite (1978) advances an hypothesis of concentration of phosphorus in seawater through evaporation in coastal basins.

The phosphorites and phosphatic sediments of marine origin in the southeastern United States are a different type from most other phosphorites. They make up a large deposit covering a large area, which occupies a different paleogeographic position from most other phosphorites, and they have a different lithologic assemblage. The Kazakov upwelling hypothesis applies with great difficulty, if at all, to this deposit. However, little agreement exists among workers on an alternative. Pevear (1966) advocated a river source of the phosphorus and an estuarine depositional environment. Rooney & Kerr (1967) hypothesized that a volcanic tuff-fall killed large numbers of organisms whose decay contributed the source for the phosphorus, an hypothesis that Cathcart (1968b) questioned because of the small amount of volcanic ash in the rock. Gibson (1967) presented evidence of a cold countercurrent to the Gulf Stream coming down along the eastern United States coast from the north at the time of the deposition of the phosphorite. Cathcart (1968a) accepted and expanded Gibson's hypothesis, attributing the cause of phosphate deposition to turbulent mixing of the cool water from the north with the warm water from the south around structural highs. Freas (1968) postulated local upwelling currents along the coast in conjunction with the ancient northward-flowing Gulf current. Riggs (1979b) speculated that hydrothermal solutions rich in phosphorus "exhaust onto the sea floor" and "supercharge the marine environment" and then "oceanic current systems and associated upwellings . . . . mix and move the supercharged waters up into the shallow-water environments." Differing views have been expressed not only on the source of the phosphorus but also on the stratigraphic relations, the sedimentation processes, and the timing of events. Because studies of these rather unique phosphate deposits have resulted in many basic disagreements, it seems evident that more research must be carried out before a widely accepted picture will emerge.

PRESENT KNOWLEDGE OF MARINE PHOSPHOGENESIS

Most researchers on the origin of marine phosphorites agree at present on many basic processes of phosphogenesis. In the following section, these basic processes are listed and the scope of agreement as well as disagreement and problems arising in interpreting ancient deposits are discussed.

Marine phosphogenesis consists of oceanographic, sedimentational, diagenetic, and erosional processes. Seven basic processes are involved: (a) supply of phosphorus to the ocean, (b) storage of dissolved phosphorus in the deeper waters of the ocean, (c) circulation of deep water to shallow levels, (d) concentration of phosphorus on the sea bottom, (e) solubility changes of ocean-derived waters with respect to apatite, (f) precipitation of apatite, and (g) concentration of apatite. These are reviewed below.

Supply of Phosphorus to the Ocean

Chemical erosion of rocks and soils on land releases phosphorus, most of which enters the biologic cycle. Eventually some phosphorus in dissolved organic and inorganic chemical species reaches the ocean. About 1.5×10^6 metric tons of phosphorus are estimated to be added to the ocean each year by rivers (Baturin 1978), of which two thirds is organic phosphorus and one third is inorganic. In addition, about 15×10^6 tons per year are added as phosphorus in mineral grains or sorbed phosphorus on mineral grains as a part of the terrigenous clastic sediments.

Phosphorus is added to the oceans by volcanos. The gases from fumaroles of the Japanese volcano Usu contain as much as 2.4 mg P/1. Volcanic gases and fumarolic products from volcanoes in Japan and Indonesia contain small amounts of phosphorus (Koritnig 1970). Some volcanic hydrothermal solutions contain significant amounts of dissolved phosphorus, which enter surface or groundwaters on land or go directly to the ocean. It is difficult to show, however, that phosphorus is a primary component and not dissolved from country rocks. Other hydrothermal solutions result from circulation of seawater through basalts at spreading centers, where they are heated and returned to the ocean. Leaching of basalt by this process may explain the metal deposits found in association with hydrothermal solutions at spreading centers (Anderson et al 1979). Although it is conceivable that these waters might add phosphorus to the ocean, Froelich et al (1980) have shown that hydrothermal solutions at the Galapagos Rift do not contain more phosphorus than seawater. Baturin (1978) and most other workers have concluded that the amount of phosphorus added to seawater from volcanic sources is small compared to the amount from rivers. However, Riggs (1979b, 1980) and many others have hypothesized that hydrothermal phosphorus is important as a local source accounting for large phosphate deposits.

The total amount of reactive phosphorus added to the ocean from all known sources was estimated by Baturin (1978), based on the available data, to be 1.57×10^6 metric tons of phosphorus per year, of which 1.5×10^6 tons is from rivers and 0.07×10^6 tons is from volcanoes. These estimates are subject to perhaps major change as new information is collected.

Storage of Phosphorus in Deep-Ocean Water

Settling organisms transport incorporated phosphorus downward, thereby depleting ocean surface water and enriching deeper water. More than 90 percent of the phosphorus in this organic debris is regenerated to the ocean before sedimentation. From about several hundred meters down, the water contains more or less constant amounts of phosphorus but reaches maximum values at about 1000 meters. Gulbrandsen & Roberson (1973) hypothesized that the solubility of apatite controls the deep-ocean concentration. Apatite is more soluble in the colder, deeper water because of decreasing temperature and increasing pressure as discussed below. Thus, if apatite precipitation does control over time the phosphorus concentration of the deep-ocean phosphorus sink, it would be through precipitation of apatite from water at shallower depths, which would be closer to saturation with respect to apatite than deep-ocean water. However, no apatite has been shown to be precipitating chemically directly out of seawater today. Broecker (1974) has suggested a kinetic model for regulating composition of the deep-ocean water, where the input of phosphorus by rivers is balanced by removal of phosphorus by organic fixation in areas of upwelling. If the turnover time were varied, it seems logical that the phosphorus content of the deep-ocean sink would vary and would increase with longer turnover times, a process suggested by Fischer & Arthur (1977) and amplified by Sheldon (1980). Whether phosphorus content increased to the point where it was controlled by apatite solubility is an interesting question. In any event, variations of turnover rates would result in variations in rate of phosphorite sedimentation, according to Broeker's model as well as Gulbrandsen & Roberson's model.

Circulation of Deep Water to Shallow Levels

The ocean is strongly density stratified. Less dense water is formed by solar and minor geothermal warming and by addition of fresh water from rain, rivers, and melting ice. Density is increased by cooling and by ice formation or evaporation, both of which increase salinity and thereby raise the specific gravity of the residual water. Dense water is generally formed in the polar areas of cooling and ice formation and in silled basins, such as the Mediterranean and Red Seas where dry trade winds cause evaporation resulting in lower-level flow over the sill into the ocean. This denser water from either source sinks in the ocean, and spreads out at levels of corresponding density to form identifiable widespread lenses of water within the ocean. Circulation and mixing of these deep water masses occur very slowly.

Vigorous vertical circulation occurs in several ways in the ocean and mostly affects water in the upper several hundred meters. The trade winds and westerly winds at ocean level create shallow ocean currents that form major gyres that circulate clockwise in the northern hemisphere and counterclockwise in the southern. On the eastern sides of these gyres (along the west coasts of continents) in the trade-wind belt, the trade winds push offshore the equatorward-flowing coastal currents. This offshore move-ment is reinforced by the Coriolis force but in decreasing amount (as the sine of the latitude) towards the equator, where it is zero. Deeper water from the upper few hundred meters wells up behind this offshore flow to replace it, resulting in coastal divergent upwelling. Divergence also occurs near the equator, where the eastward-flowing equatorial countercurrent flows between the westward-flowing currents of the northern and southern gyres. Because of friction, two zones of divergence parallel with the equatorial countercurrent form, one along the boundary of the counter-current with the northern gyre and the other within the southern gyre at about the equator. These divergences are commonly called equatorial upwelling. A third important type of divergence occurs in the Antarctic, where cold, saline waters sink along the bottom of the Antarctic con-tinental shelf and slope and are replaced by divergent upwelling water from 3000–4000 meters deep. This upwelling water in part flows back to the north to the Antarctic convergence where it sinks and flows northward beneath subantarctic water. These water movements are superimposed on the weak, eastward-flowing Antarctic circumpolar current. In some parts of the world local divergent upwelling is caused by monsoon winds, blowing offshore, for example, along the southern shore of the Arabian peninsula. In the trade-wind belt, water of high density formed by evapora-tion in coastal, semirestricted basins sinks in the ocean and is replaced by upward-flowing, deeper water. Finally, significant vertical circulation occurs from the effect of the ocean-floor relief on deep currents, a phenom-enon described as dynamic upwelling (McKelvey 1967).

The equatorial and trade-wind upwelling as well as upward-flowing water displaced by downward-flowing evaporite brines occur in latitudes below about 40°, and the Antarctic divergence occurs between 60° and 70°S. The main shallow-water areas covered by upwelling are (a) conti-nental shelves and rises along the west coasts of continents in the trade-wind belt, (b) submarine rises, seamounts, and the offshore slopes of land masses in equatorial regions, and (c) the continental slope and rise of Antarctica.

The present distribution of the oceans and their bathymetric configura-tion is unique to the Holocene, and therefore reconstruction of oceanic circulation becomes highly speculative for ancient oceans, the more remote the time the more speculative the reconstruction. Paleontologic and geologic evidence has shown that circumglobal equatorial currents existed in the past, that sea levels were higher in that much more shallow marine water existed on the continents (Hallam 1978), that world climates and intensity of oceanic circulation were variable, and that the number, configuration, and location of continents have changed as has the depth and bathymetry of the oceans themselves. In view of the complexity of the currents in the present ocean and the difficulty of studying them, the formulation of general principles, and, in turn, the prediction of the physical oceanography of poorly defined ancient oceans has been extremely difficult.

Concentration of Phosphorus on the Sea Bottom

Nutrient-rich water from the deep ocean flows over shallow-water areas in areas of upwelling. These bottom waters are richer in phosphorus than normal shallow water, which has been depleted of its phosphorus by phytoplankton. The phosphorus content of deeper ocean water on the average is 72 μg/1 P, but it can contain several times that amount in more restricted seas (Baturin 1978). The phosphorus in this water would be available for any chemical or biochemical reactions at the sediment-water interface. Settling planktonic organisms decay during settling, and the phosphorus they contain is regenerated to the water, thereby increasing the phosphorus content of deeper water. A mechanism long recognized for further concentration of phosphorus on the sea bottom is the settling and accumulation of biogenic debris, including both soft and hard parts of organisms. All plant and animal soft parts contain phosphorus, and the most common marine sediment of such material since the Cretaceous Period is diatom ooze, which contains as much as two percent phosphorus by dry weight (Baturin 1978). Piper & Codispoti (1975) proposed a process whereby this percentage could be increased. The atom ratio of total nitrogen to total phosphorus dissolved in the ocean is approximately 15:1 and equal to the nitrogen:phosphorus ratio of plankton. These two nutrients, of course, are both known to limit primary productivity. Lowering of the combined-nitrogen concentration in the deep ocean reservoir by denitrification would lower the nitrogen:phosphorus ratio below the ratio required by phytoplankton. Upwelling of this deep water into the photic zone would then allow the surface water to maintain a relatively high phosphate concentration. In other words, all of the phosphate in the upwelled water would not be removed by phytoplankton productivity. This excess phosphorus may be removed by plankton in luxury amounts and stored in the solid parts of the organism. Piper & Codispoti (1975) concluded that an extraordinary flux of phosphorus to the sediments may thus be established, even at lower productivity levels. An ooze of diatoms or other such small organisms that accumulate on the ocean floor is composed of as much as 98 percent water and is very easily moved by bottom currents, so that it collects in quiet water traps on the seafloor, where decay of organic matter continues.

Another source of phosphorus in bottom waters is interstitial water in sediments. J. R. Morse (written communication in Burnett & Sheldon

1979) reported from preliminary analysis of a few areas of marine continental shelf sediments that phosphate fluxes by diffusion from the sediments into the overlying water column reach amounts of about 5×10^{12} μg P/yr. These fluxes were measured by concentration gradients between bottom water and the upper two centimeters of cores. Compaction of sediments would also return interstitial or pore water to bottom water, particularly in the first few hundred meters of burial. The initial porosity of mud is 50 percent or more, and this porosity is mostly filled with water. By the time mud is buried by a few thousand feet of sediment it has 5–10 percent porosity; pore water is continually squeezed out during compaction and returned by paths of least pressure to the bottom water. Finally, phosphorus in pore water may be released back into the bottom water by periodic storms, which would disturb the quiet water environment and stir up organic ooze. Baturin (1978) pointed out that near-bottom water may be enriched in phosphorus due to roiling of sediments, decay of organic matter, diffusion, desorption, and leaching.

Apatite Solubility Changes of Ocean-Derived Water

The precipitation of apatite requires the concentration of PO_4^{3-} ion to be increased to the point that the solubility product (SP) of the following simplified reaction is exceeded:

$$Ca_5(PO_4)_3F = 5Ca^{2+} + 3PO_4^{3-} + F^-,$$
$$SP = [Ca^{2+}]^5 [PO_4^{3-}]^3 [F^-].$$

The PO_4^{3-} concentration is affected by its reactions with H^+ according to the following reactions:

$$H_3PO_4 = H_2PO_4^- + H^+,$$
$$H_2PO_4^- = HPO_4^{2-} + H^+,$$
$$HPO_4^{2-} = PO_4^{3-} + H^+.$$

Thus the PO_4^{3-} concentration is dependent on the pH of the system, and the higher the pH the greater the percentage of PO_4^{3-} ion to the total dissolved phosphate. The pH of ocean water is controlled primarily by the reactions of CO_2 and water to give carbonic acid, so that the more dissolved CO_2 the lower the pH at any given alkalinity. PO_4^{3-} forms a number of complex ions, of which perhaps the most important to seawater phosphorus reactions is the $MgHPO_4$ ion (Atlas 1975). Any such complex ion formation reduces the concentration of PO_4^{3-} ion. The solubility product of the reaction forming apatite is affected by temperature and pressure. A decrease of temperature and an increase of pressure both increase the solubility product or make apatite more soluble. In addition, lower temperatures and higher pressures allow increased partial pressure of dissolved CO_2, which decreases pH in ocean water. Thus, for several reasons, decreases of temperature and increases of pressure tend to change the seawater chemistry toward undersaturation with respect to apatite. On the contrary, increases of temperature and decreases of pressure tend to change it toward saturation.

Much research has been conducted to determine whether or not seawater is saturated with respect to apatite (Kramer 1964, Roberson 1966, Gulbrandsen 1969, Atlas 1975). One conclusion of the solubility studies is that apatite solubility in seawater is affected by crystallinity, crystal size, and composition (Atlas 1975) in addition to equilibrium-solution factors. Seawater appears to be close to saturation with respect to apatite, but studies show that phosphorus complex-ion formation and extremely low nucleation rates may preclude apatite precipitation, at least under present ocean conditions. Based on Atlas's data, a rise of either PO_4^{3-}, Ca^{2+}, or F^- and a decrease of Mg^{2+} would result in decreased apatite solubility.

Pore water in organic-rich sediments differs significantly in chemical composition from ocean water (Bushinski 1966a, Baturin 1978, Burnett 1977). Concentrations of H_2S, H_2CO_3, ammonia, silicic acid, and phosphorus increase in pore waters due to the degrading of organic matter by sulfate-reducing bacteria. If this change is carried far enough, the waters become saturated with respect to apatite. Apatite has been shown to be forming in sediments on the Namibian Shelf (Baturin 1978) and on the Peru Shelf (Burnett 1977). W. C. Burnett pointed out (written communication 1980) that it is not clear why apatite precipitates in these areas and does not in some others where PO_4^{3-} concentration is also high, but perhaps it is due to variations in F^- concentrations.

These results of physical-chemical analysis of the solubility of ocean-derived waters with respect to apatite to some extent are not supported by geologic data. Many phosphorites exposed on the present seafloor have been shown to be ancient (Kolodny 1980; Baturin 1978, Figure 8). Most are Tertiary in age, and some are as old as Cretaceous; all are exposed in areas of no sedimentation. If the bottom water in these areas is under-saturated with respect to phosphorus, as proposed by most marine geochemists, we must explain how these ancient phosphorites, particularly the seamount phosphorites, have survived their tens of millions of years' exposure to undersaturated waters without being dissolved. Protective coatings and metastability may be all that is necessary to account for this, and some dissolution has taken place (W. C. Burnett, written communication 1980). Even so, it remains a problem. Also, as discussed later, petrographic and stratigraphic evidence of ancient phosphorites indicates that some phosphorites were deposited chemically on the seafloor in a high-energy environment as accretionary rings on oolites. Thus, geologic evidence seems to indicate that seawater was and perhaps is saturated with respect to phosphorite and that apatite has been chemically precipitated.

Precipitation of Apatite

Apatite is known to precipitate in two ways. Apatite is precipitated biogenically within the ocean-water environment as hard parts of some nektonic and benthonic organisms. It also precipitates chemically as nodules and pellets in organic-rich muds or oozes during diagenesis, as discussed above, as well as during a number of other diagenetic chemical processes. A third mechanism of precipitation, chemical precipitation on the seafloor surface, has long been proposed but has recently been rejected by some workers for physical-chemical reasons, as discussed above. In the following sections, these three types of processes are discussed.

PRIMARY BIOGENIC PRECIPITATION   Primary biogenic precipitation of apatite in sedimentary rocks has been well documented (Rhodes & Bloxam 1971). Many animals precipitate apatite to form their hard parts. These parts include the shells of inarticulate brachiopods, bones and teeth of vertebrates, fish scales, mollusc kidney stones (Doyle et al 1978), conodonts, and tubes and opercula of some hyolithelminths. Accumulations of debris of this type form a definite phosphorite facies in some deposits. For example, in the Phosphoria Formation of Permian age in southeastern Idaho, a phosphorite that is composed of apatite shells of the brachiopod Orbiculoidea and minor amounts of apatite pellets and oolites and averages 33 percent P_2O_5 (85 percent apatite) extends for several kilometers along outcrop as a biostromal lens 0.5–4.2 meters thick (Sears 1955). This bed continues for much greater distances as a thin bed several centimeters to several meters thick (Sheldon 1963). Other thin, individual, synchronous beds of primary biogenic phosphorite deposited over several thousands of square kilometers also occur in the Phosphoria Formation. McKelvey (McKelvey et al 1959, p. 23) reported, for example,

> Throughout southeastern Idaho and adjacent areas the lowest bed of the Meade Peak is a thin phosphorite bed containing abundant fish scales, bones, and small nodules.

These primary biogenic phosphorite beds were deposited in oxygenated water in a moderate-energy environment and form the shallowest water and most shoreward element of the phosphorite facies (Sheldon 1963, p. 146). They grade into the pelletal and oolitic facies (Cressman &

Swanson 1964), which is deposited in a more basinward facies position. These biogenic phosphorite beds are ordinarily sandy and grade shoreward into thin, well-sorted sandstone beds that clearly were deposited in a relatively high-energy environment.

DIAGENETIC PRECIPITATION   Diagenetic precipitation of apatite has also been known for many years and has been well documented for many deposits. Many, but not all, phosphorites are cemented by apatite. Apatite commonly fills voids of fossils such as zooecia of bryozoans, the minute holes in echinoderm plates, axial canals of siliceous-sponge spicules, interiors of shells such as gastropods and foraminifera, and the space previously occupied by the soft organic matter of bones. Apatite also forms microconcretions or pellets in muds, as reported by Baturin (1978) and Burnett (1977) for recent muds, as well as by Sheldon (1957) for ancient mudstones. During diagenesis, apatite also replaces limestone and calcium carbonate shells and grains within phosphorites. An excellent review of the chemistry and petrology of diagenetic apatitic sediments is given by Baturin (1978), who credited Strakhov (1960) with pointing out that the driving force behind the diagenetic process is organic matter. Because of the decay of organic matter and sulfate reduction, as pointed out earlier, the chemistry of the pore waters changes. The concentrations of phosphorus typically reach 1–2 mg/l and have been recorded as high as 8–9 mg/l. At such high concentrations, the pore waters are substantially supersaturated with calcium phosphate, which begins to precipitate on material of different origin and composition. This precipitation depletes the phosphorus in the pore waters. He further pointed out that possibly the "indiscrimination" of calcium phosphate with respect to the composition of the centers of deposition is caused by the presence on the surface of the sediment particles of microcenters with high pH values due to alkalization, produced in particular by the action of bacteria, including sulfate-reducing ones, liberation of ammonia from organic matter, and solution of carbonates. He suggested that self-purging from apatite areas of nonphosphatic components with further diagenesis could increase P_2O_5 content to 20–32 percent. These nonphosphatic components would include coprolites, which originally contain only small percentages of phosphorus.

PRECIPITATION AT THE SEDIMENT-WATER INTERFACE OF APATITE   Precipitation at the sediment-water interface of apatite cannot be explained with present knowledge of the physical-chemical processes occurring in the ocean. Yet petrographic evidence clearly indicates that such precipitation has occurred in the past. Perhaps the most convincing evidence comes from sandstones that are mixtures of apatite oolites and quartz sand grains [see, for example, Figure 112*A* of Cressman & Swanson (1964)]. The oolites have nuclei of quartz sand or primary apatite shell fragments and show concentric layering indicative of accretionary growth. All evidence points to the oolite formation and deposition in a moderate-energy, oxygenated environment, and no evidence can be adduced to show a diagenetic history in muds to allow formation of the oolites. Stratigraphic evidence supports this petrographic interpretation (Cressman & Swanson 1964).

Such rocks grade basinward in the Phosphoria Formation into oolitic phosphorites with only a minor quartz-sand constituent, then grade into phosphorite which is rich in organic matter and formed of pellets with no obvious concentric banding and which commonly contains traces of very fine quartz sand in the matrix of the rock. Sheldon (1957) and Cressman & Swanson (1964) concluded that such slightly sandy pelletal phosphorite was formed in a quieter environment, as evidenced by the inclusions of clay and silt in the pellets, but was winnowed by currents strong enough to wash out the mud matrix, leaving only the sand residue and, of course, the much larger apatite pellets. These currents were not strong enough to move the apatite pellets, as indicated by lack of any cross bedding or ripple bedding. Cook (1967) concluded that the same process accounts for phosphorites of Ordovician age in central Australia. This type of phosphorite grades into organic-rich interlaminated mudstone and phosphorite, which in turn grades even farther basinward into nonphosphatic mudstone. Cressman & Swanson (1964) concluded that the phosphorites of the Phosphoria Formation in Montana could be divided into two broad types: those that were sorted and winnowed by currents and those that were not. This separation of facies is basically the same as Kazakov's (1937) platform and geosynclinal facies. The question arises that if the shoreward end of the phosphorite facies was deposited at the sediment-water interface, was the rest of the phosphorite so deposited also? Cressman & Swanson (1964) concluded that "although the evidence indicates that the pellets (of the nonsorted and winnowed type) were not transported but were formed at the site of accumulation, no evidence has been found to determine whether the pellets formed at the surface of the sediment or within the sediment during diagenesis."

A curious feature of many phosphorites is the formation of compound pellets, i.e. pellets made up of smaller pellets, some of which are broken. This structure, combined with concentric structure of pellets due to variations of organic matter and the evidence of winnowing of pellets that formed in quieter environments, indicates alternating conditions of lower and higher energy and variations in the rate of deposition of organic matter and perhaps variations between periods of oxidation and reduction. Also, thin, fossiliferous carbonates are often interbedded with organic-rich phosphorite. The carbonate rocks contain a benthonic fauna, indicating an oxygenated and more alkaline environment, and the organic-rich phosphorite indicates a reducing and less alkaline environment. For example, Yochelson (1968) gave evidence for benthonic fauna in the phosphatic shale of the Phosphoria Formation. Kolodny (1980) assembled geochemical evidence of apparently contradicting conditions indicating both a reducing and an oxidizing environment for the same rock. The cerium depletion typical of phosphate rock is also characteristic of open-ocean water and indicates a hydrogenous origin under oxidizing conditions: europium (II) will fit a calcium space in the apatite lattice, and its possible presence in apatite would suggest reduction; based on $\delta^{34}S$ data, seawater-derived sulfate occurs within the apatite lattice whereas sulfides coexist with phosphorites; both oxidized and reduced forms of uranium occur in phosphorites. He concluded that the variety is due to an episodic or sequential origin and attributed this to changing conditions from high and low sea-level stands. All of this evidence shows that the environment of deposition of many phosphorites was not constant but was fluctuating between an aerated, moderate-energy environment and a reducing, low-energy environment. Hundreds to thousands of such fluctuations must have occurred for each meter of sediment, a rate much too rapid to be accounted for entirely by sea-level changes.

The hypothesis of phosphorite formation at the sediment-water interface is enhanced when phosphorus sediment flux is taken into account, an idea suggested by W. C. Burnett (written communication 1980). In an area of 21,500 square kilometers in southeastern Idaho, the Meade Peak Phosphatic Shale Member of the Phosphoria Formation contains a more or less evenly distributed 1.46×10^{11} metric tons of phosphorus (P), calculated from data given in McKelvey et al (1953). B. R. Wardlaw (oral communication 1980) estimated on paleontologic evidence that the Meade Peak was deposited in 2–6 million years during a part of the Roadian Stage (or upper part of the Early Permian Epoch) of the Permian Period. Thus, the annual sediment flux of P during Meade Peak time was 113 to 340 $\mu g\ P/cm^2/yr$. This amount compares to a sediment flux of 0.3–28 $\mu g\ P/cm^2/yr$ in areas of modern upwelling in offshore Peru and Namibia (W. C. Burnett, written communication 1980). Thus the sediment flux is one to four orders of magnitude greater in the areas and times of maximum phosphorus sedimentation in the Permian of the Rocky Mountains of the United States than it is in the areas of maximum phosphorus sedimentation in the ocean today. This rate suggests that Permian apatite sedimentation required withdrawal of phosphorus available at the seafloor, and not just from phosphorus available in pore water in sediments during diagenesis. Also, if the apatite were formed only from phosphorus available in pore water, the amounts of pore water and mud required would be extremely large. For example, the Phosphoria Formation originally contained 0.74×10^{12} metric tons of P (McKelvey et al 1953), an amount more than five times that dissolved in the total ocean today. Other phosphogenic provinces are also large. For example, Davidson (1980) gave the reserves of Cretaceous to Eocene phosphate deposits in the North Africa–Middle East phosphogenic province as 48.3×10^9 metric tons of phosphate rock (about 20×10^9 tons of P). The total phosphorus amount deposited in rock of all grades of phosphorus content, whether they are reserves or not, has never been calculated for this province, but assuming a reserve:total phosphorus ratio the same as for the Phosphoria Formation, the total amount of phosphorus in the North African–Middle East (Gondwana) province would be twenty times that of the Phosphoria Formation. This figure may be too large, but does indicate that the Phosphoria deposit is not unique in its large size. Since gigantic amounts of pore water and mud are required to yield the known amount of phosphorus we may doubt the diagenetic origin of phosphorite in large deposits such as those of the North American–Permian and Gondwanan–Cretaceous phosphogenic provinces. Deciphering the sedimentological and paleogeographic origin and disposal of such large hypothetical quantities of mud is a formidable problem.

The formation of apatite pellets and oolites has been the subject of much controversy over the years. Some workers believe that apatite pellets form as accretionary particles in a manner similar to oolites, which are known to form by accretion, the difference being that pellets, which have no concentric structure, were formed during constant depositional conditions, whereas the oolites were formed during alternating depositional conditions.

An assumption made by some workers is that pellets and oolites must be rolled about by currents to give them their round shape. However, this need not be the case, as shown by the well-developed apatite oolites of insular phosphate deposits probably formed diagenetically beneath guano deposits and above solution surfaces on reef limestones, which were subject to downward movement and grinding as the solution surface of the limestone lowered, but were not washed by water currents. Also, the lack of ripple or cross bedding in laminar bedded marine phosphorites indicates that little if any rolling of pellets took place. Ancient marine pelletal phosphorites have a high degree of sorting. Sorting coefficients of eleven phosphorites from Wyoming ranged from 1.23 to 2.17, with all but two being below 1.45 (Sheldon 1963, p. 78). Similar results were found for Montana phosphorites by Cressman & Swanson (1964). The high degree of sorting is comparable to that of beach sands, yet the lamination indicates that the environment of phosphorite deposition had much less energy, leading to the conclusion that the size distribution of phosphorites is not due to mechanical sorting by currents but is a primary size distribution due to processes of accretionary growth.

Freas & Riggs (1965; also reported in Freas & Riggs 1968 and Riggs 1979a) proposed that the pellets of the Florida phosphate deposit were formed by erosion of an apatite mud or microsphorite and subsequent rounding abrasion as the sediment was being reworked. Well-developed apatite pellets form diagenetically in muds, as discussed earlier.

The coprolite theory is perhaps the oldest (Buckland 1829) and is still advocated by some as the predominant mode of formation of pellets (Bushinski 1964). A coprolite origin implies that either the coprolites were replaced with apatite after formation or that animals were ingesting apatite mud. The lack of trace fossils as well as the laminated structure of many pelletal phosphorites indicates a lack of bioturbation, which would seem to rule out formation by apatite-mud ingestion. Pellet formation by phosphatization of coprolites, on the other hand, requires that the coprolites, accumulated in a quiet environment, were then phosphatized diagenetically in muds to form apatite pellets and finally were concentrated by reworking, a multistage process that probably occurs (Baturin 1978) but that will not explain pellet formation of organic-rich, laminated, winnowed phosphorites that were never reworked.

Apatite mudstone has been reported by a number of workers, and it bears on the question of primary apatite sedimentation. Freas & Riggs (1965) postulated that apatite mudstone of the Miocene-Pliocene Florida deposits forms directly from seawater in very shallow water. Russell & Trueman (1971) agreed with Freas & Riggs in their interpretation of the apatite mudstone of the Cambrian deposits of Queensland, Australia; however, they noted that the apatite mudstone does not contain organic matter or sulfide minerals, which occur in the pelletal rocks of the same formation. On the other hand, de Keyser & Cook (1972) postulated that the apatite mudstones of Queensland were deposited from "phosphate-bearing surface and groundwaters moving into the intertidal or supratidal zone, becoming more alkaline and precipitating apatite within the sediments or more commonly at the surface where they were also subject to weathering. . . ." Weathering, which de Keyser & Cook believed occurred in the Cambrian, caused leaching of the trace elements that occur in the apatite lattice. It would appear that the Baturin hypothesis of diagenetic precipitation of apatite and subsequent reworking would not apply to apatite mudstones, as no reworking previous to the formation of the mudstone has been identified, and had it occurred it would have served to wash away the apatite mud. Thus, it seems difficult to attribute the formation of the apatite mud to anything other than direct precipitation from shallow water. It should be noted, however, that not all workers accept the sedimentational origin of the apatite mud; some think that intensive weathering is responsible for destroying normal sedimentary textures.

Concentration of Apatite

The concentration of apatite to form high-grade phosphorites is the final process of phosphogenesis. Three processes of concentration are important: winnowing, reworking, and weathering.

WINNOWING   Winnowing is a process of selective size sorting by wind action which is also applied to water-current activity. It washes away smaller and lighter particles leaving larger and heavier ones behind. Winnowing can be a continuous process of the movement of fine-grained clastic sediments past slowly accumulating larger grains or it can be an intermittent process of alternating high- and low-energy environments of deposition, causing finer grained clastics to be washed out during the higher energy stages. Winnowing does not imply displacement of the larger grains. This process has been suggested for concentrating apatite pellets by a number of workers including Sheldon (1957), Altschuler et al (1964), Cressman & Swanson (1964), Bushinski (1964), and Cook (1967).

REWORKING   Reworking is a different process from winnowing. It implies removal or displacement of grains from their place of origin and incorporation in recognizable form in a younger formation. This process has been suggested as a mechanism in concentration of phosphorite deposits. Baturin (1971) used it to explain the erosion of slightly phosphatic muds containing diagenetic apatite pellets and the selective sorting of the fine and coarse fractions to yield high-grade phosphorite deposits in one place while mudstones are deposited elsewhere.

Baturin proposes falling sea level as the driving mechanism of reworking, which would make the phosphorite concentrates a part of a regressive sequence. Although it is true that some phosphorites are regressive deposits, many others are not. In fact, it long has been recognized that phosphorites are commonly deposits of transgressive sequences. This relationship has been studied in detail for the Phosphoria Formation (Sheldon 1957, 1963, McKelvey et al 1959, Cressman & Swanson 1964), in which two transgressive-regressive cycles are recognized. Phosphorite is associated with both the transgressive and regressive phases of both cycles but is better developed in the transgressive cycles. Much the same relationship was reported by Eganov (1979) for the Karatau phosphorites of Cambrian age of Kazakhstan and by Semeykin and others (in Eganov 1979) for the Hubsugul phosphorites of late Precambrian age of northern Mongolia. In addition, paleogeographic analysis indicates that most of this kind of phosphorite was outer continental shelf deposition, which in general is continuous and does not contain major erosional phases. In any event, the phosphorites are deposited in layers at the laminae and bed level of stratigraphic unit, whereas the transgressions are recorded by relations between units at the member and formation level of stratigraphic unit. Thus the transgressions and regressions were caused by processes of much larger magnitude and effect than processes that were responsible for concentration of the apatite pellets. In the Phosphoria Formation the deeper-water apatite pellets probably were concentrated by winnowing. Winnowing probably is due to storm-induced variations in wave base rather than variations due to changing sea level. It is probable also that upwelling was variable both in time and space, causing variation in the sedimentologic regime.

WEATHERING   Weathering causes major changes in phosphorites and can serve to concentrate the apatite. Some phosphate deposits such as the so-called "brown rock" of Tennessee and Alabama are residual deposits formed in the Holocene by subaerial weathering and leaching of phosphatic limestones of the Middle Ordovician Nashville Group and Upper Ordovician Maysville Stage. Phosphorites of the land-pebble type in Florida are highly weathered, first by removal of carbonate, causing a residual enrichment phosphate, and finally by lateritic weathering, forming aluminum and iron phosphates (Altschuler 1973). Weathering usually leaches away the more soluble minerals such as calcite and oxidizes organic matter, increasing the percentage of phosphorus in the rock, but, as in the case of the laterite weathering of the Florida deposits as well as the Cretaceous deposits of Colombia, phosphorus is removed, causing a substantial decrease in the phosphorus content of the rock (J. B. Cathcart 1975, and written communication 1980). In addition, trace-element concentrations are usually changed and can be either increased or decreased. Weathering of phosphorites that occurs shortly after their deposition, such as postulated by de Keyser & Cook (1972) for the microsphorite of Cambrian age in Queensland, Australia, causes changes in the rock that might be confused with depositional characteristics. The light color and low organic-matter content of many phosphorites, such as those in the Middle East and Australia, may be caused by recent post-depositional weathering rather than original sedimentational processes. Without subsurface information this problem may be difficult to resolve.

In summary, weathering and its submarine counterpart, halmyrolosis, can play an important role in the concentration of apatite in some deposits.

GEOGRAPHY AND PALEOGRAPHY OF MARINE PHOSPHORITES

Phosphorites on the seafloor are distributed in water depths generally less than 2000 meters, although some phosphorites have been dredged from deeper waters. On the continental shelves and the upper part of the continental slopes, phosphorite occurs at depths generally less than about 300 meters. In pelagic waters, the phosphorites occur on flat-topped seamounts or guyots at water depths mostly about 2000 meters but as deep as 4000 meters. The first attempt to systematically show phosphorite distribution on the seafloor was by McKelvey in 1963 and in more detail in 1969 (McKelvey & Wang 1969). More recent compilations are by Baturin & Bezrukov (1971), Garrand (1977), and Baturin (1978). Most seafloor phosphorites have been dated paleontologically and geochemically and have been shown to range from Holocene to Cretaceous in age (Baturin 1978), with the majority of the continental shelf deposits being Neogene and the majority of seamount deposits being Paleogene and Cretaceous. These deposits with a few exceptions occur at latitudes less than 45°.

Phosphorites are found on the western continental shelves of North and South America and Africa in the modern trade-wind belts where strong upwelling currents occur. Phosphorites occurring on other continental shelves are not so obviously related to modern ocean currents. The Tertiary phosphorites of the southeastern United States and Blake Plateau are where the modern Gulf Stream segment of the north Atlantic gyre impacts on the continental shelf, although an inshore countercurrent flowing from the north complicates this simple picture. The Venezuelan shelf deposits are in a comparable position further south. The phosphorites on the southeastern continental shelf of Argentina and on the Chatham Rise southeast of New Zealand occur where offshoots from the modern Antarctic Circumpolar Current are diverted to the north and dynamically well up over the continental shelves; however, the relationship of these phosphorites to oceanographic conditions at the time of their formation is not known. It is curious that all of the continental-shelf phosphorites other than those in the trade-wind belt upwelling zones lie along major paths of deep-ocean circulation presented by Stommel (1958), including the phosphorites of Florida, Venezuela, Argentina, New Zealand, and the Agulhas Banks of South Africa. This location, however, may be only a reflection that these phosphorites occur on the eastern continental shelves, which coincidentally are also just offshore the site of deep-ocean currents. They may owe their origin to dynamic upwelling associated with the shallower current systems. In any event, the relation between oceanic paleocurrents and the east-coast type of phosphorites remains enigmatic.

The phosphorites on seamounts present a different picture [see the excellent review by Baturin (1978)]. They are dominantly composed of phosphatized limestone, although some are made up of crusts on and amygdule fillings in basalt blocks. Although surveying is far from complete, major clusters of seamount phosphorite are known to occur in a region in the North Pacific between 150°E–170°W and 15°–35°N and in a region in the Tasman Sea near 30°S and 156°E. The Tasman Sea guyot phosphorites have been dated as Pliocene-Pleistocene; the North Pacific guyot shallow-water limestones, which have been phosphatized, are dated paleontologically as Late Cretaceous to Eocene. In the opinion of Heezen et al (1973), there were two stages of phosphatization on Marcus-Necker Ridge in the Pacific. In Late Cretaceous time in shallow-water conditions there was phosphatization of rudist limestones and bioclastic calcarenites, and in Eocene time in deep-sea conditions there was phosphatization of the nannoplanktonic foraminiferal limestones. After the Eocene, phosphatization apparently was not resumed, as evidenced by worm tracks in the Eocene phosphorites filled with calcareous ooze carrying a fauna of Quaternary foraminifera. From this and other data, Baturin considered that phosphatization of limestones in the northern Pacific seamounts occurred shortly after the limestones were formed. The original positions of the seamounts when the limestones were formed were close to the equator (Sheldon 1980), which led to speculation that the seamount phosphorites of the northern Pacific could be associated with the zone of equatorial divergence and its associated high productivity. Results of a study of the deep-sea phosphorite of the Annan seamount in the eastern equatorial Atlantic Ocean (Jones & Goddard 1979) fit this general hypothesis in that Paleocene-Eocene shallow-water limestones were phosphatized in Eocene time while the sediments were still in shallow water. Jones & Goddard believe that the source of the phosphorus was either guano or seawater, but in either case was related to equatorial upwelling. Thus the seamount phosphorites may be genetically related to equatorial insular guano and phosphorite deposits, which also have been shown to be related to the equatorial divergence zone (Hutchinson 1950). Baturin (1978) reported that the presence of insular phosphorites on Christmas Island in the Indian Ocean served as grounds for prospecting for phosphorites in the seamounts south of the island, where they actually were found (Bezrukov 1973). This and a similar occurrence in the Caribbean Sea

(Marlowe 1971) suggest such a probable genetic relationship. Thus, the seamount phosphorites could simply be drowned insular phosphorites; however, the geologic history of many seamounts does not support this as a general hypothesis. In addition, the petrographic character of the two types of deposits usually is different. Another factor in the origin of seamount phosphorite is the depth of maximum phosphorus concentration in the ocean, which in today's ocean is 250–2000 m, averaging about 1000 m (Gulbrandsen & Roberson 1973), and which corresponds roughly with the oxygen minimum zone (Redfield et al 1963). Fischer & Arthur (1977) postulated past periods of expanded oxygen minimum zones extending as deep as several thousand feet. The oxygen minimum zone is best developed in regions of upwelling ocean currents where the organic productivity is high. Thus, the phosphatization of seamount limestone may correspond to periods of expanded oxygen minimum zones in areas of equatorial divergence or other zones of high productivity. Cenozoic episodes of equatorial high productivity, judged by accumulation rates of equatorial biogenic silica, occurred in the middle Eocene and middle and late Miocene (Leinen 1979), and correspond with phosphogenic episodes discussed earlier. Jones & Goddard (1979) reached a similar conclusion for the early Tertiary in the Atlantic Ocean. A decrease of biologic productivity occurred during the Paleogene and perhaps was caused (Heath et al 1976) by the gradual constriction of the Tethys equatorial current system due to the northward movement of South America during Paleogene time (Berggren & Hollister 1974). One can further speculate that this reduced current may have been a factor in decreased phosphatization of Pacific seamounts. It is clear that research in this field will be exciting and highly relevant to the geologic history of phosphogenesis.

An area that one might suppose should contain phosphorites is the continental shelf area of Antarctica. Very strong divergent upwelling occurs in the region, yielding large phytoplankton and zooplankton productivity during the year except for winter. Diatomaceous ooze is abundant on the seafloor beneath this area of upwelling. However, despite much marine geologic work in the region, no phosphorite has been found. Bottom currents of cold, saline water which flow downward and northward perhaps inhibit phosphorite deposition.

Ancient phosphorites have been shown to have much the same paleogeographic distribution as present seafloor phosphorites. Kazakov (1937) made the original observation that phosphorites occurred on continental shelves, and this observation has been supported by many paleogeographic studies since. The North Africa–Middle East phosphogenic province of Paleogene-Cretaceous age is related to the Tethys Sea and the continental shelves of the Africa–Arabia–Middle East craton to the south (Sheldon

1964b, Slansky 1980, Davidson 1980). The Phosphoria Formation of Permian age was related to the continental shelf (Cordilleran geosyncline) of the North American craton (McKelvey et al 1953). The Beetle Creek phosphorite of Cambrian age in Queensland, Australia, was related to the continental shelf (Tasman geosyncline) of Gondwana (Cook & Shergold 1979). These are only a few examples of many. The paleo-oceanographic setting is more difficult to determine because of lack of paleolongitudinal data for pre-Cretaceous time. Even so, paleocontinental reconstructions indicate that the phosphogenic provinces probably occur on the margins of major oceans (Freas & Eckstrom 1968, Ziegler et al 1979).

The paleolatitudes of ancient phosphorites have a range similar to that of the latitudes of phosphorites too young to have had their positions greatly altered by plate-tectonic movements (Sheldon 1964a, Cook & McElhinny 1979). The limitation of phosphogenesis to low latitudes must have an explanation. One is tempted to account for the distribution of phosphorites on the basis of the distribution of upwelling currents alone, but, in the present ocean, high-latitude Antarctic upwelling and consequent high biologic productivity has not caused the deposition of phosphatic sediments, judging from available information. The world oceanographic maps of water temperature at 100- and 200-meter depths (Ministry of Defense, USSR Navy 1976, Ministry of Defense, USSR Navy 1977) when compared to areas of seafloor phosphorites show that both continental and seamount phosphorites occur where cold tongues of low pH, phosphorus-rich water extend into warm-water areas at the same depths. This comparison, however, depends on the analogy between present currents and Tertiary currents which were responsible for formation of the Tertiary phosphorites. This relationship, combined with the low paleolatitude distribution of phosphorites, suggests that temperature increase is a factor in phosphogenesis. Exceptions to this observation are the southeastern United States shelf phosphorites and the Tasman Sea seamount phosphorites, which are not associated with present cold-water upwelling currents.

In the present ocean, circumglobal equatorial currents do not exist because of the Isthmus of Panama blockage between the Atlantic and Pacific and the blockage between the Atlantic and Indian ocean in the Mediterranean–Middle East region. However, in Jurassic to Paleogene times, the circumglobal Tethyan current existed, creating an Upper Cretaceous–Eocene phosphogenic province on the northern continental shelf of the Arabian–North African–South American paleocontinent, at paleolatitudes of about 10°–20°N (Sheldon 1964a). Cook & McElhinny (1979) have postulated that east-west seaways were a general paleo-oceanographic type of phosphogenic province. They further postulated that this type probably existed in the Lower Paleozoic, accounting for the Cambrian phosphogenic province of Asia and Australia, but they did not emphasize circumglobal currents as essential parts of the paleo-oceanography. Sheldon (1980) speculated that equatorial divergence combined with circumglobal equatorial currents along east-west continental shelves would give the paleogeographic and paleo-oceanographic configuration necessary for the Eocene-Cretaceous phosphogenic province of Arabia–North Africa–northern South America. The Cambrian phosphogenic province of Asia and Australia may be due to a similar paleocontinental–paleo-oceanographic configuration (Ziegler et al 1979); however, the sparse data available make the Lower Paleozoic and older reconstructions extremely speculative and many other different reconstructions are possible (for example, see Boucot & Gray 1979).

EPISODICITY OF PHOSPHOGENESIS

Gimmel'farb (1958) first showed the world distribution of phosphorites according to their geologic age. Strakhov (1960) revised this distribution with additional data, and Bushinski (1966a) discussed it and added Asian data. Cook & McElhinny (1979) incorporated other new information and have published the most up-to-date age distribution. These empirical data show that phosphorite deposits occur preferentially in certain intervals of the Phanerozoic, including Cambrian, Ordovician, Permian, Upper Jurassic–Lower Cretaceous, Upper Cretaceous–Lower Tertiary, and the Miocene. Several Proterozoic intervals also preferentially contain phosphorites. The question immediately arises as to whether this distribution is caused by some sort of geologic sample bias or actually reflects different rates of deposition in geologic times. Because new data have tended to reinforce instead of diminish the variation between periods, episodicity of phosphogenesis has been generally accepted.

The cause of episodicity is not clear. Cook & McElhinny (1979), in a thorough review of the subject, considered correlations of phosphogenic episodes with other global epidosdes, including glacial activity, evaporite deposition, iron-ore deposition and orogenic activity, concluding that the only correlation was between iron ore and phosphorite deposition. They postulated that the underlying cause of phosphogenic episodicity was the changing global paleogeography and paleo-oceanography due to plate-tectonic processes. For phosphogenesis to occur the continents had to be in lower latitudes, and north-south seaways had to be wide enough or east-west seaways extensive enough to allow for the required oceanic circulation. Strakhov (1960) related phosphorites to marine transgressions, a conclusion supported by many other workers. Schlanger & Jenkyns (1976) pointed out that transgressions correlate with both the deposition of organic carbon-rich black shales in the Cretaceous and the expansion of the oxygen minimum zones. Furthermore, these black shales are generally rich in P_2O_5 (Didyk et al 1978). Piper & Codispoti (1975), as discussed earlier, attributed phosphogenesis to a process of denitrification and concluded that episodicity of phosphogenesis could be caused by episodicity of denitrification associated with expanded oxygen minimum layers. This idea was further developed by Arthur & Jenkyns (1980), who related it specifically to Cretaceous and Miocene oceans. Fischer & Arthur (1977) have advanced a general hypothesis to explain episodicity of many events in the geologic record of the pelagic realm for the last 100 million years. They gave evidence for long episodes of warm, high-level, stable oceans alternating with shorter episodes of cold, low-level, rapidly circulating oceans. They postulated expanded oceanwide oxygen minimum layers during the episodes of warm, high-level oceans. Following the model of Berger & Roth (1975), the Fischer & Arthur hypothesis states that in times of high-level warm seas, the increased turnover time of deeper waters leads to an increase in their phosphate content and possibly leads to apatite saturation. Upwelling during these times, although diminished, would deliver phosphorus-rich water to the shallow-water sites of deposition. This idea is corroborated to some extent by the study of Burnett & Veeh (1977), who suggested a correlation between episodes of high sea level and phosphate deposition for Quaternary phosphorite deposits of the continental shelf of Peru. This idea seems reasonable, considering that additions of phosphorus to the ocean would double the phosphorus content in about 100,000 years if no phosphorus were withdrawn. Sheldon (1980), following these ideas further, postulated that (a) episodes of phosphogenesis occur at the onset of episodes of oceanic vertical mixing after episodes of oceanic stability, during which the phosphorus content of the deep ocean had built up to high levels, (b) the major phosphogenic episodes of the Cretaceous to early Tertiary are due to equatorial upwelling at the time of high-level, warm seas, and (c) the major phosphogenic episodes of the Cambrian, Ordovician, and Permian Periods and the Miocene Epoch are due to trade-wind belt upwelling at the time of transition from the high-level, warm oceans to low-level, cold oceans related to glacial episodes.

Episodicity, then, appears to be a process caused by a number of factors including paleo-oceanography and paleogeography, relative sea level, paleocurrents, and paleoclimate. The underlying cause of all this would appear to be plate-tectonic processes that reconfigure oceans and continents and change ocean basin geometry and land/sea surface-area ratios.

PALEOECOLOGY OF MARINE PHOSPHORITE–BEARING FORMATIONS

The genetic relation between phosphogenesis and biologic activity of the oceans is clear, as discussed above. This relation goes even further when one analyzes the paleoecology and paleobiogeography of phosphorite-bearing sequences. Siliceous sediments are common lithologic associations of marine phosphorites. They are mainly made up of siliceous sponges and their spicules in the Paleozoic sequences and of diatoms in Cretaceous and younger sequences, and owe their origin to the same oceanic current systems that supply silica as well as phosphorus and nitrogen. Fossil fish, which were part of the life chain in areas of upwelling, are extremely common in most phosphorites. For example, they are one of the most abundant fauna of the Phosphoria Formation (Yochelson 1968). A carbonate-rock facies in the Park City Formation exists shoreward of the phosphorite and chert facies of the Phosphoria Formation, where the absence of warm-water fauna such as corals and fusilinids is matched by the abundance of cold-water fauna, giving independent evidence of upwelling ocean water (Wardlaw 1980). The basin facies of phosphorite-bearing sequences commonly contain pelagic fauna, such as ammonites, foraminifera, dinoflagellates (Fauconnier & Slansky 1980), and diatoms.

PROBLEMS OF ANCIENT PHOSPHORITE PETROLOGY

A number of problems of phosphorite petrology are presently perceived and are orienting current research. Many of these have been discussed above and some of the more important ones, phosphorite sedimentation, paleo-oceanography and paleoclimates, paleogeographic reconstructions, phosphogenic episodes, and the role of plate tectonics and volcanism in phosphogenesis, are summarized below.

Phosphorite Sedimentation   Researchers are divided on the question of whether predominant ancient phosphorite sedimentation is primary at the sediment-water interface or early diagenetic within the sediments. One school says that phosphorite sedimentation was not primary because it cannot be (from geochemical evidence), and the other says that it can be primary because it was (from petrographic evidence). As with the shallow-water dolomite petrology arguments, a solution requires more laboratory, field, and marine work.

Paleo-Oceanography and Paleoclimates   The present ocean physical and chemical system is being used to interpret, by analogy, ancient oceans.

In view of the hypothesized episodicity of phosphogenesis, such a philosophic procedure is scientifically dangerous. Further development of physical and chemical oceanography principles and their application to ancient oceans as defined by paleo-oceanographic studies is required. A major issue is the phosphorus concentration of the deep ocean; is it constant over time or is it variable? Another important consideration is the change of sea level over time. With regard to paleoclimates, which are one of the driving forces of paleo-oceanographic processes, more must be learned about the episodicity of glaciation and its causes, as well as about the episodicity of globally equitable climates.

Paleogeographic Reconstructions   The current efforts to reconstruct paleogeography on plate-tectonic principles are of great value to phosphorite petrology, and studies of phosphogenic provinces need to be integrated with paleogeographic reconstructions. For many important phosphogenic provinces such studies conflict. For example, the Cambrian phosphorites of southwestern China, southern USSR, and Mongolia occur within the Asian landmass, and present paelogeographic intracontinental sea interpretations (Bushinski 1966a) conflict with paleogeographic requirements of phosphogenesis. Also, work on paleogeographic and paleo-oceanographic reconstructions will probably lead to new discoveries of phosphorite.

Phosphogenic Episodes   A better definition of the existence of phosphogenic episodes and research into the underlying geologic causes are needed. Evidence from both onshore ancient continental shelf phosphorites and offshore continental shelf and seamount phosphorites needs to be collected. The roles of glaciation, sea-level fluctuations, variations of phosphorus supply to the ocean, and variations of oceanic processes need to be defined.

Plate-Tectonic Role in Phosphogenesis   Plate-tectonic processes must play a large but presently not well-understood role in phosphogenesis. A part of the role must be in changing the shapes and distributions of ancient oceans and continents, but another part may be in changing ocean-basin configuration and thereby the global ratio of land to ocean, due not only to marine transgressions but also to continent-continent subduction.

Role of Volcanism in Phosphogenesis   The volcanic contribution of phosphorus to the ocean is not satisfactorily resolved, partly because of the paucity of chemical data on submarine hydrothermal springs. However, an indirect role of volcanism could be that of altering ocean basin geometry and concomitant sea-level changes. Hays & Pitman (1973) argued that the global Cretaceous transgression, which as noted previously may be correlated with expanded oxygen minimum zones (Fischer & Arthur 1977, Schlanger & Jenkyns 1976), was caused by an increase in spreading rate at the world mid-ocean ridge system during Cretaceous time. Further, Schlanger et al (1980) point out that the widespread mid-plate volcanism in the Pacific Basin in Cretaceous time, about 110 to 70 m.y. B.P., could have caused the Cretaceous rise in sea level that greatly altered the shallow shelf sea:deep ocean basin ratio in the world ocean.

These and other problems will determine directions of future research on marine phosphorites. This research is being stimulated by the major advances in marine geology, oceanography, and plate tectonics and promises many advances in phosphorite petrology.

ACKNOWLEDGMENTS

I would like to thank colleagues who reviewed this manuscript. They include Z. S. Altschuler, J. B. Cathcart, W. C. Burnett, P. N. Froelich, R. A. Gulbrandsen, and V. E. McKelvey. Burnett was particularly helpful in trying to keep me on firm ground in the marine geology and geochemistry sections of the paper, although any foundering is of my own doing. Correspondence and discussions (as well as arguments) with many scientists about oceanic and phosphogenic processes have been instrumental in stimulating this review. They include W. C. Burnett, G. I. Bushinski, P. J. Cook, R. A. Gulbrandsen, Y. Kolodny, S. R. Riggs, S. O. Schlanger, J. I. Tracey, Jr., B. R. Wardlaw, and a number of other colleagues.

Apologies must be made to the researchers who have published in other than the English language. Because of my language limitations I have certainly not given equitable credit to non-Anglophone researchers, particularly the Russians. The translations of reviews by Bushinski (1966a) and Baturin (1978) and the reviews by Kolodny have helped greatly to gain insight into the Russian literature.

Finally, this review on ancient marine phosphorites would have been much more difficult to have written without the scientific exchange between phosphorite geologists throughout the world made possible by Project 156 of the International Geological Correlations Programme.

Literature Cited

Altschuler, Z. S. 1973. The weathering of phosphate deposits: geochemical and environmental aspects. In *Environmental Phosphorus Handbook*, ed. E. J. Griffith, A. Beeton, J. M. Spencer, D. T. Mitchell, pp. 33–96. New York: Wiley

Altschuler, Z. S., Cathcart, J. B., Young, E. J. 1964. *Geology and geochemistry of the Bone Valley Formation and its phos-*

phate deposits. Presented at Ann. Meet. Geol. Soc. Am., Miami Beach, Florida, 1964. 68 pp.

Anderson, R. N., Hobart, M. A., Langseth, M. G. 1979. Geothermal convection through oceanic crust and sediments in the Indian Ocean. *Science* 204:828–32

Arthur, M., Jenkyns, H. 1980. *Significance of rock phosphate and other chemical sedi-*

ments in the Cretaceous and Miocene oceans. Presented at Int. Geol. Congr., Paris. In press

Atlas, E. L. 1975. *Phosphate equilibria in seawater and interstitial waters.* PhD thesis. Oregon State Univ., Corvallis, Oreg. 154 pp.

Baturin, G. N. 1971. States of Phosphorite formation on the ocean floor. *Nature Phys. Sci.* 232(29):61–62

Baturin, G. N. 1978. *Phosphorites.* P. P. Shirshow Inst. Oceanol., Acad. Sci. USSR, Moscow: Nauka. 232 pp. (In Russian)

Baturin, G. N., Bezrukov, P. L. 1971. Phosphorite on the sea floor. In *Istoriya Mirovog. Okeana (History of the World Ocean)*, pp. 237–58. Moscow: Akad. Nauk SSSR (Incl. Engl. summary)

Berger, W. H., Roth, P. H. 1975. Oceanic micropaleontology: progress and prospect. *Rev. Geophys. Space Phys.* 13(3):561 –635

Berggren, W. A., Hollister, C. D. 1974. Paleogeography, paleobiography and the history of circulation in the Atlantic Ocean. In *Studies in paleooceanography*, ed. W. W. Hay. *SEPM Spec. Publ.* 20:126–86

Bezrukov, P. L. 1973. Principal scientific results of the 54th voyage of the Vityaz' in the Indian and Pacific Oceans (Feb.– May 1973). *Oceanology* 13(5):761–66

Blackwelder, E. 1916. Geological transformations of phosphorus. *Geol. Soc. Am. Bull.* 27:47 (Abstr.)

Boucot, A. J., Gray, J. 1979. Epilogue: A Paleozoic Pangaea? In *Historical Biogeography, Plate Tectonics and the Changing Environment*, ed. J. Gray, A. J. Boucot, pp. 465–82. Corvallis, Oreg: Oregon State Univ. Press

Brodskaya, N. G. 1974. The role of volcanism in the formation of phosphorites. Moscow: Nauka

Broecker, W. S. 1974. *Chemical Oceanography.* New York: Harcourt 214 pp.

Buckland, W. 1829. On the discovery of coprolites, or fossil faeces, in the Lias at Lyme Regis, and in other formations. *Geol. Soc. London Trans.* 2d ser. 3:223–38

Burnett, W. C. 1977. Geochemistry and origin of phosphorite deposits from off Peru and Chile. *Geol. Soc. Am. Bull.* 88:813–23

Burnett, W. C. 1980a. *Apatite-glauconite associations off Peru and Chile: paleooceanographic implications.* Presented at Geol. Soc. London, 1980, London. In press

Burnett, W. C. 1980b. Oceanic phosphate deposits. *Proc. Fert. Raw Mater. Resour. Workshop*, ed. R. P. Sheldon, W. C. Bur-

nett, pp. 119–44. Honolulu, Hawaii: East-West Cent., Resource Syst. Inst.

Burnett, W. C., Sheldon, R. P., eds. 1979. *Report of the marine phosphatic sediments workshop.* Honolulu, Hawaii: East-West Center, Resource Syst. Inst. 65 pp.

Burnett, W. C., Veeh, H. H. 1977. Uranium-series disequilibrium studies in phosphorite nodules from the west coast of South America. *Geochim. Cosmochim. Acta* 41:755–64

Bushinski, G. I. 1964. On shallow water origin of phosphorite sediments. In *Deltaic and Shallow Marine Deposits*, ed. L. M. J. U. van Stratten, pp. 62–70. Amsterdam: Elsevier

Bushinski, G. I. 1966a. *Old phosphorites of Asia and the genesis.* ISR Prog. Sci. Transl. Jerusalem, 1969. 266 pp.

Bushinski, G. I. 1966b. The origin of marine phosphorites. *Litol. Polezn. Iskop.* 3:23–48. Engl. trans. *Lithol. Miner. Resour.* 3:292–311

Cathcart, J. B. 1968a. Florida-type phosphate deposits of the United States—origin and techniques for prospecting. In *Semin. on Sources of Miner. Raw Mater. for the Fert. Ind. in Asia and the Far East. Proc. UN ECAFE Miner. Res. Dev. Ser.* 32:178–86

Cathcart, J. B. 1968b. Phosphate in the Atlantic and Gulf Coastal Plains. In *Proc. 4th Forum of Geol. of Ind. Miner.*, ed. L. F. Brown, Jr., pp. 23–34. Austin: Univ. Texas Press

Cathcart, J. B. 1975. Phosphate fertilizer materials in Columbia—imports, uses and domestic supplies. *US Geol. Surv. Res.* 3(6):659–63

Cook, P. J. 1967. Winnowing—an important process in the concentration of the stairway sandstone (Ordovician) phosphorites of Central Australia. *J. Sediment. Petrol.* 37(3):818–28

Cook, P. J. 1976. Sedimentary phosphate deposits. In *Handbook of Strata-Bound and Stratiform Ore Deposits*, ed. K. H. Wolf, Chap. II, 7:505–35. New York: Elsevier

Cook, P. J., McElhinny, M. W. 1979. A re-evaluation of the spatial and temporal distribution of phosphorites in the light of plate tectonics. *Econ. Geol.* 74:315–30

Cook, P. J., Shergold, J. H., eds. 1979. *Proterozoic-Cambrian Phosphorites.* Canberra: Canberra Publ. and Print. 206 pp.

Cornet, M. F. L. 1886. On the upper Cretaceous series and the phosphatic beds in the neighbourhood of Mons (Belgium). *Geol. Soc. London Q. J.* 42:325–40

Cressman, E. R., Swanson, R. W. 1964.

Stratigraphy and petrology of the permian rocks of southwestern Montana. *US Geol. Surv. Prof. Pap. 313-C*, pp. 275–569

Davidson, D. F. 1980. *Phosphate Deposits, and the Tethyan Trough, Africa, and the Middle East.* In press

de Keyser, F., Cook, P. J. 1972. Geology of the Middle Cambrian phosphorites and associated sediments of northwestern Queensland. *Dept. of Natl. Dev. Bur. Miner. Res. Geol. Geophys. Bull.* 138:79

Didyk, B. M., Simoneit, B. R. T., Brassell, S. C. Eglinton, G. 1978. Organic geochemical indicators of paleoenvironmental conditions of sedimentation. *Nature* 272(5650):216–22

Doyle, L. J., Blake, N. J., Woo, C. C., Yevich, P. 1978. Recent biogenic phosphorite: concretions in mollusk kidneys. *Science (AAAS)* 199(4336):1431–33

Dzotsenidze, G. S. 1969. The role of volcanism in the formation of sedimentary rocks and ores. *Izd. Nedra* Moscow. 344 pp.

Eganov, E. A. 1979. The role of cyclic sedimentation in the formation of phosphorite deposits. In *Proterozoic-Cambrian Phosphorites*, ed. P. J. Cook, J. H. Shergold, pp. 22–25. Canberra: Canberra Publ. and Print. (Abstr.)

Fauconnier, D., Slansky, M. 1980. Relations entre le développement des dinoflagelles et la sédimentation phosphatée du bassin de Gafsa (Tunisie). *Bur. Rech. Geol. Min.* In press

Fischer, A. G., Arthur, M. A. 1977. Secular variations in the pelagic realm. In *Deep-Water Carbonate Environments*, ed. H. E. Cook, P. Enos, *SEPM Spec. Publ.* 25:19–50

Fisher, O. 1873. On the phosphatic nodules of the Cretaceous rocks of Cambridgeshire. *Q. J. Geol. Soc. London* 29:52–62

Freas, D. H. 1968. Exploration for Florida phosphate deposits. In *Semin. on Sources of Miner. Raw Mater. for the Fert. Ind. in Asia and the Far East. Proc. UN ECAFE Miner. Res. Dev. Ser.* 32:187–200

Freas, D. H., Eckstrom, C. L. 1968. Areas of potential upwelling and phosphorite deposition during Tertiary, Mesozoic, and Late Paleozoic time. In *Semin. on Sources of Miner. Raw Mater. for the Fert. Ind. in Asia and the Far East. Proc. UN ECAFE Miner. Res. Dev. Ser.* 32:228–38

Freas, D. H., Riggs, S. R. 1965. *Stratigraphy and sedimentation of phsophorite in the central Florida phosphate district.* Presented at Ann. Meet. AIME, Chicago, 1965. 17 pp.

Freas, D. H., Riggs, S. R. 1968. Environments of phosphorite deposition in the central Florida phosphate district. In *Forum on Geology of Industrial Minerals. Proc. Univ. Texas Bur. Econ. Geol. Austin*, pp. 117–28

Froelich, P. N., Bender, M. L., Luedtke, N. A., Heath, G. R., DeVries, T. 1980. The Marine Phosphorus Cycle. *Am. J. Sci.* Submitted

Garrand, L. J. 1977. Ocean characteristics, occurrences, origin recovery. In *Offshore Phosphorite World Occurrences*, pp. 1-1–7-3. Salt Lake City: Garrand Corp.

Gibson, T. G. 1967. Stratigraphy and paleoenvironment of the phosphatic Miocene strata of North Carolina. *Geol. Soc. Am. Bull.* 78(5):631–49

Gimmel'farb, B. M. 1958. Regularity of the tectonic distribution of phosphorite deposits in the USSR. *Zakonomern. Razmeshcheniya Polezn. Iskop.* 1:487–516 (In Russian)

Grabau, A. W. 1919. Prevailing stratigraphic relationships of the bedded phosphate deposits of Europe, N. America and N. Africa. *Geol. Soc. Am. Bull.* 30:104 (Abstr.)

Gulbrandsen, R. A. 1969. Physical and chemical factors in the formation of marine apatite. *Econ. Geol.* 64(4):365–82

Gulbrandsen, R. A., Roberson, C. E. 1973. Inorganic phosphorus in seawater. In *Environmental Phosphorus Handbook*, ed. E. J. Griffith, A. Beeton, J. M. Spencer, D. T. Mitchell, pp. 117–40. New York: Wiley

Hallam, A. 1978. Secular changes in marine inundation of USSR and North America through the Phanerozoic. *Nature* 269 (5631):769–72

Hays, J. D., Pitman, W. C. III. 1973. Lithospheric plate motion, sea level changes and climates and ecological consequences. *Nature* 246:18–22

Heath, G. R., Moore, T. C., Jr., van Andel, Tj. H. 1976. Carbonate accumulation and dissolution in the equatorial Pacific during the past 45 million years. In *Fate of Fossil CO_2 in the Oceans* (Marine Sci. Ser., Vol. 6), ed. N. R. Anderson, A. Malahoff, pp. 627–41. New York:Plenum

Heezen, B. C., Mathews, J. L., Catalano, R., Natland, J. J., Coogan, A., Tharp, M., Rawson, M. 1973. Western Pacific guyots. *Init. Rep. DSDP, Leg 20.* Washington, DC.

Hite, R. J. 1978. Possible genetic relationships between evaporites, phosphorites, and iron-rich sediments. *Mt. Geol.* 14(3): 97–107

Hutchinson, G. E. 1950. The biogeochemistry of vertebrate excretion. *Bull. Am. Mus. Nat. Hist.* 96:172–86

Jones, E. J., Goddard, D. A. 1979. Deep-sea

phosphorite of Tertiary age from Annan seamount, eastern equatorial Atlantic. *Deep-sea Res.* 26(12):1363–79

Kassin, N. R. 1925. Phosphorites of the northern Vyatka district. *Vestn. Geol. Kom. No. 5*

Kazakov, A. V. 1937. The phosphorite facies and the genesis of phosphorites. In *Geological investigations of agricultural ores.* Leningrad, Sci. Inst. Fert. and Insecto-Fungicides Trans. 142:95–113

Kolodny, Y. 1980. Phosphorites. *The Sea* 7: In press.

Koritnig, S. 1970. Phosphorus. In *Handbook of Geochemistry II/2* (15-F), ed. K. H. Wedepohl, pp. 7–8

Kramer, J. R. 1964. Sea water—saturation with apatites and carbonates. *Science* 146(3644)637–38

Leinen, M. 1979. Biogenic silica accumulation in the central equatorial Pacific and its implications for Cenozoic paleoocean-ography: summary. *Geol. Soc. Am. Bull.* 90:1310–76

Librovich, V. L. 1966. On the agreement between the hypothesis of phosphate formation of A. V. Kazakov and actual data. *Geol. Geofiz.* SSSR 1, pp. 136–42

Marlowe, J. I. 1971. Dolomite, phosphorite, and carbonate diagenesis on a Caribbean seamount. *J. Sediment. Petrol.* 41(3):809–27

Martens, C. S., Harriss, R. C. 1970. Inhibition of apatite precipitation in the marine environment by magnesium ions. *Geochim. Cosmochim. Acta* 34:621–25

McKelvey, V. E. 1963. Successful new techniques in prospecting for phosphate deposits. In *Natural Resources* 2:163–72 US Govt. Print. Off.

McKelvey, V. E. 1967. Phosphate deposits. *US Geol. Surv. Bull. 1252-D,* pp. D-1–21

McKelvey, V. E., Wang, F. F. H. 1969. *World subsea mineral resources; preliminary map.* Washington, DC: US Geol. Surv.

McKelvey, V. E., Swanson, R. W., Sheldon, R. P. 1953. The Permian phosphate deposits of western United States. In *Origine des Gisements de Phosphates de Chaux. Int. Geol. Cong. 19th, Algiers, 1952* 11(11):45–64

McKelvey, V. E., Williams, J. S., Sheldon, R. P., Cressman, E. R., Cheney, T. M., Swanson, R. W. 1959. The Phosphoria, Park City, and Shedhorn formations in the western phosphate field, *US Geol. Surv. Prof. Pap. 313-A.* 47 pp.

Ministry of Defense, USSR Navy. 1976. Pacific Ocean. In *World Ocean Atlas,* ed. S. G. Borshkov, Vol. 1. New York: Pergamon. 302 pp.

Ministry of Defense, USSR Navy. 1977. Altantic & Indian Oceans. In *World Ocean Atlas,* ed. S. G. Borshkov, Vol. 2, pp. New York: Pergamon

Penrose, R. A. F. Jr. 1888. Introduction to nature and origin of deposits of phosphate and lime with an introduction by N. S. Shaler. *US Geol. Surv. Bull. 46.* 143 pp.

Pettijohn, F. J. 1926. Intraformational phosphate pebbles of the Twin City Ordovician. *J. Geol.* 34:361–73

Pevear, D. R. 1966. The estuarine formation of United States Atlantic coastal plain phosphorite. *Econ. Geol.* 61(2)251–56

Piper, D. Z., Codispoti, L. A. 1975. Marine phosphorite deposits and the nitrogen cycle. *Science* 188(4183):15–18

Redfield, A. C., Ketchum, B. H., Richards, F. A. 1963. The influence of organisms on the composition of sea-water. In *The Sea* 2:26–77

Rhodes, F. H., Bloxam, I. W. 1971. Phosphatic organisms in the Paleozoic and their evolutionary significance. In *Proc. North American Paleontological Convention,* ed. E. L. Yochelson, Vol. II, pp. 1485–1513. Lawrence, Kansas: Allen Press

Riggs, S. R. 1979a. Phosphorite sedimentation in Florida—a model phosphogenic system. *Econ. Geol.* 74:285–314

Riggs, S. R. 1979b. Petrology of the Tertiary phosphorite system of Florida *Econ. Geol.* 74:195–220

Riggs, S. R. 1980. Tectonic model of phosphate genesis. *Proc. Fert. Raw Mater. Resour. Workshop,* ed. R. P. Sheldon, W. C. Burnett. pp. 159–90. Honolulu, Hawaii: East-West Cent., Resource Syst. Inst.

Roberson, C. E. 1966. Solubility implications of apatite in sea water. *US Geol. Surv. Prof. Pap. 550-D,* pp. D178–85

Rooney, T. P., Kerr, P. F. 1967. Mineralogic nature and origin of phosphorite, Beaufort County, North Carolina. *Geol. Soc. Am. Bull.* 78(6):731–48

Russell, R. T., Trueman, N. A. 1971. The geology of the Duchess phosphate deposits, northwestern Queensland, Australia. *Econ. Geol.* 66(8):1186–1214

Schlanger, S. O., Jenkyns, H. C. 1976. Cretaceous oceanic anoxic events: causes and consequences. *Geol. Mijinbouw* 55:179–84

Schlanger, S. O., Premoli-Silva, I., Jenkyns, H. C. 1980. Volcanism and vertical tectonics in the Pacific Basin related to global Cretaceous transgression. *Earth Plan. Sci. Lett.* In press

Sears, R. S. 1955. Phosphate deposits in the Caribou Range, Bonneville County,

Idaho. *Geol. Soc. Am. Bull.* Vol. 66, No. 12 (Abstr.)

Seeley, H. 1866. The rock of the Cambridge greensand. *Geol. Mag.* 3:302–7

Sheldon, R. P. 1957. Physical stratigraphy of the Phosphoria Formation in northwestern Wyoming. *US Geol. Surv. Bull. 1042-E*, pp. 105–85

Sheldon, R. P. 1963. Physical stratigraphy and mineral resources of Permian rocks in western Wyoming. *US Geol. Surv. Prof. Pap. 313-B*, pp. 49–272

Sheldon, R. P. 1964a. Paleolatitudinal and paleogeographic distribution of phosphorite. *US Geol. Surv. Prof. Pap. 501-C*, pp. 106–13

Sheldon, R. P. 1964b. Exploration for phosphorite in Turkey—a case history. *Econ. Geol.* 59:1159–75

Sheldon, R. P. 1980. Episodicity of phosphate deposition and deep ocean circulation—a hypothesis. *SEPM Spec. Publ.* In press

Sheldon, R. P., Burnett, W. C., eds. 1980. *Proc. Fert. Flows Raw Mater. Resour. Workshop.* Honolulu, Hawaii: East-West Cent., Resour. Syst. Inst.

Slansky, M. 1980. Eocene phosphorite deposit around West Africa. *Proc. Fert. Raw Mater. Resour. Workshop*, ed. R. P. Sheldon, W. C. Burnett, pp. 145–58. Honolulu, Hawaii: East-West Cent., Resour. Syst. Inst.

Stommel, H. 1958. The Abyssal circulation. *Deep-Sea Res.* 5(1):80–82

Strakhov, N. M. 1960. Fundamentals of the theory of lithogenesis. *Akad. Nauk. SSR*, Vol. 1. Moscow: Geol. Inst. 212 pp.

Strakhov, N. M. 1962. *Principles of Lithogenesis. Izd. AN SSSR, Moskva*, Vol. 3

Stumm, W. 1972. The acceleration of the hydrogeochemical cycling of phosphorus with discussion. In *The Changing Chemistry of the Oceans*, ed. D. Dryssen, D. Jagner, pp. 329–46. New York:Wiley

Wardlaw, B. R. 1980. Middle-Late Permian paleogeography of Idaho, Montana, Nevada, Utah and Wyoming. In *Paleozoic Paleogeography of West Central United States*, ed. T. D. Fouch, E. R. Magathan. Rocky Mountain Sect. SEPM, West Central United States Paleogeogr. Symp. 1

Yochelson, E. L. 1968. Biostratigraphy of the Phosphoria, Park City, and Shedhorn formations. *US Geol. Surv. Prof. Pap. 313-D*, pp. 571–660

Ziegler, A. M., Parrish, J., Humphreville, R. 1979. Paleogeography, upwelling and phosphorites. In *Proterozoic-Cambrian Phosphorites*, ed. P. J. Cook, J. H. Shergold, p 21. Canberra: Canberra Publ. and Print. (Abstr.)

Ziegler, A. M., Scotese, C. R., McKerrow, W. S., Johnson, M. E., Bambach, R. K. 1979. Paleozoic paleogeography. *Ann. Rev. Earth Plant. Sci.* 7:473–502

Ann. Rev. Earth Planet. Sci. 1981. 9:285–309

DEPLETED AND FERTILE ✶ 10152
MANTLE XENOLITHS FROM
SOUTHERN AFRICAN
KIMBERLITES

P. H. Nixon
Department of Earth Sciences, The University, Leeds LS2 9JT, England

N. W. Rogers
University of London Reactor Centre, Ascot, Berkshire, SL5 7PY, England

I. L. Gibson
Department of Earth Sciences, University of Waterloo, Ontario, Canada

A. Grey
Department of Earth Sciences, The University, Leeds LS2 9JT, England

INTRODUCTION

It has been demonstrated for many years (Wagner 1928) that the rounded inclusions found in kimberlite volcanoes provide a sampling of continental portions of the earth's upper mantle and crust. The most widespread types of inclusions are garnet lherzolites, of simple primary mineralogy, with usually few hydrous phases. Although in the past these have been considered to be cognate with kimberlite magma (e.g. Williams 1932), they are now generally regarded as fragments of mantle existing at the time of kimberlite intrusion. This conclusion derives from their widespread occurrence and the fact that shallower-derived spinel-bearing varieties are found in many ultrabasic-basic alkali igneous intrusions. Certain varieties including relatively ferriferous pyroxenites and griquaites (=mantle eclogites) are, however, considered to represent mantle cumulates or fractionation products within the mantle (e.g. Cox et al 1973, Barrett 1975).

285

A third group; the discrete nodules (megacrysts) consisting of large rounded crystals of garnet, clinopyroxene, orthopyroxene, and ilmenite, have been regarded as cognate, i.e. as phenocrysts within the kimberlite magma (Nixon et al 1963) or as components of crystal mush magmas within the mantle in the vicinity of kimberlite magma formation (Nixon & Boyd 1973a). Additional data for garnet lherzolites and some related garnet harzburgites, together with their host kimberlites, are considered here, in an attempt to understand the nature of the upper mantle in southern Africa and the processes that have modified it, including those that have generated kimberlite.

TECTONIC SETTING AND FACTORS AFFECTING MANTLE GEOCHEMISTRY

The samples used in this review are from kimberlites within or near the Kaapvaal craton[1] of southern Africa (Figure 1) and it is the study of the geological evolution of this stable region that provides explanations or clues to the chemical history of the underlying mantle represented by the lherzolite nodules (Nixon & Boyd 1975).

Low geothermal gradients are inferred at the present time in the lithospheric mantle underlying this and other cratonic areas, from a consideration of heat flow measurements. The use of pyroxene geotherm techniques (Boyd 1973, Boyd & Nixon 1973) indicates that such low geothermal conditions also existed at the time of kimberlite eruption, i.e. approximately 90 m.y. for most of our samples (Davis 1977). This followed a period of high heat flow conditions as witness the earlier period of widespread Stormberg plateau lava eruptions (150–200 m.y.).

The application of pyroxene geothermometry (Boyd 1973) to individual garnet lherzolite nodules also provided a means for placing the nodules in a sequence of increasing depth of origin, thus obtaining a paleo-stratigraphic cross section of the varied structure and composition of the upper mantle for a particular kimberlite pipe. Furthermore, the existence of a high temperature perturbation in the geotherm is equated with the Low Velocity Zone that marks the lithosphere-asthenosphere boundary, the "kink" in the geotherm being taken as the base of the lithosphere (Boyd 1973, Nixon et al 1973, Boyd 1976). Recent preferred estimates place this "kink" at a depth of 170 km over most of the craton but peripherally in East Griqualand (Figure 1) the depth is about 130 km (Boyd & Nixon 1979)

[1] The term "craton" is used here to signify a stable block of the earth's crust that has not suffered regional metamorphism for at least 1000 m.y. Several cratons welded together by younger high metamorphic grade "mobile belts" constitute a "shield" which is manifest in the present-day continents.

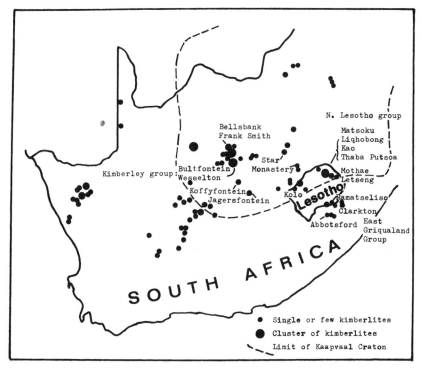

Figure 1   Kimberlite localities mentioned in the text and their relationship to the Kaapvaal craton.

The shallower-seated nodules representing the lithosphere are typically depleted in Ca, Al, Na, Ti, and Fe relative to Mg [but not always, see Cox et al (1973)]. This may be due to abstraction of partial melts represented by the widespread Stormberg Lavas (Nixon & Boyd 1973b) or a much earlier igneous episode (Danchin & Boyd 1976). A few ultradepleted nodules have been described from northern Lesotho (Carswell et al 1979) but they are much more common in the central craton Kimberley area. The lithosphere may thus have been depleted by more than one igneous event, e.g. Ventersdorp volcanism (Nixon & Boyd 1975).

The deeper-seated nodules defining the high temperature limb of the kinked geotherm and regarded as asthenospheric are relatively fertile in basaltic constituents. They are well documented from northern Lesotho (Nixon & Boyd 1973b), Jagersfontein (MacGregor 1975), and Frank Smith (Boyd 1974) where they all show deformation textures. In East Griqualand, however, some are undeformed (P. H. Nixon and F. R. Boyd, unpublished data). A much higher proportion of these asthenospheric types show deformation textures than do the depleted lithospheric types,

Table 1 Composition of garnet lherzolite and some garnet harzburgite nodules from kimberlites in southern Africa. Pyrolite quoted from Ringwood (1975).

PHN wt%	DEPLETED											METASOMATIC				FERTILE				
	1569	2492	2759	2764	2766/6	2782	2814	2823	2848	2860	2862	2713	2771	2780	2819	2829	2838	2839	3040	pyrolite
SiO_2	46.55	42.46	45.75	44.03	45.14	42.30	40.55	45.04	45.00	40.73	45.20	45.03	40.04	37.37	42.71	43.45	43.19	41.81	41.93	45.1
TiO_2	0.02	0.24	0.01	0.02	0.04	0.19	0.04	0.02	0.01	0.06	0.07	0.09	1.27	0.18	0.10	0.14	0.26	0.16	0.15	0.2
Al_2O_3	1.07	1.23	1.80	1.13	0.97	1.41	0.92	0.96	0.64	0.71	0.50	1.91	2.08	2.85	2.78	2.64	3.83	0.96	2.31	4.6
Fe_2O_3	1.35	3.00	1.52	1.82	1.99	2.18	2.14	1.88	7.26[a]	2.40	1.94	1.81	2.04	3.73	8.20[a]	8.58[a]	11.16[a]	1.69	2.60	0.3
FeO	4.79	4.14	4.32	4.36	4.18	4.15	4.33	4.22		4.07	6.86	4.90	4.51	7.24				6.30	5.49	7.6
MnO	0.11	0.11	0.11	0.11	0.11	0.10	0.10	0.09	0.11	0.09	0.12	0.12	0.11	0.12	0.12	0.12	0.14	0.12	0.13	0.1
MgO	43.77	42.22	41.53	43.24	43.57	41.82	42.97	42.72	44.17	43.32	42.80	39.82	37.05	39.29	38.81	39.95	36.59	44.32	40.05	38.1
CaO	0.84	1.16	1.00	0.95	0.95	0.83	0.45	0.54	0.54	0.29	0.46	1.58	5.68	0.94	2.31	2.80	2.71	0.88	2.26	3.1
Na_2O	0.06	0.02	0.09	0.10	0.19	0.10	0.03	0.02	0.10	0.02	0.05	0.14	0.52	0.15	0.18	0.32	0.30	0.11	0.29	0.4
K_2O	0.01	0.11	0.12	0.11	0.28	0.11	0.14	0.01	0.10	0.23	0.01	1.99	1.14	0.04	0.04	0.10	0.02	0.01	0.05	0.02
P_2O_5	0.03	0.05	0.05	0.05	0.07	0.06	0.03	0.01	0.03	0.15	0.08	0.04	0.14	0.07	0.01	0.03		0.09	0.29	0.02
H_2O^-	0.06	0.14	0.14	0.12	0.08	0.16	0.20	0.20	1.19[b]	0.24	0.01	0.14	0.22	0.22	4.37[b]	1.09[b]	1.17[b]	0.09	0.22	
H_2O^+	1.49	4.58	3.03	3.73	2.99	5.33	6.44	3.36		6.84	1.13	2.48	4.23	6.19				2.16	4.96	
CO_2	n.d.	n.d.	n.d.	n.d.	n.d.	n.d.	n.d.	n.d.		n.d.	n.d.	n.d.	1.22	0.20				n.d.	n.d.	
	100.15	99.46	99.47	99.77	100.56	98.74	98.34	99.08	99.15	99.15	99.23	100.05	100.25	98.59	99.63	99.22	99.52	98.62	100.73	

continued on next page

A	92.9	91.7	92.7	92.8	92.9	92.4	92.5	92.8	92.3	92.5	89.9	91.6	91.2	86.9	90.4	90.2	86.7	91.0	90.1	89.8
B	94.2	94.8	94.5	94.6	94.9	94.7	94.6	94.8	—	95.0	91.7	93.5		90.6				92.6	92.9	89.9
ppm																				
Cr	2826	2495	3166	3090	2716	2392	1919	2535	2356	2075	2430	3364	5734	1952	2895	2817	2919	3400	2285	2000
Co	90	96	87	95	84	86	95	96	101	99	117	88	67	126	101	111	99	102	96	1600
Ni	2009	1926	1805	2076	2061	1914	2027	2111	2201	2121	2300	1885	1518	2096	1846	1936	1781	2150	2065	
Zn	16	27	23	23	51	24	24	22	27	22	<3	29	24	61	31	35	57	36	34	
Rb	<3	5	5	4	7	5	5	2	27	8	<3	17	87	2	3	<3	9	<3	<3	
Sr	17	47	36	46	92	35	32	11	33	27	15	51	240	18	10	46	31	9	29	
Zr	7	18	18	13	15	13	<3	<3	9	4	15	24	27	19	6	12	13	8	9	
Nb	3	11	5	5	9	6	7	4	6	5	<3	5	16	3	2	4	2	1	<3	
Ba	<3	51	30	37	57	151	36	27	15	38	(Pb<3)	47	171	58	9	29	75	3	9	
PHN	1569	2492	2759	2764	2766/6	2782	2814	2823	2848	2860	2862	2713	2771	2780	2819	2829	2838	2839	3040	
ppm																				
La	1.05	5.17	2.90	3.99	2.18	3.02	4.05	1.57	10.8	2.87	1.35	3.91	5.29	0.88	—	—	0.86	1.81	1.09	
Ce	2.18	10.9	5.87	9.03	4.45	6.28	7.75	3.28	10.8	5.25	2.70	9.41	10.4	2.14	2.34	4.75	2.57	2.02	2.52	
Nd	0.93	5.60	2.69	3.91	1.70	2.56	2.36	1.28	6.9	2.05	1.33	3.53	4.40	0.81	1.66	—	1.47	1.41	1.39	
Sm	0.219	0.747	0.571	0.181	0.101	0.148	0.484	0.074	0.446	0.346	0.198	0.663	0.950	0.209	—	0.20	0.515	0.316	0.342	
Eu	0.074	0.191	0.147	0.255	0.120	0.178	0.119	0.076	1.08	0.077	0.051	0.229	0.273	0.109	0.14	—	0.191	0.086	0.123	
Dy	0.107	0.164	0.199	0.085	0.020	0.062	0.031	0.026	0.745	0.100	0.064	0.088	0.416	0.482	—	0.34	0.920	0.263	0.397	
Yb	0.024	0.095	0.061	0.014	0.006	0.007	0.011	0.006	0.120	0.022	0.018	0.015	0.085	0.230	0.56	—	0.482	0.150	0.205	
Lu	0.004	0.013	0.009	0.014	0.006	0.007	0.011	0.006	0.120	0.003	0.003	0.011	0.011	0.034	—	—	0.073	0.024	0.029	
La/Lu	236	398	322	285	363	431	368	262	90	957	450	261	481	26	—	—	12	75	38	

$100 \text{ Mg}/(\text{Mg} + \text{Fe}^{2+})$. A, with total iron as Fe^{2+}. B, with Fe^{2+} as analyzed.

Notes:

Analyses by XRF: A. Grey. $\text{Fe}^{2+}/\text{Fe}^{3+}$, H_2O^-, H_2O^+, and CO_2 by D. Richardson.
REE analysis by RNAA: N. W. Rogers (method of Steinnes; see Brunfelt et al 1974).
n.d., not detected.

[a] Total Fe as Fe_2O_3.
[b] Loss on fusion.
[c] Partial analyses of PHN 2819 and 2829 performed by I. L. Gibson.

but notable exceptions are the highly depleted varieties from the Kimber-ley pipes, which are very deformed (Dawson et al 1975, Boyd & Nixon 1978).

The origin of the deformation—whether due to stress at the base of the lithospheric plate (Boyd 1973) or diapiric upwelling of a magmatic precur-sor to kimberlite (Green & Gueguen 1974) or in some cases to localized shallow level deformation, or in connection with kimberlite intrusion—has been discussed by Dawson et al (1975), Boyd (1976), and Mercier (1979).

LHERZOLITE INCLUSIONS

A suite of mantle inclusions displaying a variety of textures, composi-tions, and depths of origin, from widespread kimberlite intrusions, are briefly described in the Appendix. Except where stated, modal propor-tions fall within the range, typical for southern African inclusions: olivine 40–75, orthopyroxene (enstatite) 5–50, clinopyroxene (chromium diop-side) 0–20, garnet (chromium pyrope) 0–15, plus variable amounts of chromite and phlogopite (see, for example, Cox et al 1973, Boyd & Nixon 1978).

The textures vary in their degree of deformation and recrystallization from coarse (grain size 2–5 mm) porphyroclastic to mosaic-porphyroclas-tic. This nomenclature (Harte 1977) incorporates the descriptions of other workers, e.g. Boullier & Nicolas (1973), and subdivides the previous "gran-ular" and "sheared" field terminology of Boyd & Nixon (1972). In thin section, olivine is traversed by serpentine stringers. It forms kink-banded grains or porphyroclasts in the deformed lherzolites and, in extreme cases, a groundmass mosaic of fine crystals (∼0.3 mm). Orthopyroxene (opx) and clinopyroxene (cpx) are less affected by chemical alteration but the latter has usually slightly turbid rims. The minerals can be subrounded or lobate but deformation produces elongated grains or mosaic textures particularly in the opx. Garnet (gt) is relatively resistant to stress but is comminuted in some nodules in the Kimberley area (disrupted texture; Harte 1977). It has an alteration "kelyphitic" rim due to late-stage hydra-tion. Coarse primary chromium-bearing spinel is present in some nodules. Phlogopite may be either primary or secondary and is occasionally ac-companied by ilmenite.

Major Elements

The lherzolite nodule compositions (Table 1) reflect the overall Mg-rich nature of the constituent minerals and a small degree of hydration (ser-pentinization and kelphitization). There is a two-fold division correspond-

Table 2 Summary of the bulk chemical characteristics of the depleted and fertile modules

Chemical parameters	Depleted	Fertile
$100 \, Mg/(Mg + Fe^{2+}$ total)	91^a–93	86–91
(range and mean)(mol %)	(92.3)	(89.2)
Al_2O_3 wt %	<2	>2
CaO wt %	⩽1	⩾1
Na_2O wt %	⩽0.15	⩾0.15
TiO_2 wt %	<0.10	>0.10
La/Lu	90–957	12–75
(range and mean)	(381)	(38)

[a] Does not include PHN 2862 (89.9 mol %).

ing to *depleted* and *fertile* nodules as was demonstrated for nodules from Thaba Putsoa (Lesotho) by Boyd & Nixon (1972). The fertile nodules approach pyrolite composition (Ringwood 1975). The relatively high contents of Fe, Al, Ca, Na, and Ti in the fertile suite are summarized in Table 2. The levels of these constituents are similar to those of fertile upper mantle spinel lherzolites (Harris et al 1972). Not all nodules, however, are sharply categorized. In addition to the metasomatized examples, which have added constituents, the depleted specimen 2492 is rather high in Ca, and 2862 is unusually enriched in Fe. Conversely, fertile 2839 is in some respects depleted, e.g. in its low abundances of Al and Ca. Localized modal variations, particularly of cpx and gt, are responsible for some chemical overlap between the two groups.

It is notable that five of the six fertile nodules in Table 1 have strong deformation textures and all represent the deeper parts of the mantle sampled by the particular host kimberlite. The Jagersfontein nodule 2819 is similar to those that have equilibrated at depths of about 190 km (MacGregor 1975). This is comparable to equilibration PT conditions of the Lesotho nodules (2829, 2838, and 2839). The nodule from Frank Smith (2790) is probably of shallower origin equivalent to 160–170 km (Boyd 1974). There is only one example here of a fertile nodule with coarse, i.e. undeformed, texture, and this is from Abbotsford pipe (3040) located outside the craton as shown in Figure 1. It should be noted that some nodules with deformation textures are very depleted especially in the Kimberley (central craton) area (i.e. 2764, 2766/6, and 2782) as recorded by Dawson et al (1975) and Boyd & Nixon (1978). Irrespective of texture all the depleted nodules are regarded as lithospheric in origin (equilibrating at shallower depths than the "kink" in the geotherm).

Two nodules from Monastery Mine (2713 and 2771) contain relatively large amounts of phlogopite represented by K_2O contents of 1.99 and 1.14 wt % respectively. Nodule 2771 contains significant amounts of Ti (ilmenite and rutile), Al (phlogophite), Ca and Na (clinopyroxene), and P (apatite) and is typical of a group of nodules that appear to have suffered metasomatism within the mantle (see, for example, Harte et al 1975).

Rare Earth Elements (REE) and Incompatible Elements

The REE contents of the suite of inclusions were determined by I. L. Gibson and those of the lherzolite inclusions were redetermined with greater precision by N. W. Rogers (Table 1). The depleted and fertile nodules show significantly contrasting REE distribution patterns [Table 1, Figures 2 and 3. The chondritic abundances of Haskin et al (1968) are used as normalizing factors.]. The depleted nodule patterns have steep slopes due to light REE (LREE) enrichment which is reflected by the high La/Lu ratios. These patterns resemble those of other garnet lherzolite nodules recorded by Philpotts et al (1972), Ridley & Dawson (1975), Shimizu (1975), and Mitchell & Carswell (1976). The fertile nodules show less-fractionated patterns (low La/Lu) but also greater relative abundances of heavy REE. These are closer to chondrite values as has been demonstrated for the "sheared" fertile deep-seated gt lherzolite 1611 from Lesotho (Nixon & Boyd 1973b) by Shimizu (1975) and Morgan et al (1979) who determined La 890 ppb and Lu 52 ppb. The differences between the depleted and the fertile groups cannot be explained by differing degrees of LREE enrichment alone. Shimizu (1975) suggested, on the basis of REE data, that the depleted nodules were originally garnet-free cumulative assemblages equilibrating with kimberlitic or equivalent liquid, and only later recrystallised below the solidus as a garnet-bearing assemblage.

The greater abundance of the heavy REE (HREE) in the fertile nodules relative to the depleted nodules can be explained by the partial-melting model mentioned in the tectonic setting section of this review. It should be noted that the tendency for slightly higher modal garnet to occur in the fertile nodules (hence the higher Al_2O_3 values in Table 1, although modal variants occur in both suites) would be expected to produce higher HREE contents in the bulk rock. This minor effect is greatly enhanced by the much higher HREE content of the garnets from the fertile nodules. These contain 2–8 times more Yb than those from the depleted nodules (Shimizu 1975).

The depletion process, however, is clearly not evident in the high LREE ranges recorded in the lithospheric depleted nodules. As in lherzolite inclusions in basaltic rocks, an additional process or contaminating fluid has to be sought to explain this (Frey & Green 1974, Frey & Prinz 1978).

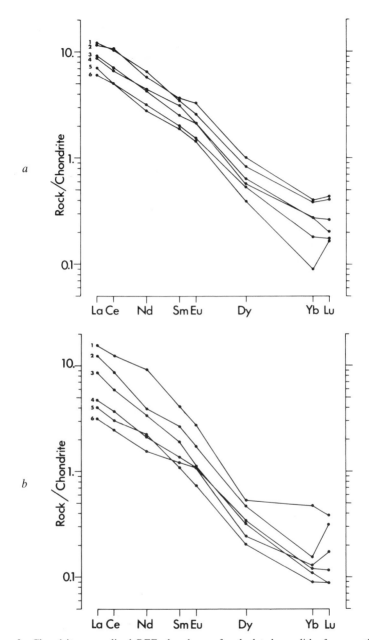

Figure 2 Chondrite normalized REE abundances for depleted xenoliths from southern African kimberlite pipes. (*a*) From Kimberley (Bultfontein), Frank Smith, and Monastery Mine. Nos. 1–6 are, respectively, PHN 2764, 2713, 2782, 2759, 2765, and 2766/6 (note 2765 is not tabulated or described in the text, but is petrographically similar to 2766/6. (*b*) From N. Lesotho and Jagersfontein. Nos. 1–6 are, respectively, PHN 2492, 2814, 2860, 2823, 2862, and 1569.

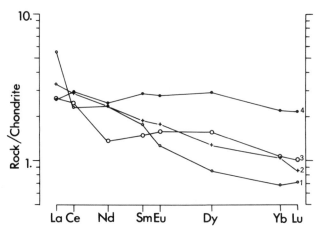

Figure 3 Chondrite normalized REE abundances for fertile xenoliths from southern African kimberlite pipes. Nos. 1–4 are, respectively, PHN 2839, 3040, 2780, and 2838.

Mysen (1979) has shown experimentally that, under high pressures equivalent to those of the lithospheric mantle, water-rich fluids constitute a "major sink" for REE as does the CO_2 vapor phase (Wendlandt & Harrison 1979). These fluids are potent metasomatizing agents, and LREE enrichment correlates with the depletion of lherzolite (Mysen 1979).

Elements that are unable to substitute in the common mantle minerals because of differences in ionic size, valency, or bond type (see Harris & Middlemost 1969) include K, Rb, Sr, Ba, U, Th, C, P, and halogens, and these tend to remain in the liquid phase during partial melting. They, thus, act like light REE under some conditions and may be concentrated by similar processes. For the lherzolite nodules Rogers (1979) has demonstrated that La and Th vary proportionally within a range La/Th = 7–10. Variations in the incompatible element data (Table 1), although poorly known, could be bound up with the regional structure. On the basis of slightly higher levels of F, Cl, K, and P in Bultfontein nodules compared with those from Lesotho, Nixon & Boyd (1975) have suggested that accretion of incompatible elements has taken place over longer periods in the central (older) part of the Kaapvaal craton lithosphere[2] compared with

[2] The mantle xenoliths from the central cratonic Premier kimberlite resemble those from Lesotho (Danchin & Boyd 1976) and may not conform to this regional picture because the Premier kimberlite is of Precambrian age. However, the Premier kimberlite could have had a more peripheral craton setting at the time of emplacement similar to that of Lesotho. The craton boundaries have since expanded and more incompatible element accretion has taken place.

the peripheral (younger) area. Unlike the data for kimberlites (see below) a comparison of the LREE (Figure 2a, 2b) does not strengthen this claim. Since C is an incompatible element it may be significant that the six lherzolite and harzburgite nodules from Lesotho in which we have observed graphite all are of the depleted type. The single recorded occurrence of diamond in garnet lherzolite from southern Africa is also of this type (Mothae; Dawson & Smith 1975).

Metasomatic processes may result from accumulations of this "background" enrichment of incompatible elements in the depleted lithosphere. They give rise to phlogopite, ilmenite, and sulphides (Harte et al 1975) as seen in nodules 2771 and 2713, in which Rb, Sr, and Ba have also been introduced (Table 1). However, there is a very high degree of enrichment in volatile and incompatible elements of specific kimberlites compared with their lherzolite nodules (Table 3; see also Harris & Middlemost 1969).

There is no incontrovertible evidence for contamination of the nodules by the enclosing kimberlite during eruption. Table 3 shows relatively low concentrations in the lherzolites of such mobile ions as P and Sr (also F and Cl; Paul et al 1975) which are abundant in kimberlite. Only in the metasomatized nodules are there comparable levels in some incompatible elements, but not REE. However, the relative degree of LREE enrichment of these nodules is similar to that of the kimberlites, as indicated by the slopes of the REE patterns (Figure 5) and La/Lu ratios. The nearly constant ratios of the relatively low REE levels in the depleted nodules, including the metasomatized varieties to the levels of the kimberlites, nearly preclude an explanation due to additions from kimberlite. It would have been expected that the cores of *some* of the larger nodules (50 cm across) would have escaped enrichment by contamination. Moreover, Shimizu's (1975) gt and cpx results demonstrate that the REE do not reside along grain boundaries.

The same degree of enrichment is not shown by the fertile, deeper-seated lherzolite nodules, which have been considered to originate close to the kimberlite magma.

These features can be accommodated in a mantle model that envisages a lithosphere and underlying relatively mobile asthenosphere having the following characteristics:

LITHOSPHERE   Mainly coarse garnet lherzolite which prior to kimberlite volcanism has been depleted by partial melting during one or more igneous events to give basaltic products. Ages greater than 2000 m.y. are indicated by Pb isotopic studies (Kramers 1979). This relatively rigid layer has been enriched in the course of geological time with mobile elements from below. The basal horizons are considered ideal sites for metasomatism and diamond/kimberlite formation, particularly beneath the older central

Table 3 Comparison of volatile and incompatible content of kimberlites and their nodules to illustrate a general lack of contamination by the magma. The deepest nodules are *not* more affected by kimberlitic fluids.

PHN	Host kimberlite pipe							Nodules in probable order of increasing depth for a particular pipe						
	K$_2$O (wt %)	P$_2$O$_5$ (wt %)	Ba (ppm)	Sr (ppm)	Nb (ppm)	Ce (ppm)		PHN	K$_2$O (wt %)	P$_2$O$_5$ (wt %)	Ba (ppm)	Sr (ppm)	Nb (ppm)	Ce (ppm)
Thaba Putsoa														
1598	1.79	0.36	600	502	36	88		2848	0.10	0.03	15	33	6	10.8
								1569	0.01	0.03	0	17	3	2.2
								2839[a]	0.01	0.01	3	9	1	2.0
								2838[b]	0.15	0.02	75	31	2	2.6
Kao														
2690 M	1.22	(data from Paul et al 1975) 0.36	—	660	—	84		2492	0.11	0.05	51	47	11	10.9
								2829	0.10	0.03	29	46	4	4.8
Jagersfontein														
2811	0.51	0.16	233	265	29	31		2814	0.14	0.03	36	32	7	7.8
								2819	0.04	0.01	9	10	2	2.3
Frank Smith														
2779	0.96	1.39	1895	1055	161	381		2780	0.04	0.07	58	18	3	2.1
2384	1.69	1.77	2156	1142	173	387		2782	0.11	0.06	151	35	6	6.3

Sample						
Wesselton						
2732	2.87	0.98	1152	865	124	217
2759	0.12	0.05	30	36	5	5.9
2771[c]	1.14	0.14	171	240	16	10.4
2764	0.11	0.05	37	46	5	9.0
2766/6[d]	0.28	0.07	57	92	9	4.5
Monastery						
2713	1.99	0.04	47	51	5	9.4
1867[e]	0.18	0.15	162	209	160	294
1870	1.78	0.94	—	—	—	—

(1870 data from Gurney & Ebrahim 1973)

Notes:
For the Wesselton kimberlite the nodules come from nearby Bultfontein pipe.
Nodules 2771 and 2713 are shallow-seated metasomatized varieties.

[a–e] F, Cl wt % (Paul et al 1975):
[a] 0.004, 0.001
[b] 0.007, 0.001
[c] 0.072, 0.002
[d] 0.023, 0.010
[e] 0.075, 0.007.

craton. The accretion is not restricted to the basal lithosphere and our evidence suggests that there are incompatible elements distributed throughout the lithosphere perhaps forming a chemical gradient, which presumably was interrupted from time to time, at various depths, by basalt magma production. On the other hand, localized concentrations of incompatible elements, at all levels in the lithosphere down to the zone of kimberlite magma formation, might produce enriched pockets represented by the metasomatized nodules.

ASTHENOSPHERE   Garnet lherzolite with fertile "pyrolitic" chemistry. Few inclusions have escaped the impact of stress arising during diapiric upwelling or conduit formation but original textures may have been coarse or poikiloblastic (P. H. Nixon and F. R. Boyd, unpublished data from East Griqualand). The low levels of incompatible elements and chondritic REE patterns indicate that these rocks have not participated in the metasomatic accreting process and strengthen the view that they represent a fundamentally distinct zone beneath the lithosphere. The garnet and pyroxene discrete nodules (megacrysts) are from this zone (see below).

KIMBERLITES

These volcanic rocks are highly serpentinized, with olivine grains in various stages of replacement, in a groundmass which typically contains phlogopite, clinopyroxene, calcite, perovskite, and spinels including magnetite (see Appendix for brief descriptions of individual rocks). Xenocrysts of ilmenite, garnet, pyroxenes, and olivine in some cases are derived from disaggregated lherzolites, or are part of the discrete nodule suite, but they normally form less than one or two percent of the rock. Xenoliths of small size (5 mm) may be included in the analyzed sample, but conspicuous fragments of lherzolite, basalt, shale, gneiss, etc, were taken out prior to fine crushing. The samples examined here are unweathered compact kimberlites (not the "yellow ground" which has disintegrated rapidly on exposure to air) and the much more resistant "hardebank" variety with groundmass cement, often calcite.

Chemical Composition

The analyses (Table 4) illustrate the well-known kimberlite characteristics, viz. highly magnesian composition, ultrabasic nature, $K_2O \gg Na_2O$, and high levels of both compatible elements, e.g. Cr and Ni, and of incompatible elements, e.g. Nb, Sr, and Ba. The high volatile content consists mainly of H_2O and CO_2.

Table 4 Composition of kimberlites from Southern Africa

PHN	1334	1598	1725	1867	2201	2257	2384	2732	2779	2796	2811
wt%											
SiO_2	31.75	31.05	26.57	31.18	35.31	31.97	31.77	32.70	30.76	35.58	37.57
TiO_2	1.98	4.48	1.84	3.74	1.06	3.03	2.72	1.82	2.06	0.79	0.35
Al_2O_3	2.40	3.57	3.22	2.43	2.74	3.62	3.26	4.60	2.18	2.08	2.35
Fe_2O_3	4.92	5.87	4.86	8.82	3.59	6.97	6.07	5.44	5.33	3.90	4.07
FeO	4.04	5.66	4.16	3.73	4.63	5.83	3.33	3.30	3.33	3.58	3.26
MnO	0.17	0.16	0.20	0.20	0.16	0.24	0.18	0.15	0.15	0.13	0.10
MgO	29.50	24.39	24.15	30.04	34.88	30.09	30.99	26.68	29.34	32.89	33.73
CaO	8.89	7.47	15.24	6.71	4.98	9.82	6.81	8.40	8.81	4.13	2.12
Na_2O	0.01	0.17	0.02	0.23	0.06	0.07	0.27	0.60	0.11	0.08	0.03
K_2O	1.11	1.79	0.82	0.18	2.49	0.56	1.69	2.87	0.96	2.82	0.51
P_2O_5	0.69	0.36	0.56	0.15	0.64	0.60	1.77	0.98	1.39	0.48	0.16
H_2O^-	0.56	1.63	0.25	0.19	0.60	0.09	0.61	0.79	3.47	0.45	1.75
H_2O^+	9.05	8.51	7.28	10.03	5.62	5.10	9.32	7.82	6.96	7.76	11.63
CO_2	5.98	3.76	10.22	1.99	3.03	2.81	1.36	4.24	4.23	4.97	1.64
	101.05	98.87	99.39	99.62	99.79	100.80	100.15	100.39	99.08	99.64	99.27
ppm											
Cr	1151	1076	1287	954	2259	1207	1778	1178	1482	2173	979
Co	65	54	49	72	70	67	63	53	50	64	77
Ni	1080	810	702	961	1274	825	1128	831	1186	1254	1512
Zn	34	46	42	63	41	63	32	56	38	37	34
Rb	63	101	26	9	99	45	93	131	59	151	33
Sr	733	502	927	209	1193	801	1141	865	1055	917	265
Zr	159	213	215	462	157	215	293	265	242	201	56
Nb	145	36	82	160	106	243	173	124	161	128	29
Ba	1104	600	1370	162	3564	844	2156	1152	1895	2894	233
Nd	95.5	35.3	45.1	127.0	161.6	317.9	148.1	104.6	169.0	148.0	13.4
Eu	3.54	1.75	1.90	5.34	4.24	11.20	5.38	4.12	5.74	3.28	0.56
Gd	9.68	-	3.77	15.75	-	-	10.48	11.50	12.80	-	2.72
Tb	1.03	0.51	-	1.54	1.01	3.55	1.26	1.27	1.71	1.36	0.20
Tm	0.35	-	-	0.57	-	-	0.58	-	0.54	-	-
Yb	-	-	1.04	1.82	-	4.64	1.21	-	-	-	0.36
Ta	-	-	8.1	26.3	-	42.5	26.6	18.3	-	25.4	2.9
Th	21.1	7.7	6.8	22.9	37.0	57.6	25.5	20.6	30.1	19.7	3.4
Hf	4.4	5.8	4.5	12.0	5.6	9.6	7.3	6.2	7.2	6.7	1.0

Notes:
Analyses by XRF: A. Grey. Fe^{2+}/Fe^{3+}, H_2O^-, H_2O^+, and CO_2 by D. Richardson.
REE analyses by INAA; estimated 2σ error: Ce Eu 2%, Nd Tb 5%, Gd Tm Yb 20%. Analyst: I. L. Gibson.

The REE pattern (Figure 4) emphasizes the incompatible element enrichment, with most LREE abundances falling within the range of 50–1000 × chondrite established by Frey et al (1971), Fesq et al (1975), Mitchell & Brunfelt (1975), Paul et al (1975), and Frey et al (1977). However, it is interesting to note that the Jagersfontein kimberlite has an exceptionally low total REE abundance. It could be significant that other peripheral craton kimberlites also have low REE abundances and a tendency to lower LREE enrichment (Table 5). There is also an indication that the peripheral craton kimberlites contain relatively low amounts of

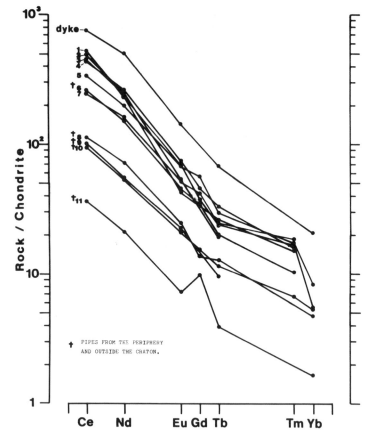

Figure 4 Chondrite normalized REE abundances of some southern African kimberlites. (*1*) 2796 (Bellsbank). (*2*) 2201 (Star, dyke but with breccia texture as found in pipes). (*3* and *4*) 2384 and 2779 (Frank Smith). (*5*) 1867 (Monastery). (*6*) 1334 (Kolo). (*7*) 2732 (Wesselton). (*8*) 1725 (Ramatseliso). (*9*) 1598 (Thaba Putsoa). (*10*) (Kao pipes, P. H. Nixon, D. K. Paul, and P. Potts, unpublished data) (*11*) 2811 (Jagersfontein). There is a tendancy for pipes from the margin and outside the craton to have lower REE abundances. However, nonbrecciated (aphanitic) kimberlite dykes, possibly representing a closed system, are very enriched in LREE. The dyke in the figure is 2257 from near Mothae, Lesotho.

incompatible elements which may or may not be the case with their inclusions (see above). A comparison of kimberlite pipes from the craton margin with those from the craton center (Tables 4 and 5) gives the following mean values (craton center values in brackets): K_2O wt % 0.83 (2.17), Rb ppm 46 (107), Ba ppm 719 (2332), Sr ppm 573 (1034), and Ce ppm 233 (1034). Gurney & Ebrahim (1973) showed that a mean of 25 Lesotho kimberlites (peripheral) compared with 80 South Africa kimberlites (mostly

Table 5 Kimberlite data arranged according to location in Kaapvaal craton (Figure 1)

Craton	Kimberlites PHN	Type of intrusion[a]	Diamond "grade"[b]	Ce/Tb	La/Lu[c]
Central	2201 Star mine	/	A	424	1641
	2796 Bellsbank	O	A	338	2367
	2384 Frank Smith	O	A	307	1250
	Bellsbank (6) (Fesq et al 1975)	o	A	259	—
	Swartruggens (4) (Mitchell & Brunfelt 1975)	/	A	226	—
	2779 Frank Smith	O	A	223	2068
	Wesselton (6) (Mitchell & Brunfelt 1975)	O	A	177	—
	2732 Wesselton type 3	O	A	171	748
Margin	1334 Kolo	O	B	220	1200
	Kao dykes (4) (P. H. Nixon et al, unpubl.)	o/	C	209	—
	1867 Monastery, east end	o	A/B	191	1028
	Monastery (Mitchell & Brunfelt 1975)	o	A/B	188	—
	2257 Dyke NW of Mothae	/	C	185	—
	1598 Thaba Putsoa	o	C	173	591
	2811 Jagersfontein	o	A	157	246
	Koffyfontein (10) (Fesq et al 1975)	o	A	123	—
	Kao pipe (7) (P. H. Nixon et al, unpubl.)	O	B	118	—
Outside	1725 Ramatseliso	o	C	145	757

[a] Type of intrusion: pipe = O, small pipe or "blow" = o, dyke = /.
[b] Diamond grades: mineable = A, marginal = B, poor to barren = C.
[c] Rogers (1979).

central) had K_2O wt % 0.79 (1.52). We have found corresponding differences in levels of Nb, Zr, and Th, but they have a lower contrast because kimberlite dyke 2257 from Mothae (peripheral craton) has high incompatible element levels. Dykes and other hypabyssal (closed system) kimberlite intrusions are relatively enriched in incompatible elements compared with associated pipes (P. H. Nixon, D. K. Paul, and P. Potts, unpublished data; Figure 4). Although our data refer to pipes, more analyses are needed to confirm incompatible element variation within the craton, and the apparent sympathetic relationship with diamond content indicated in Table 5.

Kimberlite—Diamond Genesis and Tectonic Implications

It has been shown (e.g. Mitchell & Brunfelt 1975) that a garnet lherzolite mantle source can account for much of the observed major and compatible elements (e.g. Cr, Ni, Ca) in kimberlite by variable amounts of partial melting and/or fractional crystallization. The latter process, involving formation of eclogite, may not be important (Frey et al 1977). Moreover,

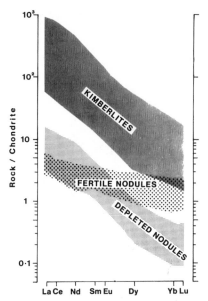

Figure 5   Summary of the ranges of REE patterns of depleted and fertile mantle nodules compared with those of host kimberlites.

a small degree of partial melting of moderately LREE-enriched garnet peridotite adequately explains the highly fractionated REE patterns of the kimberlites (Frey et al 1977). From this standpoint the depleted nodules best represent the parental mantle material (Figure 5). The silicate inclusions in natural diamonds are usually of highly depleted peridotite mineralogy (Meyer & Boyd 1972, Meyer & Svisero 1975, Meyer & Tsai 1976,

Figure 6   Schematic section through continental upper mantle illustrating the depleted lherzolite lithosphere overlying fertile asthenosphere. Accumulations of incompatible elements, especially at the base of the lithosphere, are indicated. Metasomatism is most likely to take place in this accretion zone. Kimberlite magma results from admixture of the products of two processes: (*a*) partial melting due to depressurizing in a rising diapir (Green & Gueguen 1974) in the asthenosphere to produce a "proto-kimberlite" magma and (*b*) partial melting in the lower lithosphere at an unspecified date, either long before eruption or in contact with the diapiric thermal aureole. The resulting kimberlite has characteristics that can be identified with both environments, viz the fertile (asthenosphere) rocks are the most likely source of a primitive Sr isotopic melt fraction for kimberlite, and a parent for the discrete nodule (megacryst) suite. The depleted (lithosphere) rocks and melt products are the most probable source for the incompatible elements and residual ultra-depleted diamond inclusion minerals. Deformation, illustrated by the "sheared" nodules, is mainly associated with the skin of the moving diapir or with kimberlite conduit formation especially in the asthenosphere but also at the base of the lithosphere, depending on the amount of diapir penetration (see text).

Gurney & Switzer 1973). Danchin & Boyd (1976) have described garnet and chromite harzburgite nodules from Premier Mine, South Africa, which have very similar mineral chemistry to that of inclusions in diamond. Nixon & Boyd (1975) have suggested that diamond inclusion

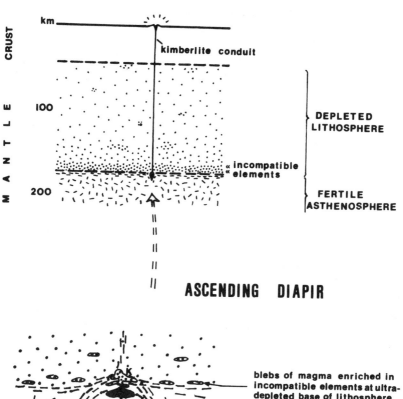

MANTLE MODEL

ASCENDING DIAPIR

blebs of magma enriched in incompatible elements at ultra-depleted base of lithosphere

proto-kimberlite magma associated with rising diapir — crystallisation of discrete nodule (megacryst) suite

zones of deformation

K,   point at which kimberlite is generated by coal-escence of magmas mineralogy has resulted from "double depletion" of mantle and Harte et al (1979) have argued that the high depletion, with respect to CaO, can be produced by further melting of depleted mantle to produce kimberlite. This view is supported by the REE data. However, there is evidence against this direct origin from Sr isotopic data. Barrett (1975) and J. D. Kramers (personal communication) have pointed out that $^{87}Sr/^{86}Sr$ ratios of the depleted types of nodules fall in the range 0.704–0.706+ and are unlikely to be parental to kimberlite with typical ranges of 0.703–0.705. On the other hand, asthenospheric fertile types of nodules from Lesotho have low isotopic ratios (N. Shimizu, personal communication).

These facts are consistent with a two-stage model in which a magma derived from the deeper fertile mantle, perhaps in connection with a rising diapir (Green & Gueguen 1974), admixes with, or is contaminated by, a melt enriched in incompatibles at the base of the lithosphere. At this point, diapiric upwelling ceases (Boyd 1976) and conduit propagation, leading to kimberlite eruption, begins (Figure 6).

The magma associated with the diapir ("proto-kimberlite") is thought to have contained large phenocrysts—the discrete nodule suite—which include high temperature subcalcic cpx with $^{87}Sr/^{86}Sr$ ratios lower than those of kimberlite (Barrett 1975). The proximity of the diapir to the base of lithosphere (enriched in incompatible elements) may have induced partial melting (thermal aureole effect, cf Mitchell 1978). Alternatively, blebs of magma may have been in existence at the base of the lithosphere for a long time. Coalescence of these blebs and intermingling with proto-kimberlite magma resulted in the formation of kimberlite. It is likely that the physical effects of an upwelling diapir accelerated coalescence into magma chambers but shearing stresses due to plate movement at the base of the lithosphere (Boyd 1973) would have the same effect.

On the basis that deformation of the mantle took place in the vicinity of the rising diapir, or immediately above it at the point of conduit propagation (Mercier 1979), it can be argued that there were varying degrees of penetration of the asthenosphere-lithosphere boundary. This appears to be greatest in the central craton Kimberley area pipes where there is a scarcity or absence of deformed asthenospheric lherzolites, but where there are many deformed depleted lithospheric types. In the Frank Smith pipe and in the pipes from more peripheral craton areas of Jagersfontein and northern Lesotho most, but not all, deformation is observed in the asthenospheric lherzolites implying conduit formation immediately below the boundary. In the extra-cratonic pipes of East Griqualand there is a range of textural types within the asthenospheric lherzolites and coarse undeformed textures occur (3040, Table 1, P. H. Nixon and F. R. Boyd, unpublished data). It is suggested that conduit propagation took place at even greater depth in the asthenosphere to allow for incorporation of these xenoliths into the magma. However, since the asthenosphere-lithosphere boundary shelves outwards from the craton center, the actual depth of conduit propagation and kimberlite formation is less in these outermost areas.

SUMMARY

Lherzolite inclusions, with some harzburgites, from kimberlites are divided into depleted and fertile groups according to Mg/Fe ratio, Al_2O_3, CaO, Na_2O, and TiO_2 content. The depleted (lithospheric) types are, however, enriched in LREE (La/Lu = 90–957) and incompatible elements, contrasting with the deeper fertile (asthenospheric) types which have near chondritic patterns (La/Lu = 12–75). It is thought that many of the nodules are not significantly contaminated by the host kimberlite. However, the relative degree of kimberlite LREE enrichment, which may show a regional cratonic variation, is similar to that of the depleted inclusions. They, hence, represent the source material for a portion of the kimberlite magma.

Combining the data with the observations of several authors, a model is proposed which involves depletion of the lithosphere due to basaltic magma(s) generation and gradual accumulation of incompatible elements particularly under the central (? older) portion of the Kaapvaal craton. Deep-seated diapiric upwelling of relatively hot fertile asthenosphere resulted in melting to form proto-kimberlite magma in which large phenocrysts developed (discrete nodule suite). Deformation textures in mantle xenoliths provide evidence that diapiric upwelling and kimberlite conduit formation took place within the basal lithosphere in the Kimberley area but within the asthenosphere in peripheral craton areas. Melting took place at the LREE-enriched base of the lithosphere, either as a result of the proximity of the diapiric thermal aureole or much earlier, producing an ultradepleted residua (the diamond inclusion suite). Kimberlite is a hybrid of this melt with proto-kimberlite.

ACKNOWLEDGMENTS

The samples used in this research were collected over a number of years with the kind cooperation of De Beers Consolidated Mines Ltd. and the Lesotho Government. Analytical and petrographic work was carried out between 1975 and 1979 at the University of London (Bedford College, and the Reactor Centre, Berkshire) and the University of Leeds. We have greatly benefited from discussions of the project with F. A. Frey, and are grateful to F. R. Boyd, J. D. Kramers, R. H. Mitchell, D. K. Paul, and S. S. Sun who criticized the manuscript. We thank F. R. Boyd and N. Shimizu for unpublished data and comments. We also thank Mr. D. A. Richardson for help with the analyses.

APPENDIX: SUMMARY PETROGRAPHIC DESCRIPTION OF ANALYZED ROCKS

LHERZOLITE, ETC, NODULES

PHN 1569 Thaba Putsoa, Lesotho Fresh 7 kg specimen. Coarse-grained garnet lherzolite with abundant enstatite having small phlogopite inclusions. Olivine slightly serpentinized. Sporadic garnet, chromium diopside, and chromite, (Nixon & Boyd 1973b).

PHN 2492 Kao, pipe No. 2, Lesotho Coarse-grained, compact, fresh, graphite-bearing garnet harzburgite. Purple garnet with narrow kelyphitic rims.

PHN 2759 Bultfontein, S. Africa Coarse-grained, slightly deformed garnet lherzolite. Olivine and enstatite up to 1 cm show peripheral recrystallization. Coarse phlogopite has granulated margins. Garnet (4 mm) mantled by secondary phlogopite and spinel (Boyd & Nixon 1978).

PHN 2764 Bultfontein, S. Africa Porphyroclastic garnet lherzolite with recrystallized olivine-forming automorphs; enstatite is deformed and spindle shaped, garnet, deep pink-lilac, rounded, lacking kelyphitic rim. Accessory diopside present (Boyd & Nixon 1978).

PHN 2766/6 Bultfontein, S. Africa Intensely deformed garnet harzburgite with fluidal texture and disrupted garnets strung out in alternating bands of fine mosaic of olivine and enstatite. Sporadic chromite apparently underformed. No cpx observed (Boyd & Nixon 1978).

PHN 2713 Monastery Mine, S. Africa Coarse-grained, fresh; gt plus cpx slightly more abundant than usual. Some phlogopite observed.

PHN 2771 Bultfontein, S. Africa Phlogopite-rich wehrlite with abundant medium green cpx and olivine porphyroclasts forming augen in mosaic olivine base. No garnet, but ilmenite and secondary rutile observed.

PHN 2780 Frank Smith Mine, S. Africa Fresh dark blue garnet harzburgite (chromium diopside lacking) with porphyroclastic texture (Boyd 1974).

PHN 2782 Frank Smith Mine, S. Africa Porphyroclastic texture. Rounded red-purple garnets and opx porphyroclasts in dark olivine mosaic, no cpx seen.

PHN 2814 Jagersfontein, S. Africa Slightly serpentinized coarse-grained lherzolite with minor cpx and garnet.

PHN 2819 Jagersfontein, S. Africa Mosaic porphyroclastic lherzolite with deep red, rounded, undeformed garnets (up to 7 mm across) with narrow kelyphitic rinds. Other minerals much finer, < 2 mm. Clinopyroxene, yellow-green subcalcic variety.

PHN 2823 Liqhobong, Lesotho Garnet lherzolite with coarse-grained texture grading to tabular texture defined by macroscopic orientation of coarse rectangular outlined orthopyroxene. Garnet rounded; very thin kelyphite. Accessory chromium diopside and patchily distributed phlogopite plates.

PHN 2829 Kao No. 2, Lesotho Highly deformed texture similar to 2819. Wine red, rounded, 4 mm garnets and yellow-green cpx together making up more than 20% of rock in fine mosaic matrix of olivine and cpx.

PHN 2838, 2839 Thaba Putsoa, Lesotho Highly deformed nodules with excellent foliation and mosaic-textured groundmass olivine, the deformed opx and cpx are elongated to produce a lineation. Rounded garnet porphyroclasts have suffered relatively slightly.

PHN 2848 Thaba Putsoa, Lesotho   Coarse-grained, olivine a little altered, sparse purple garnet, a little spinel noted. Minor bright green chromium diopside irregularly distributed.

PHN 2860 Matsoku, Lesotho   Fresh coarse garnet lherzolite with conspicuous bright green chromium diopside and large 5 mm spherical, purple garnets with narrow kelyphitic rims containing Cr spinel and pale mica. Serpentinized olivine well in excess of enstatite.

PHN 2862 Matsoku, Lesotho   Recrystallized, granuloblastic, slightly foliated rock consisting mainly of olivine, orthopyroxene, and some clinopyroxene with small "pools" of secondary opx granules, phlogopite, and spinel after ?garnet (cf plate 27B in Cox et al 1973). Generally fine-grained (0.2 mm) but occasional coarse opx (5 mm). Of possible secondary, cumulate origin.

PHN 3040 From dyke, Abbotsford, East Griqualand, S. Africa   Fresh dark grey garnet lherzolite, veined by serpentine and magnetite stringers. Garnet, subangular, with small amounts of kelyphite including peripheral picotite. Olivine, coarse (1 cm), forms contiguous background in which is found cusped pyroxene.

KIMBERLITES

PHN 1334 Kolo, Lesotho   Fine porphyritic hardebank consisting of partially altered dark pistachio-green olivine phenocrysts (15% of rock) in carbonate and serpentine matrix.

PHN 1598 Thaba Putsoa, Lesotho   Tough dark blue-green rock consisting of small (< 1 mm) rounded olivine serpentinous pseudomorphs in chlorite-magnetite carbonate matrix. Grains of garnet, diopside, ilmenite, phlogopite, and country rock basaltic xenoliths noted.

PHN 1725 Ramatseliso, Lesotho   Dark grey blue hardebank with xenocrysts of altered olivine and lesser garnet, pyroxenes, and ilmenite scattered through an aphanitic matrix consisting of calcite, magnetite, perovskite, talc (?), and apatite.

PHN 1867 Monastery Mine, S. Africa   "East end" variety (Whitelock 1973). Pale bluish grey, composed of serpentinized olivine and later calcite. Occasional coarse fresh olivine and ilmenite. Groundmass includes phlogopite, magnetite, perovskite, chromite, and possibly monticellite.

PHN 2201 Star Mine (east), S. Africa   Fresh hard dark grey kimberlite from dyke (underground). Porphyritic with fresh olivine phenocrysts up to 1 cm in groundmass noticeably rich in phlogopite.

PHN 2257 Kimb. Dyke N. W. of Mothae, Lesotho   Aphanitic grey hardebank with microporphyritic subhedral fresh olivine (0.3 mm) forming 40% of the rock. The groundmass consists of magnetite, perovskite, chlorite, and (?) monticellite.

PHN 2384 Frank Smith, S. Africa   Tough dark grey, aphanitic kimberlite rich in olivine.

PHN 2732 Kimberlite, Wesselton, South Africa   Hard dark bluish grey with talcose xenoliths and some ultrabasic inclusions with phlogopite. 25% of rock consists of rounded olivine phenocrysts, variably serpentinized with occasional ilmenite and mica in fine groundmass of calcite and phlogopite.

PHN 2779 Frank Smith, S. Africa   Similar to 2384.

PHN 2796 Bellsbank kimberlite, S. Africa   Fresh hard rock with 40% rounded fresh green olivines (< 1 cm) and sporadic garnet, clinopyroxene, coarse phlogopite in fine calcite-phlogopite matrix.

PHN 2811 Jagersfontein, S. Africa   Compact kimberlite, greenish grey, fine grained with 10–15% olivine and orthopyroxene phenocrysts mostly < 4 mm; rare garnet.

Literature Cited

Barrett, D. R. 1975. The genesis of kimberlites and associated rocks, strontium isotopic evidence. *Phys. Chem. Earth* 9:637–53

Boullier, A. M., Nicolas, A. 1973. Texture and fabric of peridotite nodules from kimberlite at Mothae, Thaba Putsoa, and Kimberley. In *Lesotho Kimberlites*, ed. P.

H. Nixon, pp. 57–66. Maseru, Lesotho: Lesotho Natl. Develop. Corp. 350 pp.

Boyd, F. R. 1973. A pyroxene geotherm. *Geochem. Cosmochim. Acta* 37:2533–46

Boyd, F. R. 1974. Ultramafic nodules from the Frank Smith kimberlite pipe, South Africa. *Carnegie Inst. Washington Yearb.* 73:285–94

Boyd, F. R. 1976. Inflected and noninflected geotherms. *Carnegie Inst. Washington Yearb.* 75:521–94

Boyd, F. R., Nixon, P. H. 1972. Ultramafic nodules from the Thaba Putsoa kimberlite pipe. *Carnegie Inst. Washington Yearb.* 71:362–73

Boyd, F. R., Nixon, P. H. 1973. Origin of the ilmenite-silicate nodules in kimberlites from Lesotho and South Africa. In *Lesotho Kimberlites*, ed. P. H. Nixon, pp. 254–68. Maseru, Lesotho: Lesotho Natl. Develop. Corp. 350 pp.

Boyd, F. R., Nixon, P. H. 1978. Ultramafic nodules from the Kimberley Pipes, South Africa. *Geochim. Cosmochim. Acta* 42:1367–82

Boyd, F. R., Nixon, P. H. 1979. Garnet lherzolite xenoliths from the kimberlites of East Griqualand, South Africa. *Carnegie Inst. Washington Yearb.* 78:488–92

Brunfelt, A. O., Roelandts, I., Steinnes, E. 1974. Determination of rubidium, cesium, barium, and eight rare earth elements in ultramafic rocks by neutron activation analysis. *Analyst* 99: 277–84.

Carswell, D. A., Clarke, D. B., Mitchell, R. H. 1979. The petrology and geochemistry of ultramafic nodules from Pipe 200, northern Lesotho. In *Proc. 2nd Int. Kimberlite Conf.*, ed. F. R. Boyd, H. O. A. Meyer, 2:127–144. Washington, DC: Am. Geophys. Union. 423 pp.

Cox, K. G., Gurney, J. J., Harte, B. 1973. Xenoliths from the Matsoku Pipe. In *Lesotho Kimberlites*, ed. P. H. Nixon, pp. 76–100. Maseru, Lesotho: Lesotho Natl. Develop. Corp. 350 pp.

Danchin, R. V., Boyd, F. R. 1976. Ultramafic nodules from the Premier kimberlite pipe, South Africa. *Carnegie Inst. Washington Yearb.* 75:531–38

Davis, G. L. 1977. The ages and uranium contents of zircons from kimberlites and associated rocks. *2nd Int. Kimberlite Conf. Abstr., Santa Fe, New Mexico*, 1977

Dawson, J. B., Gurney, J. J., Lawless, P. J. 1975. Paleogeothermal gradients derived from xenoliths in kimberlite. *Nature* 257:299–300

Dawson, J. B., Smith, J. V. 1975. Occurrence of diamond in a mica-garnet lherzolite xenolith from kimberlite. *Nature* 254:580–81

Fesq, H. W., Kable, E. J. D., Gurney, J. J. 1975. Aspects of the geochemistry of kimberlites from the Premier mine, and other selected South African occurrences with particular reference to the rare earth elements. *Phys. Chem. Earth* 9:687–707

Frey, F. A., Green, D. H. 1974. The mineralogy, geochemistry, and origin of lherzolite inclusions in Victorian basanites. *Geochim. Cosmochim. Acta* 38:1023–95

Frey, F. A., Prinz, M. 1978. Ultramafic inclusions from San Carlos, Arizona: petrologic and geochemical data bearing on their petrogenesis. *Earth Planet. Sci. Lett.* 38:129–76

Frey, F. A., Ferguson, J., Chappell, B. W. 1977. Petrogenesis of South African and Australian kimberlite suites. *2nd Int. Kimberlite Conf. Abstr., Santa Fe, New Mexico*, 1977

Frey, F. A., Haskin, L. A., Haskin, M. A. 1971. Rare earth abundances in some ultramafic rocks. *J. Geophys. Res.* 76:1184–96

Green, H. W., Gueguen, Y. 1974. Origin of kimberlite pipes by diapiric upwelling in the upper mantle. *Nature* 249:617–20

Gurney, J. J., Ebrahim, S. 1973. Chemical composition of Lesotho kimberlites. In *Lesotho Kimberlites*, ed. P. H. Nixon, pp. 280–84. Maseru, Lesotho: Lesotho Natl. Develop. Corp. 350 pp.

Gurney, J. J., Switzer, G. S. 1973. The discovery of garnets closely related to diamonds in the Finsch Pipe, South Africa. *Contrib. Mineral. Petrol.* 39:103–16

Harris, P. G., Middlemost, F. A. 1969. The evolution of kimberlites. *Lithos* 3:77–88

Harris, P. G., Hutchinson, R., Paul, D. K. 1972. Plutonic xenoliths and their relation to the upper mantle. *Philos. Trans. R. Soc. London Ser. A* 271:313–23

Harte, B. 1977. Rock nomenclature with particular relation to deformation and recrystallization in olivine-bearing xenoliths. *J. Geol.* 85:279–88

Harte, B., Cox, K. G., Gurney, J. J. 1975. Petrography and geological history of upper mantle xenoliths from the Matsoku kimberlite pipe. *Phys. Chem. Earth* 9:447–506

Harte, B., Gurney, J. J., Harris, J. W. 1979. The origin of CaO-poor silicate inclusions in diamonds. *Extended Abstr., 2nd Kimberlite Symp., Cambridge, U.K., 1979*

Haskin, L. A., Wildeman, T. R., Haskin, M. A. 1968. An accurate procedure for determination of rare-earth elements by neutron activation. *J. Radioanal. Chem.* 1:337–48

Kramers, J. D. 1979. Lead, uranium, strontium, potassium, and rubidium in inclusion-bearing diamonds and mantle-derived xenoliths from southern Africa. *Earth Planet. Sci. Lett.* 42:58–70

MacGregor, I. D. 1975. Petrologic and thermal structure of the upper mantle beneath South Africa in the Cretaceous. *Phys. Chem. Earth* 9:455–66

Mercier, J. C. 1979. Peridotite xenoliths and the dynamics of kimberlite intrusion. In *Proc. 2nd Int. Kimberlite Conf.*, ed. F. R. Boyd, H. O. A. Meyer, 2:197–212. Washington, DC: Am. Geophys. Union. 399 pp.

Meyer, H. O. A., Boyd, F. R. 1972. Composition and origin of crystalline inclusions in natural diamonds. *Geochim. Cosmochim. Acta* 36:1225–73

Meyer, H. O. A., Svisero, D. P. 1975. Mineral inclusions in Brazilian diamonds. *Phys. Chem. Earth* 9: 797–815

Meyer, H. O. A., Tsai, H. M. 1976. The nature and significance of mineral inclusions in natural diamond: a review. *Minerals Sci. Engng.* 8:242–61

Mitchell, R. H. 1978. Garnet lherzolites from Somerset Island, Canada and aspects of the nature of perturbed geotherm. *Contrib. Mineral. Petrol.* 67:341–47.

Mitchell, R. H., Brunfelt, A. O. 1975. Rare earth geochemistry of kimberlite. *Phys. Chem. Earth* 9:671–86

Mitchell, R. H., Carswell, D. A. 1976. Lanthanum, samarium and ytterbium abundances in some Southern African garnet lherzolites. *Earth Planet. Sci. Lett.* 31:175–78

Morgan, J. W., Wandless, G. A., Petrie, R. K., Irving, A. J. 1979. Earth's upper mantle: volatile element distribution and origin of siderophile element content. *Lunar Planet. Sci. XI* (Abstr.)

Mysen, B. O. 1979. Trace-element partitioning between garnet peridotite minerals and water-rich vapour: experimental data from 5 to 30 kbar. *Am. Mineral.* 64:274–87

Nixon, P. H., Boyd, F. R. 1973a. The discrete nodule association in kimberlites from northern Lesotho. In *Lesotho Kimberlites*, ed. P. H. Nixon, pp. 67–75. Maseru, Lesotho:Lesotho Natl. Develop. Corp. 350 pp.

Nixon, P. H., Boyd, F. R. 1973b. Petrogenesis of the granular and sheared ultrabasic nodule suite in kimberlites. In *Lesotho Kimberlites*, ed. P. H. Nixon, pp. 48–56. Maseru, Lesotho:Lesotho Natl. Develop. Corp. 350 pp.

Nixon, P. H., Boyd, F. R. 1975. Mantle evolution based on studies of kimberlite nodules from Southern Africa. *19th Ann. Rep. Res. Inst. Afr. Geol., Univ. Leeds,* pp. 26–31

Nixon, P. H., von Knorring, O., Rooke, J. M. 1963. Kimberlites and associated inclusions: a mineralogical and geochemical study. *Am. Mineral.* 48:1090–1132

Nixon, P. H., Boyd, F. R., Boullier, A. M. 1973. The evidence of kimberlite and its inclusion on the constitution of the outer part of the earth. In *Lesotho Kimberlites*, ed. P. H. Nixon, pp. 312–318. Maseru, Lesotho: Lesotho Natl. Develop. Corp. 350 pp.

Paul, D. K., Nixon, P. H., Buckley, F. 1975. Fluorine and chlorine geochemistry of kimberlites. *19th Ann. Rep. Res. Inst. Afr. Geol., Univ. Leeds,* pp. 32–35

Philpotts, J. A., Schnetzler, C. C., Thomas, H. H. 1972. Petrogenetic implications of some new geochemical data on eclogitic and ultrabasic inclusions. *Geochim. Cosmochim. Acta* 36:1131–66

Ridley, W. I., Dawson, J. B. 1975. Lithophile trace element data bearing on the origin of peridotite xenoliths, ankaramite, and carbonatite from Lashaine volcano, N. Tanzania. *Phys. Chem. Earth* 9:558–69

Ringwood, A. E. 1975. *Composition and Petrology of the Earth's Mantle*, pp. 176–205. New York: McGraw-Hill. 618 pp.

Rogers, N. W. 1979. *Trace element analysis of kimberlites and associated rocks and xenoliths.* PhD thesis. Univ. London. 266 pp.

Shimizu, N. 1975. Rare earth elements in garnets and clinopyroxenes from garnet lherzolite nodules in kimberlites. *Earth Planet. Sci. Lett.* 25:26–32

Wagner, P. A. 1928. The evidence of kimberlite pipes on the constitution of the outer part of the earth. *S. Afr. J. Sci.* 25:127–48

Wendlandt, R. F., Harrison, W. J. 1979. REE partitioning between immiscible carbonate and silicate liquids and CO_2 vapour. *Contrib. Mineral. Petrol.* 64:409–19

Whitelock, T. K. 1973. The Monastery Mine kimberlite pipe. In *Lesotho Kimberlites*, ed. P. H. Nixon, pp. 214–17. Maseru, Lesotho: Lesotho Natl. Develop. Corp. 350 pp.

Williams, A. F. 1932. *The Genesis of the Diamond.* Vols. 1, 2. London: E. Benn. 636 pp.

Ann. Rev. Earth. Planet. Sci. 1981. 9:311–44

THE COMBINED USE OF OXYGEN AND RADIOGENIC ISOTOPES AS INDICATORS OF CRUSTAL CONTAMINATION

✖ 10153

David E. James
Department of Terrestrial Magnetism, Carnegie Institution of Washington,
5241 Broad Branch Road, N.W., Washington, DC 20015

INTRODUCTION

The dynamical processes of mantle convection, sea-floor spreading, sub-duction, and continental accretion may be viewed as mechanisms for the redistribution or recycling of the chemical and isotopic constituents of the crust and mantle. By studying these chemical redistribution processes we obtain a more fundamental understanding of mantle and crustal evolution through geologic time. The analysis of $^{18}O/^{16}O$ ratios is a powerful tool for tracing the geochemical cycle because of the large difference in oxygen isotopic composition between crustal rocks and rocks derived from mantle material. Rocks that have reacted with the atmosphere or hydrosphere at low temperature are typically richer in ^{18}O than are those from the mantle. We can take advantage of this disparity in $^{18}O/^{16}O$ ratios to assess the following: (*a*) contamination of mantle-derived magma through assimilation of or isotope exchange with continental crustal material, and (*b*) reflux of subducted crustal material in island arc and continental arc volcanics or in oceanic basalts. For clarity of discussion, the term "crustal contamination" will refer to (*a*) processess and the term "source contamination" to (*b*) processes.

Analyses of basalts and other volcanic rocks of presumed mantle origin reveal significant and systematic differences in Sr, Nd, and Pb isotopic ratios. These isotopic variations are generally attributed, at least in part, to differing parent/daughter ratios in mantle reservoirs that have

311

0084-6597/81/0515-0311$01.00

been isolated from one another for long periods of geologic time (e.g. Brooks et al 1976a,b, DePaolo & Wasserburg 1976a,b, Tatsumoto 1978, Sun & Hanson 1975, Richard et al 1976, O'Nions et al 1977, 1978, 1979a,b, Wasserburg & DePaolo 1979, Jacobsen & Wasserburg 1979, DePaolo 1980, Sun 1980, Hofmann & White 1980). These isotopic variations are thought to provide fundamental information for understanding crust-mantle evolution and mantle structure and dynamics. In most rocks studied, however, the possibility of involvement of crustal material either as subducted crust in the mantle source region or as assimilated country rock cannot be excluded. Crustal contamination at any stage in magma-genesis will mask the isotopic character of the mantle and increase the the ambiguity of mantle-crust evolution models. The combined use of oxygen and radiogenic isotopes provides a means for assessing the role of crustal contamination in the genesis and evolution of mantle-derived magmas.

Oxygen isotope analysis is a comparatively unambiguous means of separating the relative contribution of crust and mantle. Oxygen isotopes when analyzed together with radiogenic isotopes and trace elements in the *same* rock samples are sensitive tracers of crustal contamination because variations in stable and radiogenic isotopic ratios are produced by entirely different mechanisms. Radiogenic isotopic variations reflect differing parent/daughter ratios in mantle or crustal reservoirs isolated from one another for significant periods of time. For example, continental crust, which is enriched in incompatible elements relative to the mantle, is characterized by higher $^{87}Sr/^{86}Sr$, lower $^{143}Nd/^{144}Nd$, and more radiogenic Pb isotopic ratios than mantle rocks. Variations of $^{18}O/^{16}O$ ratios are produced by isotope fractionation, the larger part of which is the result of water-rock interactions near the earth's surface. Low temperature ($< 300°C$) isotope fractionation between surface water and rock-forming minerals has, over time, produced substantial ^{18}O enrichment in crustal rocks relative to mantle rocks. Thus, the presence of *both* oxygen and radiogenic isotopic anomalies in unaltered volcanic rocks indicates contamination by material that was at one time near the earth's surface.

Combined use of oxygen and radiogenic isotopes to measure crustal contamination is a comparatively new approach. It was first employed in a systematic way by Magaritz et al (1978), although there are a number of earlier studies of volcanic rocks in which a regional correlation between oxygen and strontium or lead isotopic ratios was observed (e.g. Taylor 1968, Taylor & Turi 1976, and Turi & Taylor 1976). Taylor et al (1979) and Cortecci et al (1979) studied both oxygen and radiogenic isotopic ratios in volcanic and intrusive rocks from Italy and presented evidence for crustal contamination. Taylor (1980) quantitatively modeled the predicted effects of crustal assimilation on the oxygen and strontium isotopic systematics in mantle-derived magmas. Magaritz et al (1978) and James (1978, 1980) correlated oxygen and strontium isotopic ratios to investigate the contribution of subducted oceanic crust and continental sediments to island arc and continental arc volcanism. No systematic investigation has yet been made into possible correlations between oxygen and radiogenic isotopic ratios of oceanic basalts.

The paucity of data correlating oxygen and radiogenic isotopic variations is an obvious limiting factor in a review of the subject. The present paper, therefore, is less a traditional review than a prospectus outlining the nature and scope of the problem and describing those aspects of oxygen and radiogenic isotopic correlations that are diagnostic of the type and extent of contamination by crustal material. Crustal involvement in the genesis or evolution of mantle-derived magmas will be reviewed as follows:

1. Partial or complete melting of subducted crustal material stored deep in the suboceanic mantle to produce some oceanic basalts.
2. Mixing of subducted crustal material with normal mantle rock to generate island arc or continental arc volcanic rocks.
3. Crustal assimilation or isotope exchange with mantle-derived magma during its ascent through the continental crust.

BACKGROUND

Oxygen is the most abundant element in the earth's crust and mantle. It comprises about 45–50 wt % of most common igneous rocks. The variation in relative abundance of oxygen between rock types is small compared to variations in other elemental abundances. Oxygen is the dominant element in almost all geologic processes in the earth and it therefore plays a central role in the evolution of igneous rocks.

There are three stable isotopes of oxygen: $^{16}O = 99.763\%$; $^{17}O = 0.0375\%$; $^{18}O = 0.1995\%$ (Garlick 1969). The only isotopic ratio normally measured in terrestrial rocks is $^{18}O/^{16}O$ and it is now always expressed relative to standard mean ocean water (SMOW) by the relationship

$$\delta^{18}O(\%_0) = \left[\frac{(^{18}O/^{16}O)_{sample} - (^{18}O/^{16}O)_{SMOW}}{(^{18}O/^{16}O)_{SMOW}}\right] \times 1000.$$

Accuracy of measurement is typically ± 0.1 to 0.2 permil.

Isotope Fractionation

Oxygen isotope fractionation between two materials in equilibrium is defined as

$$\Delta_{A-B} = 1000 \ln \alpha_{A-B} \simeq (\delta^{18}O)_A - (\delta^{18}O)_B,$$

where α_{A-B} = fractionation factor for coexisting substances A and B = $(^{18}O/^{16}O)_A/(^{18}O/^{16}O)_B$. The absolute value of α_{A-B} is a function of equilibration temperature and it decreases with increasing temperature (see Urey 1947). Pressure effects on fractionation appear to be negligible (Clayton et al 1975). Oxygen isotope fractionation factors have been compiled in a paper by Friedman & O'Neil (1977).

Large differences in $\delta^{18}O$ between different materials in the earth are the result of low-temperature fractionation which, over time, has produced naturally occurring variations of about 10% (100 permil) or more (Hoefs 1973). About half that variation occurs in rocks, and the remainder in the secular variation in meteoric water.

Oxygen isotopic variations in crustal rocks are chiefly the result of isotopic fractionation between water and minerals. Because water-mineral fractionation factors are strongly temperature dependent (see Figure 1), water-rock interactions will produce ^{18}O enrichment or depletion in the rock depending upon $\delta^{18}O$ of water and rock and the equilibration temperature. The $\delta^{18}O$ of seawater is about 6 permil lower, and that of meteoric water about 10 to 20 permil lower, than that of typical volcanic rocks. For hydrothermal alteration by seawater at temperatures above

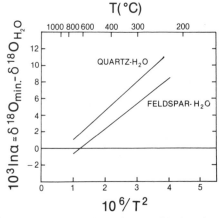

Figure 1 Experimentally determined oxygen isotope fractionation between quartz and water and feldspar and water (after Hoefs 1973).

about 300°C, the fractionation between water and rock is sufficiently small (<6 permil) that the rock is depleted in ^{18}O. At temperatures less than 300°C, the water-rock fractionation factor becomes larger (Figure 1), so that low-temperature alteration processes (e.g. weathering, diagenesis) will result in higher $\delta^{18}O$ in the altered rock.

Oxygen isotope fractionation between minerals at high temperature in igneous and metamorphic rocks is comparatively small. The effects, however, are sufficiently large (a few permil) to be important in cases where contamination is expected to produce variations in $\delta^{18}O$ that are of the same order as those that could result from natural fractionation during differentiation. A good empirical estimate for fractionation in common igneous rocks during differentiation comes from work by Matsuhisa (1979) on a highly differentiated sequence of tholeiitic lavas from Japan. He finds $\delta^{18}O$ increases from about 5.7 permil in rocks with 47% SiO_2 to about 6.7 permil in rocks with 74% SiO_2. In other regions, however, such as the Cascades or the Roman province in Italy evidence for oxygen fraction during differentiation even for rhyolites is not clear (Taylor 1968, Turi & Taylor 1976).

Numerous empirical relationships have been published for isotope fractionation between minerals (see compilation by Bottinga & Javoy 1975). The generally accepted sequence of minerals in order of *decreasing* $\delta^{18}O$ is given as quartz–alkali feldspar–muscovite–plagioclase–amphibole, pyroxene, biotite–garnet–olivine–ilmenite–magnetite (Garlick 1969, Bottinga & Javoy 1975). For fractionation coefficients determined by Anderson et al (1971) the fractionation between plagioclase and magnetite ($\delta^{18}O_{plag}-\delta^{18}O_{magnetite}$) at 1000°C is about 2.4 permil, and between plagioclase and clinopyroxene about 0.7 permil. In basalts and andesites, $\delta^{18}O$ of plagioclase is typically very close to $\delta^{18}O$ of the whole rock whereas olivine is depleted in ^{18}O. Hence, loss of olivine during magma differentiation will cause the residual liquid to be enriched in ^{18}O.

Oxygen isotope fractionation at very high (mantle) temperatures is poorly understood. On many theoretical, experimental, and empirical bases, isotope fractionation is expected to become smaller and approach zero with increasing temperature (Urey 1947, Bottinga & Javoy 1975). Little experimental work has been done, however, at temperatures above about 800°C. A small number of experimental results presented by Muehlenbachs & Kushiro (1974), and empirical studies by Kyser (1979) and Muehlenbachs & Jakobsson (1979—data presented in oral presentation but not in abstract) indicate that a crossover phenomenon may occur in some mineral pairs at very high temperatures (i.e. fractionation decreases to zero and reverses sign with increasing temperature). If correct, such behavior would greatly complicate the interpretation of oxygen isotopes in mantle-derived magmas. Other recent experimental results reported by Matthews et al (1980), however, show that the fractionation factor for common minerals extrapolates smoothly to zero at infinite temperature. Resolution of this question will depend upon considerably more data from controlled experiments at temperatures of 1000 to 1300°C.

Oxygen Isotopic Ratios in Rocks

A summary of oxygen isotopic ratios in rocks, and their relationship to radiogenic isotopic ratios and geologic setting, is useful for the discussions of crustal contamination that follow. A schematic plot of $\delta^{18}O$ versus $^{87}Sr/^{86}Sr$ given in Figure 2 shows fields of isotopic compositions that characterize common rocks.

MANTLE   Mantle compositions may be determined directly by studying unaltered ultramafic xenoliths brought rapidly to the surface in explosive volcanic vents. Such ultramafic nodules characteristically exhibit monotonously similar $\delta^{18}O$ values between about 5.4 and 5.8 permil (Reuter et al 1965, Taylor 1968, Garlick et al 1971, Kyser & O'Neil 1978). Oxygen isotopic ratios in chondritic meteorites and in lunar samples are close to ratios in ultramafic nodules (Taylor et al 1965, Reuter et al 1965, Epstein & Taylor 1975, Mayeda et al 1975). The consistency of these results is generally cited as evidence that the mantle has a uniform oxygen isotopic composition. However, Garlick et al (1971) have measured $\delta^{18}O$ values ranging between 2 and 8 permil in eclogite nodules in kimberlite. They attribute this variation to pressure-dependent fractionation in the mantle. They conclude that if their interpretation is correct, "the mean isotopic

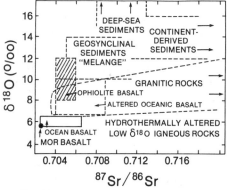

Figure 2 $\delta^{18}O$ versus $^{87}Sr/^{86}Sr$ for common igneous and sedimentary rocks (adapted from Magaritz et al 1978).

composition of the earth need not be that of the upper mantle, and the $^{18}O/^{16}O$ ratios of basaltic melts can be expected to decrease with increasing depth of origin."

OCEANIC CRUST   Evidence for the oxygen isotopic composition of oceanic crust comes from (a) the study of ocean-floor basalts collected during deep-sea dredging and drilling operations, and (b) the study of ophiolite terranes presumed to represent pieces of old oceanic crust and mantle. (There is still considerable difference of opinion, however, as to whether ophiolite bodies are actually samples of "normal" oceanic crust.)

Fresh sea-floor basalt has an average $\delta^{18}O$ of about 5.7 permil (e.g. Muehlenbachs & Clayton 1976); however, most ocean-floor basalts show evidence of oxygen isotope exchange with seawater. Weathered sea-floor basalts are enriched in ^{18}O by up to about 10 permil (Muehlenbachs & Clayton 1972, 1976, Muehlenbachs 1976a,b, Gray et al 1976). The amount of ^{18}O enrichment correlates with H_2O content of the rocks. For samples that have been in contact with seawater for most of their existence (e.g. dredged samples), $\delta^{18}O$ also shows a clear correlation with age (Muehlenbachs & Clayton 1972). Drilled samples collected as part of the deep-sea drilling program show ^{18}O enrichment of 1 to 4 permil produced by cold seawater penetration and weathering of oceanic crust to a depth of at least 600 m.

Large hydrothermal systems have been inferred beneath the mid-ocean ridges on the basis of heat flow anomalies (e.g. Wolery & Sleep 1976) and from direct observation of submarine hot springs (Corliss et al 1979). Hydrothermal alteration by seawater at temperatures in excess of about 300°C has depleted deeper parts of the oceanic crust by 1 to 2 permil.

Investigations of ophiolite complexes provide additional information on the composition of oceanic crust (see Gregory & Taylor 1980, with reference to previous work). The oxygen isotope study of the Samail ophiolite, Oman, by Gregory & Taylor (1980), together with complementary work by McCulloch et al (1980) on Nd and Sr isotopic ratios, present important results on isotopic variations in a well-preserved ophiolite section. Figure 3, modified from Gregory & Taylor, shows an oxygen isotope profile which was measured through the Samail ophiolite. Maximum $\delta^{18}O$ values (+12 permil) occur in rocks of the uppermost part of the pillows lavas. Whole rock values decrease to about 10 permil at the base of the pillow basalts (~0.5 km depth), continue decreasing through the sheeted dike complex, and reach "normal" mantle values of +5.8 permil at the base of the dike complex (~1.8 km). Below about 1.8 km, $\delta^{18}O$ values are lower than mantle values by about 1 to 2 permil down to the M-discontinuity at about 7.4 km depth.

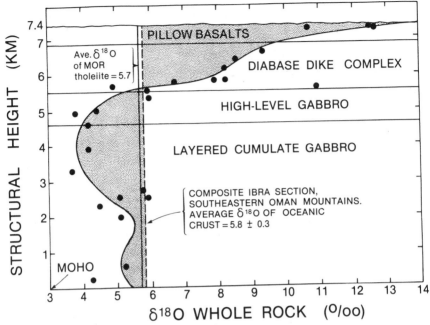

Figure 3   Whole rock $\delta^{18}O$ profile through Samail ophiolite from Gregory & Taylor (1980). Solid circles are individual whole rock determinations.

Sea-floor weathering of oceanic crust produces a marked increase in $^{87}Sr/^{86}Sr$ ratios, perhaps up to 0.706 (Hart 1971, Hart et al 1974, Dasch et al 1973, McCulloch et al 1980) but virtually no change in $^{143}Nd/^{144}Nd$ ratios (O'Nions et al 1977). In general, the increase in Sr isotopic ratios correlates with increases in $\delta^{18}O$ in weathered sea-floor basalt (e.g. Gray et al 1976, McCulloch et al 1980). For deeper parts of oceanic crust, where hydrothermal alteration is important, strontium isotopic ratios appear to be less disturbed than those of oxygen (Ito 1980, McCulloch et al 1980).

The oxygen isotopic composition of oceanic crust subducted into the mantle in the geologic past will depend upon the variation in $\delta^{18}O$ of seawater over time. Gregory & Taylor conclude that the *average* $\delta^{18}O$ of the igneous oceanic crust has not been significantly altered by interaction with seawater, and, consequently, that $\delta^{18}O$ of the oceans has not changed much with time. Much the same conclusion was reached earlier by Muehlenbachs & Clayton (1976). Perry (1967) and Perry et al (1978) suggest on the contrary that $\delta^{18}O$ of seawater has been increasing steadily through geologic time (perhaps 5 permil/b.y.). If $\delta^{18}O$ of seawater was formerly lower than at present, $\delta^{18}O$ values of ancient oceanic crust would be lower by a constant amount relative to $\delta^{18}O$ values seen in Figure 3.

ISLAND ARCS   Volcanic rocks of island arcs and continental arcs have $\delta^{18}O$ values slightly higher (perhaps 0.5 to 1.0 permil after being corrected for crystal fractionation) and more variable than $\delta^{18}O$ values of oceanic basalts (Taylor 1968, Matsuhisa 1979, James 1980). Strontium isotopic ratios of island arc volcanics are roughly comparable to those of oceanic island basalts, averaging about 0.704. Both $\delta^{18}O$ and $^{87}Sr/^{86}Sr$ ratios can be much higher in continental arc volcanic rocks (Taylor 1968, Matsuhisa 1979, James 1980, Magaritz et al 1978). One instance in which this is also true for a purely oceanic island arc is reported by Magaritz et al (1978) (see discussion below on island arc and continental arc volcanism).

CONTINENTAL CRUST   In continental igneous rocks, both Sr and O isotopic ratios tend to be higher and more variable than in oceanic rocks, although many continental flood basalts are indistinguishable from oceanic island basalts. On average, however, because continental crust is enriched in Rb relative to the mantle, $^{87}Sr/^{86}Sr$ ratios are higher in continental crustal rocks than in rocks of oceanic origin. $\delta^{18}O$ values, with a few exceptions, tend to be significantly higher than those of oceanic or island arc rocks because of extensive reworking of weathered material in the continental crust. The average $\delta^{18}O$ of crystalline rocks from the Canadian shield, for example, is about 8.0 permil, an increase of about 2 permil over average mantle values and presumed to be the result of crustal reworking (Shieh & Schwarcz 1978). Most continental granitic rocks have $\delta^{18}O$ values between $+6$ and $+9$ permil. There are, however, numerous documented examples of granitic bodies either depleted or enriched in ^{18}O (see review by Taylor 1978). Low $\delta^{18}O$ values (<6 permil) in granitic rocks and adjacent country rocks are produced by interaction with large hydrothermally convecting reservoirs of meteoric water; high $\delta^{18}O$ values ($> +9$ permil) are generally assumed to be produced by melting or massive assimilation of high $\delta^{18}O$ crustal rocks (Taylor 1978). Increases in $\delta^{18}O$ appear typically to be accompanied by increases in $^{87}Sr/^{86}Sr$.

Sedimentary and metasedimentary rocks are enriched in ^{18}O and ^{87}Sr relative to mantle material and most igneous rocks. Pelitic and argillaceous sediments are characterized by $\delta^{18}O$ values of about $+15$ to $+20$ permil (Savin & Epstein 1970, Garlick & Epstein 1967, Magaritz & Taylor 1976) and $^{87}Sr/^{86}Sr$ ratios range from about 0.71 upward. While it is well known that metasedimentary rocks may be depleted in ^{18}O by a few permil relative to the protolith (Garlick & Epstein 1967), there remains controversy over whether regional metamorphism can produce large decreases in $\delta^{18}O$. Shieh & Schwarcz (1974) studied high grade metamorphic and migmatized granulites of the Grenville Province, Canada, and concluded that $\delta^{18}O$ in metamorphosed pelitic rocks may be reduced to as low as $+8$ permil by exchange with a mantle-like reservoir. Studies by James et al (1980), however, of lower crustal xenoliths from Kilbourn Hole maar, New Mexico, show that even pelitic rocks that have undergone extensive anatexis and been subjected to high grade metamorphism for long periods of time retain $\delta^{18}O$ values of $+11$ to $+12$ permil. It is probable that large changes in $\delta^{18}O$ during metamorphism or even during partial melting require the presence of convecting fluids.

CONTAMINATION PROCESSES

The composition of magma erupted at the earth's surface reflects a complex history of partial melting of the source material, fractionation of the magma during its ascent, and exchange with or assimilation of material from the walls and roofs of the enclosing conduits and chambers. Crustal rocks may influence the composition of mantle-derived magmas by

1. Crustal contamination resulting from assimilation by, or isotopic exchange with, mantle-derived magma, or through magma mixing at crustal levels;
2. Source contamination resulting from melting of subducted material in the mantle source region.

These two contamination processes are fundamentally different. *Crustal contamination* involves physical and chemical reaction of the magma with crustal rocks it intrudes. The nature and extent of contamination depend on a variety of factors including the heat balance between assimilation and crystallization, composition and phase relations between contaminant and magma, reaction rates, and the presence or absence of fluids. Taylor (1980) points out that assimilation is a three end-member process—magma, contaminant, and cumulate—so that simple two end-member mixing calculations either are not applicable or are only approximations.

Source contamination involves closed system partial melting of crustal material or a mixture of crustal and mantle material within the mantle. Isotopic ratios in the magma will depend only on the isotopic compositions and relative proportions of materials involved in the melting. Source contamination, or hybridization, in its most elementary form should, therefore, obey two-component hyperbolic mixing relations.

Geochemical criteria are here described for identifying and distinguishing between source contamination and crustal contamination. Included in the discussion are oceanic island basalts, island and continental arc volcanic rocks, and continental volcanic rocks.

Oceanic Basalts

Fresh basalts from ocean basins exhibit systematic Sr, Nd, and Pb isotopic variations which suggest the lavas were derived from mantle reservoirs that have been separated from one another for at least 1 to 2 b.y. (see references in the introduction to this review). Although the heterogeneities appear to be ancient, it is not known whether they represent single-episodic, multiepisodic, or continuous differentiation in the mantle. Brooks et al (1976b) showed that oceanic tholeiites define a good $^{87}Sr/^{86}Sr$ versus Rb/Sr "isochron" of about 1.5 b.y., with MOR basalt occupying the low $^{87}Sr/^{86}Sr$ end of the isochron diagram and oceanic island tholeiites forming a quasilinear array toward higher $^{87}Sr/^{86}Sr$ ratios (see Figure 4a). The mantle isochron array was interpreted as due to mixing between magma originating in an undepleted intermediate or lower mantle and depleted material of the upper oceanic mantle (Brooks et al 1976b).

Compelling evidence for long-term separation of depleted and undepleted mantle reservoirs comes from the strong negative correlation observed between $^{87}Sr/^{86}Sr$ and $^{143}Nd/^{144}Nd$ ratios in oceanic basalts (see Figure 4b) (DePaolo & Wasserburg 1976a,b, O'Nions et al 1977, Richard et al 1976). The Sr-Nd anticorrelation can be interpreted as due to mixing between depleted and undepleted reservoirs (e.g. Wasserburg & DePaolo 1979).

Similar conclusions for 1 to 2 b.y. separation of depleted and undepleted mantle reservoirs were reached by Sun & Hanson (1975) and Sun (1980) on the basis of Pb isotopic measurements. Pb-Pb secondary isochrons of oceanic basalts give consistent "ages" between 1 and 2 b.y.

Because oceanic basalts are not affected by crustal contamination, the radiogenic isotopic ratios in fresh samples are almost always assumed to reflect purely mantle compositions. Indeed, the intensive study of oceanic basalts over the past several years has in large measure been aimed at describing different mantle reservoirs and relating their compositions to mantle depletion processes leading to continental accretion (e.g. O'Nions et al 1979a, Jacobsen & Wasserburg 1979, Allegre et al 1979, DePaolo 1980).

Most mantle evolution models invoke only three reservoirs: continental crust, depleted mantle, and undepleted mantle. Hofmann & White (1980) and Anderson (1979a,b) suggest that a fourth reservoir, subducted oceanic crust, must be considered. Hofmann & White emphasize that the

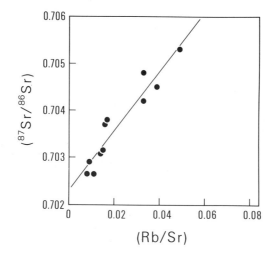

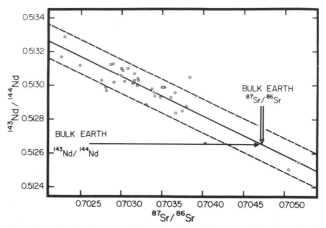

Figure 4 Isotopic variations in oceanic basalts. (*a*) Rb-Sr systematics for oceanic tholeiites (compliation from Brooks et al 1976b); (*b*) ε_{Nd} versus ε_{Sr} mantle array for oceanic basalts (compliation from O'Nions et al 1979b).

existence of such a reservoir could dramatically alter the results of the quantitative models cited above. By Anderson's model, MOR basalts are derived from subducted oceanic crust, which is presumed to accumulate above the 670 km discontinuity. Hofmann & White, however, in a comprehensive analysis of isotopic and trace element data argue that old subducted crust is *not* a plausible source for highly depleted MOR basalts, but that it could be an "undepleted" mantle reservoir for oceanic island basalts. In Hofmann & White's model, the subducted oceanic crust is assumed to reside in the lower mantle where thin (5–7 km) layers of oceanic crust accumulate to form a gradually thickening layer of material, basaltic in composition, from which oceanic island basalts are derived. According to Hofmann & White, their model provides a satisfactory explanation to the paradox of Sr and Nd isotopic time-integrated source "depletion" and Pb isotopic and incompatible trace element source "enrichment" in oceanic basalts relative to bulk mantle values.

If subducted crustal material retains its oxygen identity (i.e. does not reequilibrate with, or become homogenized with, normal mantle material), magmas derived from a deep crustal reservoir may be characterized by anomalous oxygen isotopic ratios. The correlation of these ratios with radiogenic isotopic ratios can be used to test models of oceanic volcanism.

Published oxygen isotopic studies of oceanic island basalts are almost entirely reconnaissance in nature. In no published work to date has any systematic attempt been made to correlate $\delta^{18}O$ with radiogenic isotopic ratios in the same rock samples. In fact $\delta^{18}O$ is rarely correlated with anything other than degree of differentiation or alteration. An exception to this is the work of Kyser & O'Neil (1978) and Kyser (1979) who found that alkali basalts in Hawaii are typically enriched in ^{18}O by 0.5 to 1.0 permil over tholeiites. They suggest that for Hawaii isotopically distinct sources are involved in the genesis of alkali basalts and tholeiites. Workers who have investigated other oceanic areas had previously concluded that there is no difference in $\delta^{18}O$ between alkali basalts and tholeiites.

It should be emphasized that different source materials for oceanic basalts can be distinguished reliably *only* on the basis of isotopic compositions and not on bulk chemistry, which is largely a function of degree of partial melting or differentiation. Moreover, for oxygen isotopes it is probably necessary to correct for olivine removal (see, for example, Wright et al 1975) or for other fractionation effects. Brooks et al (1976b) showed on the basis of Rb-Sr data that many alkali basalts are probably derived from the same source as their associated tholeiites but are enriched in incompatible elements as a result of smaller degrees of partial melting or through fractional crystallization. Similar conclusions may be inferred from the work of O'Nions et al (1977) who found no correlation between Sr and Nd isotopic ratios and rock type for Hawaiian lavas. The importance of oxygen isotopes in placing constraints on mantle source models comes from their *combined* use with radiogenic isotopes measured in the same rock samples. Any correlation between oxygen and radiogenic isotopic ratios in oceanic island basalts would be compelling evidence for source contamination.

Based on the comprehensive analysis by Hofmann & Hart (1978) of isotopic equilibration and homogenization processes and rates in the mantle, oceanic crust returned to the deep mantle is likely to remain out of oxygen isotopic equilibrium with the surrounding mantle. While Hofmann & Hart's arguments pertain specifically to radiogenic isotopic and trace element diffusion kinetics, their methods and conclusions are equally valid for oxygen. They consider two special cases, one where the rock is entirely solid and the other where it is incipiently melted, with melt interlocked along grain boundaries. The partial melt case is not considered here because at depths where oceanic crust is presumed to be stored (> 600 km) the mantle is completely solid (Anderson & Hart 1978, Sacks 1980). Hofmann & Hart show, however, that even in the presence of partial melt oceanic crust will not equilibrate with surrounding mantle in 10^9 years.

For temperatures of 1000 to 1200°C, the diffusivity, D, for oxygen in mantle minerals is $\sim 10^{-13}$ (see Hofmann & Hart 1978, Figure 3). The characteristic transport distance, defined by the approximate relation $X = \sqrt{Dt}$, gives a measure of the distance over which isotopic reequilibration may occur within a given time. For $t = 10^9$ years and $D = 10^{-13}$ cm^2 s^{-1}, $X = 56$ cm. Thus, while diffusion rates are sufficient to assure local equilibrium over 10^9 years, regional isotopic equilibrium between crustal layers or between crust and mantle will be completely negligible.

The above conclusions clearly depend upon numerous assumptions of varying plausibility, including the assumption that mantle convection is too slow to be effective in producing regional homogenization. A further unknown is the extent of compositional modification of oceanic crust during subduction. Even if these difficulties are ignored, however, it remains to be demonstrated that remobilization of subducted crustal rocks will produce magmas with $\delta^{18}O$ values measurably different from those of "normal" mantle. The $\delta^{18}O$ values of remobilized crust will depend on the site and rate of crustal melting and on the volumes of material involved. Partial melting of the upper parts of the crust will produce magma with high $\delta^{18}O$. If the hydrothermally altered gabbroic rocks of the crustal section melt (perhaps because of U enrichment produced during hydrothermal alteration), the melting product will have low $\delta^{18}O$. If the crustal section as a whole partially melts and rises diapirically, it is likely that mixing and equilibration will occur as a result of large shear strains in the diapir (Hofmann & Hart 1978) and the melting product could have $\delta^{18}O$ close to that of "normal mantle".

Conclusions on the validity of these speculations will have to await the gathering of much larger amounts of data. New investigations need to be undertaken specifically to test whether subducted oceanic crust ($\pm$ oceanic and/or continental sediments) plays any part in oceanic basalt magma-genesis. With growing sophistication and diversity of measurement, and with the increasing scale of effort, perhaps the time has come to coordinate studies to obtain complete isotopic measurements on comprehensive and representative suites of oceanic basalt and to synthesize the combined data into a coherent model of mantle evolution.

Island Arc and Continental Arc Volcanism

Lithospheric plates bend downward at the trenches and descend into the mantle along subduction zones. Island arcs or continental volcanic arcs overlie all of the world's subduction zones. It is widely assumed on the basis of the spatial relationship between subduction and volcanism that the magmatism is a consequence of subduction zone processes. Involve-ment of subducted continental and oceanic sediments as well as oceanic crust has been invoked by numerous workers to explain the composition-al characteristics of island arc or continental arc lavas (e.g. Armstrong 1968, 1971, Kay et al 1978, Kay 1977, Magaritz et al 1978, Whitford et al 1977, 1979, Thorpe et al 1976, James 1980, Muehlenbachs & Stern 1980). The extent to which subducted material may be recycled to the surface has a major bearing on volume estimates of continental or oceanic crustal material carried into the deep mantle, and on larger questions of conti-nental evolution.

Oceanic crust, oceanic sediments, and continental sediments may be involved in subduction zone melting. While there is considerable debate as to whether oceanic sediments are subducted or scraped off at the trench (e.g. Karig & Sharman 1975), it is clear from volume considerations that at least some of the 100 to 500 m layer of sediments that caps the oceanic crust as it nears the trench must be subducted. Oceanic sediments have radiogenic isotopic compositions similar to those of continental crust and tend to be enriched in many of the incompatible elements. $\delta^{18}O$ values for oceanic sediments are typically $+15$ to $+20$ permil (Savin & Epstein 1970), but range up to 30 permil for limestone and chert (Savin & Epstein 1970, Knauth & Epstein 1975). Where subduction zones underlie conti-nental masses (e.g. Andean and Japan arcs), large volumes of continental sedimentary material may be dumped into the trenches and consumed along with the oceanic plate (e.g. Gilluly 1971). The availability of old sialic material for subduction is one obvious distinction between purely oceanic island arcs and continental margin volcanic arcs. Studies of Franciscan graywackes (Magaritz & Taylor 1976) show them to have a narrow range of $\delta^{18}O$ between $+11$ and $+13$ permil, nearly identical to $\delta^{18}O$ of weathered oceanic basalt.

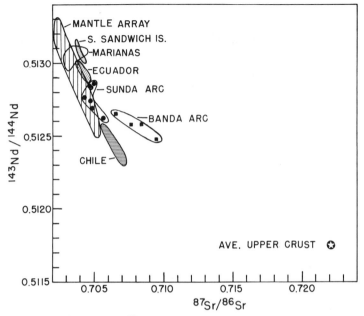

Figure 5 $^{143}Nd/^{144}Nd$ versus $^{87}Sr/86Sr$ isotopic ratios in island arc and continental volcanic arc volcanic rocks (compilation from Whitford et al 1979).

Volcanic rocks of island arcs and continental arcs exhibit a wide range of radiogenic isotopic compositions. Nd and Sr isotopic studies of some primitive island arcs show $^{87}Sr/^{86}Sr$ ratios in the volcanics to be increased relative to MOR basalts while $^{143}Nd/^{144}Nd$ ratios are the same (see Figure 5) (e.g. DePaolo & Wasserburg 1977, Hawkesworth et al 1977, DePaolo & Johnson 1979, Whitford et al 1979). In some volcanic rocks of the New Britain arc, however, neither Sr nor Nd isotopic ratios differ from those of unaltered MOR basalts (DePaolo & Johnson 1979). Where $^{87}Sr/^{86}Sr$ displacement is observed it is usually interpreted as due to involvement of altered MOR basalts, which are characterized by increased $^{87}Sr/^{86}Sr$ ratios relative to fresh MOR basalt and by undisturbed $^{143}Nd/^{144}Nd$ ratios. Pb isotopic ratios in some primitive arcs such as the Aleutians have been interpreted to require melting of small amounts of subducted oceanic sediments (e.g. Kay et al 1978).

Lavas of continental arcs and the more "evolved" island arcs are more varied in their radiogenic isotopic compositions. Relative to primitive island arc lavas, those of evolved arcs tend to have higher $^{87}Sr/^{86}Sr$, $^{207}Pb/^{204}Pb$, $^{208}Pb/^{204}Pb$, and lower $^{143}Nd/^{144}Nd$ and $^{206}Pb/^{204}Pb$ ratios. These isotopic ratios are commonly difficult to interpret because at least three reservoirs are involved: (a) oceanic crust, with or without a complement of continental sediments, (b) the overlying mantle wedge, and (c) the sialic crust through which the magma must rise. While the radiogenic isotopic composition of (b) is commonly distinct and diagnostic, the effects on radiogenic isotopic compositions of (a) and (c) cannot be separated. The discussion that follows shows that the correlated behavior of $\delta^{18}O$ and radiogenic isotopic ratios may, at least in some cases, be used to distinguish between subduction zone (i.e. source) contamination and crustal contamination.

Numerous models have been proposed for subduction zone magmatism (see Ringwood 1975 for a thorough review). For purposes of illustration, we consider here only the widely cited model of Nicholls & Ringwood (1973) for the generation of calc-alkaline magmas. The discussion below, however, may be generalized to apply to any model where melts or fluids derived from the slab percolate into and react with overlying mantle. By the Nicholls & Ringwood model, partial melts of subducted material rise into and react with overlying mantle, and diapirs of modified mantle rise to the surface to yield calc-alkaline magmas.

Subducted crustal material involved in magma genesis will influence *correlated* oxygen and radiogenic isotopic compositions in quite different ways than will crustal contamination of mantle-derived magma. This can be illustrated by considering the processes that determine the isotopic and trace element composition of the magma:

1. *Partial melts of subducted crustal material will be highly enriched in incompatible elements.* Minerals that tend to hold incompatible elements in oceanic crust or continental sediments at crustal depths (i.e. feldspar, amphibole, and biotite) are unstable at depths of calc-alkaline magma generation (~ 150 km). Major mineral phases in material of basaltic (eclogite) or graywacke (granulite) composition (Ringwood 1975, Stern & Wyllie 1978) will strongly partition incompatible elements into the melt.

2. *Mixing and equilibration of a slab-derived melt is with bulk mantle material, not with a mantle-derived melt.* Trace element concentrations in unmodified mantle are so low that trace element abundances and Sr and Nd isotopic ratios of the modified mantle will be dominated by the highly enriched slab-derived melt even if the weight proportion of melt is small. For example, if a slab-derived melt of 500 ppm Sr and $^{87}Sr/^{86}Sr$ of 0.710 equilibrates with mantle material with 25 ppm Sr and a $^{87}Sr/^{86}Sr$ ratio of 0.7030 in the weight proportion 1 part melt to 4 parts mantle material, the modified mantle will have a $^{87}Sr/^{86}Sr$ ratio of 0.7088. Regardless of the way in which the modified mantle undergoes subsequent melting, the melting product will have that $^{87}Sr/^{86}Sr$ ratio.

3. *Oxygen isotopic composition will be approximately a simple linear function of the bulk proportion of slab-derived melt to total mantle material*

with which it equilibrates and of the $\delta^{18}O$ of the two materials. Consider, for example, that the slab has a $\delta^{18}O$ value of about 12 permil and mantle rocks have a $\delta^{18}O$ value of 6 permil. If, as above, one part slab melt equilibrates with four parts mantle material, the $\delta^{18}O$ of the modified mantle will be $+7.2$ permil. Thus, while the trace element and Sr isotopic composition of partial melting products of the modified mantle are dominated by the slab-derived material, the O isotopic composition and bulk chemistry of the melt is dominated by material of mantle composition. The resulting $\delta^{18}O$ versus $^{87}Sr/^{86}Sr$ mixing hyperbolas are strongly convex downward.

The effects of subduction zone contamination on magmatic compositions may be visualized by considering the mixing relationships shown in Figure 6. It is assumed that the slab melt has $^{87}Sr/^{86}Sr = 0.710$ and $\delta^{18}O = 12$ permil, that mantle rock has $^{87}Sr/^{86}Sr = 0.703$ and $\delta^{18}O = 6$ permil, and that oxygen abundances in both end members are equal. Because the ratio of Sr concentration in slab melt (Sr_c) to Sr concentration in mantle rock (Sr_m) is probably at least 5:1 and perhaps 20:1 or more, the hyperbolic mixing curves are strongly convex *downward* as shown by vertical hatching in Figure 6. The shape of the mixing curve reflects the disproportionately large changes in $^{87}Sr/^{86}Sr$ produced by relatively small amounts of contaminant (the weight proportion of contaminant to

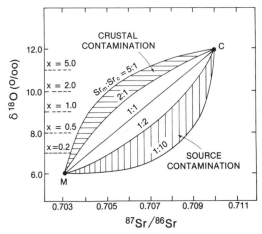

Figure 6 Theoretical two-component mixing curves for $\delta^{18}O$ versus $^{87}Sr/^{86}Sr$ for end members with equal oxygen concentrations. M is composition of hypothetical mantle end member, C is composition of hypothetical contaminant. Ratios shown with each curve denote proportion of Sr content of M (Sr_m) to Sr content of C (Sr_c). x denotes weight proportion of component C to component M. Region with vertical hatching indicates plausible field for source contamination (curves convex downward) and region with horizontal hatching indicates plausible field for crustal contamination (curves convex upward).

melt is shown by x in Figure 6). Oxygen isotopic ratios simply reflect the *bulk* mixing proportions of material.

Strong convex downward curvature on O- versus Sr-isotopic mixing diagrams is diagnostic of subduction zone contamination. In many instances of crustal contamination (see following section), Sr concentrations in the magma (Sr_m) and in the assimilated material (Sr_c) will either be about the same or higher in the magma. Thus Sr_c/Sr_m will in general be < 1 and the mixing curves will be straight or slightly convex *upward*. (This is not strictly true, as we shall see in the next section, but the statement is approximately correct.) Only rarely would crustal contamination produce a strongly convex downward mixing curve.

The example presented above also serves to demonstrate the importance of oxygen isotopic ratios in giving information on the bulk contribution of the slab to magma production. Radiogenic isotopic and trace element behavior depends on largely unknown processes of partial melting in the slab, element partitioning between phases, and reaction relations between slab-derived melt and mantle minerals. Apart from ^{18}O enrichment during crystal fractionation and possible shallow crustal contamination, oxygen isotopic ratios in calc-alkaline volcanic rocks should, within the assumptions of the simple model considered, reflect the proportion of material derived from the slab.

Very few studies have been done on island arc or continental arc volcanics that combine oxygen and radiogenic isotopic analyses of the same rock samples. Magaritz et al (1978) showed that there is a good hyperbolic correlation between $^{87}Sr/^{86}Sr$ and $\delta^{18}O$ in lavas of the Banda arc. This correlation, which is convex downward (see Magaritz et al 1978, Figure 3), is in strikingly good agreement with behavior predicted on the basis of the model considerations above. Moreover, because the Banda arc is apparently on purely oceanic crust (see Magaritz et al 1978), the contamination indicated by the $\delta^{18}O$ vs $^{87}Sr/^{86}Sr$ correlation appears to be possible only through subduction zone processes. Magaritz et al interpreted their data as showing that sediments derived from the Australian continent are being dragged down the subduction zone, melted, and recycled to the earth's surface. This conclusion, if correct, suggests that wherever significant volumes of continental sediment are being subducted large effects may be observed in both the oxygen and the radiogenic isotopic composition of the lavas (see also Whitford et al 1979).

Late Cenozoic calc-alkaline volcanic rocks of the central Andes have been studied by Magaritz et al (1978) and by James (1978, 1980). These rocks are characterized by anomalously high $^{87}Sr/^{86}Sr$ ratios and incompatible element abundances but are of normal andesitic bulk chemical composition. The very thick (~ 70 to 75 km) crustal section provides circumstantial evidence that the anomalous isotopic ratios are due to crustal contamination, but other factors argue against that mechanism (see James et al 1976, James 1978, 1980). $\delta^{18}O$ values in the volcanic rocks average close to 7.2 permil over the entire range of $^{87}Sr/^{86}Sr$ ratios (0.705–0.708). The range in $^{87}Sr/^{86}Sr$, however, is sufficiently small that it would be extremely difficult to detect a hyperbolic relationship of the kind shown in Figure 6, where Sr_c/Sr_m is large. The very low $\delta^{18}O$ (compared to the high $^{87}Sr/^{86}Sr$ ratios) and the absence of any systematic variation in $\delta^{18}O$ over the range of $^{87}Sr/^{86}Sr$ ratios are consistent with source contamination and are *not* consistent with crustal contamination (James 1980). Muehlenbachs & Stern (1980) measured an average $\delta^{18}O$ of 7.3 permil for hornblende andesites of southernmost Chile. They also concluded that the ^{18}O enrichment is consistent with source contamination.

There are many reported analyses of oxygen isotopes in island arc and continental arc volcanics, but the relationship of $\delta^{18}O$ to radiogenic isotopic ratios is either unknown or must be inferred from measurements made on similar rock samples collected from the same area (e.g. Taylor 1968, Matsuhisa 1979). In a comprehensive study of volcanic rocks from the east Japan island arcs, Matsuhisa (1979) finds that tholeiitic rocks in an immature island arc close to the trench have $\delta^{18}O$ values about 5.7 permil, indistinguishable from mantle values. He concludes that if fluids sweated from the slab were involved in magmagenesis, the fluid volume would have to be extremely small. On the continent side of the arc, $\delta^{18}O$ values tend to be 0.5 to 1.7 permil higher at a given SiO_2 content than in rocks from the immature arc. Matsuhisa concludes that the ^{18}O enrichment could be due to crustal contamination or, at least in the case of calc-alkaline volcanics, to source contamination. Oxygen isotopic ratios have been measured for volcanic rocks of the Cascades. The rocks have $\delta^{18}O$ values which lie between 6 and 7.5 permil, but unlike the rocks from eastern Japan, they show no systematic pattern of increasing $\delta^{18}O$ with increasing SiO_2 (Taylor 1968).

The two studies—of Japan and the Cascades—point up the difficulty of interpreting the origin of island arc or continental arc volcanic rocks by measuring $\delta^{18}O$ alone. The correlation between $\delta^{18}O$ and radiogenic isotopic ratios enables a clear distinction to be made between fractionation effects, crustal contamination, and source contamination. Thus, changes in $\delta^{18}O$ that are accompanied by systematic changes in radiogenic isotopic ratios cannot be the result of isotope fractionation alone; changes in radiogenic isotopic ratios without corresponding changes in $\delta^{18}O$ could be caused by source contamination but are probably not caused by crustal contamination; changes in $\delta^{18}O$ that occur without any corresponding change in radiogenic isotopic ratios could be the result of crystal fractionation.

Continental Crustal Contamination

Continental volcanic rocks exhibit much greater variability in radiogenic isotopic ratios and trace element abundances than do oceanic volcanic rocks. In general, continental volcanic rocks are characterized by higher $^{87}Sr/^{86}Sr$, lower $^{143}Nd/^{144}Nd$, and more radiogenic Pb isotopic ratios. These characteristics are commonly assumed to result from contamination by ancient, radiogenic sialic crust which the magma must traverse on its way to the surface (see Brooks et al 1976a for references). In surprisingly few cases, however, are results from radiogenic isotopes alone diagnostic of crustal contamination (e.g. Carter et al 1978b). For most mafic to intermediate continental volcanic rocks the evidence from radiogenic isotopes is insufficient to discriminate between crustal contamination and heterogeneity in the subcontinental mantle.

Brooks et al (1976a), on the basis of Sr isotopic evidence, proposed that magmas of highly diversified radiogenic isotopic composition could be derived from ancient, heterogeneous, subcontinental lithospheric mantle. Others, on the basis of Rb-Sr and Sm-Nd systematics, suggest that many continental volcanic rocks are derived from the same undepleted deep mantle reservoir that supplies oceanic island basalts or from some other enriched reservoir (e.g. Cox et al 1976, Hawkesworth & Vollmer 1979, Wasserburg & DePaolo 1979, Jacobsen & Wasserburg 1979, O'Nions et al 1978). It has become increasingly evident, however, that any resolution of the controversy regarding the nature of mantle heterogeneity is going to depend in large measure upon the ability to assess possible crustal contamination on a case-by-case basis.

There are few examples of combined oxygen and radiogenic isotopic studies on continental volcanics. Brooks & James (1978) showed that radiogenic Mesozoic tholeiites from Tasmania ($(^{87}Sr/^{86}Sr)_i \simeq 0.710$) are characterized by mantle-like $\delta^{18}O$ values of $+6.2$ to $+6.4$ permil. Leeman & Whelan (1976) found that Snake River Plain basalts enriched in ^{87}Sr have normal mantle oxygen isotopic ratios. Studies by Magaritz et al (1978) and James (1980) on continental arc volcanics are described above. By far the most thorough series of isotopic investigations into the question of crustal contamination versus enriched mantle are those reported over the past several years on the Italian volcanics (e.g. Taylor et al 1979, 1980, Hawkesworth & Vollmer 1979, Vollmer & Hawkesworth 1980, Turi & Taylor 1976, Taylor & Turi 1976, Cox et al 1976, Carter et al 1978a). A short review of the results of these investigations will be presented later.

THEORETICAL CONSIDERATIONS   Continental crustal contamination of mantle-derived magmas may in general be caused either by assimilation of crustal rock or by isotopic exchange between magma and wall rocks. In rare circumstances contamination may occur by magma mixing, but that phenomenon is easily modeled.

For most basic volcanic rocks isotopic exchange processes between magma and solid crustal rocks are likely to be too inefficient to be important in producing large changes in isotopic composition of the magma. This assertion derives from consideration of volume diffusion rates. For example, plagioclase has one of the highest rates of oxygen diffusion of any common mineral. Its diffusion coefficient, D, is approximately 5×10^{-12} cm^{2}s^{-1} at $T = 1000°C$ (extrapolated from Giletti et al 1978). Characteristic transport distances, estimated from the relationship $X = \sqrt{Dt}$, where $t = $ time in seconds, will be 0.4 cm in 10^3 years and 12.6 cm in 10^6 years (Hofmann & Hart 1978). Where blocks of assimilated material are much larger than a few centimeters in diameter, it is unlikely that they will equilibrate with the magma during reasonable crustal residence times except by melting. Similarly, equilibration is likely to be limited to no more than a few centimeters of wall rock along magma conduits. Volume diffusion is, of course, the slowest or rate-limiting exchange process. In the presence of melt where there is a large coherency stress within the crystal between initial and equilibrated compositions, the mechanism is likely to be dissolution and reprecipitation (Petrovic 1973, Hofmann & Hart 1978). If this process were many times faster than volume diffusion it could be an important mechanism for crustal contamination.

Crustal assimilation or isotope exchange during partial melting are probably the most efficient means of magma contamination. Taylor (1980) sets out some elementary principles for determining the effects of assimilation and isotope exchange on oxygen and radiogenic isotopic ratios in magma. He notes that assimilation is not a simple bulk mixing process but must involve at least three end members—the magma, the contaminant, and a cumulate phase, which crystallizes to provide the heat required to dissolve the assimilated material.[1] Thus, contamination by assimilation cannot *in general* be treated as a two-component mixing process.

---

[1] In terms of the three end-member model discussed here, assimilation and isotopic exchange should produce similar isotopic correlations. It remains an unresolved problem in magma genesis, however, to describe the specific mechanisms by which magma interacts with country rock. Especially relevant is the role of fluid and vapor phases, grain boundary melts, disequilibrium melting of crustal phases, and various scavenging mechanisms. These processes all require considerably improved quantitative understanding before their importance for crustal contamination can be assessed.

Taylor emphasizes that the liquid line of descent of magmatic differentiates will not be drastically altered by assimilation of common rock types, although fractional crystallization may be speeded up. Assimilation promotes the formation of cumulus phases already forming as a result of normal fractional crystallization. Thus, although the proportion of late differentiates may be significantly increased by assimilation, effects of assimilation may be quite difficult to detect either in the differentiation sequence or in major element compositions.

On the other hand, isotopic compositions may be strongly affected by assimilation. The following discussion considers in some detail the behavior of O, Sr, and Nd isotopic ratios and Sr and Nd abundances during assimilative contamination. The approach is that outlined by Taylor (1980) and includes the following assumptions:

1. Crystallization of a cumulate phase provides the heat necessary to dissolve the assimilated rock;
2. The crystal cumulate is immediately removed from the magma body (i.e. Rayleigh crystallization);
3. The crystal cumulate has the same $\delta^{18}O$ as the magma (i.e. there is no fractionation of oxygen isotopes) and all three components—magma, contaminant, and cumulate—have the same oxygen concentrations.

In the hypothetical models considered by Taylor and in this paper, plagioclase is the cumulate phase. The low density of plagioclase, however, makes it a particularly difficult phase to remove from a magma; indeed, its abundance and ubiquitous presence in mafic and intermediate rocks indicate that a large proportion of crystallized plagioclase remains in the melt. If plagioclase crystallizes but remains in the melt and equilibrates with it, the mixing relations will be simple two component. In fact, Rayleigh crystallization is widely assumed to be common in basic volcanic suites (e.g. Allegre et al 1977), but the cumulate phase is more commonly taken to be olivine.

In the model considered by Taylor, the assumption that the cumulate phase is plagioclase fulfills two requirements: (a) that $\delta^{18}O$ of the cumulate be the same as that of the magma; and (b) that the solid-liquid partition coefficient for strontium, K_{Sr}, be greater than 1.0. In the first instance, if $\delta^{18}O$ of the cumulate is significantly different from that of the melt (e.g. olivine), $\delta^{18}O$ of the magma will change in response to fractionation. In the second instance, unless the cumulate phase is feldspar, with $K_{Sr} > 1$, Sr abundances in the differentiation sequence will not decrease with increasing degree of differentiation. No other cumulate mineral is likely to have $K_{Sr} > 1$, so that extraction of any other phase will produce an increase than decrease in Sr abundances in the differentiates.

Despite these obvious difficulties, oxygen and radiogenic isotopic variations in rocks such as those of the Italian volcanic provinces suggest that to a first approximation plagiocalse may be considered the dominant cumulate phase, at least in upper crustal assimilation. The methodology outlined here is perfectly general, however, and may be applied to any cumulate phase or phases and to other isotopes (e.g. Pb).

Simple two-component mixing is discussed in the previous section and $\delta^{18}O$ versus $^{87}Sr/^{86}Sr$ hyperbolic mixing curves shown in Figure 6. For source contamination, the hyperbolic mixing curves will be strongly convex *downward*. For crustal contamination, the curves will in general

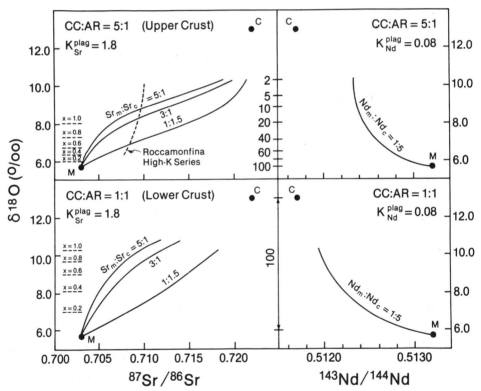

Figure 7 $\delta^{18}O$ versus $^{87}Sr/^{86}Sr$ and $^{143}Nd/^{144}Nd$ crustal contamination mixing curves for assimilation accompanied by Rayleigh crystallization. Subscript m denotes composition of mantle-derived melt, subscript c denotes composition of crustal contaminant. Sr and Nd concentrations ratios between starting magma and contaiminant are expressed as $Sr_m:Sr_c$ and $Nd_m:Nd_c$ respectively. Ratio CC:AR denotes weight proportion of crystal cumulate (CC) removed from melt to weight proportion of assimilated rock (AR). K is solid-liquid partition coefficient. Scale shown on line separating Sr and Nd diagrams gives percent magma remaining. x denotes wt proportion assimilated material to 1 part starting magma. Sr and Nd concentration ratios are shown on the mixing trajectories.

be convex *upward* or weakly convex downward. The addition of a cumulate phase as an end member can produce a significant change in the mixing trajectories. Examples of three-component (magma, contaminant, and cumulate) mixing relations are shown in Figures 7 and 8. The composition of the contaminant is assumed to have Sr and Nd isotopic ratios (0.722 and 0.5117, respectively) for average upper continental crust as given by DePaolo & Wasserburg (1979). $\delta^{18}O$ is arbitrarily chosen to be a typical metasedimentary value of $+13.0$ permil. Abundance ratios for Sr and Nd between magma and contaminant are given as $Sr_m:Sr_c$ and $Nd_m:Nd_c$, where m and c denote magma and contaminant respectively. In the example shown, $Nd_m:Nd_c$ is assumed to be 1:5, the ratio of Nd concentration in MOR basalt to Nd in the North American shale composite. Solid-liquid partition coefficients for both upper and lower crust are assumed to be $K_{Sr}^{plag} = 1.8$ and $K_{Nd}^{plag} = 0.08$ (Arth 1976). It is important to note that in terms both of magma to contaminant elemental abundances and plagioclase partition coefficients, Sr and Nd exhibit opposite behavior: $Sr_m:Sr_c > 1$ whereas $Nd_m:Nd_c < 1$ and $K_{Sr}^{plag} > 1$ whereas $K_{Nd}^{plag} < 1$. The correlated behavior of the isotopic ratios and concentrations of the two elements with each other and with $\delta^{18}O$ can place important constraints on the composition of cumulate phases.

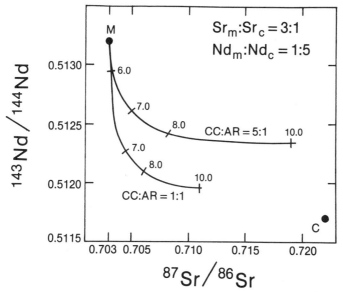

Figure 8 $^{143}Nd/^{144}Nd$ versus $^{87}Sr/^{86}Sr$ mixing trajectories for end members shown in Figure 9. $Sr_m/Sr_c = 3.0$ and $Nd_m/Nd_c = 0.2$. Other compositions of end members as given in Figure 7. $\delta^{18}O$ values are shown on each curve for reference.

In Figure 7, the weight proportion of cumulate removed to the starting weight of the magma is given by the quantity x. The scale marked along the axis separating Sr and Nd diagrams represents the weight proportion (in percent) of melt remaining at any given $\delta^{18}O$. Sr and Nd concentrations are given along each mixing trajectory. The quantity CC:AR is the weight ratio of crystal cumulate (CC) to assimilated rock (AR) dissolved. For CC:AR = 1:1, the weight ($\sim$volume) of magma remains unchanged during assimilation.

All of the mixing trajectories shown in Figures 7 and 8 are for crystal cumulate to assimilated rock ratios of 1:1 and 5:1, the former for hot lower crustal rocks already near melting and the latter for cool upper crustal rocks where large amounts of heat are necessary to raise temperatures to the melting point (Taylor 1980). All other parameters including composition of end members and partition coefficients are assumed to be the same for both the upper crust and lower crust.

Observations and conclusions to be drawn from the mixing relations shown may be summarized as follows:

1. For hypothetical lower crustal contamination (CC:AR = 1:1), isotopic trajectories for both Sr and Nd versus O (Figure 7) will be similar to those of simple two-component hyperbolic mixing (Figure 6). For a cumulate to assimilation ratio of 1:1, the amount of magma remains unchanged along the entire mixing trajectory and isotopic compositions vary smoothly between end members.

2. For hypothetical upper crustal contamination (CC:AR = 5:1), mixing trajectories deviate markedly from those of the two-component case. For neither Sr nor Nd do trajectories trend toward the contaminant end member (Figure 7). Sr trajectories are flatter and curve to the right of the two-component mixing curves (i.e. toward higher $^{87}Sr/^{86}Sr$ ratios for any give $\delta^{18}O$), reflecting decreasing Sr concentrations in the residual melt and progressively smaller $Sr_m:Sr_c$ ratios. Nd trajectories behave in the opposite way, reflecting *increasing* Nd concentrations in the residual melt and progressively increasing $Nd_m:Nd_c$ ratios. As a result, $^{143}Nd/^{144}Nd$ ratios reach a limiting value somewhere between magma and contaminant end members while $\delta^{18}O$ continues to increase with progressive contamination.

3. $^{143}Nd/^{144}Nd$ versus $^{87}Sr/^{86}Sr$ trajectories are shown in Figure 8. Corresponding $\delta^{18}O$ values are indicated on each mixing curve. For both CC:AR ratios given for this example, the mixing curves have a steep slope near the parent magma end point, with large decreases in $^{143}Nd/^{144}Nd$ ratios accompanied by comparatively small increases in $^{87}Sr/^{86}Sr$ ratios. The curves flatten markedly with increasing degree of contamination so that $^{87}Sr/^{86}Sr$ and $\delta^{18}O$ increase rapidly with little corresponding

decrease in ^{143}Nd/^{144}Nd ratios. For CC:AR = 5:1, the mixing trajectory for late differentiates terminates at a limiting ^{143}Nd/^{144}Nd ratio considerably higher than that of the contaminant but at a ^{87}Sr/^{86}Sr ratio close to that of the contaminant.

One conclusion to be drawn from these simple model calculations is that oxygen and radiogenic isotopic variations resulting from crustal contamination will occur in systematic and predictable ways that may be quite different from two end-member mixing. Although it is likely that in most volcanic suites mixing trajectories will be too short to define clear trends on isotopic correlation diagrams, the *relative* differences in O-Sr, O-Nd, and O-Pb isotopic correlations may commonly be diagnostic of the nature of the contamination.

THE ITALIAN VOLCANICS   The late Cenozoic alkaline volcanic province of west-central Italy has for decades been the subject of intensive geological and geochemical research. Over the past several years probably more oxygen and radiogenic isotopic studies have been done of these rocks than on any other group of continental volcanics in the world (see Taylor et al 1979, Taylor 1980, Hawkesworth & Vollmer 1979, Vollmer & Hawkesworth 1980, Cox et al 1976, Carter et al 1978a, Turi & Taylor 1976, Taylor & Turi 1976). The study of the Italian volcanics has emerged as a classic example of the difficulties involved in distinguishing between crustal contamination, source contamination, and enriched mantle. The debate between proponents of crustal contamination and those of enriched mantle (or source contamination) is thoroughly summarized in four papers: Taylor et al (1979) and Taylor (1980) argue for crustal contamination; Hawkesworth & Vollmer (1979) and Vollmer & Hawkesworth (1980) argue for source enrichment.

The volcanic rocks of central Italy are usually divided into two groups— the Tuscany igneous province to the north and the Roman and Campanian provinces to the south. It is generally agreed that volcanics of the Tuscan province were probably produced by crustal anatexis and that the lavas of the northern part of the Roman province could reflect some crustal contamination. The controversy over crustal contamination versus source enrichment centers on the volcanics of the southern part of the Roman province (south of Rome), with emphasis on the Roccamonfina volcano. For simplicity, the discussion that follows concerns only results from Roccamonfina. The reader is referred to the papers cited above for a comprehensive reading of other results from Italian volcanics.

The Roccamonfina lavas form a low-K series (olivine basalt–trachy-basalt–trachyte) and a high-K series (tephritic phonolite–phonolitic leucite tephryte–leucitite) (Hawkesworth & Vollmer 1979). As shown by

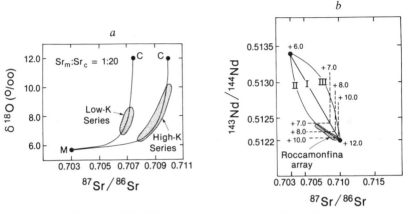

Figure 9 (a) $\delta^{18}O$ versus $^{87}Sr/^{86}Sr$ in Roccamonfina lavas. Theoretical two-component source mixing hyperbolas with $Sr_m:Sr_c = 1:20$ are shown for comparison. (b) Two-component (source) mixing relations between $^{143}Nd/^{144}Nd$ and $^{87}Sr/^{86}Sr$. End-member compositions as in (a) and $(^{143}Nd/^{144}Nd)_c = 0.5122$. I. $Sr_m/Sr_c = Nd_m/Nd_c$; II. $Sr_m/Sr_c = 4 \times Nd_m/Nd_c$; III. $Nd_m/Nd_c = 4 \times Sr_m/Sr_c$. $\delta^{18}O$ values are shown for reference. Roccamonfina sample array from Hawkesworth & Vollmer (1979).

Taylor et al (1979), the two groups are totally distinct. The lavas show wide variation in $\delta^{18}O$ and Sr, Nd, and Pb isotopic ratios and are particularly amenable to geochemical modeling.

Consider first the $\delta^{18}O$ versus $^{87}Sr/^{86}Sr$ correlation shown in Figure 9a adapted from Taylor et al (1979). The high-K series shows the greatest dispersion and forms a convex downward hyperbolic array which can be fitted by a simple two-component *source* contamination hyperbola where $Sr_m:Sr_c = 1:20$. A similar curve, but with a lower $^{87}Sr/^{86}Sr$ contaminant end member can also be fitted through the low-K series. The $\delta^{18}O$ versus $^{87}Sr/^{86}Sr$ correlation for the high-K series is very similar to that observed in Banda arc lavas (Magaritz et al 1978).

The point to be emphasized here is that the $\delta^{18}O$ versus $^{87}Sr/^{86}Sr$ data arrays for either series do *not* follow the kinds of trajectories shown in Figure 7. The very high Sr concentrations observed in the Roccamonfina lavas and the positive correlation between Sr concentrations and $^{87}Sr/^{86}Sr$ ratios are also inconsistent with crustal contamination models developed earlier.

A possible explanation for the $\delta^{18}O$ versus $^{87}Sr/^{86}Sr$ correlation is source mixing whereby subducted crustal material enriched in ^{87}Sr and ^{18}O and in incompatible elements is partly melted and percolated into and equilibrated with overlying mantle rock to produce enriched modified mantle. Such a process will produce the convex downward mixing curve shown by the Roccamonfina lavas.

The Roccamonfina lavas exhibit increasing Sr with increasing $^{87}Sr/^{86}Sr$ ratios and increasing $\delta^{18}O$. For crustal contamination with plagioclase as the cumulate phase, Sr concentrations should decrease with increasing $^{87}Sr/^{86}Sr$. Thus either plagioclase is not the cumulate phase or the compositional variations are not due to crustal contamination.

Finally, Figure 9b shows the field of Nd versus Sr isotopic ratios for the Roccamonfina lavas as published by Hawkesworth & Vollmer (1979). Superimposed are hypothetical source contamination mixing curves for different $Sr_m:Sr_c$ and $Nd_m:Nd_c$ ratios. $\delta^{18}O$ values are shown for reference. $\delta^{18}O$ and $^{87}Sr/^{86}Sr$ of the contaminant are as in Figure 9a and $^{143}Nd/^{144}Nd = 0.5122$. For that composition of contaminant, O, Sr, and Nd isotopic mixing relations are satisfied by a model in which $Sr_m:Sr_c$ is four times greater than $Nd_m:Nd_c$. The observed $^{87}Sr/^{86}Sr$ versus $^{143}Nd/^{144}Nd$ correlation can also be modeled somewhat less satisfactorily by crustal contamination (see Figure 8), but the mixing trajectories will not embrace the contaminant composition.

The above observations lead to the following conclusions:

1. The high $\delta^{18}O$ values of the Roccamonfina lavas *require* the participation of crustal material.
2. The pronounced convex downward shape of the $\delta^{18}O$ versus $^{87}Sr/^{86}Sr$ isotopic mixing curve is diagnostic of *source* contamination rather than crustal contamination.
3. The shape of the $^{143}Nd/^{144}Nd$ versus $^{87}Sr/^{86}Sr$ isotopic mixing curve is consistent with either source contamination or crustal contamination.
4. Observed $^{143}Nd/^{144}Nd$ versus $1/Nd$ and $^{87}Sr/^{86}Sr$ versus $1/Sr$ are both straight line relationships (Hawkesworth & Vollmer 1979, Figure 9) consistent within the scatter of data with either two-component mixing (i.e. source contamination), or with crustal contamination.
5. Despite the large amount of data overall, the paucity of combined oxygen and radiogenic isotopic measurements (particularly of Nd and Pb) on the *same* samples from Roccamonfina remains a serious impediment to interpretation.

CONCLUSIONS

The rapidly increasing number of radiogenic isotopic measurements on mantle-derived volcanic rocks over the past several years and the resulting proliferation of crust-mantle evolution models have accentuated how little is known about the role of crustal material in the genesis and evolution of mantle-derived magmas. Subducted crustal rocks may be recycled to the earth's surface by partial melting in the subduction zone or in a deep mantle reservoir. Contamination may also occur when mantle-derived magmas ascend through old sialic crust.

Systematic correlations between $\delta^{18}O$ and radiogenic isotopic ratios that *require* involvement of crustal material have been well documented for many island arc and continental arc volcanic associations and for some continental volcanic rocks. No such correlations have been found for oceanic basalt but the data are far too sparse to say that none exist.

Model studies of source contamination and crustal contamination suggest that each may have its own distinctive oxygen versus radiogenic isotopic "signature." Source contamination characteristically obeys two-component mixing relationships. Where partial melts from subducted crustal material are infiltrated into mantle rock, $\delta^{18}O$ versus $^{87}Sr/^{86}Sr$ isotopic mixing curves are strongly convex downward.

Crustal contamination is a three end-member process involving magma, assimilated material, and crystal cumulates. Oxygen versus radiogenic isotopic mixing curves for upper crustal contamination generally follow trajectories that depart significantly from those of two-component mixing and do not trend toward the composition of the contaminant end member. Isotopic mixing behavior is controlled largely by Sr, Nd, and Pb partition coefficients of the crystallizing phase(s) and by the weight ratio of assimilated material to cumulate. For lower crustal contamination where that ratio may be close to 1, mixing trajectories will be similar to those for simple two-component mixing. $\delta^{18}O$ versus $^{87}Sr/^{86}Sr$ mixing curves for crustal contamination are typically convex upward, distinguishing them from the strongly convex downward curves for source contamination.

Studies relating $\delta^{18}O$ values to radiogenic isotopic ratios in island arc or continental arc volcanics where crustal contamination is suspected invariably exhibit correlated isotopic compositions consistent with source contamination rather than crustal contamination. Far more data are needed, however, before firm conclusions can be drawn.

Lavas from Roccamonfina volcano in the alkaline volcanic province of Italy have been widely interpreted as owing their oxygen and radiogenic isotopic composition to crustal contamination. Observed $\delta^{18}O$ versus $^{87}Sr/^{86}Sr$ mixing curves, however, fail to follow predicted trajectories for crustal contamination but do exhibit good agreement with source contamination two-component mixing relationships.

ACKOWLEDGMENTS

I thank D. J. Whitford and A. W. Hofmann for many stimulating and thought-provoking discussions on the role of subducted crustal material in the genesis of mantle-derived magmas. Special thanks go to E. Ito for an exhaustive review of parts of the manuscript.

Literature Cited

Allegre, C. J., Treuil, M. Minster, J.-F., Minster, B., Albarede, F. 1977. Systematic use of trace element in igneous process. Part I: fractional crystallization process in volcanic suites. *Contrib. Mineral. Petrol.* 60:57–75

Allegre, C. J., Othman, D. B., Polve, M., Richard, P. 1979. The Nd-Sr isotopic correlation in mantle materials and geodynamic consequences. *Phys. Earth Planet. Inter.* 19:293–306

Anderson, A. T., Clayton, R. N., Mayeda, T. K. 1971. Oxygen isotope thermometry of mafic igneous rocks. *J. Geol.* 79:715–29

Anderson, D. L. 1979a. Irreversible differentiation of the earth. *Geol. Soc. Am. Abstr. with Programs* 11(7):378

Anderson, D. L. 1979b. Chemical stratification of the mantle. *J. Geophys. Res.* 84:6297–98

Anderson, D. L., Hart, R. S. 1978. The Q of the earth. *J Geophys. Res.* 83:5869–82

Armstrong, R. L. 1968. A model for the evolution of strontium and lead isotopes in a dynamic earth. *Rev. Geophys.* 6:175–99

Armstrong, R. L. 1971. Isotopic and chemical constraints on models of magma genesis in volcanic arcs. *Earth Planet. Sci. Lett.* 12:137–42

Arth, J. G. 1976. Behavior of trace elements during magmatic processes—a summary of theoretical models and their applications. *J. Res. US Geol. Surv.* 4(1):41–47

Bottinga, Y., Javoy, M. 1975. Oxygen isotope partitioning among the minerals in igneous and metamorphic rocks. *Rev. Geophys. Space Phys.* 13:401–18

Brooks, C., James, D. E. 1978. Strontium and oxygen isotopes in Mesozoic tholeiites and the case for heterogeneous mantle. *Geol. Soc. Am. Abstr. with Programs* 10(7):372

Brooks, C., James, D. E., Hart, S. R. 1976a. Ancient lithosphere: its role in young continental volcanism. *Science* 193:1086–94

Brooks, C., Hart, S. R., Hofmann, A., James, D. E. 1976b. Rb-Sr mantle isochrons from oceanic regions. *Earth Planet. Sci. Lett.* 32:51–61

Carter, S. R., Evensen, N. M., Hamilton, P. J., O'Nions, R. K. 1978a. Continental volcanics derived from enriched and depleted source regions: Nd- and Sr-isotope evidence. *Earth Planet. Sci. Lett.* 37:401–08

Carter, S. R., Evensen, N. M, Hamilton, P. J., O'Nions, R. K. 1978b. Neodymium and strontium isotope evidence for crustal contamination of continental volcanics. *Science* 202:743–47

Clayton, R. N., Goldsmith, J. R., Karel, K. J. Mayeda, T. K., Newton, R. C. 1975. Limits on the effect of pressure on isotopic fractionation. *Geochim. Cosmochim. Acta* 39:1197–1201

Corliss, J. B., Dymond, J., Gordon, L. I., von Herzen, R. P., Ballard, R. D., Green, K., Williams, D., Bainbridge, A., Crane, K., Van Andel, H. 1979. Submarine thermal springs on the Galapagos Rift. *Science* 203:1073–83

Cortecci, G., Del Moro, A., Leone, G., Pardini, G. C. 1979. Correlation between strontium and oxygen isotopic compositions of rocks from the Adamello Massif (northern Italy). *Contrib. Mineral. Petrol.* 68:421–27

Cox, K. G., Hawkesworth, C. J., O'Nions, R. K. 1976. Isotopic evidence for the derivation of some Roman region volcanics from anomalously enriched mantle. *Contrib. Mineral. Petrol.* 56:173–80

Dasch, E. J., Hedge, C. E., Dymond, J. 1973. Effect of sea-water interaction on strontium isotope composition of deep-sea basalts. *Earth Planet. Sci. Lett.* 19:177–83

DePaolo, D. J. 1980. Earth structure and crustal evolution models inferred from neodymium isotopes. *Trans. Am. Geophys. Union* 61:207

DePaolo, D. J., Johnson, R. W. 1979. Magma genesis in the New Britain island-arc: Constraints from Nd and Sr isotopes and trace element patterns. *Contrib. Mineral. Petrol.* 70:367–79

DePaolo, D. J., Wasserburg, G. J. 1976a. Nd isotopic variations and petrogenetic models. *Geophys. Res. Lett.* 3:249–52

DePaolo, D. J., Wasserburg, G. J. 1976b. Inferences about magma sources and mantle structure from variations of $^{143}Nd/^{144}Nd$. *Geophys. Res. Lett.* 3:743–46

DePaolo, D. J., Wasserburg, G. J. 1977. The sources of island arcs as indicated by Nd and Sr isotopic studies. *Geophys. Res. Lett.* 4:465–68

DePaolo, D. J., Wasserburg, G. J. 1979. Petrogenetic mixing models and Nd-Sr isotopic patterns. *Geochim. Cosmochim. Acta* 43:615–27

Epstein, S., Taylor, H. P. Jr. 1975. Investigation of the carbon, hydrogen, oxygen, and silicon isotope and concentration relationships on the grain surfaces of a variety of lunar soils and in some Apollo 15 and 16 core samples. *Proc. Lunar Sci. Conf. 6th*, pp. 1771–98

Friedman, I., O'Neil, J. R. 1977. Compilation of stable isotope fractionation factors

342   JAMES

of geochemical interest. In *Data of Chemistry*, Chapter KK, ed. M. Fleischer. *US Geol. Surv. Prof. Pap. 440-KK*. 12 pp. 48 figs. 6th ed.

Garlick, G. D. 1969. The stable isotopes of oxygen. In *Handbook of Geochemistry*, ed. K. H. Wedepohl, pp. 8B(1–27). Berlin/Heidelberg/New York: Springer

Garlick, G. D., Epstein, S. 1967. Oxygen isotope ratios in coexisting minerals of regionally metamorphosed rocks. *Geochim. Cosmochim. Acta* 31:181–214

Garlick, G. D., MacGregor, I. D., Vogel, D. E. 1971. Oxygen isotope ratios in eclogites from kimberlites. *Science* 172:1025–27

Giletti, B. J., Semet, M. P., Yund, R. A. 1978. Studies in diffusion—III. Oxygen in feldspars: an ion microprobe determination. *Geochim. Cosmochim. Acta* 42(1):45–57

Gilluly, J. 1971. Plate tectonics and magmatic evolution. *Geol. Soc. Am. Bull.* 82:2383–96

Gray, J., Cumming, G. L., Lambert, R. St. J. 1976. Oxygen and strontium isotopic compositions and thorium and uranium contents of basalt from DSDP 37 cores. In *Initial Reports of the Deep-Sea Drilling Project*, 37:607–9. Washington, DC: Govt. Print. Off.

Gregory, R. T., Taylor, H. P. Jr. 1980. An oxygen isotope profile in a section of Cretaceous oceanic crust, Samail ophiolite, Oman: evidence for δ^{18} O-buffering of the oceans by deep (<5 km) seawater-hydrothermal circulation at mid-ocean ridges. *J. Geophys. Res.* In press

Hart, S. R. 1971. Dredge basalts: some geochemical aspects. *Trans. Am. Geophys. Union* 52:376

Hart, S. R., Erlank, A. J., Kable, E. J. D. 1974. Sea-floor basalt alteration: some chemical and Sr-isotopic effects. *Contrib. Mineral. Petrol.* 44:219–230

Hawkesworth, C. J., Vollmer, R. 1979. Crustal contamination versus enriched mantle: $^{143}Nd/^{144}Nd$ and $^{87}Sr/^{86}Sr$ evidence from the Italian volcanics. *Contrib. Mineral. Petrol.* 69:151–65

Hawkesworth, C. J., O'Nions, R. K., Pankhurst, R. J., Hamilton, P. J., Evensen, N. M. 1977. A geochemical study of island-arc and back-arc tholeiites from the Scotia Sea. *Earth Planet. Sci. Lett.* 36:253–62

Hoefs, J. 1973. *Stable Isotope Geochemistry*. Berlin:Springer. 140 pp.

Hofmann, A. W., Hart, S. R. 1978. An assessment of local and regional isotopic equilibrium in the mantle. *Earth Planet. Sci. Lett.* 38:44–62

Hofmann, A. W., White, W. M. 1980. Role of oceanic crust in mantle differentiation.

Carnegie Inst. Washington Yearb. 79: In press

Ito, E. 1980. K, Rb, Sr, and Ba concentrations and Sr isotopic composition of plagioclase from submarine gabbro. *Carnegie Inst. Washington Yearb.* 79: In press

Jacobsen, S. B., Wasserburg, G. J. 1979. The mean age of mantle and crustal reservoirs. *J. Geophys. Res.* 84:7411–27

James, D. E. 1978. Origin of high $^{87}Sr/^{86}Sr$ ratios in central Andean calc-alkaline lavas. *Short Papers, 4th Int. Conf. Geochron., Cosmochron., and Isotope Geol.*, Aspen, Colo. USGS Open-File Rep. No. 78-701, pp. 199–201

James, D. E. 1980. Role of subducted continental material in the genesis of calc-alkaline volcanics of the central Andes. *Geol. Soc. Am. Nazca Plate Mem.* In press

James, D. E., Brooks, C., Cuyubamba, A. 1976. Andean Cenozoic volcanism: magma genesis in the light of strontium isotopic composition and trace-element geochemistry. *Geol. Soc. Am. Bull.* 87:529–600

James, D. E., Padovani, E. R., Hart, S. R. 1980. Preliminary results on the oxygen isotopic composition of the lower crust, Kilbourne Hole maar, New Mexico. *Geophys. Res. Lett.* 7:321–24

Karig, D. E., Sharman, G. F. III 1975. Subduction and accretion in trenches. *Geol. Soc. Amer. Bull.* 86:377–89

Kay, R. W. 1977. Geochemical constraints on the origin of Aleutian magmas. In *Island Arcs, Deep Sea Trenches and Back Arc Basins*, ed. M. Talwani, W. C. Pitman III. *Maurice Ewing Ser. 1*, pp. 229–42. Washington, DC: AGU

Kay, R. W., Sun, S.-S., Lee-Hu, C.-N. 1978. Pb and Sr isotopes in volcanic rocks from the Aleutian Islands and Pribilof Islands, Alaska. *Geochim. Cosmochim. Acta* 42:263–73

Knauth, L. P., Epstein, S. 1975. Hydrogen and oxygen isotope ratios in silica from the Joides Deep Sea Drilling Project. *Earth Planet. Sci. Lett.* 25:1–10

Kyser, T. K. 1979. The temperature dependence of oxygen isotope distributions and the origin of basalts and ultramafic nodules. *Geol. Soc. Am. Abstr. with Programs* 11(7):462

Kyser, T. K., O'Neil, J. R. 1978. Oxygen isotope relations among oceanic tholeiites, alkali basalts, and ultramafic nodules. In *Short Papers, 4th Int. Conf. Geochron., Cosmochron., and Isotope Geol.*, Aspen, Colo. USGS Open-Fire Rep. 78-701, pp. 237–40

Leeman, W. P., Whelan, J. F. 1976. Oxygen isotopic studies of lavas from the Snake

River Plain. *Geol. Soc. Am. Abstr. with Programs* 8:975

Magaritz, M., Taylor, H. P. Jr. 1976. Oxygen hydrogen and carbon isotope studies of the Franciscan formation, Coast Ranges, Calif. *Geochim. Cosmochim. Acta* 40:215–34

Magaritz, M., Whitford, D. J., James, D. E. 1978. Oxygen isotopes and the origin of high $^{87}Sr/^{86}Sr$ andesites. *Earth Planet. Sci. Lett.* 40:220–30

Matsuhisa, Y. 1979. Oxygen isotopic compositions of volcanic rocks from the east Japan island arcs and their bearing on petrogenesis. *J. Volcanol. Geotherm. Res.* 5:271–96

Matthews, A., Goldsmith, J. R., Clayton, R. N. 1980. $^{18}O/^{16}O$ and $^{17}O/^{16}O$ fractionation studies on Ca-Mg silicate minerals. *Trans. Am. Geophys. Union* 61:403

Mayeda, T. K., Shearer, J., Clayton, R. N. 1975. Oxygen isotope fractionation in Apollo 17 rocks. *Proc. 6th Lunar Sci. Conf.*, pp. 1799–1802

McCulloch, M. T., Gregory, R. T., Wasserburg, G. J., Taylor, H. P. Jr. 1980. A neodymium, strontium, and oxygen isotopic study of the Cretaceous Samail ophiolite and implications for the petrogenesis and seawater-hydrothermal alteration of oceanic crust. *Earth Planet. Sci. Lett.* 46:201–11

Muehlenbachs, K. 1976a. Oxygen isotope geochemistry of DSDP leg 34 basalt. In *Initial Reports of Deep-Sea Drilling Project*, 34:337–39. Washington, DC: US Gov't. Print. Off.

Muehlenbachs, K. 1976b. Oxygen isotope geochemistry of DSDP leg 37 rocks. In *Initial Reports of Deep-Sea Drilling Project*, 37:617–19. Washington, DC: US Govt. Print. Off.

Muehlenbachs, K., Clayton, R. N. 1972. Oxygen isotope studies of fresh and weathered submarine basalt. *Can. J. Earth Sci.* 9:172–84

Muehlenbachs, K., Clayton, R. N. 1976. Oxygen isotope composition of the oceanic crust and its bearing on seawater. *J. Geophys. Res.* 81(23):4365–69

Muehlenbachs, K., Kushiro, I. 1974. Oxygen isotope exchange equilibrium of silicates with CO_2 or O_2. *Carnegie Inst. Washington Yearb.* 72:232–36

Muehlenbachs, K., Jakobsson, S. P. 1979. The $\delta^{18}O$ anomaly in icelandic basalts related to their field occurrence and composition. *Trans. Am. Geophys. Union* 60:408

Muehlenbachs, K., Stern, C. R. 1980. $\delta^{18}O$ variations among calc-alkaline volcanic rocks from S. Chile: implications for the effects of ridge subduction on the source region of orogenic volcanism. *Trans. Am. Geophys. Union* 1:401

Nicholls, I. A., Ringwood, A. E. 1973. Effect of water on olivine stability in tholeiites and the production of silica-saturated magmas in the island arc environment. *J. Geol.* 81:285–300

O'Nions, R. K., Hamilton, P. J., Evensen, N. M. 1977. Variations in $^{143}Nd/^{144}Nd$ and $^{87}Sr/^{86}Sr$ ratios in oceanic basalts. *Earth Planet. Sci. Lett.* 34:13–22

O'Nions, R. K., Evensen, N. M., Hamilton, P. J., Carter, S. R. 1978. Melting of the mantle past and present: isotope and trace element evidence. *Philos. Trans. R. Soc. London* 258:547–59

O'Nions, R. K., Evensen, N. M., Hamilton, P. J. 1979a. Geochemical modeling of mantle differentiation and crustal growth. *J. Geophys. Res.* 84:6091–6101

O'Nions, R. K., Carter, S. R., Evensen, N. M., Hamilton, P. J. 1979b. Geochemical and cosmochemical applications of Nd isotope analysis. *Ann. Rev. Earth Planet. Sci.* 7:11–38

Perry, E. C. Jr. 1967. The oxygen isotope chemistry of ancient cherts. *Earth Planet. Sci. Lett.* 3:62–66

Perry, E. C. Jr., Ahmad, S. N., Swulius, T. M. 1978. The oxygen isotope composition of 3,800 m.y. old metamorphosed chert and iron formation from Isukasia, West Greenland. *J. Geol.* 86: 223–39.

Petrović, R. 1973. The effect of coherency stress on the mechanism of the reaction albite + K^+ ⇄ K-feldspar + Na^+ and on the mechanical state of the resulting feldspar. *Contrib. Mineral. Petrol.* 41:151

Reuter, J. H., Epstein, S., Taylor, H. P. 1965. O^{18}/O^{16} ratios of some chondritic meteorites and terrestrial ultramafic rocks. *Geochim. Cosmochim. Acta* 29:481–88

Richard, P., Shimizu, N., Allegre, C. J. 1976. $^{143}Nd/^{146}Nd$, a natural tracer, an application to oceanic basalts. *Earth Planet. Sci. Lett.* 31:269–78

Ringwood, A. E. 1975. *Composition and Petrology of the Earth's Mantle*. New York: McGraw-Hill. 618 pp.

Sacks, I. S. 1980. Q_s of the lower mantle—a bodywave determination. *Carnegie Inst. Washington Yearb.* 79: In press

Savin, S. M., Epstein, S. 1970. The oxygen and hydrogen isotope geochemistry of ocean sediments and shales. *Geochim. Cosmochim. Acta* 34:43–64

Shieh, Y.-N., Schwarcz, H. P. 1974. Oxygen isotope studies of granite and migmatite, Grenville province of Ontario, Canada. *Geochim. Cosmochim. Acta* 38:21–45

Shieh, Y.-N., Schwarcz, H. P. 1978. The oxygen isotope composition of the surface crystalline rocks of the Canadian Shield. *Can. J. Earth Sci.* 15:1773–82

Stern, C. R., Wyllie, P. J. 1978. Phase compositions through crystallization intervals in basalt-andesite-H_2O at 30 kb with implications for subduction zone magmas. *Am. Mineral.* 63:641–63

Sun, S.-S. 1980. Lead isotopic study of young volcanic rocks from mid-ocean ridges, ocean islands and island arcs. *Philos. Trans. R. Soc. London.* In press

Sun, S.-S., Hanson, G. N. 1975. Evolution of the mantle: geochemical evidence from alkali basalt. *Geology* 3:297–302

Tatsumoto, M. 1978. Isotopic composition of lead in oceanic basalt and its implication to mantle evolution. *Earth Planet. Sci. Lett.* 38:63–87

Taylor, H. P. Jr. 1968. The oxygen isotope geochemistry of igneous rocks. *Contrib. Mineral. Petrol.* 19:1–71

Taylor, H. P. Jr. 1978. Oxygen and hydrogen isotope studies of plutonic granitic rocks. *Earth Planet. Sci. Lett.* 38:177–210

Taylor, H. P. Jr. 1980. The effects of assimilation of country rocks by magmas on $^{18}O/^{16}O$ and $^{87}Sr/^{86}Sr$ systematics in igneous rocks. *Earth Planet. Sci. Lett.* 47:243–54

Taylor, H. P. Jr., Turi, B. 1976. High-^{18}O igneous rocks from the Tuscan magmatic province, Italy. *Contrib. Mineral. Petrol.* 55:33–54

Taylor, H. P. Jr., Duke, M. B., Silver, L. T., Epstein, S. 1965. Oxygen isotope studies of minerals in stony meteorites. *Geochim. Cosmochim. Acta* 29:489–512

Taylor, H. P. Jr., Giannetti, B., Turi, B. 1979. Oxygen isotope geochemistry of the potassic igneous rocks from the Roccamonfina volcano, Roman comagmatic region, Italy. *Earth Planet. Sci. Lett.* 46:81–106

Thorpe, R. S., Potts, P. J., Francis, P. W. 1976. Rare earth data and petrogenesis of andesite from the north Chilean Andes. *Contrib. Mineral. Petrol.* 54:65–78

Turi, B., Taylor, H. P. Jr. 1976. Oxygen isotope studies of potassic volcanic rocks of the Roman Province, Central Italy. *Contrib. Mineral. Petrol.* 55:1–31

Urey, H. C. 1947. Thermodynamic properties of isotopic substances. *J. Chem. Soc.*, pp. 562–81

Vollmer, R., Hawkesworth, C. J. 1980. Lead isotopic composition of the potassic rocks from Roccamonfina (South Italy). *Earth Planet. Sci. Lett.* 47:91–101

Wasserburg, G. J., DePaolo, D. J. 1979. Models of earth structure inferred from neodymium and strontium isotopic abundances. *Proc. Natl. Acad. Sci. USA* 76(8):3594–98

Whitford, D. J., Compston, W., Nicholls, I. A. 1977. Geochemistry of the late Cenozoic lavas from eastern Indonesia: role of subducted sediments in petrogenesis. *Geology* 5:571–75

Whitford, D. J., White, W. M., Jezek, P. A., Nicholls, I. A. 1979. Nd isotopic composition of recent andesites from Indonesia. *Carnegie Inst. Washington Yearb.* 78:304–8

Wolery, T. J., Sleep, N. H. 1976. Hydrothermal circulation and geochemical flux at mid-ocean ridges. *J. Geol.* 84:249–75

Wright, T. L., Swanson, D. A., Duffield, W. A. 1975. Chemical compositions of Kilauea east-rift lava, 1968–1971. *J. Petrol.* 16:110–33

Ann. Rev. Earth Planet. Sci. 1981. 9:345–83
Copyright © 1981 by Annual Reviews Inc. All rights reserved

INTERVALENCE TRANSITIONS IN MIXED-VALENCE MINERALS OF IRON AND TITANIUM

✖ 10154

Roger G. Burns

Department of Earth and Planetary Sciences, Massachusetts Institute of
Technology, Cambridge, Massachusetts 02139

> Since ferrous iron usually colors minerals green, and ferric iron yellow or brown, it may
> seem rather remarkable that the presence of both together should give rise to a blue
> color, as in the case of vivianite. . . . . Other instances may perhaps be discovered, should
> this subject ever be investigated as it deserves to be.
>
> E. T. Wherry, *Am. Mineral.* 3:161 (1918)

> This term (charge transfer) is used as a catch phrase for a wide variety of electronic
> processes which give rise to strong absorption in the visible and ultraviolet and there
> is a great danger in merely assigning any unexplained feature in a spectrum as a "charge
> transfer band."
>
> W. B. White, *Am. Mineral.* 52:555 (1967)

INTRODUCTION

Mixed-valence minerals containing an element in two different oxidation
states frequently have unusual physical properties, one of which is intense
coloration originating from the transfer of electrons between the two
valence states. Interpretations of such "charge transfer" or intervalence
transitions in absorption spectra of minerals have often been controversial.
The two statements cited in the preface above convey the anticipation,
excitement, and frustrations of research on mixed-valence transition
metal-bearing minerals, which have been compounded during the past
decade as a result of spectral measurements on extraterrestrial materials
from the Moon and meteorites.

345

0084-6597/81/0515-0345$01.00

Nevertheless, light-induced electron transfer processes between neighboring elements in a crystal structure continue to have important applications in the earth and planetary sciences. Not only do they affect color and pleochroism, visible-region and nuclear gamma resonance (Mössbauer) spectra, magnetism, electrical and thermal conductivities of minerals, but they also manifest themselves in such wide-ranging phenomena as geophysical properties of the Earth's interior, remote-sensed spectra of planetary surfaces, and transition metal geochemistry. Recent investigations of mixed-valence compounds (Brown 1980) have greatly clarified electron transfer processes in minerals (Burns et al 1980). This review focusses on results and applications of such measurements for silicates and oxides of iron and titanium, the most abundant transition elements.

Mixed-valence minerals have attracted attention since antiquity. The subtle blue hues of the gems sapphire, aquamarine, and iolite, the intense colorations of certain amphiboles, micas, tourmalines, and pyroxenes, and the opacity of magnetite, ilmenite, etc, were generally attributed to the transfer of electrons between Fe^{2+} and Fe^{3+} ions (Watson 1918, MacCarthy 1926, Weyl 1951, Martinet & Martinet 1952). The highly directional character of such "charge transfer" interactions in minerals was demonstrated by absorption spectral measurements in polarized light (Faye et al 1968) and electrical conductivity measurements (Littler & Williams 1965). While the spectral features due to $Fe^{2+} \rightarrow Fe^{3+}$ charge transfer transitions could usually be distinguished from crystal field transitions within individual Fe^{2+} and Fe^{3+} ions, complexities arose when titanium was present. The color and spectra of blue kyanites and sapphire, for example, have sparked considerable debate (White & White 1967, Faye & Nickel 1969a, Lehmann & Harder 1970, Faye 1971, Smith 1978b). Subsequently, the occurrence of Ti^{3+} ions in natural and synthetic pyroxenes and glasses in terrestrial and extraterrestrial rocks led to further problems of distiguishing Ti^{3+} crystal field transitions from $Ti^{3+} \rightarrow Ti^{4+}$ and $Fe^{2+} \rightarrow Ti^{4+}$ charge transfer transitions (Dowty & Clark 1973a,b, Burns & Huggins 1973, Mao & Bell 1974). Some of these opposing spectral assignments of Fe-Ti-bearing silicates and oxide minerals are addressed in this review.

Another area of interest in mixed-valence minerals concerns the nature and extent of electron transfer processes in the crystal structures. Here the question is whether electron hopping is localized on adjacent cations leading to discrete valencies (e.g. distinguishable Fe^{2+} and Fe^{3+} ions), or whether electrons are delocalized along chains of cations leading to indistinguishable valencies. These phenomena are affected not only by crystal chemical and structural factors, but also by variations of pressure and temperature. This review highlights pertinent crystal structure data for mixed-valence minerals of iron and titanium and temperature-pressure variations of their electronic absorption and Mössbauer spectra.

TERMINOLOGY

In mixed-valence minerals, we are primarily interested in the excitation of electrons *between* adjacent ions and not within individual cations. The latter intra-electronic or crystal field transitions take place between d or f orbital energy levels of single transition metal ions. Electronic transitions between neighboring ions are broadly called charge transfer (CT) transitions and include anion → cation as well as cation → cation transitions. In oxides and silicates, oxygen → cation CT transitions are generally excited by higher energy ultraviolet radiation, but absorption edges may extend into the visible region and induce dark brown colors in a mineral (e.g. biotite, hornblende). For cations most frequently encountered in terrestrial and lunar minerals, oxygen → cation CT energies for octahedrally coordinated cations are calculated or observed to decrease in the order (Loeffler et al 1974)

$$Cr^{3+} > Ti^{3+} > Fe^{2+} > Ti^{4+} > Fe^{3+}.$$

The focus of this review, however, is on electron transfer between adjacent cations which momentarily change valence during the lifetime of the transition. Such electron exchange processes are referred to as intervalence transitions. The cation → cation intervalence transitions are classified as either homonuclear (e.g. $Fe^{2+} \rightarrow Fe^{3+}$; $Ti^{3+} \rightarrow Ti^{4+}$) or heteronuclear (e.g. $Fe^{2+} \rightarrow Ti^{4+}$), and are typically induced by incident light in the visible–short wave infrared region (wavelengths 400–2000 nm or wave numbers 25,000–4,000 cm^{-1}). Light-induced intervalence transitions are sometimes referred to as optically excited electron exchange transitions, energies of which are denoted by E_{OP}.

In order for intervalence transitions to occur the interacting cations must be adjacent to one another in the mineral crystal structure. Therefore, they are generally observed between transition metal ions in coordination sites sharing edges or faces. The majority of $Fe^{2+} \rightarrow Fe^{3+}$ and $Fe^{2+} \rightarrow Ti^{4+}$ CT transitions identified in silicate and oxide minerals involve cations in edge-shared octahedra, although examples involving face-shared octahedra (sapphire), edge-shared octahedra-tetrahedra (cordierite) and edge-shared cube-tetrahedra (garnet) are known.

A classification scheme for mixed valence compounds has been developed (Robin & Day 1967, Day 1976) based on the degree to which the sites occupied by the cations of differing valence can be distinguished in the ground state and the consequent ease or difficulty of transferring an electron from one site to another. When the two sites are very different (e.g. Fe^{3+}-bearing octahedral and Fe^{2+}-bearing cube sites in the garnet structure), the transfer of electrons is difficult so that the properties of the mineral are essentially the sum of those of the individual cations in the two sites. These are termed Class I compounds (Day 1976). If the two sites are identical, the electron can be transferred from one site to the other with no expenditure of energy. If the structure contains continuous arrays of sites, the mineral will have metallic properties. Because the valencies of the individual ions are "smeared out" the characteristic properties of the single valence states may not be found. Materials of this kind are called Class IIIB compounds. Magnetite is the classic example of a Class IIIB compound. If the structure is not completely continuous, electron delocalization only takes place within a finite cluster of equivalent cations. Although the individual ion properties may not be seen the structure does not conduct electrons. This type of substance is called a Class IIIA compound.

Intermediate between these two extremes are minerals classified as Class II compounds in which the two sites are similar but distinguishable (i.e. both are octahedral sites, but with slightly different metal-oxygen distances, or ligand orientation or bond-type, such as the mica M1 and M2 sites). Such materials still exhibit ions with discrete valencies, but have low energy intervalence CT bands and are semiconductors. The physical properties of each class of mixed-valence compound are summarized in Table 1.

This approach to mixed-valence compounds represents an essentially static model and takes no account of the dynamical behavior of either the nucleus or the electrons. Class III compounds will be observed, for example, only if electrons are transferred from cation to cation more rapidly than the ligand atoms constituting the coordination site can relax to equilibrium positions corresponding to the individual discrete valencies. Fe^{2+} ions, for example, require a larger site than Fe^{3+} ions, but there may be insufficient time for the coordination polyhedra to expand each time an electron is momentarily transferred from Fe^{3+} to Fe^{2+}. The frequencies of such lattice relaxation effects are temperature dependent, but at elevated temperatures electrons may overcome activation energies to transfer from cations of discrete valencies to become electron-delocalized species. The energy required for the adiabatic delocalization of an electron between two sites is denoted E_{AD}, and an approximate relationship between this thermal activated electron energy and the optically excited electron exchange energy, E_{OP} is (Hush 1967): $E_{AD} = E_{OP}/4$.

Table 1 Physical properties of classes of mixed-valence compounds (after Day, 1976).

Property	Class I	Class II	Class III A	Class III B
Nature of sites in crystal structure	Vastly different, e.g. cube-octahedron, octahedron-tetrahedron	Similar, e.g. octahedra with slightly different bond lengths or ligands	Identical, finite clusters, e.g. edge-shared octahedral dimers	Identical, continuous chains, e.g. chains of edge-shared octahedra
Optical absorption spectroscopy	No intervalence transitions in visible region	One mixed-valence transition in visible region; absorption bands intensify at elevated pressures and low temperatures	One or more mixed-valence transition in visible region; temperature lowers intensity of absorption bands, pressure intensifies	Opaque; metallic reflectivity in visible region
Electrical conductivity	Insulator; resistivity $> 10^{12}$ ohm cm	Semiconductor; resistivity $10–10^8$ ohm cm	Insulator or high resistance semiconductor	Metallic conductor; resistivity $10^{-2}–10^{-6}$ ohm cm
Magnetic properties	Diamagnetic or paramagnetic to very low temperature	Magnetically dilute; either ferro- or anti-ferromagnetic at low temperatures	Magnetically dilute	Pauli paramagnetism or ferro-magnetic with high Curie temperature
Mössbauer spectra	Spectra of constituent Fe ions; discrete Fe^{2+} and Fe^{3+} ions	Spectra of constituent Fe ions; discrete Fe^{2+} and Fe^{3+} ions	Electron-delocalized species contribute to spectra; higher contributions at elevated temperatures and pressures	Electron delocalized species contribute to spectra; higher contributions at elevated temperatures and pressures

One consequence of thermally activated electron delocalization behavior is that techniques such as Mössbauer spectroscopy, which might be expected to distinguish discrete Fe^{2+} and Fe^{3+} valencies from electron delocalized Fe cation species, will detect Class III compound behavior only if electrons are transferred between Fe^{2+} and Fe^{3+} ions in neighboring sites more rapidly than the lifetime of the Mössbauer transition $(10^{-7}$ s) between the 14.41 keV ground and excited nuclear energy levels of ^{57}Fe. Thus, Mössbauer spectrum profiles of Class III compounds are expected to show significant temperature variations.

CRYSTAL STRUCTURE FEATURES

It is apparent that the classification and properties of mixed-valence compounds depend greatly on the nature and extent of coordination sites containing adjacent cations (Loeffler et al 1975, Burns et al 1980). Many of the electronic properties can be correlated with crystal structure features, including metal-oxygen (M-O) bond strength as evidenced by M-O distances; metal-oxygen-metal overlap indicated by metal-metal (M-M) distances; symmetries of coordination polyhedra, e.g. regular or distorted octahedra, tetrahedra, cubes; types of polyhedral linkages, e.g. face-shared, edge-shared, corner-shared; and orientation and extent of linked polyhedra, e.g. isolated pairs, chains, bands, sheets. Such structural features of mixed-valence oxide and silicate minerals of Fe and Ti are summarized in Table 2. Note that cation → cation intervalence transitions generally produce intense absorption bands in visible-region spectra only when light is polarized along the metal-metal directions in the crystal structures.

MINERALS WITH FE^{2+}-FE^{3+} OCTAHEDRAL CLUSTERS

Vivianite

Perhaps the best-known mixed-valence compound exhibiting a $Fe^{2+} \rightarrow Fe^{3+}$ intervalence transition is vivianite, $Fe_3(PO_4)_2 \cdot 8H_2O$. Although the formula of this mineral is not indicative of a mixed-valence compound, freshly cleaved vivianite crystals or newly precipitated ferrous phosphate which are pale green turn blue when exposed to air. Such intense blue colorations are atypical of pure Fe(II) or Fe(III) compounds containing Fe^{2+} or Fe^{3+} ions octahedrally coordinated to oxygen ligands, and are indicative of an intervalence transition (Allen & Hush 1967, Hush 1967, Robin & Day 1967).

Table 2 Structural data for mixed-valence minerals of Fe and Ti

Mineral (formula)	Sites (coord. no.)	M-M distances (Å)	Polyhedral linkages (type of sharing; extent of M-M cluster)	References
vivianite $Fe_3(PO_4)_2 \cdot 8H_2O$	Fe_A (6) Fe_B (6)	$Fe_B\text{-}Fe_B = 2.85$	edge: isolated clusters	Mori & Ito 1950
sapphire $Al_2O_3/Fe,Ti$	Al (6)	$\perp c = 2.79$ $\|c = 2.65$	edge: finite clusters; face: isolated clusters	Newnham & de Haan 1962
kyanite $Al_2SiO_5/Fe,Ti$	Al_1 (6) Al_2 (6) Al_3 (6) Al_4 (6)	2.75–2.78	edge: infinite $Al_1\text{-}Al_2$; chains: isolated $Al_1\text{-}Al_2$, $Al_2\text{-}Al_3$, $Al_1\text{-}Al_4$, $Al_2\text{-}Al_4$ clusters; substitutional blocking by Al^{3+}	Burnham 1963
aquamarine (beryl) $Be_3Al_2Si_6O_{18}$	Al (6)	$\|c = 4.60$	isolated octahedra; cations possibly in channels	Gibbs et al 1968
cordierite (iolite) $Al_3(Mg,Fe^{2+})_2$ $(Al,Fe^{3+})Si_5O_{18}$	M (6) T_1 (4)	$M\text{-}T_1 = 2.74$	edge: isolated octahedral-tetrahedral clusters; cations possibly in channels	Gibbs 1966
osumilite $K(Mg,Fe^{2+})_2(Al,Fe^{3+})_3$ $(Si,Al)_{12}O_{30}$	similar to cordierite	comparable to cordierite	similar to cordierite	Brown & Gibbs 1969
tourmaline $Na(Mg,Fe,Mn,Li,Al)_3$ $(Al,Fe)_6(Si_6O_{18})$ $(BO_3)_3(OH,F)_4$	Mg (6) Al (6)	$Mg\text{-}Mg = 3.04$ $Mg\text{-}Al = 2.97$ $Al\text{-}Al = 2.80$	edge: trigonal planar clusters of Mg-Mg and Mg-Al $\perp c$; infinite spiral clusters of Al-Al $\|c$; substitutional blocking by Mg^{2+} and Al^{3+}	Buerger et al 1962

Table 2 (*continued*)

Mineral (formula)	Sites (coord. no.)	M-M distances (Å)	Polyhedral linkages (type of sharing; extent of M-M cluster)	References
yoderite (Al,Mg,Fe,Mn)$_8$ Si$_4$O$_{18}$(OH)$_2$	Al$_1$ (6) Al$_2$ (5) Al$_3$ (5)	Al$_1$-Al$_1$ = 2.90 Al$_1$-Al$_2$ = 3.50 Al$_1$-Al$_3$ = 3.28	edge: infinite Al$_1$-Al$_1$ chains; isolated Al$_1$-Al$_2$ and Al$_1$-Al$_3$ clusters; substitutional blocking by Mg^{2+} & Al^{3+}	Fleet & Megaw 1962
orthopyroxenes (Mg,Fe)$_2$Si$_2$O$_6$	M1 (6) M2 (6)	M1-M1 = 3.15 M1-M2 = 3.08 and 3.27	infinite M1-M1 and M1-M2 chains ‖c; substitutional blocking by Mg^{2+}	Burnham et al 1971
augite Ca(Mg,FeAl) (Si,Al)$_2$O$_6$	M1 (6) M2 (8)	M1-M1 = 3.11 M1-M2 = 3.21	same as orthopyroxene	Clark et al 1969
Allende fassaite Ca(Mg,Ti^{3+},Ti^{4+},Al) (Si,Al)$_2$O$_6$	M1 (6)	M1-M1 = 3.15	same as augite	Dowty & Clark 1973a
Angra dos Reis fassaite (Ca,Fe^{2+})(Fe^{2+},Ti^{4+},Mg,Al) (Si,Al)$_2$O$_6$	M1 (6)	M1-M1 = 3.13	same as augite	Hazen & Finger 1977
omphacite (Na,Ca)(Al,Fe,Ti) Si$_2$O$_6$	M1 (6)	M1-M1 = 2.90	similar to orthopyroxene	Curtis et al 1975

Mineral	Sites	M-M distances	Description	Reference
glaucophane $Na_2(Mg,Fe^{2+})_3(Al,Fe^{3+})_2Si_8O_{22}(OH)_2$	M1 (6) M2 (6) M3 (6) M4 (8)	M1-M1 = 3.22 M1-M2 = 3.09 M1-M3 = 3.10 M2-M3 = 3.31 M2-M4 = 3.15 M1-M4 = 3.31	infinite bands of edge-shared octahedra ||c; substitutional blocking by Mg^{2+} and Al^{3+}	Papike & Clark 1968
riebeckite, crocidolite $Na_2Fe_3^{2+}Fe_2^{3+}Si_8O_{22}(OH)_2$	as for glaucophane	comparable to glaucophane	infinite bands of edge-shared $[FeO_6]$ octahedra	Hawthorne 1976
actinolite-hornblende $Ca_2(Mg,Fe,Al)_5(Si,Al)_8O_{22}(OH)_2$	as for glaucophane	M1-M1 = 3.23 M1-M2 = 3.11 M1-M3 = 3.10 M2-M3 = 3.20 M2-M4 = 3.20 M1-M4 = 3.42	similar to glaucophane	Mitchell et al 1971
biotite (annite-phlogopite) $K(Mg,Fe^{2+},Fe^{3+})_{2-3}(Si_3AlO_{10})(OH)_2$	M1 (6) M2 (6)	M1-M1 and M1-M2 $\approx$ 3.10–3.20	infinite 2-D sheets of edge-shared octahedra	Hazen & Burnham 1973
babingtonite $Ca_2Fe^{2+}Fe^{3+}Si_5O_{14}(OH)$	Fe_1 (6) Fe_2 (6)	Fe_1-Fe_1 = 3.37 Fe_1-Fe_2 = 3.30	clusters of edge-shared $[FeO_6]$ octahedra four octahedra wide	Araki & Zoltai 1972 Kosoi 1976
ilvaite $CaFe_2^{2+}Fe^{3+}Si_2O_8(OH)$	Fe_A (6) Fe_B (6)	Fe_A-Fe_A = 2.83 and 3.03 (||c) and 3.01 (⊥c), Fe_A-Fe_B = 3.15 and 3.25	infinite double chains of edge-shared Fe_A octahedra ||c	Beran & Bittner 1974 Haga & Takéuchi 1976
deerite $Fe_6^{2+}Fe_3^{3+}O_3(Si_6O_{17})(OH)_5$	cations in nine octahedral sites	range of M-M distance 3.11–3.31	infinite bands, six octahedra wide, of edge-shared $[FeO_6]$ octahedra	Fleet 1977

Table 2 (*continued*)

Mineral (formula)	Sites (coord. no.)	M-M distances (Å)	Polyhedral linkages (type of sharing; extent of M-M cluster)	References
howieite $NaFe^{2+}_{10}Fe^{3+}_{3}(Si_{12}O_{34})(OH)_{10}$	cations in six octahedral sites	range of M-M distance 3.16–3.29	infinite bands, four octahedral wide of edge-shared [FeO₆] octahedra	Wenk 1974
sapphirine $(Mg,Fe^{2+},Al,Fe^{3+})_7(Si,Al)_6O_{20}$	cations in seven octahedral sites	range of M-M distance 3.18–3.35	infinite bands, four and three octahedral wide of edge-shared octahedra	Moore 1969 Higgins et al 1979 Higgins & Ribbe 1979 Merlino 1980
aenigmatite $Na_2Fe_5TiSi_6O_{20}$	similar to sapphirine	comparable to sapphirine	similar to sapphirine	Cannillo et al 1971
magnetite Fe_3O_4	Fe_A (4) Fe_B (6)	Fe_B-Fe_B = 2.97 Fe_A-Fe_B = 3.48	infinite single chains of edge-shared [FeO₆] running in 3-D	Hamilton 1958
Ti andradite (schorlomite) $(Ca,Fe^{2+})_3(Fe^{3+},Fe^{2+},$ $Ti^{4+},Ti^{3+},Al)_2$ $(Si,Fe^{3+},Al)_3O_{12}$	{X} (8) [Y] (6) (Z) (4)	X-Z = 3.01 X-Y = 3.37 Y-Z = 3.37	edge: chains of {XO₈} cubes and (ZO₄) tetrahedra; corner-shared [YO₆] octahedra and (ZO₄) or {XO₈}	Novak & Gibbs 1971

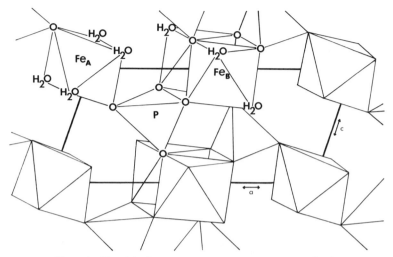

Figure 1   The vivianite crystal structure viewed along the *b* axis.

The crystal structure of vivianite illustrated in Figure 1 contains two distinct octahedral sites, designated Fe_A and Fe_B. The Fe_A octahedra are isolated, but pairs of Fe_B octahedra share a common edge across which the Fe^{2+} ions are separated by only 2.85 Å along the *b* axis. Polarized absorption spectra of vivianites (Hush 1967, Faye et al 1968) show an intense absorption band centered at 15,200 cm^{-1} when light is polarized along the *b* axis (the Fe_B-Fe_B direction) which effectively masks all but blue light when radiation is transmitted through the crystals. The polarization dependence of the 15,200 cm^{-1} band is the key to its assignment; it originates from a $Fe^{2+} \rightarrow Fe^{3+}$ intervalence transition between adjacent Fe^{2+} and Fe^{3+} ions in the edge-shared Fe_B octahedra. The intensity of this charge transfer band increases significantly with decreasing temperature (Smith & Strens 1976) and rising pressure (Mao 1976), supporting its assignment as an intervalence transition between Fe^{2+} and Fe^{3+} ions.

Inherent in this assignment, however, are the assumptions that Fe^{3+} ions are present and that some occur in the Fe_B sites, which are not implied by the stoichiometry of vivianite. Confirmation comes from recent Mössbauer measurements of vivianites (McCammon & Burns 1980). The Mössbauer spectra of samples mechanically ground in air, such as the profiles illustrated in Figure 2, show that Fe^{2+} ions in both Fe_A and Fe_B octahedra are oxidized to Fe^{3+} ions. However, Fe_A^{2+}/Fe_B^{2+} ratios increase with rising Fe^{3+} concentration, suggesting that there is a greater probability of oxidizing remaining Fe_A^{2+} ions than the second Fe^{2+} ion of a Fe_B^{2+}-Fe_B^{3+} pair.

Vivianite has the potential of being a Class IIIA mixed-valence compound with respect to the Fe_B octahedra. However, the Mössbauer spectra measured as high as the thermal decomposition temperature of vivianite (56°C) resolve only octahedral Fe^{2+} and Fe^{3+} ions indicating that individual iron cations have discrete valencies in the vivianite structure. Thus, although $Fe^{2+} \rightarrow Fe^{3+}$ intervalence transitions occur when light is polarized along the short Fe_B-Fe_B axis, the electron exchanged between each Fe_B^{2+}-Fe_B^{3+} pair is effectively localized on an individual

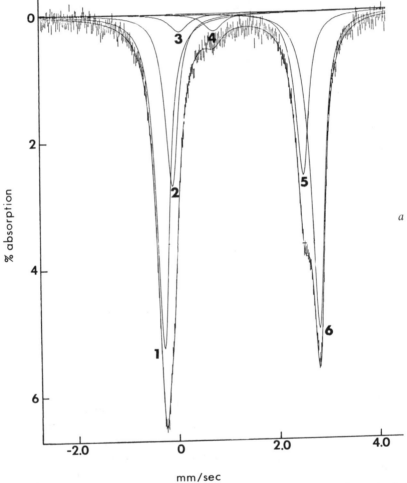

Figure 2 Mössbauer spectra of vivianite (a) before and (b) after oxidation by mechanical grinding in air. Peaks 1 and 6: Fe^{2+} in the Fe_B sites; peaks 2 and 5: Fe^{2+} in the Fe_A sites; peaks 3 and 4: octahedral Fe^{3+}. From McCammon & Burns (1980).

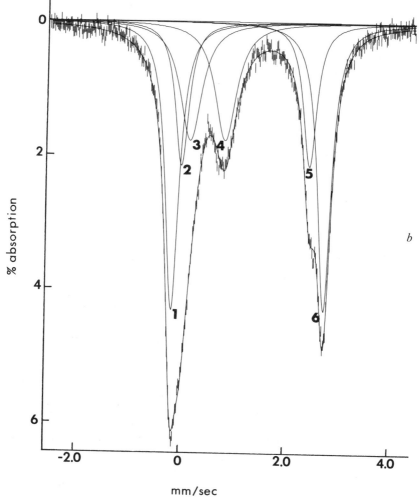

Figure 2   (continued)

cation during the lifetime (10^{-7} s) of the Mössbauer transition. Vivianite is thus a Class II mixed-valence compound.

Magnetite

Another mineral cited as a typical example of a mixed-valence compound (Robin & Day 1967, Day 1976) is magnetite, Fe_3O_4 or $Fe^{2+}Fe_2^{3+}O_4$. It is considered to be a Class IIIB mixed-valence compound because the magnetite structure contains infinite chains of Fe^{2+}-Fe^{3+} octahedra exhibiting electron delocalization. However, a phase change in magnetite at

119 K complicates interpretations of electron delocalization in this structure type.

The magnetite structure, an inverse spinel, contains Fe^{3+} ions in tetrahedral A sites, and Fe^{2+} and Fe^{3+} ions separated by 2.97 Å in three-dimensional infinite chains of edge-shared octahedral B sites extending along [100] directions. Above 119 K, electron delocalization involving the B-site cations occurs in magnetite, resulting in opacity and high electrical and thermal conductivities. The Mössbauer spectra show two cation species: tetrahedral Fe^{3+} and mixed-valence octahedral iron (Sawatzky et al 1969, Lotgering & Diepen 1977, Kundig & Hargrove 1969). At 119 K a phase change occurs (Hamilton 1958) which doubles the unit cell, leads to magnetic ordering of Fe^{2+} and Fe^{3+}, and produces distinguishable Fe^{2+} and Fe^{3+} in octahedral sites with attendant drop in electrical conductivity and changes in the Mössbauer spectra (Verwey & Haayman 1941, Verble 1974). The transition in magnetite, which is called the Verwey transition, is interpreted as an ionic order-disorder transition, but controversy has arisen recently concerning the number of disordering parameters necessary and the validity of the order-disorder model versus a band model or polaron model (Cullen & Callen 1971, Verble 1974). Thus, although magnetite is classified as the type-example of a Class IIIB mixed-valence compound, application of a Verwey-type model of quenched electron delocalization to other systems not showing a phase change should be made with caution.

Ilvaite

Many of the attributes of electron delocalization in a mixed-valence compound are displayed by ilvaite, $CaFe_2^{2+}Fe^{3+}Si_2O_8(OH)$. The ilvaite structure, which is illustrated in Figure 3, consists of a framework of infinite double chains of edge-shared Fe(A) octahedra linked by corner-shared double tetrahedral Si_2O_7 groups. Six-coordinated Fe(B) sites and seven-coordinated Ca sites are bonded to oxygens of this framework. Neutron diffraction measurements have demonstrated that Fe^{2+} and Fe^{3+} ions are located in the Fe(A) sites while Fe(B) sites accomodate Fe^{2+} ions (Haga & Takéuchi 1976). Many $Fe^{2+} \rightarrow Fe^{3+}$ interactions are possible between the edge-shared octahedra shown in Figure 4. Thus, Fe(A)-Fe(A) couples occur along the chains in the c direction (Fe-Fe distances of 2.83 Å and 3.03 Å), and across the chain in the a-b plane (Fe-Fe = 3.01 Å). Four Fe(B) $\rightarrow$ Fe(A) interactions occur per Fe(B) site with Fe-Fe distances ranging from 3.15 Å to 3.25 Å.

Electron delocalization behavior has been deduced from the temperature variations of the Mössbauer spectra of ilvaite (Gérard & Grandjean 1971, Grandjean & Gérard 1975, Nolet 1978, Nolet & Burns 1979). The spectral profiles show significant variations between 80 K and 390 K (for

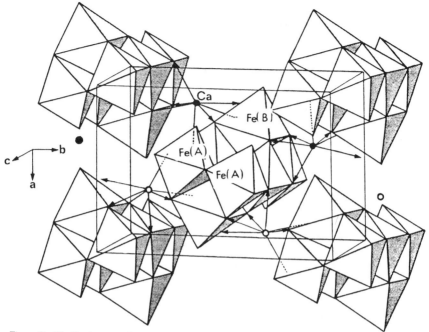

Figure 3   The ilvaite crystal structure, showing the arrangement of infinite double chains of edge-shared Fe(A) octahedra extending along the *c* axis. From Nolet & Burns (1979).

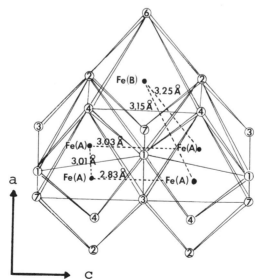

Figure 4   Projection of one unit of the Fe(A) double chain showing possible Fe(A)-Fe(A) and Fe(A)-Fe(B) interactions across shared edges and appropriate metal-metal distances. From Nolet & Burns (1979).

example, see Figure 5) which was earlier attributed to a Verwey-type transition in ilvaite at about 320 K. Recently, the Mössbauer spectra of a suite of ilvaite specimens were successfully fitted to five component doublets (Nolet & Burns 1979), the assignments of which are shown in Figure 5. In addition to one doublet for octahedral Fe^{3+} and two Fe^{2+} doublets representing cations in the Fe(A) and Fe(B) sites, two additional doublets were resolved representing delocalized $[Fe^{2+}(A) \rightarrow Fe^{3+}(A)]$ species parallel to and perpendicular to the chains of edge-shared Fe(A) octahedra (Figure 4). The latter two assignments explain the strong absorption and dark colors of ilvaite in all crystallographic directions. The temperature variations of the ilvaite Mössbauer spectra are therefore not due to a localized-delocalized order-disorder transition, but rather to conditions where peaks due to $[Fe^{2+}(A) \rightarrow Fe^{3+}(A)]$ delocalized species become resolvable at elevated temperatures. Certain electronic levels associated with discrete valencies become depopulated with increasing temperature contemporaneous with population of new delocalized levels associated with intermediate valencies.

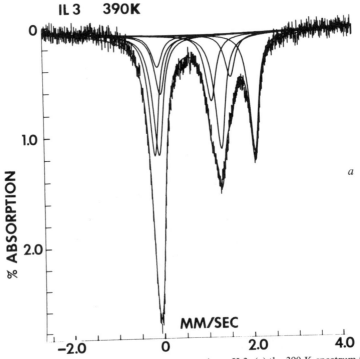

Figure 5 Fitted Mössbauer spectra for ilvaite specimen IL3: (*a*) the 390 K spectrum fitted to eight peaks or four doublets, $Fe^{2+}(A)$ is below resolution of spectrometer; (*b*) the 80 K spectrum showing assignments of component doublets. From Nolet & Burns (1979).

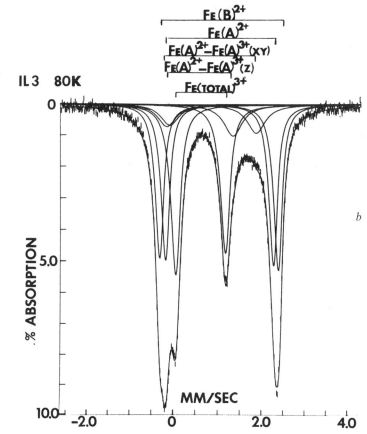

Figure 5 (continued)

The effect of pressure to 60 kb on the Mössbauer spectra of ilvaite are similar to temperature, but are highly orientation dependent (Evans & Amthauer 1980). Thus, the emergence of delocalized $[Fe^{2+}(A) \to Fe^{3+}(A)]$ species is most pronounced for sections cut parallel to (001), in which the compression axis is c, corresponding to the direction of infinite chains of linked octahedra (Figure 3).

Ilvaite with its infinite chains of edge-shared Fe^{2+}-Fe^{3+} octahedra is a model example of a Class IIIB mixed-valence compound with respect to the Fe(A) octahedra. The short Fe(A)-Fe(A) distances, 2.83–3.03 Å, are particularly conducive to electron delocalization. The Fe(A)-Fe(B) octahedral clusters, however, constitute a Class II compound, the somewhat larger Fe(A)-Fe(B) distances (3.15–3.25 Å) being less conducive to electron delocalization. Ilvaite is representative of several end-member Fe^{2+}-Fe^{3+} silicate minerals in which electron delocalization is predicted or observed along chains of linked octahedra. Further examples are discussed below.

Deerite

Another mixed-valence mineral recently demonstrated by Mössbauer spectroscopy to exhibit electron delocalization is deerite, $Fe_6^{2+}Fe_3^{3+}O_3$ $Si_6O_{17}(OH)_5$. The crystal structure of this chain silicate mineral contains bands of edge-shared octahedra, six octahedra wide, extending along the c axis (Fleet 1977). Although the structure contains nine distinct Fe atom positions, they are grouped into three sets of three virtually equivalent octahedral sites. The Fe-Fe distances across the edge-shared octahedra range from 3.10 Å to 3.31 Å. By analogy with ilvaite, the profiles of the Mössbauer spectra of deerite are strongly temperature dependent (Frank & Banbury 1974, Amthauer et al 1980). The spectra indicate that at 77 K discrete Fe^{2+} and Fe^{3+} ions exist in deerite, but at 298 K about 20% of the iron cations take part in thermally activated electron delocalization processes across the edge-shared octahedra (Amthauer et al 1980).

Babingtonite

Babingtonite, $Ca_2Fe^{2+}Fe^{3+}Si_5O_{14}(OH)$, has a formula resembling ilvaite and deerite and might be expected to display electron delocalization. Edge-shared octahedra also occur in the babingtonite structure. However, they do not form infinite chains as exist in ilvaite and deerite. There are two distinct Fe sites in babingtonite, designated Fe(1) and Fe(2), and clusters of edge-shared Fe(2)-Fe(1)-Fe(1)-Fe(2) octahedra, four wide, link two chains of corner-shared SiO_4 tetrahedra (Araki & Zoltai 1972). The Fe(2)-Fe(1) and Fe(1)-Fe(1) distances are about 3.30 Å and 3.37 Å, respectively, and average Fe-O distances are approximately 2.17 Å and 2.05 Å, respectively, leading to the assignment of Fe^{2+} and Fe^{3+} ions in the larger Fe(1) and smaller Fe(2) octahedra, respectively. Mössbauer measurements between 30 K and 600 K confirmed this highly ordered cation distribution and failed to detect any electron-delocalized iron cation species (Amthauer 1980). Babingtonite, therefore, resembles vivianite by containing discrete Fe^{2+} and Fe^{3+} ions. Evidently, the limited extent of edge-shared octahedra and large Fe-Fe distances are not conducive to electron delocalization behavior in babingtonite.

Glaucophane-Riebeckite

Ilvaite and deerite represent end-member mixed-valence silicates the crystal structures of which contain infinite chains of edge-shared $[FeO_6]$ octahedra permitting electron delocalization. Several other silicate minerals also have structures in which infinite chains of edge-shared octahedra exist, but atomic substitution of non–transition metal ions such as Mg^{2+} and Al^{3+} with ionic radii similar to those of Fe^{2+} and Fe^{3+} limit the extent of $Fe^{2+} \rightarrow Fe^{3+}$ interactions by substitutional blocking.

The model mineral showing intervalence transitions and substitutional blocking of electron delocalization is glaucophane, $Na_2(Mg,Fe^{2+})_3$ $(Al,Fe^{3+})_2Si_8O_{22}(OH)_2$. The Fe^{2+}-Fe^{3+} end-member mineral is riebeckite, $Na_2Fe_3^{2+}Fe_2^{3+}Si_8O_{22}(OH)_2$, the asbestiform variety of which is called crocidolite. The glaucophane structure, like other amphiboles, contains bands of octahedral sites, designated M1, M2, and M3, two or three octahedra wide, extending along the c axis (Papike & Clark 1968). In highly ordered blue-schist facies glaucophanes, the larger M1 and M3 octahedra are occupied preferentially by Mg^{2+} and Fe^{2+} ions, while the smaller Al^{3+} and Fe^{3+} ions are enriched in the smaller M2 octahedra. Metal-metal distances across the edge-shared octahedra are in the range 3.09–3.31 Å. Intervalence $Fe^{2+} \rightarrow Fe^{3+}$ transitions, primarily Fe^{2+}(M1) $\rightarrow$ Fe^{3+}(M2) and Fe^{2+}(M3) $\rightarrow$ Fe^{3+}(M2), account for the blue-violet pleochroism of glaucophane and the polarization dependence of the absorption bands in the region 16,130–18,520 cm^{-1} when light is polarized only in the plane of edge-shared octahedra (Burns 1970). Similar charge transfer bands have been measured in the optical spectra of several alkali amphiboles and shown to have intensities proportional to the product of donor Fe^{2+} and acceptor Fe^{3+} concentrations (Littler & Williams 1965, Smith & Strens 1976). A similar concentration dependence was found for the electrical conductivity of crocidolite (Littler & Williams 1965) which is significantly high along the fiber axis (the direction of infinite chains of edge-shared $[FeO_6]$ octahedra).

Although the amphibole structure is conducive to electron delocalization by having short enough metal-metal distances and infinite chains of edge-shared octahedra, the Mössbauer spectra of several compositions along the glaucophane-riebeckite solid solution series have resolved only doublets attributable to discrete Fe^{2+} and Fe^{3+} cations. Negligible electron delocalization in glaucophane is certainly attributable to Mg^{2+} and Al^{3+} blocking $Fe^{2+} \rightarrow Fe^{3+}$ interactions along the chains of edge-shared octahedra. In the riebeckite end-member composition, however, electron delocalization may also be limited by the slightly larger Fe-Fe distances (compared to ilvaite and magnetite) and dissimilarities between the M1, M2, and M3 octahedra which render alkali amphiboles Class II mixed-valence compounds. Riebeckite might acquire electron delocalization detectable by the Mössbauer effect if cations were more randomly distributed in the crystal structure. Such disordering may account for the unusually high electrical conductivity of heated crocidolites (Littler & Williams 1965), while additional peaks in the Mössbauer spectra of pegmatitic riebeckites (Bancroft & Burns 1969) might represent delocalized $Fe^{2+} \rightarrow Fe^{3+}$ species instead of Fe^{2+} ions disordered into M2 sites.

Other Examples

Substitutional blocking typified by glaucophane is an important factor controlling the electronic properties of minerals. Several iron-bearing silicates contain infinite chains, bands, or sheets of edge-shared octahedra, but electron delocalization is prevented by atomic substitution of Mg^{2+} and Al^{3+} in solid solution series which block extended $Fe^{2+} \rightarrow Fe^{3+}$ interactions. However, intense intervalence bands are frequently observed in polarized spectra of such minerals measured in the visible region. Examples of minerals showing discrete valencies for Fe^{2+} and Fe^{3+} ions as a result of atomic substitution, to which $Fe^{2+} \rightarrow Fe^{3+}$ intervalence transitions across edge-shared octahedra have been assigned, are listed in Table 3. They include micas of the phlogopite-biotite series, amphiboles of the actinolite-hornblende series, many pyroxenes, omphacite, certain tourmalines, chlorite, kyanite, and potentially sapphirine.

On the other hand, end-member Fe^{2+}-Fe^{3+} silicates expected to display electron delocalization behavior, in addition to deerite, ilvaite, and perhaps riebeckite, include howieite, aenigmatite, annite, and stilpnomelane.

$Fe^{2+} \rightarrow Fe^{3+}$ INTERACTIONS IN OTHER COORDINATION SYMMETRIES

The majority of mixed-valence iron-bearing minerals displaying $Fe^{2+} \rightarrow Fe^{3+}$ interactions contain cations in edge-shared octahedra. However, intervalence transitions and electron delocalization behavior have been documented between iron cations in other coordination symmetries.

Cordierite

The best known example of Fe^{2+}(octahedral) $\rightarrow Fe^{3+}$(tetrahedral) interactions is in the mineral cordierite, $Al_3(Mg,Fe^{2+})_2[(Al,Fe^{3+})Si_5O_{18}]$. The crystal structure of cordierite contains M octahedra which share edges with two T_1 and one T_2 tetrahedra (Gibbs 1966). The Fe^{3+} ions are believed to be concentrated in the T_1 tetrahedra, while Fe^{2+} ions occupy M octahedra. The M-T_1 distances across the isolated M octahedron-T_1 tetrahedron clusters is only 2.74 Å. The absorption spectra of cordierite show a broad intense band at about 17,000 cm^{-1} polarized in the plane of the edge-shared M and T_1 polyhedra, the intensity of which increases at low temperatures (Smith & Strens 1976, Faye et al 1968, Goldman & Rossman 1978). The Mössbauer spectra of such cordierites, however, show

Table 3   Spectral data for mixed-valence minerals of Fe and Ti

Mineral	CT energy (cm^{-1}) and assignment	Comments (Möss. spect.; elect. cond.; T variations, etc)	References
vivianite	$15,200$: $Fe_B^{2+} \rightarrow Fe_B^{3+}$	M/S spectra: discrete valencies; inverse T depend. of CT; P intensifies CT.	Hush 1967, Faye et al 1968, Smith & Strens 1976, Mao 1976, McCammon & Burns 1980
sapphire	$17,000$ ($\perp c$): $Fe^{2+} \rightarrow Ti^{4+}$ $11,150$ ($\perp c$): $Fe^{2+} \rightarrow Ti^{4+}$ $12,900$ ($\|$): $Fe^{2+} \rightarrow Fe^{3+}$ $9,700$ ($\|c$): $Fe^{2+} \rightarrow Fe^{3+}$	No M/S; inverse T depend. of CT; debate over assignment	Townsend 1968, Lehmann & Harder 1970, Faye 1971, Ferguson & Fielding 1972, Eigenmann et al 1972, Smith & Strens 1976, Smith 1978b
kyanite	$16,500$: $Fe^{2+} \rightarrow Ti^{4+}$ $11,500$: $Fe^{2+} \rightarrow Fe^{3+}$	M/S confirms Fe^{2+}; inverse T dependence of CT; debate over assignment	White & White 1967, Faye & Nickel 1969a, Faye 1971, Smith & Strens 1976, Parkin et al 1977
aquamarine (beryl)	$16,100$ ($\|c$): $Fe^{2+} \rightarrow Fe^{3+}$	M/S confirms Fe^{2+}; debate over assignment	Wood & Nassau 1968, Loeffler & Burns 1976, Parkin et al 1977, Goldman et al 1978
cordierite (iolite)	$17,500$ ($\perp c$): $Fe_M^{2+} \rightarrow Fe_{T_1}^{3+}$	M/S detects Fe^{3+} at 77 K only; inverse T dependence of CT	Farrell & Newnham 1967, Faye et al 1968, Smith & Strens 1976, Pollak 1976, Parkin et al 1977; Goldman et al 1978
osumilite	$15,480$ cm^{-1}($\perp c$): Fe^{2+}(oct) $\rightarrow Fe^{3+}$(tet); or Fe^{2+}(oct) $\rightarrow Fe^{3+}$(channel)	M/S detects Fe^{2+} in oct. and channel sites; Fe^{3+} in tet.; discrete valencies	Faye 1972, Goldman & Rossman 1978
tourmaline	$18,000$ ($\perp c \gg \|c$): Fe^{2+}(Mg) $\rightarrow Fe^{3+}$(Mg,Al)	M/S detects Fe^{2+} in Mg and Al sites; discrete valencies; inverse T dependence of CT; considerable debate over assignment	Faye et al 1968, Townsend 1970, Wilkins et al 1969, Burns 1972, Hermon et al 1973, Faye et al 1974, Smith & Strens 1976, Smith 1977, 1978a

Table 3 (*continued*)

Mineral	CT energy (cm^{-1}) and assignment	Comments (Möss. spect.; elect. cond.; T variations, etc)	References
yoderite	13,800: $Fe^{2+}(Al_1) \rightarrow Fe^{3+}(Al_1)$ 16,500: $Mn^{2+}(Al_1) \rightarrow Mn^{3+}(Al_1)$ 21,000: $Mn^{2+}(Al_1) \rightarrow Mn^{3+}(Al_1)$	M/S detects Fe^{2+} and Fe^{3+}: discrete valencies	Abu-Eid et al 1978
orthopyroxenes	14,500 (‖c): $Fe^{2+}(M2,M1)$ $\rightarrow Fe^{3+}(M1)$	M/S detects Fe^{2+} and Fe^{3+}; Fe^{2+} strongly ordered in distorted M2 site	Burns 1970, Burnham et al 1971, Goldman & Rossman 1977a, Annersten et al 1978
augite	13,000: $Fe_{M1}^{2+} \rightarrow Fe_{M1}^{3+}$	M/S detects Fe^{2+} and Fe^{3+}; discrete valencies	Burns et al 1976
Allende fassaite	16,000: $Ti_{M1}^{3+} \rightarrow Ti_{M1}^{4+}$	P intensifies CT, but negligible shift in energy	Dowty & Clark 1973a,b, Burns & Huggins 1973, Mao & Bell 1974
Angra dos Reis fassaite	20,6000: $Fe_{M1}^{2+} \rightarrow Ti_{M1}^{4+}$	M/S detects Fe^{2+} only; P intensifies CT and shifts to lower energy	Bell & Mao 1976, Mao et al 1977, Hazen et al 1977
omphacite	15,040: $Fe^{2+} \rightarrow Fe^{3+}$ or $Fe^{2+} \rightarrow Ti^{4+}$	Blue Ti variety; M/S detects Fe^{2+} and Fe^{3+} discrete valencies; P variations of position and intensity of CT	Abu-Eid 1976, Strens et al 1980, Aldridge et al 1978, Bancroft et al 1969
glaucophane	18,520 (‖b): $Fe^{2+}(M1,M3)$ $\rightarrow Fe^{3+}(M2)$ 16,130 (‖c): $Fe^{2+}(M1)$ $\rightarrow Fe^{3+}(M2)$	M/S detects Fe^{2+} and Fe^{3+}; discrete valencies; Fe^{2+} ordered in M1 and M3; inverse T dependence of CT	Bancroft & Burns 1969, Ernst & Wai 1970, Smith & Strens 1976

Mineral	Spectral feature	M/S observations	References
riebeckite, crocidolite	15,000–18,000: $Fe^{2+} \rightarrow Fe^{3+}$	High elect cond.; M/S detects Fe^{2+} and Fe^{3+}; discrete valencies	Littler & Williams 1965, Hush 1967, Bancroft & Burns 1969, Manning & Nickel 1969, Faye & Nickel 1969b, Borg & Borg 1980
actinolite-hornblende	14,200 ($\|b$) and 13,700 ($\perp c$): $Fe^{2+} \rightarrow Fe^{3+}$	M/S detects Fe^{2+} in M1, M2, M3, and oct. Fe^{3+}; discrete valencies	Burns 1970, Burns & Greaves 1971, Bancroft & Brown 1975, Goldman & Rossman 1977b, Goldman 1979
biotite (annite-phlogopite)	13,650 and 16,400: $Fe^{2+} \rightarrow Fe^{3+}$ (M1 $\rightarrow$ M2 and M2 $\rightarrow$ M2)	M/S resolved into four doublets; Fe^{2+} and Fe^{3+} in M1 and M2; discrete valencies; inverse T dependence of CT	Faye 1968, Robbins & Strens 1972, Annersten 1974, Bancroft & Brown 1975, Smith & Strens 1976, Smith 1977, 1978b, Faye & Hogarth 1969
babingtonite	opaque—very dark green	M/S detects Fe^{2+} and Fe^{3+}, each on only one oct. site	Araki & Zoltai 1972, Amthauer 1980
ilvaite	green–dark brown	M/S detects discrete valencies Fe^{2+} in Fe_A and Fe_B, and Fe^{3+} in Fe_A; two ED species detected assigned to $Fe^{2+}_A \rightarrow Fe^{3+}_A$ $\|c$ and $\perp c$.	Gérard & Grandjean 1971, Grandjean & Gérard 1975, Heilmann et al 1977, Paques-Ledent et al 1977, Nolet 1978, Nolet & Burns 1978, 1979, Evans & Amthauer 1980, Yamanaka & Takéuchi 1979.
deerite	opaque	low mag. susc.; M/S detects discrete valencies Fe^{2+} and Fe^{3+} each in two or more oct. at least two ED species detected	Carmichael et al 1966, Bancroft et al 1968, Frank & Banbury 1974, Bancroft 1979, Amthauer et al 1980, Pollak et al 1979

Table 3 *(continued)*

Mineral	CT energy (cm^{-1}) and assignment	Comments (Möss. spect.; elect. cond.; T variations, etc)	References
howieite	dark-green; no electronic spectra	M/S detects discrete valencies Fe^{2+} and Fe^{3+} ions; evidence of some ED species	Bancroft et al 1968
sapphirine	blue-green; no electronic spectra	M/S detects discrete valencies Fe^{2+} and Fe^{3+}	Bancroft et al 1968
aenigmatite	dark-blue; no electronic spectra	M/S detects discrete valencies Fe^{2+} and Fe^{3+}	Osborne & Burns unpubl. results
magnetite	opaque	high elect. cond. above 119 K; M/S resolves Fe_A^{3+} (tet.) and ED species in Fe_B	Verwey & Haayman 1941, Sawatzky et al 1969, Kundig & Hargrove 1969, Verble 1974, Cullen & Callen 1971, Lotgering & van Diepen 1977
Ti garnet	$5,280: Fe^{2+}\{X\} \rightarrow Fe^{3+}(Z)$	M/S detects discrete valencies Fe^{3+} in [Y] and (Z), Fe^{2+} in $\{X\}$ and [Y]; and ED species $Fe^{2+}\{X\} \rightarrow Fe^{3+}(Z)$; inverse T dependence of CT	Huggins et al 1977a,b, Amthauer et al 1977, Schwartz et al 1980, Moore & White 1971

no evidence of Fe^{3+} at room temperature, even in specimens reported to contain as much as 22% of the iron in the ferric state (Parkin et al 1977). It is only at 77 K that ferric iron is detected in the Mössbauer spectra, but then accounting for smaller proportions of the chemically analyzed Fe^{3+} ions. The failure to detect Fe^{3+} ions in the room temperature spectra might be explained if the electron transfer between adjacent Fe^{2+} and Fe^{3+} ions is faster than the Mössbauer transition. Ferric ions could appear in the 77 K spectra because electron hopping is partially quenched. The phenomena would be comparable to electron delocalization observed in magnetite and ilvaite in which $Fe^{2+} \rightarrow Fe^{3+}$ interactions produce species with Mössbauer parameters intermediate between those for discrete Fe^{2+} and Fe^{3+} ions. Such "averaging" is not observed in the room-temperature Mössbauer spectra of cordierite, however, even in specimens containing Fe^{3+}/total Fe ratios as high as 0.25 (Parkin et al 1977). Another possible explanation is that the $Fe^{2+} \rightarrow Fe^{3+}$ intervalence transition involves iron cations located in the channels enclosed by the hexagonal rings of corner-shared $[SiO_4]$ tetrahedra in the cordierite structure (Goldman et al 1977). Iron in a loosely bound channel site would have an anomalously low recoil-free fraction and not contribute to the room temperature Mössbauer spectra of cordierite. The problem remains unresolved at present.

Another example of $Fe^{2+} \rightarrow Fe^{3+}$ interactions between cations in octahedral and tetrahedral sites is osumilite which is related to cordierite. Here, the $Fe^{2+} \rightarrow Fe^{3+}$ intervalence transition occurs at 15,480 cm^{-1} and the room temperature Mössbauer spectra detect coexisting discrete Fe^{2+} and Fe^{3+} ions (Goldman & Rossman 1978, Faye 1972).

Fe-Ti Garnets

Garnets of the andradite-melanite-schorlomite series represent another mineral group containing Fe^{2+}-Fe^{3+} clusters of different coordination symmetries. Many andradites contain high titanium contents, leading to subsilicic compositions and unusual crystal chemistries of iron and titanium. In the garnet structure, {X} cations are coordinated to eight oxygens at the vertices of a distorted cube. The [Y] cations are octahedrally coordinated and (Z) cations are in tetrahedral coordination. The only edge-sharing interaction involves the {XO_8} cubes. In particular, three-dimensional chains of edge-shared {XO_8} cubes and (ZO_4) tetrahedra are present in the garnet structure, and in andradite the X-Z interatomic distance is only 3.01 Å (Novak & Gibbs 1971).

Mössbauer measurements of several titanium-bearing andradites (i.e. melanites and schorlomites) have revealed the presence of coexisting Fe^{2+} and Fe^{3+} ions, and have demonstrated that Fe^{3+} ions, and not Ti^{4+}, make up for the deficiencies of silicon in the tetrahedral sites (Burns

1972, Huggins et al 1977a,b, Schwartz et al 1980). Up to five component doublets have been fitted to the Mössbauer spectra, four of which may be unambigously assigned to Fe^{3+} ions in octahedral [Y] and tetrahedral (Z) sites and to Fe^{2+} ions in the cubic {X} and octahedral [Y] sites. A fifth doublet showing unusual temperature variations (Amthauer et al 1977) may be resolved in the Mössbauer spectra of Ti-rich garnets, and its assignment has aroused considerable debate. Its parameters are suggestive of tetrahedral Fe^{2+} ions (Amthauer et al 1977, Huggins et al 1977b), but the substitution of large Fe^{2+} ions in the small Si^{4+} sites is considered unlikely (Schwartz et al 1980), particularly when octahedral Fe^{3+} ions predominate. An alternative assignment of the fifth doublet to $[Fe^{2+}\{X\} \rightarrow Fe^{3+}(Z)]$ electron delocalization has been proposed (Schwartz et al 1980). This assignment argues against tetrahedral Fe^{2+} replacing Si^{4+} in the garnet structure, and also correlates with the short X-Z separation of 3.01 Å between edge-shared $\{XO_8\}$ cubes and (ZO_4) tetrahedra, as well as electronic spectral data. The near infrared spectra of titanian garnets show an absorption band at 5,280 cm^{-1} (Burns 1972, Huggins et al 1977b, Moore & White 1971), which was originally assigned to a crystal field transition in tetrahedral Fe^{2+} ions (Manning & Harris 1970). However, the temperature dependence of the 5,280 cm^{-1} band, which deceases in intensity with rising temperature (Moore & White 1971) is not consistent with a crystal field transition. Instead, it confirms with the temperature variations found for intervalence transitions in other minerals (Smith & Strens 1976, Smith 1977, 1978b). Furthermore, electrical conductivity measurements on garnets show that pure andradite is an insulator, but becomes a semiconductor as the Ti content increases (Moore & White 1971). The activation energy for conduction, estimated to be about 3,800 cm^{-1}, is close to the onset of the intervalence transition centered at 5,280 cm^{-1}. These results support the assignment of a doublet to a $[Fe^{2+}\{X\} \rightarrow Fe^{3+}(Z)]$ electron-delocalized species in the Mössbauer spectra of Fe-Ti garnets (Schwartz et al 1980).

Although the garnet structure contains infinite chains of edge-shared $\{XO_8\}$ cubes and (ZO_4) tetrahedra, the extent of $[Fe^{2+}\{X\} \rightarrow Fe^{3+}(Z)]$ interactions is limited by substitutional blocking because Ca^{2+} and Si^{4+} ions are the major constituents of the {X} and (Z) sites, respectively, of andradite. It has been suggested that the $[Fe^{2+}\{X\} \rightarrow Fe^{3+}(Z)]$ inter-action is facilitated by the similar relative energy levels of $t_{2g}(t_2)$ and e_g (e) molecular orbitals in cubic and tetrahedral coordinations (Schwartz et al 1980). The $\{XO_8\}$-(ZO_4) coordination clusters in the garnet structure suggest that this mineral should be a Class I mixed-valence compound (Table 1). However, the semiconductor and electron delocalization properties of $[Fe^{2+}\{X\} \rightarrow Fe^{3+}(Z)]$ clusters in Fe-Ti garnets detected by Mössbauer spectroscopy are more indicative of a Class IIIA compound.

MINERALS WITH Ti^{3+}-Ti^{4+} OCTAHEDRAL CLUSTERS

The Ti(III) oxidation state is rare in terrestrial minerals due to the comparatively high redox conditions on Earth. However, Ti^{3+} ions may exist in Fe-Ti garnets (Burns 1972, Huggins et al 1977b, Schwartz et al 1980) and in some high pressure phases in the Mantle (Burns 1976). Trivalent titanium does occur in extraterrestrial materials, however, including titanian pyroxenes and glasses from certain meteorites and the Moon (Burns et al 1972, Sung et al 1974, Bell et al 1976, Dowty & Clark 1973a).

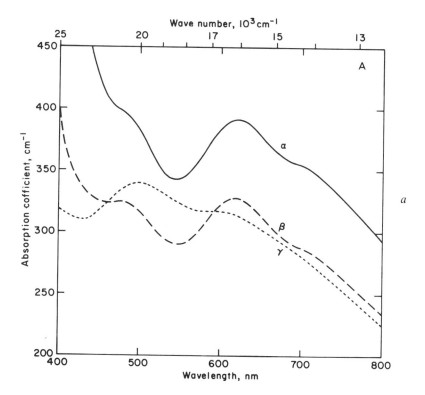

Figure 6   Spectra of the Ti^{3+}-Ti^{4+} pyroxene from the Allende meteorite (from Mao & Bell 1974). (*a*) polarized spectra of the Allende pyroxene at 1 atmosphere; (*b*)—next page— polarized spectra at 1 bar, 20 kbar, and 40 kbar. Note that the doublet at 650 nm splits with increasing pressure, the higher energy peak shifting to higher energy and the lower energy peak remaining stationary with rising pressure. (*c*) spectra of polycrystalline pyroxene at very high pressures. The CF bands shift to higher energy and the CT band gains intensity but remains stationary with increasing pressure.

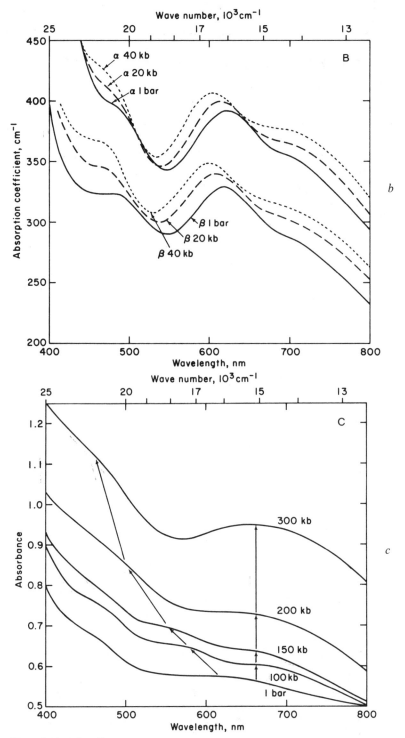

Figure 6 (continued)

One such pyroxene occurs in the Allende meteorite. Its chemical analysis revealed that it is an iron-free fassaite with coexisting Ti^{3+} and Ti^{4+} ions, having the chemical formula (Dowty & Clark 1973a): $Ca_{1.01}Mg_{0.38}Ti^{3+}_{0.34}Ti^{4+}_{0.14}Al_{0.87}Si_{1.26}O_6$.

The crystal structure of the Allende fassaite, like other calcic clino-pyroxenes, contains chains of edge-shared M1 octahedra extending along the c axis, and M1-M1 distances are 3.15 Å. Absorption bands in the polarized spectra illustrated in Figure 6 occur around 20,000 cm^{-1} and 16,000 cm^{-1}, and were originally assigned to Ti^{3+} crystal field and $Ti^{3+} \rightarrow Ti^{4+}$ intervalence transitions, respectively (Dowty & Clark 1973a). An alternative assignment, which proposed that both absorption bands originated from crystal field transitions in Ti^{3+} ions in the distorted M1 octahedral sites, was debated (Burns & Huggins 1973, Dowty & Clark 1973b). One method for distinguishing between crystal field and inter-valence transitions is to measure pressure-induced variations of the spectra, since increased pressure markedly shifts crystal field bands to higher energies and appreciably intensifies intervalence bands (Abu-Eid 1976, Mao 1976). The conflicting assignments of the Allende pyroxene spectra were subsequently resolved by measurements of the polarized spectra at elevated pressures (Mao & Bell 1974). The spectra illustrated in Figures 6b and c show that in addition to the bands at 16,000 and 20,000 cm^{-1}, another occurs around 15,000 cm^{-1} and intensifies at high pressures. Furthermore, the position of the 15,000 cm^{-1} band is insensitive to pressure, whereas the 16,000 and 20,000 cm^{-1} bands shift to higher energies with rising pressure, which is consistent with their assignment to Ti^{3+} crystal field transitions. Such pressure measurements demonstrate conclusively that the $Ti^{3+} \rightarrow Ti^{4+}$ intervalence transition in clinopyrox-enes occurs at 15,000 cm^{-1}.

$Ti^{3+} \rightarrow Ti^{4+}$ intervalence transitions have been assigned to a number of other natural minerals and synthetic phases (Burns & Vaughan 1975), including andalusite at 20,800 cm^{-1} (Faye & Harris 1969) and tourmaline (Manning 1969). Pressure- and temperature-dependent studies of the absorption spectra of these minerals are required to confirm these assignments.

MINERALS EXHIBITING $Fe^{2+} \rightarrow Ti^{4+}$ INTERACTIONS

The Ti(IV) oxidation state is far more common in terrestrial minerals than Ti(III), and the existence of Fe^{2+}-Ti^{4+} assemblages is potentially very common (Burns & Vaughan 1975). In fact, Fe^{2+}-Fe^{3+} and Fe^{2+}-Ti^{4+} coordination clusters may coexist in the same structure, and give rise to homonuclear $Fe^{2+} \rightarrow Fe^{3+}$ and heteronuclear $Fe^{2+} \rightarrow Ti^{4+}$ intervalence transitions in the visible region, thereby complicating assignments of absorption bands. As a result some of the bands originally assigned to $Fe^{2+} \rightarrow Fe^{3+}$ intervalence transitions have been reassigned recently to $Fe^{2+} \rightarrow Ti^{4+}$ intervalence transitions. The role of Fe^{2+}-Ti^{4+} clusters in mixed-valence iron minerals is illustrated by the following examples.

Titanian Pyroxenes

One mineral in which Fe^{2+}-Ti^{4+} octahedral clusters occur and Fe^{3+} ions are absent is the pyroxene from the Angra dos Reis meteorite. The composition of the Angra dos Reis fassaite, $Ca_{0.97}Fe_{0.22}Mg_{0.58}Ti_{0.06}Al_{0.43}Si_{1.79}O_6$, together with measurements of its crystal structure and Mössbauer spectrum, indicate that iron and titanium are predominantly in the M1 octahedral sites and that no ferric iron is detectable (Hazen & Finger 1977, Mao et al 1977). Light polarized in the plane of the M1 cations gives rise to a broad intense absorption band centered around $20,600 \text{ cm}^{-1}$ (Bell & Mao 1976, Mao et al 1977), which may be assigned to a $Fe^{2+} \rightarrow Ti^{4+}$ intervalence transition. Pressure not only intensifies this band, but also results in a systematic shift of it to *lower* energies, so that at 52 kb it is centered at $19,200 \text{ cm}^{-1}$ (Hazen et al 1977). Although such a pressure-induced shift for the $Fe^{2+} \rightarrow Ti^{4+}$ intervalence transition contrasts with the negligible shift observed for the $Ti^{3+} \rightarrow Ti^{4+}$ (Allende pyroxene, Figure 6c) and $Fe^{2+} \rightarrow Fe^{3+}$ (vivianite) intervalence transitions, the intensification of the $19,200 \text{ cm}^{-1}$ band at elevated pressures is consistent with trends observed for other intervalence transitions.

The $Fe^{2+} \rightarrow Ti^{4+}$ intervalence transition has also been identified in the spectra of titanian pyroxenes from the Moon and in crustal rocks (Burns et al 1976, Dowty 1978). Absorption spectra of terrestrial titanaugites illustrated in Figure 7 are particularly complex. Mössbauer measurements show that Fe^{2+} ions occur in the M1 and M2 sites of the clinopyroxene structure, and also reveal that discrete Fe^{3+} ions occur in both octahedral and tetrahedral coordinations (Burns et al 1976). As a result, polarized absorption spectra of titanaugites contain spin-allowed and spin-forbidden crystal field transitions in multiple-site Fe^{2+} and Fe^{3+} ions in addition to the $Fe^{2+} \rightarrow Fe^{3+}$ and $Fe^{2+} \rightarrow Ti^{4+}$ intervalence transitions (Figure 7). The latter are superimposed on Fe^{3+} crystal field bands which are somewhat intensified due to Fe^{3+} ions in non-centrosymmetric tetrahedral sites.

Alternative assignments of an intervalence transition have been proposed for the spectra of an unusual blue titanian omphacite (Curtis et al 1975). An intense absorption band centered around $15,000 \text{ cm}^{-1}$ which shifted to lower energies at 40 kb was originally interpreted as a

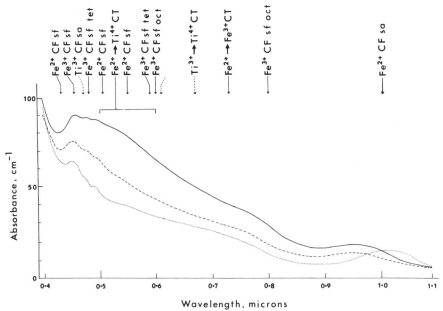

Figure 7   Polarized absorption spectra of a terrestrial titanaugite (from Burns et al 1976).
————, ――――, and ····· correspond to Z, Y, and X polarizations, respectively. Assignments of peaks and inflections in the absorption spectra are shown: CF sa: crystal field,
spin-allowed; CF sf: crystal field, spin-forbidden; CT: intervalence charge transfer; oct:
octahedral; tet: tetrahedral.

$Fe^{2+} \rightarrow Fe^{3+}$ intervalence transition (Abu-Eid 1976). Recently, however,
the same band has been assigned to a $Fe^{2+} \rightarrow Ti^{4+}$ transition (Strens et
al 1980). The dual assignments remain unresolved at this time.

Kyanite

The major feature in the visible-region spectra of blue kyanites is an
intense polarization-dependent band at 16,500 cm^{-1} with a prominent
shoulder in the region 11,500–12,500 cm^{-1} (Faye & Nickel 1969a, White
& White 1967). The 16,500 cm^{-1} band was assigned to a $Fe^{2+} \rightarrow Fe^{3+}$
intervalence transition (Faye & Nickel 1969a, Faye 1971) between Fe^{2+}
and Fe^{3+} ions located in chains of edge-shared $[AlO_6]$ octahedra in the
kyanite structure, in which Al-Al distances are 2.76–2.88 Å (Burnham
1963). The shoulder at 12,500 cm^{-1} was attributed to a spin-allowed
crystal field transition in octahedral Fe^{2+} ions. Mössbauer spectroscopy
subsequently confirmed that octahedral Fe^{2+} and Fe^{3+} ions coexist in
blue kyanites (Parkin et al 1977). However, reported correlations of the

intensity of the blue color with Ti contents of kyanites (White & White 1967, Rost & Simon 1972) led to the suggestion that the 16,500 cm^{-1} band is due instead to the $Fe^{2+} \to Ti^{4+}$ intervalence transition (Smith & Strens 1976). A coupled substitution of Fe^{2+} and Ti^{4+} for two Al^{3+} ions was assumed, since this maintains local charge balance within the structure. It also allows for a high probability of Fe^{2+}-Ti^{4+} couples to exist in adjacent sites, which is necessary to explain the intensity of absorption with rather low concentration of Ti. The shoulder at 11,500–12,500 cm^{-1}, like the 16,500 cm^{-1} band, intensifies at low temperatures, leading the shoulder to be reassigned to a $Fe^{2+} \to Fe^{3+}$ intervalence transition (Smith & Strens 1976).

Sapphire

A similar assignment to that suggested for the kyanite spectra has been proposed for absorption bands in sapphire (Smith & Strens 1976). The corundum structure consists of hexagonal close-packed oxygen ions in which $[AlO_6]$ octahedra share faces parallel to the c axis and share edges perpendicular to c. Thus, two types of cation $\to$ cation interactions are possible in the corundum structure. Crystal growth studies have established that minor amounts of both Fe and Ti must be added to Al_2O_3 to generate the blue coloration of sapphire. In polarized spectra of synthetic sapphires (Ferguson & Fielding 1972, Eigenmann et al 1972), bands at 17,000 cm^{-1} and 12,900 cm^{-1} occur only when Fe and Ti are both present, while a band at about 11,500 cm^{-1} occurs when Fe alone is present. These bands intensify at low temperatures, indicating that they are intervalence transitions (Smith & Strens 1976, Smith 1978b). A fourth band has been identified at 9,700 cm^{-1} (Lehmann & Harder 1970). It is apparent that Fe-Ti interactions in the corundum structure are important factors determining the color and spectra of sapphire. The following assignments are generally accepted for the intervalence transitions in the sapphire spectra:

17,000 cm^{-1}:	$Fe^{2+} \to Ti^{4+}$	perpendicular to c,
12,900 cm^{-1}:	$Fe^{2+} \to Ti^{4+}$	parallel to c,
11,150 cm^{-1}:	$Fe^{2+} \to Fe^{3+}$	perpendicular to c,
9,700 cm^{-1}:	$Fe^{2+} \to Fe^{3+}$	parallel to c.

Hematite-Ilmenite Solid Solutions

The Fe-Ti octahedral clusters producing the color of sapphire are also responsible for electron delocalization in the hematite-ilmenite solid solution series, $(1-x)Fe_2O_3 \cdot xFeTiO_3$, which is isostructural with corundum. The Fe^{3+} ions in hematite occupy face-shared and edge-shared $[FeO_6]$

octahedra as in corundum, with Fe-Fe distances 2.89 Å and 2.97 Å parallel and perpendicular, respectively, to the c axis (Blake et al 1970). In ilmenite Fe^{2+} and Ti^{4+} ions are ordered so as to be coupled in the face-shared octahedra (Fe-Ti = 2.94 Å), while planes of edge-shared $[FeO_6]$ and $[TiO_6]$ octahedra alternate along the c axis (Fe-Fe = 3.00 Å; Ti-Ti = 2.99 Å) (Raymond & Wenk 1971). A Mössbauer study of the ilmenite-hematite series demonstrated that all Fe^{2+} ions in hematite-rich samples participate in electron delocalization with an equal number of Fe^{3+} ions (Warner et al 1972). Electron delocalization continues for values of x as great as 0.60, but no delocalization is observed in the composition range $0.75 < x < 1.00$. The cations are believed to be completely disordered in the range $0 < x < 0.6$, so that $Fe^{2+} \rightarrow Fe^{3+}$ electron delocalization occurs both between face-shared and edge-shared octahedra. Cation ordering increases in the range $0.60 < x < 0.75$, with the result that $Fe^{2+} \rightarrow Fe^{3+}$ electron delocalization between face-shared octahedra predominates. Between $0.75 < x < 1.00$, Ti effectively blocks electron delocalization between Fe^{2+} and Fe^{3+} ions. Nevertheless, a prominent broad band centered around 16,000 cm^{-1} in the diffuse reflectance spectrum of ilmenite (Adams 1975) has been attributed to a $Fe^{2+} \rightarrow Ti^{4+}$ intervalence transition.

Other Examples

The $Fe^{2+} \rightarrow Ti^{4+}$ intervalence transition is believed to contribute to visible–near ultraviolet spectra of lunar regolith and synthetic Fe-Ti silicate glasses (Adams 1975, Bell et al 1976, Wells & Hapke 1977, Osborne et al 1978, Nolet et al 1979). Other minerals to which $Fe^{2+} \rightarrow Ti^{4+}$ intervalence CT has been assigned (Burns & Vaughan 1975) include andalusite (Faye & Harris 1969, Smith 1977), micas (Faye 1968, Smith 1978b), tourmaline (Manning 1969, Smith 1977, 1978a), and vesuvianite (Manning 1975). They might also occur in neptunite, aenigmatite, and taramellite.

Besides schorlomites (Schwartz et al 1980) and hematite-ilmenite solid-solutions (Warner et al 1972), electron delocalization behavior in Fe-Ti minerals has been observed in magnetite-ulvöspinel solid solutions (Banerjee et al 1967, Jensen & Shive 1973, O'Donovan & O'Reilly 1980). The latter have important implications in geomagnetism.

CONCLUSIONS

Intervalence transitions and electron delocalization phenomena between coexisting cations of Fe and Ti continue to be conspicuous in several rock-forming minerals. In many instances, confusion over previous assignments of visible-region and Mössbauer spectra has been clarified by recent controlled spectral measurements spanning wider ranges of temperatures, pressures, and mineral specimens. Continued critical measurements are still required for outstanding controversial minerals, and additional examples should be sought of phases exhibiting light-induced electron transfer phenomena between coexisting transition metal ions. Understanding the mechanism of such electronic transitions will broaden the scope of their applications in the earth and planetary sciences beyond explanations of color, pleochroism, and magnetism of minerals; deductions about geomagnetic, geoelectric, and radiative heat transfer properties of the Earth's interior; and interpretations of remote-sensed reflectivity spectral profiles of regoliths on the Moon, Mars, and Mercury.

ACKNOWLEDGMENTS

Many of the results and interpretations of intervalence transitions in minerals have resulted from research by and discussions with Dr. F. E. Huggins, Dr. R. M. Abu-Eid, Dr. C. M. Sung, Dr. K. M. Parkin, Margery Osborne, Kenneth Schwartz, Daniel Nolet, and Catherine McCammon at M.I.T., and with Dr. G. R. Rossman, Dr. D. S. Goldman, Dr. P. M. Bell, Dr. H. K. Mao, Dr. G. Smith, and Dr. G. Amthauer. Mrs. Virginia Mee Burns assisted with bibliographic research. The research is supported by a grant from the National Aeronautics and Space Administration (grant no. NSG-7604).

Literature Cited

Abu-Eid, R. M. 1976. Absorption spectra of transition metal bearing minerals at high pressures. In *The Physics and Chemistry of Minerals and Rocks*, ed. R. G. J. Strens, pp. 641–75. London:Wiley

Abu-Eid, R. M., Langer, K., Seifert, F. 1978. Optical absorption and Mössbauer spectra of purple and green yoderite, a kyanite-related mineral. *Phys. Chem. Miner.* 3:271–89

Adams, J. B. 1975. Uniqueness of visible and near-infrared diffuse reflectance spectra of pyroxenes and other rock-forming minerals. Chapter 1. In *Infrared and Raman Spectroscopy of Lunar and Terrestrial Minerals*, ed. C. Karr, Jr., pp. 1–38. New York: Academic

Aldridge, L. P., Bancroft, G. M., Fleet, M. E., Herzberg, C. T. 1978. Omphacite studies, II. Mössbauer spectra of C2/c and P2/n omphacites. *Am. Mineral.* 63:1107–15

Allen, G. C., Hush, N. S. 1967. Intervalence absorption. Part I. Qualitative evidence for intervalence transfer absorption in inorganic systems in solution and in the solid state. In *Progr. Inorg. Chem.* 8:357–90

Amthauer, G. 1980. ^{57}Fe Mössbauer study of babingtonite. *Am. Mineral.* 65:157–62

Amthauer, G., Annersten, H., Hafner, S. S. 1977. The Mössbauer spectrum of ^{57}Fe in titanium-bearing andradites. *Phys. Chem. Miner.* 1:399–413

Amthauer, G., Langer, K., Schliestedt, M. 1980. Thermally activated electron delocalization in deerite. *Phys. Chem. Miner.* 6:19–30

Annersten, H. 1974. Mössbauer studies of natural biotites. *Am. Mineral.* 59:143–51

Annersten, H., Olesch, M., Seifert, F. A. 1978. Ferric iron in orthopyroxene: a Mössbauer spectroscopic study. *Lithos* 11:301–10

Araki, T., Zoltai, T. 1972. Crystal structure of babingtonite. *Z. Krist.* 135:355–75

Bancroft, G. M. 1979. Mössbauer spectroscopic studies of the chemical state of iron in silicate minerals. *J. Phys.* 40:C2-464–71

Bancroft, G. M., Brown, J. R. 1975. A Möss-bauer study of coexisting hornblendes and biotites: quantitative Fe^{3+}/Fe^{2+} ratios. *Am. Mineral.* 60:265–72

Bancroft, G. M., Burns, R. G. 1969. Möss-bauer and absorption spectral studies of alkali amphiboles. *Miner. Soc. Am. Spec. Pap.* 2:137–48

Bancroft, G. M., Burns, R. G., Stone, A. J. 1968. Applications of the Mössbauer effect to silicate mineralogy. II. Iron silicates of unknown and complex crystal structures. *Geochim. Cosmochim. Acta* 32:547–59

Bancroft, G. M., Williams, P. G. L., Essene, E. J. 1969. Mössbauer spectra of ompha-cites. *Miner. Soc. Am. Spec. Pap.* 2:59–65

Banerjee, S. K., O'Reilly, W., Gibb, T. C., Greenwood, N. 1967. The behavior of ferrous ions in iron-titanium spinels. *J. Phys. Chem. Solids* 28:1323–35

Bell, P. M., Mao, H. K. 1976. Crystal-field spectra of fassaite from the Angra dos Reis meteorite. *Ann. Rep. Geophys. Lab. Carnegie Inst. Yearb.* 75:701–5

Bell, P. M., Mao, H. K., Weeks, R. A. 1976. Optical spectra and electron paramagnetic resonance of lunar and synthetic glasses: a study of the effects of controlled atmosphere, composition, and temperature. *Proc. 7th Lunar Sci. Conf., Suppl. 7, Geochim. Cosmochim. Acta* 3:2543–59

Beran, A., Bittner, H. 1974. Untersuchungen zur Kristallchemie des Ilvaits. *Tschermaks Mineral. Petrog. Mitt.* 21:11–29

Blake, R. L., Zoltai, T., Hessevick, R. E., Finger, L. W. 1970. Refinement of hematite crystal structure. *US Bur. Mines Rep. No. 7384*, pp. 1–20

Borg, R. J., Borg, I. Y. 1980. Mössbauer study of behavior of oriented single crystals of riebeckite at low temperatures and their magnetic properties. *Phys. Chem. Miner.* 5:219–34

Brown, D. B., ed. 1980. *Mixed Valence Compounds. Theory and Applications in Chemistry, Physics, Geology, and Biology* Dordrecht: Reidel. 519 pp.

Brown, G. E., Gibbs, G. V. 1969. Refinement of the crystal structure of osumilite. *Am. Mineral.* 54:101–16

Buerger, M. J., Burnham, C. W., Peacor, D. R. 1962. Assessment of several structures proposed for tourmaline. *Acta Crystallog.* 15:583–90

Burnham, C. W. 1963. Refinement of the structure of kyanite. *Z. Kristallogr.* 118:337–60

Burnham, C. W., Ohashi, Y., Hafner, S. S., Virgo, D. 1971. Cation distribution and atomic thermal vibrations in an iron-rich orthopyroxene. *Am. Mineral.* 56:850–76

Burns, R. G. 1970. *Mineralogical Applications of Crystal Field Theory.* Cambridge Univ. Press. 224 pp.

Burns, R. G. 1972. Mixed valencies and site occupancies of iron in silicate minerals from Mössbauer spectroscopy. *Can. J. Spectrogr.* 17:51–59

Burns, R. G. 1976. Partitioning of transition metals in mineral structures of the mantle. In *The Physics and Chemistry of Minerals and Rocks,* ed. R. G. J. Strens, pp. 555–72. London:Wiley

Burns, R. G., Greaves, C. J. 1971. Correlations of infrared and Mössbauer site population measurements of actinolites. *Am. Mineral.* 56:2010–33

Burns, R. G., Huggins, F. E. 1973. Visible absorption spectra of a Ti^{3+} fassaite from the Allende meteorite: A discussion. *Am. Mineral.* 58:955–61

Burns, R. G., Vaughan, D. J. 1975. Polarized electronic spectra. Chapter 2. In *Infrared and Raman Spectroscopy of Lunar and Terrestrial Minerals,* ed. C. Karr, Jr., pp. 39–72. New York:Academic

Burns, R. G., Abu-Eid, R. M., Huggins, R. E. 1972. Crystal field spectra of lunar pyroxenes. *Proc. 3rd Lunar Sci. Conf., Suppl. 3, Geochim. Cosmochim. Acta* 1:533–43

Burns, R. G., Parkin, K. M., Loeffler, B. M., Abu-Eid, R. M., Leung, I. S. 1976. Visible-region spectra of the moon: progress toward characterizing the cations in Fe-Ti bearing minerals. *Proc. 7th Lunar Sci. Conf., Suppl. 7, Geochim. Cosmochim. Acta* 3:2561–78

Burns, R. G., Nolet, D. A., Parkin, K M., McCammon, C. A., Schwartz, K. B. 1980. Mixed-valence minerals of iron and titanium: correlations of structural, Mössbauer, and electronic spectral data. In *Mixed Valence Compounds. Theory and Applications in Chemistry, Physics, Geology, and Biology,* ed. D. B. Brown. Dordrecht: Reidel. 295–336

Cannillo, E., Mazzi, F., Fang, J. H., Robinson, P. D., Ohya, Y. 1971. The crystal structure of aenigmatite. *Am. Mineral.* 56:427–46

Carmichael, I. S. E., Fyfe, W. S., Machin, D. J. 1966. Low spin ferrous iron in the iron silicate deerite. *Nature* 211:1389

Clark, J. R., Appleman, D. E., Papike, J. J. 1969. Crystal-chemical characterization of clinopyroxenes based on eight new structure refinements. *Mineral. Soc. Am. Spec. Pap.* 2:31–50

Cullen, J. R., Callen, E. 1971. Band theory of multiple ordering and the metal-semiconductor transition in magnetite. *Phys. Rev. Lett.* 26:236–38

Curtis, L., Gittins, J., Kocman, V., Rucklidge, J. C., Hawthorne, F. C., Ferguson, R. B. 1975. Two crystal structure refinements of a P2/n titanium ferro-omphacite. *Can. Mineral.* 13:62–67

Day, P. 1976. Mixed valence chemistry and metal chain compounds. In *Low Dimensional Cooperative Phenomena*, ed. H. J. Keller, pp. 191–214. New York Plenum

Dowty, E. 1978. Absorption optics of low symmetry crystals—Application to titanian clinopyroxene spectra. *Phys. Chem. Miner.* 3:173–81

Dowty, E., Clark, J. R. 1973a. Crystal structure refinement and visible-region absorption spectra of a Ti^{3+} fassaite from the Allende meteorite. *Am. Mineral.* 58:230–242

Dowty, E. Clark, J. R. 1973b. Crystal-structure refinement and optical properties of a Ti^{3+} fassaite from the Allende meteorite: Reply. *Am. Mineral.* 58:962–64

Eigenmann, K., Kurtz, K., Gunthard, H. H. 1972. Solid state reactions and defects in doped Verneuil sapphire. *Helv. Phys. Acta* 45:452–80

Ernst, W. G., Wai, C. M. 1970. Mössbauer, infrared, X-ray, and optical study of cation ordering and dehydrogenation in natural and heat treated sodic amphiboles. *Am. Mineral.* 55:1226–58

Evans, B. J., Amthauer, G. 1980. The electronic structure of ilvaite and the pressure and temperature dependence of its ^{57}Fe Mössbauer spectrum. *J. Phys. Chem. Solids.* 41:985–1001

Farrell, E. F., Newnham, R. E. 1967. Electronic and vibrational absorption spectra in cordierite. *Am. Mineral.* 52:380–88

Faye, G. H. 1968. The optical absorption spectra of iron in six-coordinate sites in chlorite, biotite, phlogopite and vivianite. Some aspects of pleochroism in the sheet silicates. *Can. Mineral.* 9:403–25

Faye, G. H. 1971. On the optical spectra of di- and tri- valent iron in corundum: a discussion. *Am. Mineral.* 56:344–50

Faye, G. H. 1972. Relationship between crystal-field splitting parameter, "Δ_{VI}" and M_{host}-O bond distance as an aid in the interpretation of absorption spectra of Fe^{2+}-bearing materials. *Can. Mineral.* 11:473–87

Faye, G. H., Harris, D. C. 1969. On the origin and pleochroism in andalusite from Brazil. *Can. Mineral.* 10:47–56

Faye, G. H., Hogarth, D. D. 1969. On the origin of "reverse pleochrosim" of a phlogopite. *Can. Mineral.* 10:25–34

Faye, G. H., Nickel, E. H. 1969a. On the origin of color and pleochroism of kyanite. *Can. Mineral.* 10:35–46

Faye, G. H., Nickel, E. H. 1969b. The effect of charge transfer processes on the colour and pleochroism of amphiboles. *Can. Mineral.* 10:616–35

Faye, G. H., Manning, P. G., Nickel, E. H. 1968. The polarized optical absorption spectra of tourmaline, cordierite, chloritoid, and vivianite: ferrous-ferric electronic interaction as a source of pleochroism. *Am. Mineral.* 53:1174–1201

Faye, G. H., Manning, P. G., Gosselin, J. R., Tremblay, R. J. 1974. Optical absorption spectra of tourmaline: importance of charge transfer processes. *Can. Mineral.* 12:370–80

Ferguson, J., Fielding, P. E. 1972. The origins of the colours of natural yellow, blue, and green sapphires. *Aust. J. Chem.* 25:1371–85

Fleet, M. E. 1977. The crystal structure of deerite. *Am. Mineral.* 62:990–98

Fleet, S. G., Megaw, H. D. 1962. The crystal structure of yoderite. *Acta Crystallogr.* 15:721–28

Frank, E., Banbury, D. St. P. 1974. A study of deerite by the Mössbauer effect. *J. Inorg. Nucl. Chem.* 36:1725–30

Gérard, A., Grandjean, F. 1971. Observations by the Mössbauer effect of an electron hopping process in ilvaite. *Solid State Comm.* 9:1845–49

Gibbs, G. V. 1966. The polymorphism of cordierite. I. The crystal structure of low cordierite. *Am. Mineral.* 51:1068–87

Gibbs, G. V., Breck, D. W., Meagher, E. P. 1968. Structural refinements of hydrous and anhydrous synthetic beryl, $Al_2(Be_3 Si_6)O_{18}$ and emerald $Al_{1.9}Cr_{0.1}(Be_3Si_6) O_{18}$. *Lithos* 1:275–85

Goldman, D. S. 1979. A re-evaluation of Mössbauer spectroscopy of calcic amphiboles. *Am. Mineral.* 64:109–18

Goldman, D. S., Rossman, G. R. 1977a. The spectra of iron in orthopyroxene revisited: the splitting of the ground state. *Am. Mineral.* 62:151–57

Goldman, D. S., Rossman, G. R. 1977b. The identification of Fe^{2+} in the M(4) site of calcic amphiboles. *Am. Mineral.* 62:205–16

Goldman, D. S., Rossman, G. R. 1978. The site distribution of iron and anomalous biaxiality in osumilite. *Am. Mineral.* 63:490–98

Goldman, D. S., Rossman, G. R., Dollase, W. A. 1977. Channel constituents in cordierite. *Am. Mineral.* 62:1144–57

Goldman, D. S., Rossman, G. R., Parkin, K. M. 1978. Channel constituents in beryl. *Phys. Chem. Miner.* 3:225–35

Grandjean, F., Gérard, A. 1975. Analysis by Mössbauer spectroscopy of the electron hopping process in ilvaite. *Solid State Comm.* 16:553–56

Haga, N., Takéuchi, Y. 1976. Neutron diffraction study of ilvaite. *Z. Kristallogr.* 144:161–74

Hamilton, W. C. 1958. Neutron diffraction study of the 119K transition in magnetite. *Phys. Rev.* 110:1050–57

Hawthorne, F. C. 1976. The crystal chemistry of the amphiboles. V. The structure and chemistry of arfvedsonite. *Can. Mineral.* 14:346–56

Hazen, R. M., Burnham, C. W. 1973. The crystal structures of one-layer phlogopite and annite. *Am. Mineral.* 58:889–900

Hazen, R. M., Finger, L. W. 1977. Crystal structure and compositional variation of Angra dos Reis fassaite. *Earth Planet. Sci. Lett.* 35:357–62

Hazen, R. M., Bell, P. M., Mao, H. K. 1977. Polarized absorption spectra of Angra dos Reis fassaite to 52 k bar. *Ann. Rep. Geophys. Lab., Carnegie Inst. Washington Yearb.* 76:515–16

Heilmann, I. U., Olsen, N. B., Olsen, J. S. 1977. Electron hopping and temperature dependent oxidation states of iron in ilvaite studied by Mössbauer effect. *Phys. Scr.* 15:285–88

Hermon, E., Simkin, D. J., Donnay, G., Muir, W. B. 1973. The distribution of Fe^{2+} and Fe^{3+} in iron-bearing tourmalines: a Mössbauer study. *Tschermaks Mineral. Petrogr. Mitt.* 19:124–32

Higgins, J. B., Ribbe, P. H. 1979. Sapphirine II. A neutron and x-ray diffraction study of $(Mg-Al)^{VI}$ and $(Si-Al)^{VI}$ ordering in monoclinic sapphirine. *Contrib. Mineral. Petrol.* 68:357–68

Higgins, J. B., Ribbe, P. H., Herd, R. K. 1979. Sapphirine I. Crystal chemical contributions. *Contrib. Mineral. Petrol.* 68:349–56

Huggins, F. E., Virgo, D. Huckenholz, H. G. 1977a. Titanium-containing silicate garnets. I. The distribution of Al, Fe^{3+} and Ti^{4+} between octahedral and tetrahedral sites. *Am. Mineral.* 62:475–90

Huggins, F. E., Virgo, D., Huckenholz, H. G. 1977b. Titanium-containing silicate garnets. II. The crystal chemistry of melanites and schorlomites. *Am. Mineral.* 62:646–65

Hush, N. S. 1967. Intervalence-transfer absorption. Part 2. Theoretical considerations and spectroscopic data. *Progr. Inorg. Chem.* 8:391–444

Jensen, S. D., Shive, P. N. 1973. Cation distribution in sintered titano-magnetites. *J. Geophys. Res.* 78:8474–80

Kosoi, A. L. 1976. The structure of babingtonite. *Sov. Phys. Crystallogr.* 20:446–51

Kundig, W., Hargrove, R. S. 1969. Electron hopping in magnetite. *Solid State Comm.* 7:223–27

Lehmann, G., Harder, H. 1970. Optical spectra of di- and trivalent iron in corundum. *Am. Mineral.* 55:98–105 (See also: *Am. Mineral.* 56:349–50)

Littler, J. G. F., Williams, R. J. P. 1965. Electrical and optical properties of crocidolite and some other iron compounds. *J. Chem. Soc.*, pp. 6368–71

Loeffler, B. M., Burns, R. G. 1976. Shedding light on the color of gems and minerals. *Am. Sci.* 64:636–47

Loeffler, B. M., Burns, R. G., Tossell, J. A., Vaughan, D. J., Johnson, K. H. 1974. Charge transfer in lunar materials: interpretation of ultra-violet-visible spectral properties of the moon. *Proc. 5th Lunar Sci. Conf., Suppl. 5, Geochim. Cosmochim. Acta* 3:3007–16

Loeffler, B. M., Burns, R. G., Tossell, J. A. 1975. Metal-metal charge transfer transitions: interpretations of visible-region spectra of the moon and lunar materials. *Proc. 6th Lunar Sci. Conf., Suppl. 6, Geochim. Cosmochim. Acta* 3:2663–76

Lotgering, F. K., van Diepen, A. M. 1977. Electron exchange between Fe^{2+} and Fe^{3+} ions on octahedral sites in spinels studied by means of paramagnetic Mössbauer spectra and suceptibility measurements. *J. Phys. Chem. Solids* 38:565–72

MacCarthy, G. R. 1926. Colors produced by iron in minerals and the sediments. *Am. J. Sci.* 12:16–36

Manning, P. G. 1969. An optical absorption study of the origin of colour and pleochroism in pink and brown tourmalines. *Can. Mineral.* 9:678–90

Manning, P. G. 1975. Charge-transfer processes and the origin of colour and pleochroism of some titanium-rich vesuvianites. *Can. Mineral.* 13:110–16

Manning, P. G., Harris, D. C. 1970. Optical-absorption and electron microprobe studies of some high-Ti andradites. *Can. Mineral.* 10:260–71

Manning, P. G., Nickel, E. H. 1969. A spectral study of the origin of colour and pleochroism of a titanaugite from Kaiserstuhl and of a riebeckite from St. Peter's Dome, Colorado, *Can. Mineral.* 10:71–83

Mao, H. K. 1976. Charge-transfer processes at high pressure. In *The Physics and Chemistry of Minerals and Rocks*, ed. R. G. J. Strens, pp. 573–81

Mao, H. K., Bell, P. M. 1974. Crystal-field effects of trivalent titanium in fassaite from the Pueblo de Allende meteorite. *Ann. Rep. Geophys. Lab., Carnegie Inst. Washington Yearb.* 73:488–92

Mao, H. K., Bell, P. M., Virgo, D. 1977. Crystal-field spectra of fassaite from the the Angra dos Reis meteorite. *Earth Planet. Sci. Lett.* 35:352–56

Martinet, J., Martinet, A. 1952. Pleochroism and structure of natural silicates. *Bull. Soc. Chem. France* 19:563–65

McCammon, C. A., Burns, R. G. 1980. The oxidation mechanism of viviante as studied by Mössbauer spectroscopy. *Am. Mineral.* 65:361–66

Merlino, S. 1980. Crystal structure of sapphirine-1Tc *Z. Kristallogr.* 151:91–100

Mitchell, J. T., Bloss, F. D., Gibbs, G. V. 1971. Examination of the actinolite structure and four other C2/m amphiboles in terms of double bonding. *Z. Kristallogr.* 133:273–300

Moore, P. B. 1969. The crystal structure of sapphirine. *Am. Mineral.* 54:31–49

Moore, R. K., White, W. B. 1971. Intervalence electron transfer effects in the spectra of the melanite garnets. *Am. Mineral.* 56:826–40

Mori, H., Ito, T. 1950. The structure of vivianite and symplesite. *Acta Crystallogr.* 3:1–6

Newnham, R. E., de Haan, Y. E. 1962. Refinement of the α-Al_2O_3, Ti_2O_3, V_2O_3, and Cr_2O_3 structures. *Z. Kristallogr.* 117:235–37

Nolet, D. A. 1978. Electron delocalization observed in the Mössbauer spectrum of ilvaite. *Solid State Comm.* 20:719–22

Nolet, D. A., Burns, R. G. 1978. Temperature dependent Fe^{2+}-Fe^{3+} electron delocalization in ilvaite. *Geophys. Res. Lett.* 4:821–24

Nolet, D. A., Burns, R. G. 1979. Ilvaite: a study of temperature dependent electron delocalization by the Mössbauer effect. *Phys. Chem. Miner.* 4:221–34

Nolet, D. A., Burns, R. G., Flamm, S. L., Besancon, J. R. 1979. Spectra of Fe-Ti silicate glasses: implications to remote-sensing of planetary surfaces. *Proc. Lunar Planet. Sci. Conf., 10th, Suppl. II, Geochim. Cosmochim. Acta* 2:1775–86

Novak, G. A., Gibbs, G. V. 1971. The crystal chemistry of the silicate garnets. *Am. Mineral.* 56:791–825

O'Donovan, J. B., O'Reilly, W. 1980. The temperature dependent cation distribution in titanomagnetites. *Phys. Chem. Miner.* 6:235–43

Osborne, M. D., Parkin, K. M., Burns, R. G. 1978. Temperature-dependence of Fe-Ti spectra in the visible region: implications to mapping Ti concentrations of hot planetary surfaces. *Proc. Lunar Planet. Sci. Conf., 9th, Suppl. 9, Geochim. Cosmochim. Acta* 3:2949–60

Papike, J. J., Clark, J. 1968. The crystal structure and cation distribution of glaucophane. *Am. Mineral.* 53:1156–73

Paques-Ledent, M. T., Grandjean, F., Gerard, A. 1977. Chemical formula of ilvaite: infrared and Mössbauer data. *Bull. Soc. R. Sci. Liege* 46:337–42

Parkin, K. M., Loeffler, B. M., Burns, R. G. 1977. Mössbauer spectra of kyanite, aquamarine, and cordierite showing intervalence charge transfer. *Phys. Chem. Miner.* 1:301–11

Pollak, H. 1976. Charge transfer in cordierite. *Phys. Stat. Sol.* 74:K31–K34

Pollak, H., Quartier, R., Bruyneel, W., Walter, P. 1979. Electron relaxation in deerite. *J. Phys.* 40:C2–455

Raymond, K. N., Wenk, H. R. 1971. Lunar ilmenite (refinement of the crystal structure). *Contrib. Mineral. Petrol.* 30:135–40

Robbins, D. W., Strens, R. G. J. 1972. Charge transfer in ferromagnesian silicates: the polarized electronic spectra of trioctahedral micas. *Mineral. Mag.* 38:551–63

Robin, M. B., Day, P. 1967. Mixed valence chemistry–a survey and classification. *Adv. Inorg. Chem. Radiochem.* 10:247–423

Rost, F., Simon, E. 1972. Zur Geochemie und Färbung des Cyanits. *Neues Jahrb. Mineral. Monatsh.* 9:383–95

Sawatzky, G. A., Coey, J. M. D., Morrish, A. H. 1969. Mössbauer study of electron hopping in the octahedral sites of Fe_3O_4. *J. Appl. Phys.* 40:1402–3

Schwartz, K. B., Nolet, D. A., Burns, R. G. 1980. Mössbauer spectroscopy and crystal chemistry of natural Fe-Ti garnets. *Am. Mineral.* 65:142–53

Smith, G. 1977. Low temperature optical studies of metal-metal charge-transfer transitions in various minerals. *Can. Mineral.* 15:500–7

Smith, G. 1978a. A reassessment of the role of iron in the 5000–30000 cm^{-1} range of the electronic absorption spectra of tourmaline. *Phys. Chem. Miner.* 3:343–73

Smith, G. 1978b. Evidence for absorption by exchange-coupled Fe^{2+}-Fe^{3+} pairs in the near infrared spectra of minerals. *Phys. Chem. Miner.* 3:375–83

Smith, G., Strens, R. G. J. 1976. Intervalence transfer absorption in some silicate, oxide and phosphate minerals. In *The Physics and Chemistry of Minerals and Rocks*, ed. R. G. J. Strens, pp. 583–612. New York: Wiley

Strens, R. G. J., Mao, H. K., Bell, P. M. 1980. Quantitative spectra and optics of some meteoritic and terrestrial titanian clinopyroxenes. *Phys. Chem. Miner. In press*

Sung, C-M., Abu-Eid, R. M., Burns, R. G. 1974. Ti³⁺/Ti⁴⁺ ratios in lunar pyroxenes: implications to depth of origin of mare basalt magma. *Proc. 5th Lunar Sci. Conf., Suppl. 5, Geochim. Cosmochim. Acta* 1:717–26

Townsend, M. G. 1968. Visible charge transfer band in blue sapphire. *Solid State Comm.* 6:81–83

Townsend, M. G. 1970. On the dichroism of tourmaline. *J. Phys. Chem. Solids* 31:2481–88

Verble, J. L. 1974. Temperature-dependent light scattering studies of the Verwey transition and electronic disorder in magnetite. *Phys. Rev. B* 9:5236–48

Verwey, É. J., Haayman, P. W. 1941. Electronic conductivity and transition point of magnetite (Fe₃O₄). *Physica* 8:979–87

Warner, B. N., Shive, P. N., Allen, J. L., Terry, C. 1972. A study of the hematite-ilmenite series by the Mössbauer effect. *J. Geomag. Geoelectr.* 24:353–67

Watson, T. L. 1918. The color change in vivianite and its effect on the optical properties. *Am. Mineral.* 3:159–61

Wells, E., Hapke, B. 1977. Lunar soil: iron and titanium bands in the glass fraction. *Science* 195:977–79

Wenk, H. R. 1974. Howieite, a new type of chain silicate. *Am. Mineral.* 59:86–97

Weyl, W. A. 1951. Light absorption as a result of the interaction of two states of valency of the same element. *J. Phys. Colloid Chem.* 55:507–12

White, E. W., White, W. B. 1967. Electron microprobe and optical absorption study of colored kyanites. *Science* 158:915–17

Wilkins, R. W. T., Farrell, E. F., Naimen, C. S. 1969. The crystal field spectra and dichroism of tourmaline. *J. Phys. Chem. Solids* 30:43–56

Wood, D. L., Nassau, K. 1968. The characterization of beryl and emerald by visible infrared absorption spectroscopy. *Am. Mineral.* 53:777–800

Yamanaka, T., Takéuchi, Y. 1979. Mössbauer spectra and magnetic features of ilvaites. *Phys. Chem. Miner.* 4:149–59

Ann. Rev. Earth Planet. Sci. 1981. 9:385–413

FREE OSCILLATIONS OF   ✖ 10155
THE EARTH[1]

Ray Buland
US Geological Survey, Denver, Colorado 80225

INTRODUCTION

All of seismology depends on an ability to model wave propagation within the Earth. We may think about small oscillations within the Earth, either as traveling waves or as standing waves (Goldstein 1965). Traditional seismology takes the former approach while free oscillation (or normal mode) seismology takes the latter. The mathematical equivalence of these formalisms may be most satisfyingly demonstrated by direct numerical computation. The upper panel of Figure 1 shows one component of ground motion of an earthquake in Fiji recorded in Australia. The dominant period of this record is about 40 s. The numerous sharp pulses are due to the arrival of body waves. The lower panel is a synthetic seismogram constructed by summing over 10,000 normal modes. The few body wave arrivals are formed by interference among all of the free oscillations.

Because our intuition tends to be very good for traveling waves and very poor for free oscillations, a great deal of effort has been expended in studying the connection between the two formalisms. The so called "ray-mode duality" has been examined by Jeans (1923), Gilbert & MacDonald (1960), Brune (1964), Ben-Menahem (1964), and Gilbert (1976). The closely related topic of the asymptotic relationships among eigenperiods of the Earth has been examined by Anderssen & Cleary (1974), Anderssen et al (1975), Gilbert (1975), and Woodhouse (1978). These results allow us to speak of a free oscillation as being "equivalent" to a given type of traveling wave.

Using modern instrumentation, it is possible to observe free oscillations of the Earth in the period range of 1 hour to 80 seconds from most earthquakes larger than magnitude 6.0. For convenience, this range will be

385

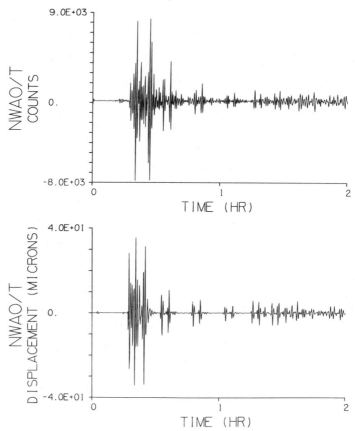

Figure 1   The top panel is a transverse (rotated) component recording of a deep earthquake in Fiji made by a Seismic Research Observatory (SRO) instrument in Narrogin, Australia. An additional low pass filter has been applied with a corner of 40 s period. The bottom panel is a synthetic seismogram with the same source-receiver geometry and filtering constructed by summing 10,000 toroidal modes. The initial large amplitude pulses result from inter-ference among S, sS, ScS, and sScS. The three well-separated doublets are ScS and sScS pairs for two, three, and four core reflections. The rattling late in the seismogram is due to multiply reflected body wave energy which traveled away from the receiver and around the Earth.

referred to as the normal mode (period) band. Figure 2 shows seismograms of a very large earthquake in Indonesia recorded by four normal mode band seismographs. The only traveling waves visible in these records are the fundamental mode Rayleigh mantle waves. The irregular, slowly decaying background is characteristic of time domain free oscillation observations. Figure 3 shows Fourier amplitude spectra of the same four records. The low frequency portion of each spectrum is dominated by

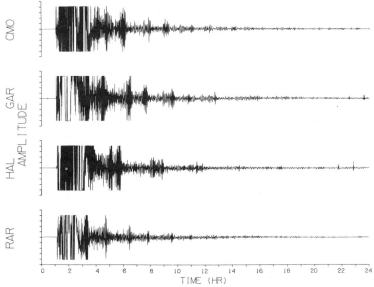

Figure 2 The four time series are one day each of raw data recorded by stations of the IDA network from a very large event in Indonesia, located at (from top to bottom) Fairbanks, Alaska; Garm, Soviet Union; Halifax, Canada; and Rarotonga, Cook Islands. The first four to six hours of each record is clipped. Following the clipped portions the large pulses of energy are fundamental mode surface waves. They repeat in a regular pattern since they circumnavigate the Earth in about three hours. The slowly decaying background is due to overtones. The sharp pulses late in some of the records are from small earthquakes, local to the station.

sharp free oscillation spectral peaks. The large, regularly spaced peaks are equivalent to the fundamental Rayleigh waves. The smaller peaks are equivalent to body waves, which are not well developed in this period band. At the higher frequencies, the normal mode peaks are becoming so numerous as to obscure one another.

In less than thirty years, free oscillation observations have progressed from a curiosity to a vital constraint on any Earth model. About 1500 modes have been observed and at least tentatively identified (roughly 80% of all modes with periods longer than 80 s). Spherically averaged Earth models based on any significant subset of these modes are good enough that few surprises are left in the gross spectral structure of the normal mode band. Attention has turned, instead, to the fine structure of free oscillation spectral peaks. Current research topics in this direction include the study of the very long period nature of seismic sources, the investigation of the spherically averaged, anelastic structure of the Earth, and the elucidation of lateral elastic variations of continental scale.

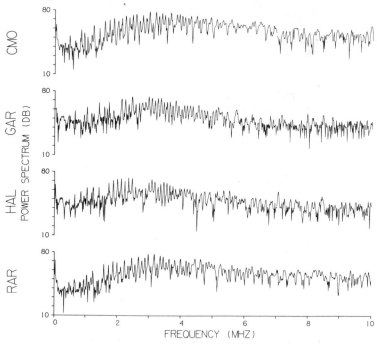

Figure 3   The four power spectra were made from the time series in Figure 2 after removing the clipped portion, other earthquakes, and the solid Earth tides. The frequency scale is in milli-Hertz (10 mHz corresponds to 100 s period). The "forest" of regularly spaced peaks (below 5 mHz) are mostly fundamental spheroidal modes.

THE STUDY OF FREE OSCILLATIONS THROUGH 1975

The Spherically Averaged Earth

The first step in the development of free oscillations theory, computation, observation, and inference is understanding the elastic, nonrotating, spherical, laterally homogeneous Earth model. For purposes of inference the model is considered to be an estimate of the Earth's true laterally heterogeneous structure averaged over concentric spherical shells. As long as self-gravitation and hydrostatic equilibrium are incorporated into this model, the theory of small oscillations and Hooke's Law provide an acceptable zeroth-order approximation to reality for periods in the normal mode band.

The restoring forces for oscillations in the normal mode band are predominantly elastic although self-gravitation cannot be neglected, particularly at the longer periods. Therefore, these modes will be referred to as elastic-gravitational free oscillations. Oscillations at periods shorter than 80 s are essentially acoustic waves and self-gravitation may be neglected. Oscillations at periods longer than one hour, with the exception of the Slichter mode (Slichter 1961, Smith 1976), correspond to energy trapped in the Earth's fluid outer core. The theory and interpretation of these modes, sometimes called undertones, is very different from that of the elastic-gravitational modes. Although no undertone has ever been observed, their theory is well developed (Smith 1974, Johnson & Smylie 1977, Crossley & Rochester 1980). Because they are so different from the elastic modes, undertones will not be considered further here.

The equations of motion for the spherically averaged Earth are conveniently developed in spherical coordinates: radius r, colatitude ϑ, and longitude φ. The Earth model itself consists of three functions of radius alone: rigidity μ, bulk modulus κ, and density ρ. The equations of motion take the form of four second-order differential equations: one for each of three components of displacement and one for the pertubation to the gravitational potential resulting from the motion. The equations for radial displacement U, one component of lateral displacement V, and the potential perturbation P are coupled. Solutions of this system are called spheroidal modes and are analogous to coupled compressional and vertically polarized shear waves (P-SV motion). Solutions of the remaining equation, for the other component of lateral displacement W, are called toroidal modes and are analogous to horizontally polarized shear waves (SH motion).

The functional dependence of both problems on the spatial coordinates is partially separable. In fact, the lateral dependences of U, V, W, and P are represented by various simple functions of spherical harmonics: $Y_l^m(\vartheta, \varphi)$. The angular order, l, and the azimuthal order, m, are integers such that $l = 0, 1, 2, \ldots$ and $m = -1, -l+1, -l+2, \ldots, +l$. Radial solutions are found by solving the systems of coupled differential equations numerically for U, V, and P in the spheroidal mode case and for W in the toroidal mode case. The radial equations depend on l, but not on m. For each l there are an infinite number of discrete solutions. By convention these solutions are assigned integer radial order numbers, n. Beginning from zero, n values are assigned sequentially to the solutions in order of decreasing period. Asymptotically, for short periods, $l + 1/2$ is proportional to horizontal wavenumber and n is related to vertical wavenumber.

Notationally, spheroidal and toroidal modes are designated $_nS_l^m$ and $_nT_l^m$ respectively. Functionally, each free oscillation is of the form

$$\hat{r} \cdot {}_nS_l^m(r, \vartheta, \varphi) = {}_nU_l(r) Y_l^m(\vartheta, \varphi) \cos(\omega t) e^{-\alpha t}, \tag{1}$$

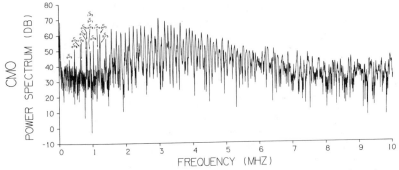

Figure 4 This power spectrum is the same as the top panel of Figure 3 except that a longer time series (two days) has been used to emphasize the longer period normal modes. Some of the longest period modes are identified for reference.

where $\hat{r}$ is the unit vector in the radial direction, ω is the radian eigenfrequency, α is the damping parameter, and t is time. The radial component of spheroidal motion has been used for simplicity. Other components have similar representations (e.g. Gilbert & Dziewonski 1975). Because m does not enter into the differential equations, modes with the same type, n, and l will all have the same eigenperiod. Such a group is called a multiplet and is said to have a $2l + 1$ fold degeneracy. Spheroidal and toroidal multiplets are designated $_nS_l$ and $_nT_l$ respectively. In a similar vein, each member or individual free oscillation of a multiplet is called a singlet. Because the spherically averaged model is an idealization of the Earth, real multiplets are not quite degenerate. The multiplets are said to be split. For the vast majority of observable modes, splitting is so small that observed multiplet center periods are reasonable estimates of the hypothetical degenerate multiplet eigenperiod. Under this assumption large numbers of multiplet center periods have been used to constrain spherically averaged Earth structure. For reference, Figure 4 shows a normal mode spectrum with some of the longest period modes identified.

Splitting

The real Earth is slightly anelastic, slowly rotating, roughly elliptical, and laterally heterogeneous. Each of these effects is important, to first order, in at least some portion of the normal mode band. Each can and has been treated by perturbation theory. As we have seen, the primary effect of anelasticity is to slightly damp each free oscillation. On the other hand, rotation, ellipticity, and lateral variations, as well as anelasticity, result in a slight shift in the eigenperiod of each normal mode. In general, the shift is different for each singlet, splitting each multiplet apart in frequency.

This splitting is usually extremely small compared with the average frequency spacing between different multiplets.

A theoretical treatment of anelasticity may be found in a number of continuum mechanics texts (e.g. Malvern 1969). The application of this theory to free oscillations is straightforward (e.g. Gilbert & Buland 1977). The effect of a spherically symmetric anelastic structure would be to modify the spectral shape and degenerate eigenperiod of each multiplet. In other words, each singlet in a multiplet would be affected in exactly the same way. The result of a general, laterally varying anelastic structure would be to modify the spectral shape and eigenperiod of each singlet in a multiplet in a slightly different manner. While anelasticity has been studied for many years, even the spherically averaged anelastic structure of the Earth is not yet very well constrained.

Splitting due to the rotation of the Earth and ellipticity of figure has a simple form. In both cases the splitting depends on powers of azimuthal order, m, times constants called splitting parameters. The splitting parameters are functions of the radial part of each eigenfunction. Both perturbations are relatively easy to deal with because of their first-order axisymmetric nature. Early observations of rotational splitting were reported by Smith (1961) and Ness et al (1961). The theoretical development of first-order rotational splitting soon followed (Backus & Gilbert 1961, Pekeris et al 1961a). Rotational splitting has been investigated in more detail by Dahlen (1968) and Dahlen & Smith (1975). The theory of splitting due to ellipticity has been developed by Dahlen (1968, 1975, 1976a), Woodhouse (1976), and Dziewonski & Sailor (1976). Extensive calculations of rotation and ellipticity splitting parameters are given by Dahlen & Sailor (1979). The first completely resolved multiplet (all singlets observed) was reported by Buland et al (1979). This is the only direct measurement to date of splitting due to both rotation and ellipticity.

Lateral variation in the Earth's elastic structure is by far the most complicated perturbation to treat of any discussed here. This is because of a lack of symmetry properties of any kind. Lateral elastic variations are the dominant cause of splitting over most of the normal mode band. The effects of this splitting are very commonly observed on modern instruments (e.g. Jobert et al 1978). However, because of the extreme difficulty in interpreting these observations, the study of the laterally varying elastic structure of the Earth has not progressed very far.

Exploring the Spherically Averaged Earth

Through the mid-1970s the bulk of free oscillation observation and inference was directed towards constructing ever more precise spherically averaged Earth models. This process has been iterative. The longest period free oscillations are well separated and may be unambiguously identified even using rather crude models. At shorter period, the number of modes per frequency band increases rendering positive identification more difficult. The correctly identified longer period modes were used to construct more accurate Earth models. Better models enabled the identification of more, shorter period modes and so on. This process has been used very successfully to bootstrap mode identification to periods as short as 80 s.

Most of the theory for elastic oscillations in a sphere was presented by Lamb (1882). The most realistic of the early free oscillation computations was performed by Love (1911). Using a uniform Earth model, he found a surprisingly accurate eigenperiod for the gravest (longest period) spheroidal mode, $_0S_2$. Benioff et al (1954) reported a possible observation of the same mode excited by the 1952 great Kamchatka earthquake. The excitement generated by this finding resulted in a number of theoretical investigations (Matumoto & Satô 1954, Jobert 1957). However, the study of free oscillations, as we know it today, dates from the theoretical and computational work using realistic Earth models reported by Pekeris & Jarosch (1958) and from the convincing observations by a variety of instruments of a number of normal modes excited by the 1960 great Chilean earthquake.

Most of the early observations were made using recordings from exotic, one (or two) of a kind instruments (Smith 1961, 1966, MacDonald & Ness 1961, Ness et al 1961, Connes et al 1962, Takeuchi et al 1962, Nowroozi 1965, Slichter 1967). These observations were supplemented with data from the more traditional seismic instruments of the World Wide Standardized Seismograph Network (WWSSN) (Brune 1964, Alsop 1964, Nowroozi & Alsop 1968). In either case, reliable observation required modes excited by the largest earthquakes of the century. This work was so successful that by the end of the decade Derr (1969a) reports the observation and tentative identification of 265 normal mode multiplets.

Identification requires modeling. The theory needed to compute normal mode eigenfunctions and eigenperiods given an elastic nonrotating, spherically averaged Earth model was essentially complete in the work of Pekeris & Jarosch (1958). Early computations used existing seismological models derived mainly from body wave travel times (Alterman et al 1959, Takeuchi 1959, Gilbert & MacDonald 1960, Pekeris et al 1961b, Alsop 1963). By the mid-1960s there was sufficient information to go beyond comparisons among existing models. In fact the construction of more accurate models became of paramount importance in order to identify newly observed normal modes. Early attempts generally involved hand "fiddling" of model parameters taken one at a time (Landisman et al 1965, Bullen & Haddon 1967) or inversion for one model component (e.g. density with the others held constant; Pekeris 1966). The ultimate in this trial-and-error modeling were the Monte-Carlo inversions of Press (1970). Meanwhile, formal linear inverse theory and a new class of Earth models were evolving hand in hand (Backus & Gilbert 1967, 1968, 1970, Gilbert & Backus 1968, Derr 1969b, Wiggins 1972). These models are characterized by the simultaneous adjustment of all model parameters and a well-developed theory of the resolving power of the data (see Parker 1977).

During the early 1970s, the precision and detail of spherically averaged Earth models increased dramatically, due to a rapid increase in the number of observational constraints available. One factor in this increase was the explicit incorporation of both body wave travel times and free oscillation eigenperiods in the same inversion (Johnson & Gilbert 1972). In addition there was an order-of-magnitude increase in the number of free oscillation observations, made possible by imaginative use of the global, three-component coverage of the WWSSN. Although the WWSSN seismographs are most sensitive to periods shorter than the normal mode band, it has proven possible to observe free oscillation eigenperiods as long as 500 s from a single WWSSN record, given a sufficiently large earthquake. It has been found that appropriate combinations of WWSSN records have the ability to enhance a desired normal mode and to suppress some of its neighbors.

Dziewonski & Gilbert (1972, 1973) used simple coincidence of spectral peaks among many WWSSN records to discover new modes and observed polarizations to identify them. Mendiguren (1973) advanced the data processing a giant step by using the WWSSN as an array. By knowing the geometry of an earthquake source, he was able to make linear combinations of Fourier spectra to enhance a particular sequence of modes. This process, called stacking, not only aides in mode observation but also contributes towards positive identification. Brune & Gilbert (1974) used Brune's (1964) method to detect many new toroidal modes by correlating direct and surface-reflected mantle shear waves. Finally, Gilbert & Dziewonski (1975) used both stacking and a new technique called stripping, which compensates for nonuniform global station coverage, to observe even more normal modes. An integral part of their method was the estimation of the source time function of the earthquake from the mode data. Their improved source representation enabled stacking and stripping to work even more effectively. By mid-decade, Gilbert & Dziewonski (1975) report that 1461 free oscillation multiplets with periods longer than 80 s had been observed and tentatively identified. Theoretically this is about 80% of all multiplets in the normal mode band.

As new modes were observed and identified, a whole series of new models evolved (Dziewonski & Gilbert 1973, Gilbert et al 1973, Jordan & Anderson 1974, Gilbert & Dziewonski 1975, Dziewonski et al 1975). These models satisfied from several hundred to over a thousand free oscillation eigen-periods and as many as several hundred travel time constraints. Because of a high degree of redundancy in information content among the less completely observed shorter period modes, it seemed that the problem of modeling the spherically averaged Earth was, for all practical purposes, solved except for the baseline problem. The baseline problem is a nearly constant discrepancy for each body wave type between the traveling and standing wave data sets. In part, the baseline results from bias in body wave travel times resulting from the fact that most seismic stations are on continents. However, this bias could only account for a small portion of the observed baseline. The rediscovery (Randall 1976, Liu & Archambeau 1976) that anelasticity affected normal mode eigenperiods to first order finally explained the remainder of the discrepancy (Kanamori & Anderson 1977). Anderson & Hart (1976) show that correctly treating anelasticity results in 1% changes in Earth models. Saying that the spherically averaged Earth is understood from a normal mode point of view does not mean that better spherically averaged Earth models will not be found. It does mean that changes will be in the details of the structure (e.g. Burdick & Helmberger 1978), not in its gross properties.

The triumphs of this period of free oscillations research are a very complete understanding of the spectral structure of the normal mode band, at least on a multiplet level, and the ability to constrain spherically averaged, elastic Earth structure much more tightly than has been possible using travel time data alone. In fact, the longest period free oscillations provide the only known direct constraints on the Earth's density structure other than the total mass and moment of inertia of the Earth. Pre-normal-mode density models (e.g. Bullen 1950, Bullard 1957) necessarily depended on indirect arguments and many assumptions. Further, a subset of the longest period normal modes has provided the best evidence to date for the solidity of the Earth's inner core (Dziewonski & Gilbert 1971, 1972, 1973). This was possible because at extremely long periods, energy trapped in the inner core can leak enough to be observable at the Earth's surface.

THE STUDY OF FREE OSCILLATIONS AFTER 1975

New Instrumentation, New Directions

As we have seen, pre-1975 normal mode instrumentation consisted of instruments designed for the purpose and of seismic instruments used for the purpose despite their design. The former have always been few in number, poorly distributed geographically, and diverse in type and sensitivity: strain meters (Smith 1961), tiltmeters (Connes et al 1962), and accelerometers (Ness et al 1961) have all been used with some success. The latter are numerous and provide reasonable geographic coverage by nearly identical instruments. In spite of being peaked near 30 s period, the array nature of the WWSSN has proven powerful indeed. It was clear by the early 1970s that a marriage of these two concepts would be optimal for ever more detailed normal mode observations.

It was already well known in the early 1960s that very long period, vertical component accelerometers, also known as gravity meters or gravimeters, were capable of providing the highest signal-to-noise ratio, teleseismic free oscillations recordings. Reasons for this involve instrumental limitations, characteristics of Earth noise, and the nature of the signal itself. Further, gravimeters have improved considerably over the years, as a result of superior design, modern electronics, and automatic control theory. With the addition of feedback, they have proven to be extremely stable, linear, and quiet (Block & Moore 1966, Moore & Farrell 1970, Prothero & Goodkind 1972). Instrumental noise has been lowered considerably below Earth noise throughout the normal mode band and, in fact, to longer than tidal periods. As a consequence, Block et al (1970) reported that normal modes may be observed from about 20 large events per year rather than from a few great earthquakes per century.

The International Deployment of Accelerometers (IDA) network (Agnew et al 1976) finally combined force feedback gravimeters, digital recording, and global distribution (Figure 5). The combination of high signal-to-noise ratio, wide spatial distribution, and accurate calibration (Farrell & Berger 1979, Berger et al 1979) is already being used to advantage in many areas of free oscillations research. In general, modern research is directed towards the observation and interpretation of spectral parameters other than multiplet eigenperiod. Of these. the three most important are source studies (amplitude and phase information), anelasticity studies (spectral width information), and lateral variations studies (multiplet splitting information). Each subject will be discussed separately below.

Source Studies

For our purposes, source studies will imply the examination of teleseismically recorded elastic waves to infer the geometry and time history of the source that generated them. Two kinds of sources are commonly found to generate observable seismic waves: earthquakes and explosions. Both are concentrated in space and impulsive in time. An ideal explosion is purely isotropic. That is, the source results in a volumetric change rather than a distortion of the source region. Seismic waves generated from a

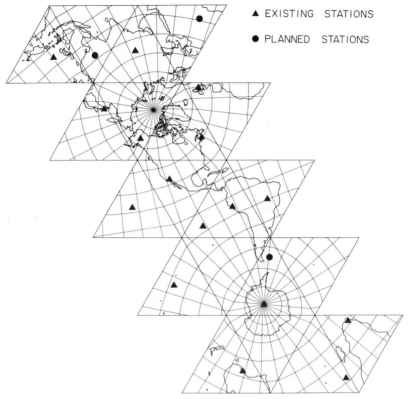

Figure 5   This novel view of the Earth was made by projecting a spherical Earth onto an icosahedron and then unfolding it. The triangles mark operating IDA stations and the circles mark stations to be installed over the next few years. Figure courtesy of Jon Berger.

purely isotropic source have the same amplitude and phase in every direction. An ideal earthquake is purely deviatoric. That is, the source results in a distortion of the medium, but no volumetric change. Seismic waves generated from a purely deviatoric source have a pronounced radiation pattern. The various types of seismic waves have distinct patterns of nodal lines and amplitude variations, functions of the direction in which the waves emerge from the source region.

It is well known that most earthquakes are the result of long term plate tectonic stresses, relieved quickly by slip on a nearly planar fault surface. The simplest mathematical description of an earthquake is as a couple, that is, as two equal forces acting in opposed directions at either end of a lever arm. A couple is intuitively satisfying because it describes forces that should result in slip in opposite directions on either side of a fault surface and conserves linear momentum. However, a single couple is not physically realizable because it doesn't conserve angular momentum. The rotation of a fault plane induced by the tectonic driving couple is countered dynamically by the strength of the rock bounding the fault surface. Mathematically, a second couple with smaller forces and a longer lever arm acts at right angles to the first to conserve angular momentum (Knopoff & Gilbert 1960). A point, double couple source correctly explains the observed radiation pattern of first arriving compressional body waves, which has been long used to constrain faulting geometry (Byerly 1955).

Given the enormous range in fault plane area over which slip occurs during an earthquake, it may seem surprising that an equivalent point source is a useful concept. Point sources are appropriate from the point of view of a body wave because generally only the first arriving wavelet is examined. It appears that the short period arrival is mainly sensitive to properties of the nucleation point of a rupture event. Free oscillations, on the other hand, are sensitive to the entire rupture process. However, the shortest normal mode wavelength of interest is very long compared with all but the largest earthquake source dimensions. Therefore, both body wave and normal mode seismology may safely assume that most earthquakes are point double couples to first order.

Source modeling has progressed in several directions. The surface and body wave approach is to constrain the faulting geometry, assuming a point source, and if the data warrant it to constrain the speed, direction, and duration of rupture as well (assuming a finite slip area; e.g. Kanamori 1970, Boatwright 1980). Usually, this approach involves a complicated nonlinear parameter search aided by considerable geophysical intuition. The normal mode approach is to constrain the first term in a multipolar expansion about the source of the stress excess, or glut, which results in the material failure. This first term is known as the moment (density) tensor (Gilbert 1970). Being a three-dimensional, second-order tensor, the moment tensor has nine elements, each a function of time. In order to conserve momentum it must be symmetric with only six independent elements. To be deviatoric it must have a zero trace. To be double-couple equivalent it must have a zero determinant. For purposes of inversion, all properties except the last may be guaranteed by linear constraints. To force a moment tensor inversion to yield a double couple is possible, but requires nonlinear iteration (Strelitz 1978).

Each source model has advantages and disadvantages, and each has seen application in all areas of seismology. For instance the parameter search has been used with normal mode data (Mendiguren & Aki 1978) and the moment tensor formalism has been applied to both surface (Patton & Aki 1979) and body waves (Strelitz 1978, Ward 1980). The parameter search method has the advantage of describing the commonly observed phenomenon of asymmetry in wave radiation patterns due to source finiteness. The advantage of the moment tensor formalism is that it is linear and that it includes the most general possible point source. Linearity allows a high degree of automation in moment tensor inversions. Generality allows the modeling of sources forbidden by faulting models. However, asymmetry can be included in the moment tensor formalism only by using higher-order moments. Unfortunately, this involves an impractically large number of free parameters.

The prerequisite for using free oscillation amplitudes to constrain source parameters is a theory of coupling between them. This problem has been studied by Saito (1967), Gilbert (1970), Abe (1970), Dahlen (1972, 1973), and Phinney & Burridge (1973). Even earlier Benioff et al (1961) used qualitative amplitude information to constrain the geometry of the 1960 great Chilean earthquake. Ben-Menahem et al (1972) and Mendiguren (1973) were able to constrain some source parameters of two great earthquakes by observing normal mode amplitudes. However, Dziewonski & Gilbert (1974) and Gilbert & Dziewonski (1975) were the first to invert observed multiplet amplitudes to determine complete time-dependent moment tensors.

Dziewonski & Gilbert (1974) reported finding an energetic, long, slow, precursive implosion to the July 31, 1970 Colombian deep earthquake. This result touched off a controversy unparalleled in the history of free oscillations research. For several reasons their result is very plausible. They find the deviatoric part of their solution to be consistent with the origin time and faulting geometry determined by short period body waves. Also, the use of their source for stacking and stripping resulted in the observation of 500 more multiplets than their use of a source geometry determined by body wave observations alone. Finally, an implosive component for a deep earthquake is not totally unexpected. The mechanism of deep earthquakes is not well understood. It is at least conceivable that they are associated with mantle phase changes which result in a denser packing of atoms on a short time scale (i.e. an implosion; see, for example, Bridgman 1945, Dennis & Walker 1965).

Hart & Kanamori (1975) examined WWSSN records for independent time domain evidence of the precursor without success. Kennett & Simons (1976) found corroborative evidence in one strain recording. Okal & Geller (1979) argued that because of the relatively small coupling of most normal modes with isotropic sources that the large inferred implosive component was based on a few relatively noisy compressional wave equivalent modes. However, Buland & Gilbert (1976) and Buland (1976) essentially duplicated the result of Dziewonski & Gilbert (1974) using a more robust technique and a small subset of the same data. Further,

Mendiguren & Aki (1978) find that if the isotropic component is an artifact of the processing then its cause must be coherent from record to record. All in all, it seems likely that evidence for an implosive component of the Colombian earthquake does exist in the WWSSN records. However, both its size and precursory nature are poorly constrained because the implosive component is extremely long period compared with the pass band of the instruments.

This is not to say that the implosive component of the Colombian earthquake is necessarily real. There are, at least, two factors that could result in an erroneous conclusion. First, it is possible that the assumed theoretical very long period calibration of the WWSSN seismometers is systematically in error. Because of the mechanical nature of these instruments, none of them has ever been reliably tested at periods longer than about 100 s. Second, it is possible that the signal is valid, but that the interpretation is incorrect. Self-gravitation is neglected in the normal mode excitation equations used by Dziewonski & Gilbert (1974). Given the inverse distance fall off of the gravitational potential perturbation it is possible that self-gravitation is important, particularly for isotropic sources. Various aspects of this problem have been investigated by Vlaar (1976) and Backus & Mulcahy (1976a,b). Although the nature of the Colombian earthquake remains in doubt, possibly related phenomena have been reported for other events by Kanamori & Anderson (1975) and Kanamori & Stewart (1979).

With the advent of the IDA network, researchers have moved on to the study of other events. By eliminating the question of instrument response it is hoped that the observation of other large deep events will eventually shed light on the controversy over the Colombian earthquake. However, recent normal mode source studies have concentrated on the much more common large shallow earthquakes. H. Kanamori (personal communication, 1979) has used surface-wave-equivalent normal modes near 250 s period to constrain the geometry of several recent earthquakes. He has found good agreement with faulting geometry derived from short period body waves. G. Masters and F. Gilbert (personal communication, 1980) have had similar success using complete normal mode band seismograms. As it uses all available information, their method is very robust. However, they find that great care must be taken to minimize the effects of lateral variations in elastic structure.

Anelasticity

The most commonly, indeed universally observed violation of the elastic, nonrotating, spherically symmetric Earth model is anelasticity. Although the Earth's anelastic structure is often observed to be laterally varying (e.g. Solomon & Toksöz 1970, Der & McElfresh 1977, Sipkin

& Jordan 1980), the data are not currently precise enough to give more than a rough idea of its spherically averaged nature. Moreover, in spite of frequent speculation on the subject (Knopoff 1964, Jackson & Anderson 1970, O'Connell & Budiansky 1977), the mechanisms of seismic attenuation in the Earth are not understood. However, for our purposes, anelasticity may be modeled by representing the elastic parameters' rigidity and bulk modulus as complex numbers. The real part of each represents the elastic response of the Earth. The imaginary part represents the anelastic response. It is known that most, if not all, attenuation can be modeled using the imaginary part of rigidity alone, i.e. the bulk modulus is purely real.

A complex Earth model results in complex free oscillation eigenfrequencies. The real part of the eigenfrequency corresponds to the frequency or period of the oscillation. The imaginary part corresponds to the time constant of attenuation or damping of the mode. The mode decays exponentially by a factor $1/e$ in a time of $1/\alpha$ where α, also known as the damping parameter, is the imaginary part of the radian eigenfrequency. $\alpha = \omega/2Q$ where ω is the real part of the radian eigenfrequency and Q is the quality factor of the mode. In the frequency domain, normal mode spectral peaks of an elastic Earth model would have zero width and infinite amplitude (Dirac delta functions). In an anelastic Earth, normal mode spectral peaks have both finite width and finite amplitude. In fact, the half width at half power of each spectral peak is just α.

It is important to understand anelasticity in order to understand the composition and mechanical behavior of the interior of the Earth and the structure of the normal mode spectrum. However, anelasticity also affects our ability to understand other aspects of Earth structure and the nature of seismic sources. A poorly resolved anelastic structure will result in systematic errors in determinations of source parameters and laterally varying elastic structure. Furthermore, Jeffreys (1965, 1967) and Davies (1967) pointed out that anelasticity has a first-order effect on eigenperiods as well. That is, ignoring anelasticity will result in systematic errors in inferring spherically averaged elastic Earth structure. This shift in eigenperiods, called physical dispersions, was ignored until the effect was "rediscovered" by Liu & Archambeau (1975, 1976) and Randall (1976). Kanamori & Anderson (1977) have shown that accounting for physical dispersion results in small, but significant changes in elastic Earth models. Physical dispersion is particularly important when combining normal mode and traveling wave data in one inversion (the baseline problem).

It may seem surprising that a phenomenon as fundamental and commonly observed as anelasticity is not better understood. There are several reasons for this. The most important is that Q is intrinsically difficult to measure. Second, anelastic structure is not only laterally varying, but apparently frequency dependent as well (e.g. Der & McElfresh 1977, Sipkin & Jordan 1979) greatly complicating modeling efforts. The first problem is particularly serious for traveling wave data. Traveling wave amplitudes are sensitive to many effects other than anelasticity, such as dispersion, geometrical spreading, and interaction with internal caustics. Therefore, traveling wave amplitude data usually require many assumptions before their attenuation characteristics may be extracted. Also traveling wave amplitudes (especially at the shorter periods) are quite sensitive to poorly understood, short scale, laterally varying structures in the Earth. Free oscillations are little affected by lateral variations due to their very long wave lengths. Also, many free oscillations are individually observable. These modes exhibit a simple exponential decay regardless of observation point.

Traveling wave Q's are always determined by examining either the time domain or the spectral amplitude decay of body or surface waves with distance. Analogously, a free oscillation Q may be measured by examining either the time domain or the spectral amplitude decay of a mode with time at one station. This technique has been applied in the time domain by Stein & Geller (1978) and Geller & Stein (1980) and in the frequency domain by Alsop et al (1961), Nowroozi (1968), and Sailor & Dziewonski (1978). However, a free oscillation Q may also be determined by measuring the width of its spectral peak. This property, unique to normal modes, has led to a variety of specialized techniques for estimating Q.

The most obvious means of estimating Q, after amplitude decay, is to simultaneously determine all parameters of a free oscillation spectral peak by least squares. Very early, it was recognized that this method can be misleading because the width and amplitude of a resonance peak are highly correlated. Gilbert & Dziewonski (1975) integrated across spectral peaks to eliminate the effect of Q on peak amplitude estimates. Unfortunately, their method guarantees the inclusion of a good deal of noise in the integral. Buland & Gilbert (1978) extended this idea by using an integration width estimator. Buland & Gilbert (1978) also experimented with Fourier spectra which were analytically continued into the complex plane. This method has the curious property of sharpening spectral lines when the imaginary part of the complex Fourier frequency is near the mode's damping parameter. The utility of this method for estimating Q and for the resolution of interfering spectral peaks is limited by its propensity for amplifying noise. Bolt & Brillinger (1979) advocate a method called complex demodulation because of its ability to provide error estimates. They shift a resonance peak to zero frequency, narrow band filter it, and examine its exponential behavior directly. Although several of these methods show promise, each requires special circumstances for successful application. The majority of values in the current Q data set have, in fact, been determined by more traditional methods.

The observation of free oscillation Q's has proceeded by two parallel paths. The majority of Q estimates have been derived from 20 to 30 of the most commonly observed modes using standard methods. On the other hand, a great deal of effort has been expended in studying the Q's of a few rather exotic modes using very sophisticated techniques. The former data set, which is dominated by fundamental surface wave equivalent modes, is sensitive to the anelastic properties of the upper mantle. Information about attenuation in the rest of the Earth comes almost exclusively from the latter data set. Both data sets are necessary to construct even the most elementary attenuation model for the whole Earth.

The exotic modes may be divided into three categories: radial modes, Stoneley modes, and core modes. Radial modes are spheroidal modes of angular order zero (zero phase velocity). They are equivalent to compressional waves which travel straight through the center of the Earth. The radial modes, particularly $_0S_0$ and $_1S_0$, are the most extensively studied of the three categories (Ness et al 1961, Slichter 1967, Buland et al 1979, Knopoff et al 1979, Riedesel et al 1980). The measured Q's of the radial modes, while high compared with other modes, are slightly lower than predicted by simple Q models. This fact is currently the only evidence requiring bulk dissipation (a complex bulk modulus; Sailor & Dziewonski 1978) or viscous dissipation in the fluid outer core (D. Anderson, personal communication, 1979).

Stoneley modes are spheroidal modes which are equivalent to traveling waves trapped on an internal discontinuity in material properties. Generally, Stoneley mode amplitudes die out so quickly with distance away from the trapping interface that they are quite unobservable at the Earth's surface. However, at sufficiently long period, Stoneley mode energy can "leak" away from the interface enough to be detectable. The only normal mode band Stoneley modes observed to date are representative of energy trapped on the core-mantle interface: $_0S_2$, $_0S_3$, $_0S_4$, and $_0S_5$. These modes are very sensitive to the structure of the lowermost mantle and of the core-mantle interface itself. The Q's of these modes have been studied by Buland & Gilbert (1978), Stein & Geller (1978), and Sailor & Dziewonski (1978). While these Stoneley modes constrain averaged structure in the lower mantle, their wavelengths are so long that they are insensitive to features with radial extents less than several hundred kilometers.

Core modes are spheroidal and toroidal modes which are trapped in the Earth's inner core. The toroidal modes are not directly observable at the Earth's surface although Dahlen & Sailor (1979) find that they can affect some splitting parameters. However, like the Stoneley modes, some of the longer period spheroidal core modes "leak" enough to be observable. To date the Q of only one core mode, $_2S_2$, has been measured (Buland & Gilbert 1978). This one number represents the most direct free oscillation constraint on the anelasticity of the inner core.

The set of more commonly observed modes is dominated by the fundamental radial order zero spheroidal free oscillations with periods between 500 and 200 s. Fundamental modes are always better excited than any others (the others are collectively referred to as overtones) due to large amplitudes throughout the upper mantle and very efficient coupling with earthquake sources. The longer period fundamental modes are usually at least 10 db larger than any overtone even for very deep (650 km depth) earthquakes. Unfortunately, toroidal modes are less well observed for instrumental reasons than the spheroidal modes. Fundamental spheroidal modes are less often observed at periods longer than 500 s because they begin to behave like Stoneley modes. For periods shorter than 200 s these modes decay too quickly to be individually observed in the frequency domain.

Sailor & Dziewonski (1978) measured Q's of fundamental modes with periods as short as 80 s and many overtones by stacking and stripping WWSSN data in the manner of Gilbert & Dziewonski (1975). In this way, they were able to measure the Q's of many more modes than would be possible using single records. On the other hand, stacking tends to broaden the spectral peaks of multiplets split by lateral variations. Therefore, their results must be considered as lower bounds on the true Q's of these modes. Geller & Stein (1980) used the more straightforward method of time domain amplitude decay. They were able to estimate the Q's of well-separated, fundamental spheroidal modes from individual IDA records. Curiously, they found that their Q estimates scattered much more from record to record than one would expect from theoretical arguments (Dahlen 1976b).

Normal mode Q data have been used to constrain models of spherically averaged anelasticity since the 1960s (Anderson & Archambeau 1964, Anderson 1967). The recent increase in the number and quality of free oscillation Q measurements has prompted many new modeling efforts. Sailor & Dziewonski (1978) inverted normal mode data alone. They found that all data could be satisfied by a smooth low resolution model. However, their model predicts unacceptably high attenuation for one-second period body waves penetrating into the lower mantle. In an effort to reconcile both free oscillation and body wave data, Anderson & Hart (1978) decreased the model anelasticity of the lower mantle compensated by a dramatic increase in attenuation at the base of the mantle. Recent studies (e.g. Choy & Boatwright 1980) indicate that even this model is not extreme enough to satisfy shorter period body wave data. The resolution of this problem has already been mentioned. That is, there seems to be a strong frequency dependence in anelasticity in the period range 10 to 1 s. A number of people are now working on frequency-dependent modeling (e.g. Anderson & Minister 1979).

Lateral Variations

The existence of plains, plateaus, mountain ranges, and ocean basins makes it clear that the Earth is anything but spherically symmetric. It is fortunate for seismology that the most severe lateral structural variations are in the crust and uppermost mantle. Since the spherically symmetric portion of the Earth's elastic structure is now fairly well understood, at least from a normal mode point of view, the study of lateral elastic variations is a natural next step. The term lateral variations or heterogeneity will be used to mean elastic variations since variations in lateral anelastic structure are currently beyond the scope of free oscillations research.

To review, lateral variations result in a splitting of free oscillations multiplets. In fact, for periods shorter than about 500 s this is the dominant cause of splitting. For most free oscillations the spacing between adjacent singlets is small compared with the width or damping parameter of each singlet. At high enough frequency, the singlets in a multiplet combine according to the splitting, dictated by the lateral variation, and their relative amplitudes and phases, dictated by the source-receiver geometry, to form one peak. The center period of this peak will shift in a manner that is simply related to the mean phase velocity of the great circle path passing through both source and receiver (Jordan 1978, Dahlen 1979a,b). Similar shifts will occur in all multiplets that are equivalent to a particular traveling wave. As a result, the superposition of these multiplets will form a phase arrival with the correct travel time through the laterally varying medium between the source and receiver. Singlets belonging to a multiplet, split dominantly by lateral variations, will rarely, if ever, be observable individually. However, a slight shifting in multiplet peaks for different source-receiver geometries is commonly observed (Madariaga & Aki 1972, Jobert et al 1978, Silver & Jordan 1980).

Unfortunately, the study of a laterally varying elastic perturbation is at least an order-of-magnitude more formidable than the study of spherically averaged Earth structure or axisymmetric perturbations like rotation and ellipticity. To understand why this should be true, it is necessary to consider the mathematical form of an eigenfunction of various Earth models. For a spherically symmetric Earth model, we have seen that eigenfunctions are represented laterally by spherical harmonics. The position of the spherical harmonic axis ($\vartheta = 0$ in spherical coordinates) is arbitrary. Gilbert & Dziewonski (1975) have used this fact to simplify excitation calculations by making the eigenfunction axis pass through the source. The presence of axisymmetric perturbations about the same axis changes matters very little. In this case, eigenfunctions are represented by spherical harmonics about the axis of symmetry (Dahlen 1968). Thus, eigenfunctions of a rotating, elliptical Earth model have an axis that is coincident with the Earth's spin axis. Lateral variations are different, however. By their very nature they have no symmetry properties of any kind. Madariaga (1971) has shown that eigenfunctions of a rotating, elliptical, laterally varying Earth model are linear combinations of eigenfunctions of a rotating, elliptical Earth model. Unfortunately, the linear combination depends on the unknown lateral perturbation.

Some work has been done on the nature of large scale lateral variations using traveling waves (Julian & Sengupta 1973, Sipkin & Jordan 1976, Sengupta & Toksöz 1976, Dziewonski et al 1977). However, none of these investigations has had enough resolution or precision to compute even one eigenfunction of the laterally varying Earth with any confidence. Without knowing the form of singlet eigenfunctions, array-processing tools like stacking cannot be used. Singlets belonging to a multiplet split dominantly by lateral variations are generally so close together in frequency that no one of them will ever be observable by itself. If stacking cannot be used to isolate individual singlets there is little hope of ever directly measuring splitting due to lateral variations.

However, this is not the only problem encountered in studying lateral heterogeneity in the Earth. There is a serious difficulty in representing the perturbation itself. The theory required for computing splitting due to lateral variations has been developed by Zharkov & Lyubimov (1970a,b), Madariaga (1971), Saito (1971), Luh (1973, 1974), Woodhouse (1976), Dziewonski & Sailor (1976), and Woodhouse & Dahlen (1978). The theory has been made tractable by representing the lateral perturbation by a spherical harmonic expansion

$$\delta m = \sum_{s=1}^{\infty} \sum_{t=-s}^{+s} \delta m_s^t(r) Y_s^t(\vartheta, \varphi) \qquad (2)$$

where δm is the perturbation in rigidity, bulk modulus, or density and s and t are the angular and azimuthal orders of the lateral perturbation. Madariaga (1971) has shown that the splitting may be computed as the product of a complicated radial integral and a surface integral over a product of three spherical harmonics (called a Wigner 3-j symbol). Exclusion properties of the 3-j symbol immediately provide several rather peculiar properties. First, a multiplet of angular order l will be sensitive to angular orders of the heterogeneity no greater than $2l$. Second, no multiplet interacts with odd angular orders of the heterogeneity. That is, there is a component of the Earth's lateral variation that is never expressed in perturbations to seismic waves. Unfortunately, even if one is willing to truncate the spherical harmonic expansion of the Earth's lateral variation at angular order $2S$, there are still $S(2S + 3)$ different radial components of the model $(\delta m_s^t(r))$ to be determined, neglecting odd angular orders. If, for instance, the perturbation could be truncated at angular order 20, then 230 functions of radius would be needed to represent each of the three model parameters.

While the lateral variations problem is extremely difficult it is not impossible. Good results have been obtained on a regional basis by Forsyth (1975), Nolet (1976), and Chou (1979). For this work, teleseismic sources from one area are examined on a regional array of receivers. Normal mode techniques are used to separate overtone dispersion branches to periods as short as 20 s. The coherent perturbation in dispersion is then used to infer the structure of the source-receiver path. Some progress has been made in the whole Earth problem as well. Since the splitting due to lateral variations is not, for practical purposes, an observable quantity, several researchers have proposed the use of aggregate properties of split multiplets as data (Dahlen 1974, Jordan 1978). Dahlen (1974) proposed a direct inversion scheme for the lowest terms in Equation (2) based on such data. While this method has not proven very useful for the real Earth, Dahlen (1976b) was able to develop several models consistent with the data then available.

Jordan (1978) proposes a greatly simplified, computationally realizable inversion scheme using the center periods of laterally split multiplets as data. He postulates that the lateral perturbation is separable

$$\delta m_s^t(r) = \delta m(r)f(s, t). \tag{3}$$

Since there is evidence (e.g. Sipkin & Jordan 1976) that large scale upper mantle structure is correlated with surface geology, f may be computed directly from a model like the one given by Jordan (1979). It would then be a relatively simple matter to invert for the single radial dependence of the perturbation.

Silver & Jordan (1980) describe the first results of this effort. Using IDA records they have obtained over 2000 data from fundamental spheroidal modes. So far they have been able to draw some preliminary conclusions about the nature of the Earth's longest wavelength lateral variations. However, it remains to be seen whether the Earth will really permit simplifications like Equation (3). Because of the enormous complexity of the lateral variations problem, this field is still in its infancy. While there are some signs that researchers are beginning to find the right track, it will probably be many years before even continental dimension lateral variations are well resolved.

Some Problems

As investigations into the very long period properties of seismic sources, spherically averaged anelastic Earth structure, and laterally varying elastic Earth structure progress new puzzles will inevitably emerge. In fact, a few anomalies have apparently already been found. For instance, Q estimates for some well-observed modes seem to scatter more from record to record than they should based on signal-to-noise arguments (Geller & Stein 1980). Also, apparent center periods for some multiplets seem to scatter more from record to record than others although they should behave similarly given reasonable models of lateral heterogeneity (Silver & Jordan 1980).

These observations make it likely that one or more factors are operating which are not yet sufficiently well understood. Several possibilities have been proposed. The best studied is quasi-degenerate coupling (Dahlen 1969, Luh 1973, Woodhouse 1980). In general, the singlets of one multiplet are orthogonal to one another and will not couple no matter how close together they are in eigenperiod. This is also true for singlets belonging to different multiplets for most free oscillations which are likely to be close together in period. However, certain perturbations, most notably rotation of the Earth, result in the development of an affinity between special pairs of multiplets. If the members of one of these pairs are also close together in period, then quasi-degenerate coupling may occur. The coupling results in substantial alteration of the properties of both multiplets. The most likely candidates for quasi-degenerate coupling in the normal mode band are $_0S_l$ and $_0T_{l+1}$ pairs for $l = 11, 12, 13, \ldots, 18$ which fits well with the findings of Silver & Jordan (1980). Chao & Gilbert (1980) have rather direct evidence of coupling between $_0S_{11}$ and $_0T_{12}$. Unfortunately, these few modes are disproportionately important in most data sets because they are so easily observed.

Another mechanism which may contribute towards one or more of the observed anomalies is nonasymptotic interference among the singlets of a multiplet. At short periods, Jordan (1978) and Dahlen (1979a) have shown that singlet interference behaves in a simple manner. The singlets interfere to produce one spectral peak whose width and center period are simply related to the structure of the great circle path passing through both the source and the receiver. Buland (1980) has shown that in much of the normal mode band singlet interference may be more complicated. The singlets still interfere to produce one spectral peak. However, the width and center period of this peak are not necessarily simply related to Earth structure.

One or both of these mechanisms is probably contributing towards at least some of the observed anomalies. Indeed, it is likely that more anomalies and complications will be discovered as the free oscillations spectrum is examined in ever finer detail in the years to come.

SUMMARY

The first three decades of the study of free oscillations as a subfield of geophysics has been successful indeed. Concurrent developments in the theory of free oscillations of the Earth and long period instrumentation have encouraged rapid advances in the science. This has resulted today in a very complete understanding of the gross features of the observed free oscillations spectrum for periods longer than 80 s. This understanding has led in turn to rather precise models of the spherically averaged internal elastic structure of the Earth. Further, there have been comprehensive theoretical developments of perturbations to this spectrum due to the rotation of the Earth, ellipticity of figure, spherically averaged anelastic structure, and laterally varying elastic structure.

Current topics of research include using amplitude and phase information to study seismic sources, measuring normal mode Q's to estimate anelastic structure, and studying the effects of lateral elastic variations. The first two topics are better understood and more completely developed than the third. In fact, each has been developed far enough to encounter some unexpected and not yet satisfactorily explained difficulties.

ACKNOWLEDGMENTS

Many thanks to Freeman Gilbert who taught me much of what I know about free oscillations and to Jon Berger, Adam Dziewonski, Tom Jordan, and Tony Dahlen who filled in some of the gaps. Thanks also to Jim Dewey and Stu Sipkin who critically read and to Barbara Sloan and Madeleine Zirbes who prepared the manuscript.

Literature Cited

Abe, K. 1970. Determination of seismic moment and energy from the Earth's free oscillations. *Phys. Earth Planet. Inter.* 4:49–61.

Agnew, D., Berger, J., Buland, R., Farrell, W., Gilbert, F. 1976. International deployment of accelerometers: a network for very long period seismology. *EOS, Trans. Am. Geophys. Union* 57:180–88

Alsop, L. E. 1963. Free spheriodal vibrations of the Earth at very long periods—I. calculation of periods for several Earth models. *Bull. Seismol. Soc. Am.* 53:483–501

Alsop, L. E. 1964. Spheriodal free periods of the Earth observed at eight stations around the world. *Bull Seismol. Soc. Am.* 54:755–76

Alsop, L. E., Sutton, G. H., Ewing, M. 1961. Measurement of Q for very long period free oscillations. *J. Geophys. Res.* 66: 2911–15

Alterman, Z., Jarosch, H., Pekeris, C. L. 1959. Oscillations of the Earth. *Proc. R. Soc. London Ser. A* 252: 80–95

Anderson, D. L. 1967. The anelasticity of the mantle. *Geophys. J. R. Astron. Soc.* 14: 135–64

Anderson, D. L., Archambeau, C. B. 1964. The anelasticity of the Earth. *J. Geophys. Res.* 69: 2071–84

Anderson, D. L., Hart, R. S. 1976. An Earth model based on free oscillations and body waves. *J. Geophys. Res.* 81: 1461–75

Anderson, D. L. Hart, R.S. 1978. The Q of the Earth. *J. Geophys. Res.* 83: 5869–82

Anderson, D. L., Minister, J. B. 1979. The frequency dependence of Q in the Earth and implications for mantle rheology and Chandler wobble. *Geophys. J. R. Astron. Soc.* 58: 431–40

Anderssen, R. S., Cleary, J. R. 1974. Asymptotic structure in torsional free oscillations of the Earth—I. overtone structure. *Geophys. J. R. Astron. Soc.* 39: 241–68

Anderssen, R. S., Cleary, J. R., Dziewonski, A. M. 1975. Asymptotic structure in eigenfrequencies of spheriodal modes of the Earth. *Geophys. J. R. Astron. Soc.* 43: 1001–5

Backus, G., Gilbert, F. 1961. The rotational splitting of the free oscillations of the Earth. *Proc. Nat. Acad. Sci. USA* 47: 362–71

Backus, G. E., Gilbert, J. F. 1967. Numerical application of a formalism for geophysical inverse problems. *Geophys. J. R. Astron. Soc.* 13: 247–76

Backus, G., Gilbert, F. 1968. The resolving power of gross Earth data. *Geophys. J. R. Astron. Soc.* 16: 169–205

Backus, G., Gilbert, F. 1970. Uniqueness in the inversion of inaccurate gross Earth data. *Philos. Trans. R. Soc. London Ser. A* 266: 123–92

Backus, G. Mulcahy, M. 1976a. Moment tensors and other phenomenological descriptions of seismic sources—I. continuous displacements. *Geophys. J. R. Astron. Soc.* 46: 341–61

Backus, G., Mulcahy, M. 1976b. Moment tensors and other phenomenological descriptions of seismic sources—II. discontinuous displacements. *Geophys. J. R. Astron Soc.* 47: 301–29

Benioff, H., Gutenberg, B., Richter, C. F. 1954. Progress report, Seismological Laboratory, California Institute of Technology, 1953. *Trans. Am. Geophys. Union* 35: 979–87

Benioff, H., Press, F., Smith, S. 1961. Excitation of the free oscillations of the Earth by earthquakes. *J. Geophys. Res.* 66: 605–19

Ben-Menahem, A. 1964. Mode-ray daulity. *Bull. Seismol. Soc. Am.* 54: 1315–21

Ben-Menahem. A., Rosenman, M., Israel, M. 1972. Source mechanism of the Alaskan earthquake of 1964 from amplitudes of free oscillations and surface waves. *Phys. Earth Planet. Inter.* 5: 1–29

Berger, J., Agnew, D. C., Parker, R. L., Farrell, W. E. 1979. Seismic system calibration: 2. cross-spectral calibration using random binary signals. *Bull. Seismol. Soc. Am.* 69: 271–88

Block, B., Moore, R. D. 1966. Measurements in the Earth mode frequency range by an electrostatic sensing and feedback gravimeter. *J. Geophys. Res.* 71: 4361–75

Block, B., Dratler, J., Moore, R. D. 1970. Earth normal modes from a 6.5 magnitude earthquake. *Nature* 226: 343–44

Boatwright, J. L. 1980. Preliminary body-wave analysis of the St. Elias, Alaska earthquake of February 28, 1979. *Bull. Seismol. Soc. Am.* 70: 419–36

Bolt, B. A., Brillinger, D. R. 1979. Estimation of uncertainties in eigenspectral estimates from decaying geophysical time series. *Geophys. J. R. Astron. Soc.* 59: 593–603

Bridgman, P. W. 1945. Polymorphic transitions and geological phenomena. *Am. J. Sci.* 243A: 90–97

Brune, J. N. 1964. Travel times, body waves, and normal modes of the Earth. *Bull. Seismol. Soc. Am.* 54: 2099–2128

Brune, J. N., Gilbert, F. 1974. Torsional overtone dispersion from correlations of S waves to SS waves. *Bull. Seismol. Soc. Am.* 64: 313–20

Buland, R. P. 1976. *Retrieving the seismic moment tensor.* PhD thesis. Univ. Calif., San Diego. 66 pp.

Buland, R. 1980. On interference among free oscillations of the Earth. *Geophys. J. R. Astron. Soc.* In press

Buland, R., Gilbert, F. 1976. Matched filtering for the seismic moment tensor. *Geophys. Res. Lett.* 3: 205–6

Buland, R., Gilbert, F. 1978. Improved resolution of complex eigenfrequencies in analytically continued seismic spectra. *Geophys. J. R. Astron. Soc.* 52: 457–70

Buland, R., Berger, J., Gilbert, F. 1979. Observations from the IDA network of attenuation and splitting during a recent earthquake. *Nature* 277: 358–62

Bullard, E. C. 1957. The density within the Earth. *Verh. K. Ned. Geol. Mijnbouwkd. Genoot. Geol. Ser.* 18: 23–41

Bullen, K. E. 1950. An Earth model based on the compressibility-pressure hypothesis. *Mon. Not. R. Astron. Soc. (Geophys. Suppl.)* 6:50–59

Bullen, K. E., Haddon, R. A. W. 1967. Derivation of an Earth model from free oscillation data. *Proc. Nat. Acad. Sci. USA* 58:846–52

Burdick, L. J., Helmberger, D. V. 1978. The upper mantle P velocity structure of the Western United States. *J. Geophys. Res.* 83:1699–1712

Byerly, P. 1955. Nature of faulting as deduced from seismograms. In *Crust of the Earth, Geol. Soc. Am. Special Paper*, ed. A. Poldervaart. 62:75–85. Baltimore: Waverly. 762 pp.

Chao, B. F., Gilbert, F. 1980. Autoregressive estimation of complex eigenfrequencies in low frequency seismic spectra. *Geophys. J. R. Astron. Soc.* Submitted

Chou, T-A., 1979. *Continental and Oceanic Upper Mantle Structure from Dispersion of Higher Modes of Surface Waves.* PhD thesis. Harvard Univ., Cambridge. 79 pp.

Choy, G. L., Boatwright, J. L. 1980. The rupture characteristics of two deep earthquakes inferred from broad-band GDSN data. *Bull. Seismol. Soc. Am.* Submitted

Connes, J., Blum, P. A., Jobert, G., Jobert, N. 1962. Observation des oscillations propres de la Terre. *Ann. Geophys.* 18:260–68

Crossley, D. J., Rochester, M. G. 1980. Simple core undertones. *Geophys. J. R. Astron. Soc.* 60:129–61

Dahlen, F. A. 1968. The normal modes of a rotating, elliptical Earth. *Geophys. J. R. Astron. Soc.* 16:329–67

Dahlen, F. 1969. The normal modes of a rotating, elliptical earth, II: near-resonance multiplet coupling. *Geophys. J. R. Astron. Soc.* 18:397–436

Dahlen, F. A. 1972. Elastic dislocation theory for a self-gravitating elastic configuration with an initial static stress field. *Geophys. J. R. Astron. Soc.* 28:357–83

Dahlen, F. A. 1973. Elastic dislocation theory for a self-gravitating elastic configuration with an initial static stress field—II: energy release. *Geophys. J. R. Astron. Soc.* 31:469–84

Dahlen F. A. 1974. Inference of the lateral heterogeneity of the Earth from the eigenfrequency spectrum: a linear inverse problem. *Geophys. J. R. Astron. Soc.* 38:143–67

Dahlen, F. A. 1975. The correction of great circle surface wave phase velocity measurements for the rotation and ellipticity of the Earth. *J. Geophys. Res.* 80:4895–4903

Dahlen, F. A. 1976a. Reply to comments by A. M. Dziewonski and R. V. Sailor. *J. Geophys. Res.* 81:4951–56

Dahlen. F. A. 1976b. Models of lateral heterogeneity of the Earth consistent with eigenfrequency splitting data. *Geophys. J. R. Astron. Soc.* 44:77–105

Dahlen, F. 1979a. The spectra of unresolved split normal mode multiplets. *Geophys. J. R. Astron. Soc.* 58:1–33

Dahlen, F. A. 1979b. Exact and asymptotic synthetic multiplet spectra on an ellipsoidal Earth. *Geophys. J. R. Astron. Soc.* 59:19–42

Dahlen, F. A., Sailor, R. V. 1979. Rotational and elliptical splitting of the free oscillations of the Earth. *Geophys. J. R. Astron. Soc.* 58:609–23

Dahlen, F. A., Smith, M. L. 1975. The influence of rotation on the free oscillations of the Earth. *Philos. Trans. R. Soc. London Ser. A* 279:583–629

Davies, D. 1967. On the problem of compatibility of surface wave data, Q and body wave travel times. *Geophys. J. R. Astron. Soc.* 13:421–21

Dennis, J. G., Walker, C. T. 1965. Earthquakes resulting from metastable phase transitions. *Tectonophysics* 2:401–7

Der, Z. A., McElfresh, T. W. 1977. The relationship between anelastic attenuation and regional amplitude anomalies of short period P waves in North America. *Bull. Seismol. Soc. Am.* 67:1303–17

Derr, J. S. 1969a. Free oscillation observations through 1968. *Bull. Seismol. Soc. Am.* 59:2079–99

Derr, J. S. 1969b. Internal structure of the Earth inferred from free oscillations. *J. Geophys. Res.* 74:5202–20

Dziewonski, A. M., Gilbert, F. 1971. Solidity of the inner core of the Earth inferred from normal mode observations. *Nature* 234:465–66

Dziewonski, A. M., Gilbert, F. 1972. Observations of normal modes from 84 recordings of the Alaskan earthquake of 1964 March 28. *Geophys. J. R. Astron. Soc.* 27:393–446

Dziewonski, A. M., Gilbert, F. 1973. Observations of normal modes from 84 recordings of the Alaskan earthquake of 1964 March 28—II. Further remarks based on new spheriodal overtone data. *Geophys. J. R. Astron. Soc.* 35:401–37

Dziewonski, A. M., Gilbert, F. 1974. Temporal variation of the seismic moment tensor and the evidence of precursive compression for two deep earthquakes. *Nature* 247:185–88

Dziewonski, A. M., Sailor, R. V. 1976. Comments on 'The correction of great circular surface wave phase velocity measurements for the rotation and ellipticity of the Earth' by F. A. Dahlen. *J. Geophys. Res.* 81:4947–50

Dziewonski, A. M., Hales, A. L., Lapwood, E. R. 1975. Parametrically simple Earth models consistent with geophysical data. *Phys. Earth Planet. Inter.* 10:12–48

Dziewonski, A. M., Hager, B. H., O' Connell, R. J. 1977. Large-scale heterogeneities in the lower mantle. *J. Geophys. Res.* 82:239–55

Farrell, W. E., Berger, J. 1979. Seismic system calibration: 1. parametric models. *Bull. Seismol. Soc. Am.* 69:251–70

Forsyth, D. W. 1975. A new method for the analysis of multi-mode surface-wave dispersion: application to Love-wave propagation in the East Pacific. *Bull. Seismol. Soc. Am.* 65:323–42

Geller, R. J., Stein, S. 1980. Time domain attenuation measurements for fundamental spheroidal modes $(_0S_6-_0S_{28})$ for the 1977 Indonesian earthquake. *Bull. Seismol. Soc. Am.* 69:1671–91

Gilbert, F. 1970. Excitation of the normal modes of the Earth by earthquake sources. *Geophys. J. R. Astron. Soc.* 22:223–26

Gilbert, F. 1975. Some asymptotic properties of the normal modes of the Earth. *Geophys. J. R. Astron. Soc.* 43:1007–11

Gilbert, F. 1976. The representation of seismic displacements in terms of traveling waves. *Geophys. J. R. Astron. Soc.* 44:275–80

Gilbert, F., Backus, G. E. 1968. Approximate solutions to the inverse normal mode problem. *Bull. Seismol. Soc. Am.* 58:103–31

Gilbert, F., Buland, R. 1977. Dissipation of energy radiated by earthquakes. *Ann. Geofis.* 30:471–89

Gilbert, F., Dziewonski, A. M. 1975. An application of normal mode theory to the retrieval of structural parameters and source mechanisms from seismic spectra. *Philos. Trans. R. Soc. London Ser. A* 278:187–269

Gilbert, F., MacDonald, G. J. F. 1960. Free oscillations of the Earth, I. toroidal oscillations. *J. Geophys. Res.* 65:675–93

Gilbert, F., Dziewonski, A., Brune, J. 1973. An informative solution to a seismological inverse problem. *Proc. Nat. Acad. Sci. USA* 70:1410–13

Goldstein, H. 1965. Small oscillations. In *Classical Mechanics*, pp. 318–38. Dallas: Addison-Wesley. 399 pp.

Hart, R. S., Kanamori, H. 1975. Search for compression before a deep earthquake. *Nature* 253:333–36

Jackson, D. D., Anderson, D. L. 1970. Physical mechanisms of seismic-wave attenuation. *Rev. Geophys. Space Phys.* 8:1–63

Jeans, J. H. 1923. The propagation of earthquake waves. *Proc. R. Soc. London Ser. A* 102:554–74

Jeffreys, H. 1965. The damping of S waves. *Nature* 208:675–75

Jeffreys, H. 1967. Radius of the Earth's core. *Nature* 215:1365–66

Jobert, N. 1957. Sur la période propre des oscillations sphéroïdales de la Terre. *C. R. Acad. Sci.* 245:1941–43

Jobert, N., Leveque, J. J., Roult, G. 1978. Evidence of lateral variations from free oscillations and surface waves. *Geophys. Res. Lett.* 5:569–72

Johnson, I. M., Smylie, D. E. 1977. A variational approach to whole-Earth dynamics. *Geophys. J. R. Astron. Soc.* 50:35–54

Johnson, L. E., Gilbert, F. 1972. A new datum for use in the body wave travel time inverse problem. *Geophys. J. R. Astron. Soc.* 30:373–80

Jordan, T. H. 1978. A procedure for estimating lateral variations from low-frequency eigenspectra data. *Geophys. J. R. Astron. Soc.* 52:441–55

Jordan, T. H. 1979. The deep structure of the continents. *Sci. Am.* 240:92–107

Jordan, T. H., Anderson, D. L. 1974. Earth structure from free oscillations and travel times. *Geophys. J. R. Astron. Soc.* 36:411–59

Julian, B. R., Sengupta, M. K. 1973. Seismic travel time evidence for lateral inhomogeneity in the deep mantle. *Nature* 242:443–47

Kanamori, H. 1970. The Alaska earthquake 1964: radiation of long-period surface waves and source mechanism. *J. Geophys. Res.* 75:5029–40

Kanamori, H., Anderson, D. L. 1975. Amplitude of the Earth's free oscillations and long-period characteristics of the earthquake source. *J. Geophys. Res.* 80:1075–78

Kanamori, H., Anderson, D. L. 1977. Importance of physical dispersion in surface wave and free oscillation problems: review. *Rev. Geophys. Space Phys.* 15:105–12

Kanamori, H., Stewart, G. S. 1979. A slow earthquake. *Phys. Earth Planet. Inter.* 18:167–75

Kennett, B. L. N., Simons, R. S. 1976. An implosive precursor to the Colombia earthquake 1970 July 31. *Geophys. J. R. Astron. Soc.* 44:471–82

Knopoff, L. 1964. Q. *Rev. Geophys. Space Phys.* 2:625–60

Knopoff, L., Gilbert, F. 1960. First motions from seismic sources. *Bull. Seismol. Soc. Am.* 50:117–34

Knopoff, L. Zurn, W., Rydelek, P. A., Yogi, T. 1979. Q of mode $_0S_0$. *J. Geophys. Res.* 46:89–95

Lamb, H. 1882. On the vibrations of an elastic sphere. *Proc. London Math. Soc.* 13:189–212

Landisman, M., Satô, Y., Nafe, J. 1965. Free vibrations of the Earth and the properties of its deep interior regions, part 1: density. *Geophys. J. R. Astron. Soc.* 9:439–502

Liu, H.-P., Archambeau, C. B. 1975. The effect of anelasticity on periods of the Earth's free oscillations (toroidal modes). *Geophys. J. R. Astron. Soc.* 43:795–814

Liu, H.-P., Archambeau, C. B. 1976. Correction to 'The effect of anelasticity on periods of the Earth's free oscillations (toroidal modes)'. *Geophys. J. R. Astron. Soc.* 47:1–7

Love, A. E. H. 1911. Vibrations of a gravitating compressible planet. In *Some Problems in Geodynamics*, pp. 126–143. New York: Dover. 180 pp.

Luh, P. C. 1973. Free oscillations of the laterally inhomogeneous Earth: quasi-degenerate multiplet coupling. *Geophys. J. R. Astron. Soc.* 32:187–202

Luh, P. C. 1974. Normal modes of a rotating, self-gravitating inhomogeneous Earth. *Geophys. J. R. Astron. Soc.* 38:187–224

MacDonald, G. J. F., Ness, N. F. 1961. A study of the free oscillations of the Earth. *J. Geophys. Res.* 66:1865–1911

Madariaga, R., 1971. *Free oscillations of the laterally heterogeneous earth.* PhD thesis. Mass. Inst. Tech., Cambridge. 105 pp.

Madariaga, R., Aki, K. 1972. Spectral splitting of toroidal-free oscillations due to lateral heterogeneity of the Earth's structure. *J. Geophys. Res.* 77:4421–31

Malvern, L. E. 1969. General principles. In *Introduction to the Mechanics of a Continuous Medium*, pp. 197–272. Englewood Cliffs: Prentice-Hall. 713 pp.

Matumoto, T., Satô, Y. 1954. On the vibration of an elastic globe with one layer. The vibration of the first class. *Bull. Earthquake Res. Inst. Tokyo Univ.* 32:247–58

Mendiguren, J. A. 1973. Identification of free oscillation spectral peaks for 1970, July 31, Colombian deep shock using the excitation criteria. *Geophys. J. R. Astron. Soc.* 33:281–321

Mendiguren, J. A., Aki, K. 1978. Source mechanism of the deep Colombian earthquake of 1970 July 31 from the free oscillation data. *Geophys. J. R. Astron. Soc.* 55:539–56

Moore, R. D., Farrell, W. E. 1970. Linearization and calibration of electrostatically fedback gravity meters. *J. Geophys. Res.* 75:928–32

Ness, N. F., Harrison, J. C., Slichter, L. B. 1961. Observations of the free oscillations of the Earth. *J. Geophys. Res.* 66:621–29

Nolet, G. 1976. *Higher Modes and the Determination of Upper Mantle Structure.* PhD thesis. Univ. Utrecht, Netherlands. 90 pp.

Nowroozi, A. A. 1965. Eigenvibrations of the Earth after the Alaskan earthquake. *J. Geophys. Res.* 70:5145–56

Nowroozi, A. A. 1968. Measurement of Q values from the free oscillations of the Earth. *J. Geophys. Res.* 73:1407–15

Nowroozi, A. A., Alsop, L. E. 1968. Torsional free periods of the Earth observed at six stations around the Earth. *Nuovo Cimento* 6:133–46

O'Connell, R. J., Budiansky, B. 1977. Viscoelastic properties of fluid-saturated cracked solids. *J. Geophys. Res.* 82:5719–35

Okal, E. A., Geller, R. J. 1979. On the observability of isotropic seismic sources: the July 31, 1970 Colombian earthquake. *Phys. Earth Planet. Inter.* 18:176–96

Parker, R. L. 1977. Understanding inverse theory. *Ann. Rev. Earth Planet. Sci.* 5:35–64

Patton, H., Aki, K. 1979. Bias in the estimate of seismic moment tensor by the linear inversion method. *Geophys. J. R. Astron. Soc.* 59:479–95

Pekeris, C. L. 1966. The internal constitution of the Earth. *Geophys. J. R. Astron. Soc.* 11:85–132

Pekeris, C. L., Jarosch, H. 1958. The free oscillations of the Earth. In *Contributions in Geophysics in Honor of Beno Gutenberg*, *Int. Ser. Monogr. Earth Sci..* ed. E. Ingerson, 1:171–92. Los Angeles: Pergamon. 244 pp.

Pekeris, C. L., Alterman, Z., Jarosch, H. 1961a. Rotational multiplets in the spectrum of the Earth, *Phys. Rev.* 122:1692–1700

Pekeris, C. L., Alterman, Z., Jarosch, H. 1961b. Comparison of theoretical with observed values of the periods of free oscillation of the Earth. *Proc. Nat. Acad, Sci. USA* 47:91–98

Phinney, R. A., Burridge, R. 1973. Representation of the elastic-gravitational excitation of a spherical Earth model by generalized spherical harmonics. *Geophys. J. R. Astron. Soc.* 34:451–87

Press, F. 1970. Earth models consistent with geophysical data. *Phys. Earth Planet. Inter.* 3:3–22

Prothero, W. A., Goodkind, J. M. 1972. Earth-tide measurements with the superconducting gravimeter. *J. Geophys. Res.* 77:926–37

Randall, M. J. 1976. Attenuative dispersion and frequency shifts of the Earth's free oscillations. *Phys. Earth Planet. Inter.* 12:P1–P4

Riedesel, M. A., Agnew, D., Berger, J., Gilbert, F. 1980. Stacking for the frequencies and Q's of $_0S_0$ and $_1S_0$. *Geophys. J. R. Astron. Soc.* 62:457–71

Sailor, R. V., Dziewonski, A. M. 1978. Measurements and interpretation of normal mode attenuation. *Geophys. J. R. Astron. Soc.* 53:559–81

Saito, M. 1967. Excitation of free oscillations and surface waves by a point source in a vertically heterogeneous Earth. *J. Geophys. Res.* 72:3689–99

Saito, M. 1971. Theory for the elastic-gravitational oscillation of a laterally heterogeneous Earth. *J. Phys. Earth* 19:259–70

Sengupta, M. K., Toksöz, M. N. 1976. Three dimensional model of seismic velocity variation in the Earth's mantle. *Geophys. Res. Lett.* 3:84–86

Silver, P. G., Jordan, T. H. 1980. Fundamental spheriodal mode observations of aspherical heterogeneity. *Geophys. J. R. Astron. Soc.* Submitted

Sipkin, S. A., Jordan, T. H. 1976. Lateral heterogeneity of the upper mantle determined from the travel times of multiple ScS. *J. Geophys. Res.* 81:6307–20

Sipkin, S. A., Jordan, T. H. 1979. Frequency dependence of Q_{ScS}. *Bull. Seismol. Soc. Am.* 69:1055–79

Sipkin, S. A., Jordan, T. H. 1980. Regional variations of Q_{ScS}. *Bull. Seismol. Soc. Am.* 70:1071–1102

Slichter, L. B. 1961. The fundamental free mode of the Earth's inner core. *Proc. Nat. Acad. Sci. USA* 47:186–90

Slichter, L. B. 1967. Spherical oscillations of the Earth. *Geophys. J. R. Astron. Soc.* 14:171–77

Smith, M. L. 1974. The scalar equations of infinitesimal elastic-gravitational motion for a rotating, slightly elliptical Earth. *Geophys. J. R. Astron. Soc.* 37:491–526

Smith, M. L. 1976. Translational inner core oscillations of a rotating slightly elliptical Earth. *J. Geophys. Res.* 81:3055–65

Smith, S. W. 1961. *An investigation of the Earth's free oscillations.* PhD thesis. Calif. Inst. Tech., Pasadena. 91 pp.

Smith, S. W. 1966. Free oscillations excited by the Alaskan earthquake. *J. Geophys. Res.* 71:1183–93

Solomon, S. C., Toksöz, M. N. 1970. Lateral variation of attenuation of P and S waves beneath the United States. *Bull. Seismol. Soc. Am.* 60:819–38

Stein, S., Geller, R. J. 1978. Attenuation measurements of split normal modes for the 1960 Chilean and 1964 Alaskan earthquakes. *Bull. Seismol. Soc. Am.* 68:1595–611

Strelitz, R. A. 1978. Moment tensor inversions and source models. *Geophys. J. R. Astron. Soc.* 52:359–64

Takeuchi, H. 1959. Torsional oscillations of the Earth and some related problems. *Geophys. J. R. Astron. Soc.* 2:89–100

Takeuchi, H., Saito, M., Kobayashi, N., Nakagawa, I. 1962. Free oscillations of the Earth observed on gravimeters. *J. Seismol. Soc. Jpn.* 15:122–37

Vlaar, N. J. 1976. On the excitation of the Earth's seismic normal modes. *Pure Appl. Geophys.* 114:863–75

Ward, S. N. 1980. Body wave calculations using moment tensor sources in spherically symmetric, inhomogeneous media. *Geophys. J. R. Astron. Soc.* 60:53–66

Wiggins, R. A. 1972. The general linear inverse problem: implication of surface waves and free oscillations for Earth structure. *Rev. Geophys. Space Phys.* 10:251–85

Woodhouse, J. H. 1976. On Rayleigh's principle. *Geophys. J. R. Astron. Soc.* 46:11–22

Woodhouse, J. H. 1978. Asymptotic results for elastodynamic propagator matrices in plane-stratified and spherically-stratified Earth models. *Geophys. J. R. Astron. Soc.* 54:263–80

Woodhouse, J. H. 1980. The coupling and attenuation of nearly resonant multiplets in the Earth's free oscillation spectrum. *Geophys. J. R. Astron. Soc.* 61:261–83

Woodhouse, J. H., Dahlen, F. A. 1978. The effect of a general aspherical perturbation on the free oscillations of the Earth. *Geophys. J. R. Astron. Soc.* 53:335–54

Zharkov, V. N., Lyubimov, V. M. 1970a. Torsional oscillations of a spherically asymmetrical model of the Earth. *Izv. Acad. Sci. USSR Phys. Solid Earth*, pp. 71–76

Zharkov, V. N., Lyubimov, V. M. 1970b. Theory of spheriodal vibrations for a spherically asymmetric model of the Earth. *Izv. Acad. Sci. USSR Phys. Solid Earth*, pp. 613–18

Ann. Rev. Earth Planet. Sci. 1981. 9:415–48
Copyright © by Annual Reviews Inc. All rights reserved

LONG WAVELENGTH ✖ 10156
GRAVITY AND
TOPOGRAPHY ANOMALIES

A. B. Watts and S. F. Daly[1]

Lamont-Doherty Geological Observatory of Columbia University, Palisades, New York 10964

INTRODUCTION

The principal inference about the nature of the Earth's interior drawn from most early studies of gravity anomalies and topography was the widespread occurrence of isostasy. The main aim of these studies was to determine the form of the compensating masses associated with individual geological features of the continents and oceans. The models for the compensating masses most preferred by geologists and physical geodesists were those developed by G. B. Airy and J. H. Pratt. Airy (1855) proposed that the crust was thicker beneath mountain ranges than lowlands while Pratt (1858) proposed that the average crustal density beneath mountain ranges was smaller than beneath lowlands. These models successfully explained gravity anomalies of relatively small horizontal extent, such as those associated with mountain ranges and sedimentary basins.

With the introduction of artificial satellites in 1955 it became possible to determine directly gravity anomalies of more regional extent. These anomalies were of interest to geophysicists since they could not be attributed to isostasy, at least not in the form envisaged by Airy and Pratt. Jeffreys (1959) argued that gravity anomalies characterized by full wavelength λ greater than a few thousand km were due to the finite strength of the mantle. Runcorn (1963, 1965), on the other hand, attributed them to density differences associated with convective flow in the mantle.

Barrell (1914) pointed out much earlier, however, that isostasy in the form proposed by Airy and Pratt could not exist everywhere since the

---

[1] Present address: Jet Propulsion Laboratory, California Institute of Technology, Pasadena, California 91103.

0084-6597/81/0515-0415$01.00

strength of the Earth's crust and upper mantle would prevent it. The presence of a strong outer layer of the Earth, or lithosphere, strongly modifies the gravity field and makes it difficult to directly interpret gravity anomalies in terms of density differences in the mantle (McKenzie 1967). During the past few years, however, a number of studies have been carried out of the flexural strength of the lithosphere and how this varies with age (Watts et al 1980). These studies suggest that the strength of the lithosphere contributes most to the short wavelength gravity field ($\lambda \sim 30$ to 800 km). Thus, only the long wavelength gravity field, such as derived from observations of satellite orbits, may be directly attributed to density differences in the mantle.

It is now generally believed that the Earth's lithospheric plates are driven by some form of thermal convection in the mantle. A number of studies (Runcorn 1963, 1965, McKenzie 1977, McKenzie et al 1974) have pointed out that if convection occurs in the mantle then some correlation would be expected between long wavelength gravity and topography anomalies on the Earth's surface, but not otherwise. There is presently little agreement, however, on the plan form of convection or even whether whole mantle or layered convection is dominant.

A useful approach to the problem, therefore, is to determine the correlation between gravity and topography anomalies that would be expected for different models of mantle convection (McKenzie et al 1974, McKenzie 1977). McKenzie (1977) has shown, for example, using layered convection models, that the correlation would be expected to depend little on the exact form of the heating, but strongly on the viscosity structure and depth of the convecting layer. Gravity and topography anomalies maybe either positive or negative over a rising region in the flow, depending on the viscosity structure. Thus the correlation between long wavelength gravity and topography anomalies would be expected to vary with wavelength and may provide constraints on the viscosity structure and depth of the convecting layer.

There have, unfortunately, only been a few attempts (Anderson et al 1973, Sclater et al 1975, Watts 1976, Cochran & Talwani 1977) to determine the correlation between observed long wavelength gravity and topography anomalies and none of these determined the correlation as a function of wavelength. McKenzie & Bowin (1976) attempted to define the correlation as a function of wavelength using surface ship profiles in the Atlantic ocean. Although they did find a strong correlation between gravity and topography anomalies, the correlation could be satisfactorily explained in terms of the flexural strength of the lithosphere.

The purpose of this paper is to review the current knowledge of gravity and topography anomalies on the Earth which are of large horizontal extent. In particular, we emphasize gravity and topography anomalies characterized by wavelengths in the range of a few hundred to a few thousand km since these anomalies appear to hold the most promise in determining information on the plan form of convection in the mantle. We limit the review to mainly oceanic regions for two reasons. First, in comparison to the continents, the oceans have had a relatively simple thermal and mechanical evolution. It is now possible, for example, to quantify how the oceanic lithosphere contributes to observed gravity and topography anomalies. Second, the advent of altimeters on orbiting satellites, beginning with SKYLAB in 1973, has enabled the long wavelength gravity field over the oceans to be determined with great precision. These studies of mainly oceanic regions have direct implications, however, for the interpretation of the gravity field of the continents and the other terrestrial planets.

GRAVITY ANOMALIES

The Earth's gravity field is now well known from terrestrial measurements over both continents and oceans. The continents are largely surveyed except for parts of Africa, South America, and Asia while the oceans are reasonably well surveyed except south of about 30° S. Terrestrial measurements have been used as a basis for a wide range of geologic and geodetic studies. These include the state of isostasy of individual geological features in the continents and oceans and the construction of detailed gravimetric geoids.

There has been increasing interest over the past several years in the long wavelength gravity field of continents and oceans since the long wavelength field cannot be attributed to isostasy. It is difficult, however, to obtain the long wavelength field from terrestrial measurements since some form of smoothing is required. Estimates of the mean gravity anomaly in a $1 \times 1°$ square are limited to regions of closely spaced terrestrial measurements, such as parts of the Atlantic, Indian, and Pacific oceans (Talwani et al 1972, Watts & Leeds 1977, Kahle & Talwani 1973) and North and South America, Europe, India, and Australia (Woollard 1972). These $1 \times 1°$ mean anomalies, which are generally accurate to better than a few mgal, resolve features in the Earth's gravity field with $\lambda \gtrsim 200$ km.

The advent of orbiting satellites, however, has allowed the long wavelength gravity field to be determined directly. The initial observations of satellites provided information on the zonal harmonics and therefore the long wavelength variations in the gravity field. By 1963 the tesseral harmonics could be determined by averaging over much shorter periods of time, of the order of a few days. The gravity anomaly Δg is related to these harmonics by

$$\Delta g = (GM/a_e^2) \sum_{n=2}^{\infty} \sum_{m=0}^{n} (a_e/r)^{n+2} \times (n-1)(\delta C_{nm} \cos m\lambda$$

$$+ \delta S_{nm} \sin m\lambda) P_{nm}(\sin \phi) \qquad (1)$$

where

$GM =$ the product of the gravitational constant and mass of the Earth,

$a_e =$ equatorial radius of the Earth,

$(r, \phi, \lambda) =$ the spherical coordinates of the computation point,

$P_{nm}(\sin \phi) =$ the associated Legendre's functions,

$C_{nm}, S_{nm} =$ spherical harmonic coefficients of degree n and order m,

$\delta C_{nm} =$ observed $C_{nm} -$ reference C_{nm},

$\delta S_{nm} =$ observed $S_{nm} -$ reference S_{nm}.

The reference field most frequently used is an ellipsoid of revolution so that reference $C_{nm} = S_{nm} = 0$ for $m \neq 0$. Also all odd zonal harmonics are zero. The reference surface can be defined by the 2nd and 4th zonal harmonics, since higher orders are very small. Kahn (1977a) has summarized the $C_{2,0}$ and $C_{4,0}$ coefficients which correspond to the most commonly used reference surfaces.

The steady increase in the number of orbiting satellites has led to successive improvements in gravity field models. The Smithsonian Standard Earth model SE II (Gaposchkin & Lambeck 1971) included data from 21 orbiting satellites and 935 individual $5 \times 5°$ mean gravity anomaly values and was complete to $n = 16$. The resolution of this gravity field model is about $11°$, which corresponds to $\lambda \sim 2400$ km. The SE II model was followed by the SE III model which was complete to $n = 18$ ($\lambda \gtrsim 2200$ km) with a few higher order coefficients (Gaposchkin 1974). The Goddard Space Flight Center GEM models GEM 1, 3, and 5 (Lerch et al 1974) are satellite-derived models and are complete to $n = 12$ ($\lambda \gtrsim 2600$ km). GEM 2, 4, and 6 are combination solutions of satellite-derived and terrestrial data and are complete to $n = 16$, with some higher order terms.

The SE and GEM (even numbers) gravity field models are derived mainly from the same types of satellite and terrestrial gravity data, although computational methods differ and the recent GEM models are based on more terrestrial gravity data. Khan (1977a) carried out statistical comparisons of these Earth models and showed that although they are in close agreement for $n \leq 4$ they are in poor agreement for $n > 11$. Terrestrial gravity data contributes most for $n > 11$ (Khan 1977a) and variable amounts and quality of gravity data have been used in these solutions.

Following the GEM 6 gravity field model, there has been a steady improvement in the accuracy of the harmonics for $n > 11$ (Lerch et al 1979, Wagner et al 1976). The most recent solution, GEM 9 (Lerch et al 1979), is a satellite-derived model based on improved satellite tracking techniques and observations of the perturbations of 30 satellite orbits. The GEM 10 model (Lerch et al 1979) is a combination solution of the GEM 9 satellite-derived solution and terrestrial gravity based on 1645 $5 \times 5°$ mean gravity anomalies. These models are complete to $n = 22$ and therefore resolve features in the gravity field with $\lambda \gtrsim 1800$ km.

We illustrate the GEM 10 gravity field model complete to $n = 12$ in Figure 1. The shortest wavelength resolved in this figure is about 3600 km. The most striking correlation shown in Figure 1, and one that has been pointed out in previous studies using other gravity field models (Kaula 1969), is between positive gravity anomalies (up to $+30$ mgal) and island arc–deep sea trench systems bordering the Pacific and the northeast Indian oceans. Prominent positive anomalies (up to $+30$ mgal) also occur over the North Atlantic and the southern Indian oceans. Gravity anomaly lows (up to -40 mgal) correlate with the glaciated regions of North America and with some ocean basins in the western North Atlantic and North Pacific oceans. A number of the gravity anomalies in Figure 1, however, appear to be independent of continental and oceanic structure and must therefore have their source at depth in the mantle.

Statistical studies have been carried out in order to estimate the depths of the source of the low order harmonics ($n < 12$) of the Earth's gravity field. Cook (1963) speculated that, because of their amplitude, the low order harmonics ($n \leq 3$) have their sources in the deeper parts of the mantle. Allan (1972) assumed randomly distributed density variations concentrated on a single density interface and estimated that the source depth for harmonics $n = 2$ to 6 was about 1700 km. Using similar assumptions and more refined Earth models Khan (1977b) estimated the source depth for harmonics $n = 2$ to 11 was in the range 600 to 800 km. The usefulness of these depth estimates has recently been questioned, however (McNutt 1980), since the actual distribution of density that exists in the Earth is too poorly known.

McKenzie (1967) approached the problem by estimating the stress differences in the Earth required to support observed gravity anomalies. By assuming the lithosphere could not support shear stresses greater than 200 bar, McKenzie (1967) argued that only the short wavelength components of the gravity field could be supported by the lithosphere and that the long wavelength components ($\lambda \gtrsim 300$ to 800 km), such as defined in gravity field models, must be due to density differences at depth in the

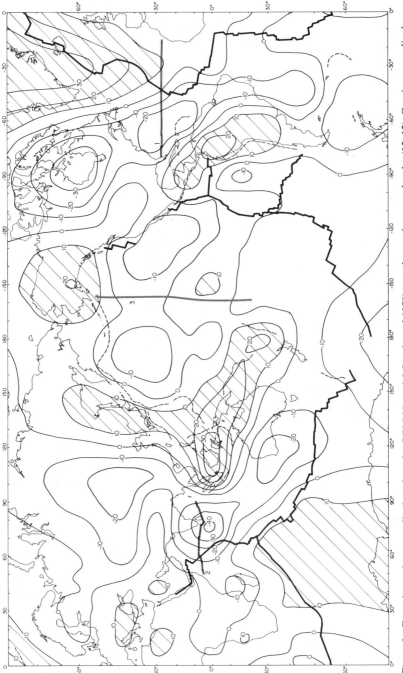

Figure 1 Free-air gravity anomalies based on the GEM 10 model (Lerch et al 1979) complete to degree and order (12, 12). Gravity anomalies have been referred to the best fitting reference ellipsoid (flattening 1/298.25). Areas greater than + 10 mgal are shaded. Profiles 1 to 3 (Figures 10 to 12) are shown as heavy broken lines.

mantle. Lambeck (1972) argued, however, that the lithosphere could support shear stresses of at least 0.8 kbar and that gravity anomalies characterized by $\lambda \gtrsim 3000$ km could be supported by the lithosphere.

More recently, forward modeling techniques have been used, in which the gravity effect of simple plate models is computed and compared to observed gravity anomalies. Studies have now been made of the gravity effect of thermal cooling models at mid-ocean ridge crests (Lambeck 1972), of dense lithospheric slabs descending beneath island arcs (Grow 1973, Grow & Bowin 1975), and of models for the structure of oceanic lithosphere at large fracture zone offsets (Sibuet et al 1974, Sibuet & Veyrat-Peinet 1980). A problem with these studies, however, is that the gravity effect of the models is characterized by small amplitudes and long wavelengths and is generally difficult to distinguish in observed free-air gravity anomaly profiles (Cochran & Talwani 1977, Watts & Talwani 1975a, Louden & Forsyth 1976).

A number of authors have emphasized the use of geoid anomalies rather than gravity anomalies since the geoid better defines the long wavelength gravity field. The geoid anomaly ΔN is related to the gravity anomaly Δg in (1) by

$$\Delta N = \frac{\Delta g}{(n-1)g}, \tag{2}$$

where g is average gravity and n has been previously defined. Equation (2) indicates that, compared to gravity anomalies, long wavelength $(n \to 0)$ geoid anomalies are amplified relative to the short wavelengths. Chase (1979) and Crough & Jurdy (1980), for example, computed the geoid effect of simple thermal models of the dense downgoing lithospheric slabs beneath island arcs and compared them to observed geoid anomalies based on the GEM 8 gravity field model. They did not, however, use the comparison as a basis for modifying the density distribution and shape of the downgoing slab. Rather, they determined the geoid anomaly associated with the downgoing slab and subtracted it from the observed geoid. The resulting "residual" geoid anomaly was then interpreted in terms of other mass distributions in the mantle.

The advent of radar altimeters in orbiting satellites, beginning with SKYLAB in 1973 (Leitao & McGoogan 1975) and GEOS-3 in 1975 (Stanley 1979), has enabled the geoid in oceanic regions to be determined directly. Radar altimeters measure the distance between the satellite and the instantaneous sea surface which, in oceanic regions, corresponds closely to the geoid. By accurately determining the satellite orbit, the geoid anomaly can be calculated directly from altimeter data. The geoid in oceanic regions, of course, includes disturbances due to oceanographic factors as well as to density differences in the Earth. The oceanographic factors, however, are known to be small (generally less than about 130 cm), enabling the gravimetric or tectonic part of the geoid to be easily separated.

Preliminary studies with data obtained by SKYLAB and GEOS-3 indicate that satellite altimeters can resolve information in the geoid for wavelengths greater than about 80 km (Wagner 1979). The altimeter data shows geoid anomalies of up to ±10 meters associated with deep sea trenches, seamounts, fracture zones, mid-ocean ridge crests, and continental margins (Haxby & Turcotte 1978, Roufousse 1978, Watts 1979, Crough 1979, Chapman 1978, Haxby 1981). The altimeter data has been interpreted in terms of the crustal and lithospheric structure at these features, although in a number of cases the interpretations substantiate the results of previous studies based on gravity data.

The most important implication of satellite altimeter data is the information it provides on long wavelength components in the gravity field. At convergent plate boundaries, GEOS-3 data has been used to provide general constraints on the density distribution in the downgoing slab (Chapman 1978). In plate interiors, GEOS-3 data has been used to suggest that the Hawaiian and Bermuda topographic swells are compensated at relatively shallow depths in the lithosphere (Haxby & Turcotte 1978, Crough 1978). Unfortunately, there have been few studies carried out that have examined, in detail, the correlation between GEOS-3 data and topography in oceanic regions.

TOPOGRAPHY ANOMALIES

The topography of the Earth's surface is now known from altimetric surveys in the continents as well as bathymetric surveys in the oceans. In general, the topography is well known for most parts of the Earth, but poorly known in Tibet, central Africa, and the southern oceans, south of about 40° S. The Earth's topography shows two distinct mean values (for example, Egyed 1969). The higher value (about 0.1 km) corresponds to the mean elevation of the continents while the lower value (about 4.7 km) corresponds to the mean depth of the oceans. This bi-model character of the Earth's topography contrasts with that of the other terrestrial planets.

There have been only a limited number of studies of the regional variations of topography on the Earth's surface (Vening Meinesz 1964, Balmino et al 1973, Bills 1978)). Vening Meinesz (1964) carried out a spherical harmonic analysis of the topography complete to $n = 30$, using data compiled by Prey (1922), and Balmino et al (1973) carried out a similar analysis complete to $n = 36$, using more recent data compiled by Lee (1966). These studies, therefore, considered features in the Earth's topography with $\lambda \gtrsim 1100$ km. The spectrum of the Earth's topography (Vening Meinesz

1964, Balmino et al 1973) shows a strong peak for $n = 1$, corresponding to the distribution of most continents in one hemisphere, and a remarkably regular decrease with increasing n. Vening Meinesz (1964) argued that this phenomenon was related to some pattern of thermal convection in the Earth's mantle. Bills (1978) pointed out, however, that since the topography spectrum for the Earth was similar to that of the other terrestrial planets it was unlikely that convection could be the only cause.

The topography of the continents on smaller scales is, in fact, very variable. The surface of the continents has been modified by complex geological processes which include mountain building, volcanism, and granite emplacement. In addition, erosion, which is very active at present, appears to have been active throughout the Phanerozoic (Garrels et al 1972). In general, the most recent orogenic belts, such as the Alpine and Himalaya (orogeny culminated in the Miocene), are associated with high relief (up to 5 km) while older orogenic belts, such as the Appalachian and Caledonian (orogeny culminated in Silurian-Devonian), are associated with low relief (up to 1 km). Seismic refraction measurements show that regional changes in the topography of the continents are generally accompanied by changes in crustal thickness, as defined by the depth to Moho discontinuity. Woollard (1972) pointed out that, at least in North America, there are indications of a correlation between depth to the Moho and geological age of the basement complex. There is, however, no simple general relationship between topography and age in the continents.

In contrast, the topography of the ocean floor shows a remarkably simple relationship with crustal age (Sclater & Francheteau 1970). The systematic increase in the depth of the ocean floor away from a mid-ocean ridge crest can be explained by simple thermal models for the evolution of oceanic lithosphere. These models, in which hot material is added at the ridge axis and subsequently cools, also explain other geophysical observations such as the variations in heat flow and free-air gravity anomaly away from a ridge crest (Sclater & Francheteau 1970, Lambeck 1972).

A detailed analysis of ocean floor topography has been carried out by Parsons & Sclater (1977) using data from relatively old sea floor in the western North Atlantic and central Pacific oceans. They showed that for sea floor of age 0 to 70 m.y.B.P. topography is described by

$$d(t) = 2500 + 350t^{1/2} \tag{3}$$

and for sea floor older than 20 m.y. by

$$d(t) = 6400 - 3200e^{-t/62.8}, \tag{4}$$

where t is the age (m.y.B.P.) and $d(t)$ is the depth (meters).

There are, however, large portions of the ocean floor whose depth cannot be explained by the simple thermal model (Menard 1973, Weissel & Hayes 1974, Sclater et al 1975, Watts 1976, Cochran & Talwani 1977,

Menard & Dorman 1977). For example, the North Atlantic ocean north of about 30° N is generally too shallow for its age and the southeast Indian ocean, between Australia and Antarctica, is too deep. These areas of anomalous topography are of interest since they cannot be explained by simple thermal models for the cooling lithosphere and thus may provide information on deep-seated processes occurring in the underlying mantle.

A useful approach to the problem is to calculate residual depth anomalies, which represent regions of the ocean floor not at the expected depth for their age (Menard 1973). The residual depth anomaly Δb is given by

$$\Delta b = d(t) + \frac{S(\rho_s - \rho_w)}{(\rho_m - \rho_s)} - b, \tag{5}$$

where $d(t)$ is the expected depth on the basis of age, S is the sediment thickness, ρ_s, ρ_w, and ρ_m are the mean densities of sediments, water, and mantle respectively, and b is the observed depth.

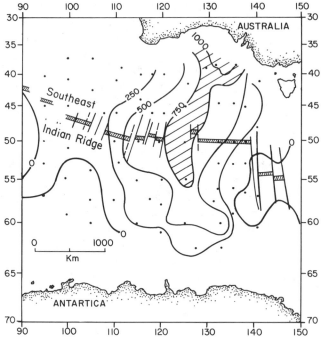

Figure 2 Residual depth anomalies between Australia and Antarctica based on Cochran & Talwani (1977). The solid dots represent the points at which residual depths were calculated. The shaded area is the southeast Indian ridge crest based on the bathymetric and magnetic studies of Weissel & Hayes (1974).

Residual depth anomaly maps have now been constructed for large regions of the ocean floor (Menard 1973, Sclater et al 1975, Watts 1976, Menard & Dorman 1977, Cochran & Talwani 1977). For example, Figure 2 illustrates the residual depth anomaly mapped by Cochran & Talwani (1977) between Australia and Antarctica. This anomaly is about 2000 km in width and reaches amplitudes of up to 1100 meters. The depth anomaly low trends generally N–S, normal to the local trend of the southeast Indian ocean ridge crest. The minimum depth anomaly correlates closely with a topographically rough zone referred to by Weissel & Hayes (1974) as the "discordance" zone.

The most obvious origin of residual depth or topography anomalies is regional changes in crustal thickness. For example, Figure 3 shows that the relatively shallow Iceland-Faeroes, Chagos-Laccadive, and Nazca aseismic ridges are associated with thick crust while the relatively deep

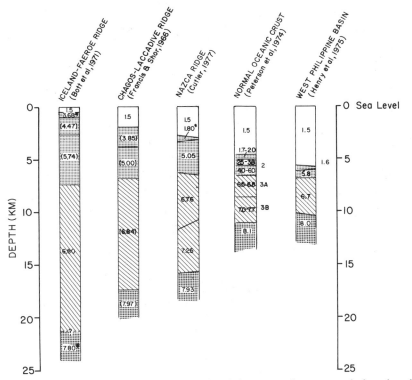

Figure 3 Summary seismic refraction profiles of the crust and upper mantle for selected bathymetric features in the Atlantic, Pacific, and Indian oceans. These profiles illustrate the variability in the thickness of the oceanic crust and how changes in water depths correlate with changes in crustal thickness.

West Philippine marginal basin is associated with thin crust. The expected change in water depth would be 230 meters for each 1 km change in crustal thickness, assuming a normal oceanic crustal section (Figure 3). Thus, since the mean crustal thickness in the Pacific ocean varies by up to 200 meters about its mean (Goslin et al 1972), depth anomalies of up to 420 meters could arise in the Pacific simply as a result of changes in crustal thickness.

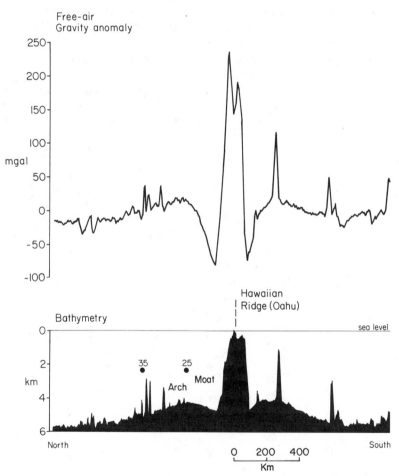

Figure 4 Free-air gravity anomaly and bathymetry profiles obtained on Vema cruise 21 leg 5 of the region around the southeastern end of the Hawaiian ridge. This profile, which extends north and south of Oahu, is located in Watts & Talwani (1975b). The age of the sea floor in this region is in the range 85 to 100 m.y.B.P. indicating a depth anomaly of about 1.2 to 1.6 km in amplitude and 1500 to 2000 km width around the southeastern end of the Hawaiian ridge.

A number of depth anomalies cannot, however, be attributed to variations in crustal thickness. For example, seismic refraction stations 25 and 35 in the central Pacific ocean (Figure 4; Watts 1976) show that the crust is similar in thickness on the Hawaiian swell as in flanking regions. Thus, factors other than variations in crustal thickness must contribute to the origin of depth anomalies. Watts (1976) interpreted the Hawaiian swell depth anomaly and its associated long wavelength gravity anomaly (Figure 4) in terms of a passive uplift of the plate in response to some form of convection in the underlying mantle. Detrick & Crough (1978), on the other hand, suggested the anomaly was caused by thinning of the Pacific plate over the Hawaiian hot spot.

Cochran & Talwani (1977) pointed out that positive depth anomalies were generally associated with volcanic regions such as Bermuda, Hawaii, Azores, and Cape Verde. Crough (1978) concluded that hot spot reheating of the lithosphere was the main cause of depth anomalies in the oceans, citing as an additional example the Cook-Austral chain. There are, however, some problems with this interpretation. For example, Bermuda is inactive and yet is presently associated with a topographic swell. In addition, the Cook-Austral chain includes active volcanoes and yet is presently associated with a topographic ridge that may, at least in part, be greater in age (Johnson & Malahoff 1971). Thus, factors other than thermal reheating of the lithosphere by hot spots may contribute to the origin of depth anomalies.

MECHANICAL PROPERTIES OF THE PLATES

The short wavelength gravity field of the Earth is determined mainly by the strength and mechanical properties of the lithosphere. This is not true of the long wavelength gravity field which cannot be explained by the strength of the lithosphere. Any discussion of the significance of the long wavelength gravity field, however, must be based on a preferred model for the mechanical properties of the lithosphere.

We show, for example, in Figure 4 a free-air gravity anomaly and topography profile of the Hawaiian ridge in the central Pacific ocean. This figure shows that the volcanic load of the Hawaiian ridge is associated with a relatively short wavelength gravity anomaly high of up to $+240$ mgal over the ridge and gravity lows of up to -75 mgal over flanking regions, which are superimposed on a long wavelength gravity high associated with the Hawaiian swell. Clearly, it is necessary to evaluate the contribution to the gravity field of the load of the Hawaiian ridge before the origin of this long wavelength gravity anomaly can be understood.

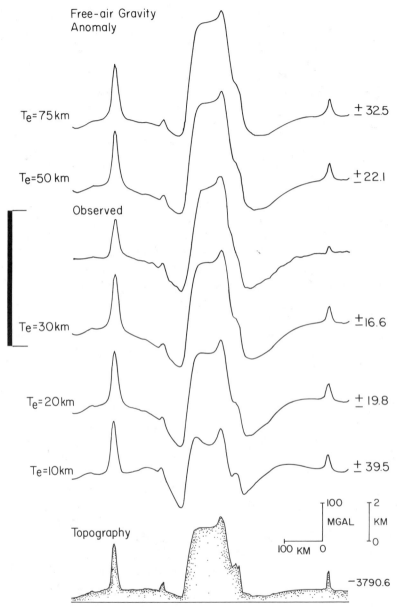

Figure 5 Comparison of observed and computed free-air gravity anomaly profiles of the Hawaiian ridge near Oahu. The observed profile is located in Watts (1978a, Figure 2) and was obtained during *R/V Vema* cruise 21 leg 5 and *R/V Robert D. Conrad* cruise 12 leg 20. The computed profiles are based on the elastic plate model and assumed values of T_e of 10, 20, 30, 50, and 75 km. The best overal fit to the amplitude and wavelength of the observed data is for $T_e = 30$ km.

A number of studies have, in fact, shown that the short wavelength gravity anomalies associated with the Hawaiian ridge can be satisfactorily explained by the flexure or plate model of isostasy (Gunn 1943, Vening Meinesz 1941, Walcott 1970, Watts & Cochran 1974, Watts 1978a). The flexure model, first proposed by Barrell (1914) and quantified by Gunn (1943), is similar to the Airy model in which the compensation is achieved by thickening of the crust, but differs in taking into account its lateral strength. In this model, the oceanic lithosphere is assumed to respond to seamount and oceanic island loads by flexure, in a similar manner as would a thin elastic plate overlying a weak fluid substratum. A useful parameter in this model is the effective flexural rigidity D, or stiffness, of the plate which is given by

$$D = ET_e^3/12(1 - \sigma^2), \tag{6}$$

where E is Young's Modulus, σ is Poisson's ratio, and T_e is the effective elastic thickness. Figure 5 shows the amplitude and wavelength of the observed gravity anomaly over the ridge can be best explained by the flexure for $T_e = 30$ km. A value of $T_e = 75$ km predicts too large an amplitude over the crest of the ridge and too long wavelengths in flanking regions while a value of $T_e = 10$ km predicts too small an amplitude over the crest of the ridge and too short wavelengths in flanking regions.

Recently, Watts et al (1980) compiled all recent estimates of T_e for the oceanic lithosphere and plotted them against the age of the lithosphere at the time it was loaded. These estimates, which are plotted in Figure 6, are based mainly on gravity and bathymetry data from mid-ocean ridge crests and flanks (solid squares), seamounts and oceanic islands (solid circles), and deep sea trench–outer rises (solid triangles). Figure 6 shows that T_e is a strong function of the age of the lithosphere at the time it is loaded. Surface loads associated with young oceanic lithosphere correlate with small values of T_e while loads associated with old lithosphere correlate with large values of T_e.

The observations in Figure 6 are generally compatible with extrapolations of data based on data from experimental rock mechanics (Goetze & Evans 1979, Anderson & Minster 1980, Bodine et al 1981). Figure 6 suggests that on loading the oceanic lithosphere, there must be a rapid relaxation from its seismic or short-term thickness (20 to 150 s) to its mechanical or long-term ($> 10^6$ years) thickness. Since the ages of loads plotted in Figure 6 are in the range 1 to 55 m.y., the mechanical thickness does not apparently change substantially with time. This behavior is explicable in terms of simple models (Goetze & Evans 1979, Anderson & Minster 1980, Bodine et al 1981) in which materials at shallow depths

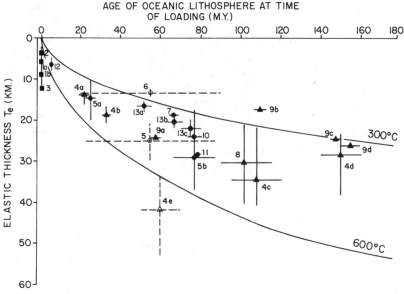

Figure 6 Plot of effective elastic thickness of the oceanic lithosphere T_e against age of the lithosphere at the time of loading. The sources of data are given in Watts et al (1980). The filled squares indicate ridge crest estimates, filled circles seamounts and oceanic islands, and filled triangles deep-sea trenches. The solid lines are the 300°C and 600°C oceanic isotherms based on a cooling plate model.

respond to surface loads by brittle failure while at great depths, where temperatures and pressures are higher, materials respond by ductile flow.

The flexure model of isostasy (Figure 6) can, in fact, explain the gravity field over a variety of different geological features on the ocean floor. For example, Figure 7 shows that free-air gravity anomaly profiles over the Line islands ridge and Nova ridge in the Pacific ocean can be best explained for $T_e = 5$ km while profiles over the Louisville ridge and Makarov guyot can be best explained for $T_e = 25$ km. This indicates (Figure 6) that the Line islands ridge and Nova ridge formed on young oceanic lithosphere at or near a mid-ocean ridge crest while the Louisville ridge and Makarov guyot formed on old lithosphere, off-ridge. These interpretations are in excellent agreement with geological evidence for the tectonic setting of these features (Watts et al 1980).

The results in Figures 6 and 7 suggest that the mechanical properties of the lithosphere can be determined for volcanic loads that form either on-ridge or off-ridge. These tectonic settings are also applicable to a number of other geological features in the world's ocean basins such as Walvis and 90° East ridges (on-ridge) and Bermuda and Azores (off-ridge).

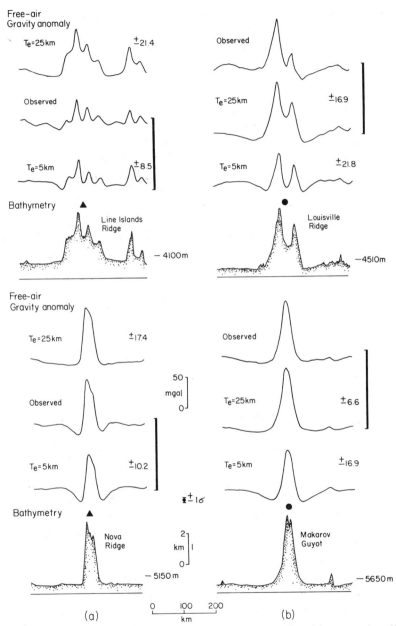

Figure 7 Comparison of observed free-air gravity anomaly profiles with computed profiles based on the flexure model for four selected bathymetric features in the Pacific ocean. Profiles of the Line islands and Nova ridge can be best explained by $T_e = 5$ km while profiles of Makarov guyot and Lousiville ridge can be best explained by $T_e = 25$ km. These estimates of T_e, which suggest an on-ridge and off-ridge origin for these features respectively, are consistent with available geological evidence on their age and the age of the underlying sea floor (Watts et al 1980).

Thus, it should be possible, by considering the flexural response function (Walcott 1976) corresponding to these tectonic settings, to quantify the contribution of the mechanical properties of the lithosphere to the gravity field as a function of wavelength.

We consider, for example, the topography of the ocean floor as a simple harmonic $b \cos kx$ of wavelength $\lambda = 2\pi/k$ and amplitude $b =$ topography above or below some mean depth. The gravity effect of distributed topography at sea level can then be expressed as a summation of Fourier components. Assuming the topography is uncompensated, and neglecting nonlinear contributions to the gravity anomaly, we have

$$GA(k) = 2\pi GB(k)(\rho_L - \rho_w)e^{-kd}, \tag{7}$$

where $GA(k)$ and $B(k)$ are the Fourier transforms of the gravity anomaly and topography respectively, ρ_L is the mean density of the oceanic crust and d is the mean depth. If, however, the topography is assumed to be compensated according to the flexure model then $GA(k)$ is given by

$$GA(k) = 2\pi GB(k)(\rho_L - \rho_w)e^{-kd}$$
$$\times \left\{ 1 - \Phi(k, D) \frac{[(\rho_3 - \rho_2)e^{-k(t-t_3)} + (\rho_m - \rho_3)e^{-kt}]}{[(\rho_m - \rho_{\text{infill}})]} \right\}, \tag{8}$$

where

$$\Phi(k, D) = \left[\frac{Dk^4}{(\rho_m - \rho_{\text{infill}})g} + 1 \right]^{-1}$$

and ρ_{infill} is the material infilling the crustal flexure, t the total crustal thickness, t_3 the thickness of oceanic layer 3, ρ_2 the mean density of oceanic layer 2, and ρ_3 the mean density of oceanic layer 3.

We see from Equations (7) and (8) that for short wavelength topography ($k \to \infty$) the effect on the gravity anomaly is the same, whether or not the topography is compensated, and is given by (7). For long wavelength topography ($k \to 0$), however, the gravity anomaly approaches zero if the topography is compensated. The quantity in the bracket at the right side of (8) is referred to as the isostatic or flexural response function, and is determined by the mechanical properties of the plate and the crustal structure prior to flexure.

The flexural response function is plotted in Figure 8 as a function of wavelength λ, for different assumed values of T_e and an assumed normal oceanic crust section (Figure 3). This figure illustrates the contribution of isostasy to the gravity field, by wavelength, of geological features which originate off-ridge, such as the Hawaiian-Emperor seamount chain, and on-ridge, such as the topography of oceanic layer 2 at the East Pacific rise crest. In the case of features that originate off-ridge the gravity anomaly is determined mainly by the topography for $\lambda \gtrsim 200$ to 300 km. The cor-

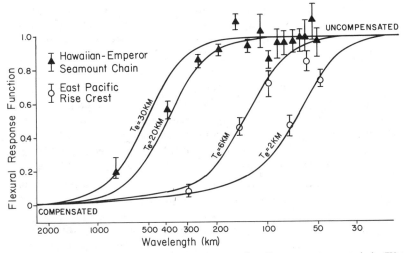

Figure 8 Flexural response functions for the Hawaiian-Emperor seamount chain (Watts 1978a) and East Pacific rise crest (Cochran 1979) plotted as a function of λ. These functions were computed from Equation (8) using observed values of $GA(k)/B(k)$ at each feature. The solid lines represent theoretical functions based on the flexure model and an assumed normal oceanic crustal structure. This figure illustrates the effect of isostasy on the gravity anomaly as a function of wavelength for geological features formed off-ridge or on-ridge.

relation between gravity and topography decreases at longer wavelengths because of isostasy and for $\lambda \gtrsim 1000$ to 1500 km there is little or no correlation between gravity and topography. For features that originate on-ridge, however, the corresponding wavelength ranges are $\lambda \gtrsim 30$ to 80 km and $\lambda \gtrsim 200$ to 300 km, respectively. Thus isostasy contributes most to the gravity field for features that form off-ridge, but these effects are small for $\lambda \gtrsim 1000$ to 1500 km.

The presence of gravity anomalies with $\lambda \gtrsim 1000$ to 1500 km is, however, a common feature of both terrestrial gravity data (for example, Figure 4) and satellite-derived gravity field models (for example, Figure 1). Thus Figure 8 suggests these anomalies cannot be attributed to the strength of the lithosphere and must be due to other causes.

CORRELATION OF GRAVITY AND TOPOGRAPHY ANOMALIES

During the past few years there has been particular interest in examining the correlation between long wavelength gravity and topography anomalies (Anderson et al 1973, Sclater et al 1975, Watts 1976, Cochran & Talwani 1977). Long wavelength gravity anomalies cannot be interpreted directly in terms of density differences in the mantle. Similarly, long wavelength topography anomalies cannot be used to directly infer vertical motions in the mantle. The correlation between long wavelength gravity and topography anomalies, however, makes a good argument for convection in the mantle. Simple models have now been constructed (McKenzie et al 1974, McKenzie, 1977) which show how convective flow may contribute to long wavelength gravity and topography anomalies. The main problem, however, has been in determining the nature of the observed correlation between long wavelength gravity and topography anomalies.

The first studies that related gravity and topography anomalies to convection in the mantle were carried out by Runcorn (1965) and Morgan (1972). Runcorn (1965) suggested that a rising region in the flow would be associated with a negative gravity anomaly due to its large temperature and density differences with the surroundings. Morgan (1972), on the other hand, suggested that a positive anomaly should be associated with a rising region since the gravity anomaly is the sum of both the mass deficiency of the rising region and the mass excess of the deformed surface layer.

The relationships between gravity anomaly and convection proposed by Morgan (1972) were confirmed by McKenzie et al (1974) using the results from several numerical experiments for convection in a constant viscosity fluid heated from below or within. More recently McKenzie (1977) has shown that the correlation between gravity and topography anomalies depends little on the amplitude of convection or the exact form of heating, but strongly on the viscosity structure and depth of the convecting layer. In addition, McKenzie (1977) has suggested the gravity anomaly and topography anomalies may be either positive or negative over a rising region in the flow when the viscosity is temperature dependent.

We show in Figure 9 an example of the gravity anomaly, geoid anomaly, and topography anomaly (surface deformation) that would be expected for a time-dependent numerical solution of convection, similar to those discussed by McKenzie (1977). The solution shown is characterized by an instability in the upper boundary layer so that cold blobs of fluid are detached from it and sink into the interior of the fluid. A similar type of behavior is characteristic of two-dimensional numerical experiments with total internal heating and constant viscosity at Rayleigh numbers $\gtrsim 10^6$.

The depth of the convecting layer was assumed to be at 700 km in Figure 9, even though there is presently little agreement on whether whole mantle or layered convection is dominant (Elsasser 1969, Davies 1977, O'Connell 1977, Richter & McKenzie 1978, Jeanloz & Richter 1979). Both the non-hydrostatic pressure and the curvature of the stream lines contribute to the surface deformation (McKenzie et al 1974). The gravity anomaly is given by the sum of the gravity effect of the deformation of the

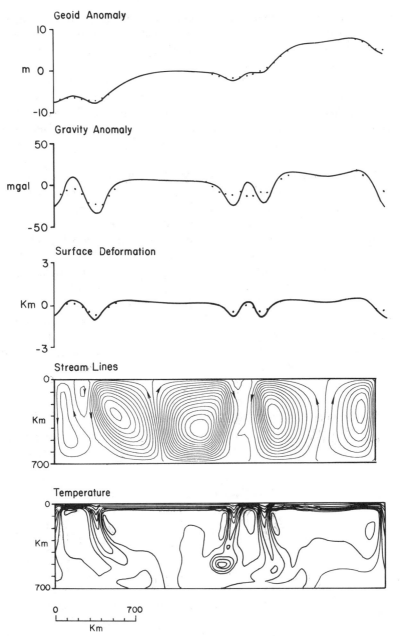

Figure 9 Surface deformation, gravity anomaly, and geoid anomaly obtained from a time-dependent, constant viscosity numerical experiment with Rayleigh number $R_a = 1.4 \times 10^6$, heated entirely from within. The temperature and stream function are also given. The dimensional values have been computed based on a depth of 700 km for the convective layer. The effect of an elastic plate overlying the convective layer is indicated by dots.

upper surface and the temperature and density differences in the interior. For simplicity, we have neglected the deformation of the bottom surface, although such effects will also contribute to the gravity anomaly (DeBremaecker 1976, McKenzie 1977).

We have included in Figure 9 the effect of an elastic plate above the convecting region using an isostatic response function, similar to that defined by (8). The effect of including an elastic plate, which is shown by a dotted line in Figure 9, is to smooth out the short wavelength components of the gravity anomaly leaving the long wavelength components unaffected.

The results shown in Figure 9, however, only illustrate those characteristics of the gravity anomaly, geoid anomaly, and surface deformation that are associated with a specific numerical convection experiment. They should not therefore be considered as a prediction of actual gravity and topography anomalies. The main result shown is that gravity highs would be expected over rising regions in the flow while geoid lows occur over sinking regions. In addition, surface deformations with wavelengths varying between 800 and 3000 km and amplitudes of up to 1.5 km would be expected over rising and sinking regions. Therefore, Figure 9, as well as other similar numerical experiments including some with temperature dependent viscosity, suggests that the correlation between gravity and topography anomalies would be expected to vary with wavelength and could provide constraints on the viscosity structure and the form of convective flow in the mantle.

The first attempt to relate the predictions of the correlation between gravity and topography anomalies based on numerical models of convection to observations was by Anderson et al (1973). They used the SE II gravity field complete to $n = 16$ to examine the correlation between free-air gravity anomaly and topography along the crest of mid-ocean ridges since the empirical relationship between oceanic depth and age predicts a uniform crustal depth everywhere. Anderson et al (1973) suggested there was a positive correlation between gravity anomaly and ridge crest depths which agreed in slope with the numerical models of McKenzie et al (1974). Weissel & Hayes (1974), who also used the SE II field, showed there was a correlation between gravity and topography on either side of the southeast Indian mid-ocean ridge crest, but they did not correct the observed bathymetry for the effects of age and sediment loading. The SE II gravity field is, however, unreliable for $n > 11$ (Kahn 1977a) and therefore is a poor indicator of the nature of the correlation between gravity anomaly and topography anomalies.

A different approach was used by Sclater et al (1975) and Watts (1976) who used mean $5 \times 5°$ free-air gravity anomalies based on terrestrial measurements rather than satellite-derived observations. Sclater et al (1975) showed there was a good correlation between gravity and topography anomalies in the North Atlantic, particularly around the Azores and Cape Verde islands. They obtained a positive correlation with a slope of 30 mgal/km, similar to the predictions by McKenzie et al (1974). Watts (1976) showed there was a good correlation between gravity and topography anomalies around the southeastern end of the Hawaiian ridge, although the slope obtained of 22 mgal/km was somewhat smaller than predicted. Both Sclater et al (1975) and Watts (1976) suggested the observed correlation was caused by some pattern of flow in the underlying mantle.

It has recently been suggested that convection in the upper mantle may occur on several scales (Richter 1973, Richter & Parsons 1975, McKenzie & Weiss 1975, Richter & Daly 1978). One scale, which is directly observable, is associated with the motion of the plates themselves (up to $\sim 10^5$ km for the Pacific plate). A second smaller scale, suggested by laboratory and numerical experiments, is one with dimensions on the order of the depth of the layer. Such a two-scale convective flow pattern was first studied in the laboratory by Richter & Parsons (1975). They showed that rolls perpendicular to the direction of motion of a moving top boundary decayed and were replaced by rolls parallel to the motion.

Marsh & Marsh (1976), using a gravity field model similar to GEM 8 (Wagner et al 1976), suggested there was a good correlation between gravity and topography anomalies in the Pacific ocean, at least for $n = 12$ to 23 ($1700 \gtrsim \lambda \gtrsim 3300$ km). Their model showed a linear pattern of positive and negative anomalies which spanned the Pacific ocean. They interpreted these anomalies as caused by rolls in the mantle, similar to those predicted earlier by Richter (1973). However, Watts (1978b) pointed out that west of about longitude $150°$ W the satellite-derived gravity field model was in poor agreement with surface ship data, due probably to inclusion of insufficient surface data in the model. Thus the GEM 8 model could not reliably be used in support of small-scale convection occurring in the mantle.

Cochran & Talwani (1977) carried out a systematic comparison of gravity and topography anomalies in the world's oceans using all available surface ship data. They concluded that there is little correlation between long wavelength gravity and topography anomalies (λ greater than a few hundred km), except over areas associated with off-ridge volcanism such as Hawaii, Azores, and Bermuda. Their study was based, however, on gravity rather than geoid data obtained on widely separated ship tracks.

We illustrate some of the main features in the correlation of gravity and topography in the world's oceans using three selected surface-ship free-air gravity and topography profiles in Figures 10–12. Since the geoid

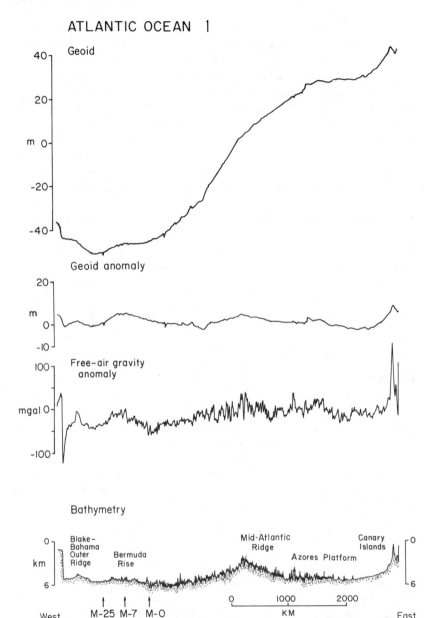

Figure 10 Free-air gravity anomaly, bathymetry, geoid, and geoid anomaly profile of the North Atlantic ocean at about 28° North (Figure 1). The gravity and bathymetry profile was obtained during the 1965 cruise of *HMS Snellius*. The geoid profile was obtained by interpolation of the adjusted GEOS-3 altimeter data (Rapp 1979) onto the actual ship track. The geoid anomaly profile was obtained by subtracting the geoid profile from the GEM 6 Earth model complete to $n = 12$. Note a geoid high occurs over a region of sea floor characterized by magnetic anomalies M-0 (112 m.y.B.P) to M-25 (155 m.y.B.P.).

INDIAN OCEAN 2

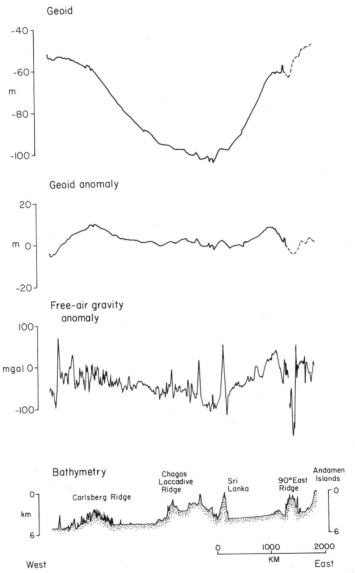

Figure 11 Free-air gravity anomaly, bathymetry, geoid, and geoid anomaly profile of the Indian ocean at about 5.5° North (Figure 1). The gravity anomaly and bathymetry profile were obtained during *Vema* cruise 19 leg 10 and *Robert D. Conrad* cruise 12 leg 16.

PACIFIC OCEAN 3

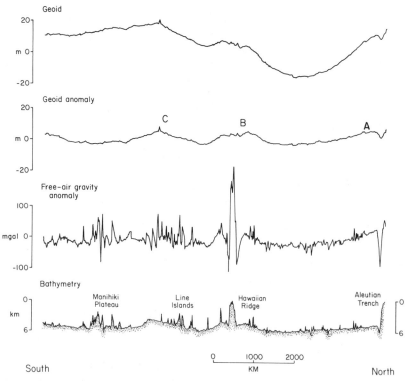

Figure 12 Free-air gravity anomaly, bathymetry, geoid, and geoid anomaly profile of the Pacific ocean along about 160°W (Figure 1). The gravity anomaly and bathymetry profile were obtained on *Robert D. Conrad* cruise 12 leg 9, *Dmitry Mendeleyev* cruise leg 5, and *Vema* cruise 21 leg 5. The geoid anomaly highs A, B, C are discussed in the text and are shown in map form in Figure 13.

better defines the long wavelength gravity field we also show the geoid based on adjusted GEOS-3 satellite altimeter data (Rapp 1979). The geoid anomaly was obtained by subtracting the GEM 6 field complete to $n = 12$ from the adjusted GEOS-3 data, since this field is relatively well determined (Khan 1977a). The adjusted GEOS-3 data and geoid anomaly were interpolated onto each ship track and then projected, along with the gravity anomaly and topography data, along either east-west (Indian, Atlantic) or north-south (Pacific) profiles.

Figure 10 shows a gravity anomaly, geoid, and topography profile of the North Atlantic ocean between the Blake escarpment and the Canary islands. This profile shows an asymmetry in the depth of the ocean basin either side of the mid-Atlantic ridge crest. The Canary basin is nearly 1 km shallower than the western Atlantic basin. This shallowing in the sea floor is associated with a gravity anomaly difference of about 20 mgal either side of the ridge crest and a geoid difference of about 60 meters. The steep increase in the geoid is almost completely removed in the geoid anomaly profile, indicating this increase is a dominant feature of the gravity field for $n < 11$ (see also Figure 1).

The main features of the geoid anomaly profile of the North Atlantic (Figure 10) are geoid highs of up to 10 meters amplitude and 1500 km widths associated with the mid-ocean ridge crest and the Bermuda rise. The geoid high over the ridge crest can be generally explained by simple cooling models for the oceanic lithosphere (Haxby 1981). Haxby & Turcotte (1978) have interpreted the geoid high over the Bermuda rise as indicating the rise is compensated at shallow depths in the lithosphere. The origin of the high is of particular interest, however, since it occurs over a region of sea floor that Parsons & Sclater (1977) included in their analysis of the variation of depth with crustal age > 80 m.y.

Figure 11 shows a gravity anomaly, geoid, and bathymetry profile of the northern Indian ocean between Africa and the Andamen islands. The profile includes crossings of the Carlsberg, Chagos-Laccadive, and 90° East ridges and the continental margin off southern India. The most striking feature of the profile is the gravity anomaly and geoid low, which reaches a minimum off southern India. The geoid low is almost completely removed in the geoid anomaly profile, indicating that it dominates the gravity field for $n < 12$ (Figure 1). The Indian ocean gravity low, however, unlike the North Atlantic gravity high, shows no obvious correlation with topography.

Geoid anomaly highs occur over the Carlsberg ridge and the sea floor west of the 90° East ridge (Figure 11). The geoid high of about 10 meters over the Carlsberg ridge is somewhat larger than that observed over the mid-Atlantic ridge (Figure 10). The geoid high west of the 90° East ridge correlates well with the buried 85° East ridge, mapped by Curry et al (1981).

Figure 12 shows a gravity anomaly, geoid, and topography profile of the Pacific ocean between the Aleutian islands and northeast of New Zealand. This profile shows a gravity anomaly and geoid low over the deep sea floor between the Aleutian trench and Hawaii and highs over the Hawaiian swell, Aleutian outer rise, and Line islands ridge.

The geoid low is removed in large part in the geoid anomaly profile in Figure 12 but the geoid highs are still clearly visible. These highs, referred to Figure 12 as geoid highs *A*, *B* and *C*, are about 1500 km in width and are associated with amplitudes of up to 7 meters. Figure 13 shows that the geoid highs *A*, *B*, *C* extend either side of the profile in

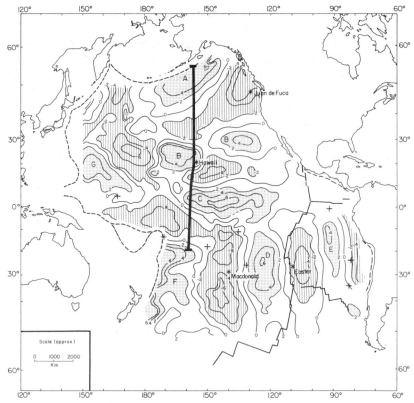

Figure 13 Geoid anomaly map of the Pacific ocean obtained by subtracting adjusted GEOS-3 satellite altimeter data (Rapp 1979) from the GEM 6 Earth model complete to $n = 12$. The heavy line shows the location of the profile in Figure 12, the continuous lines are active spreading centers, and the asterisks indicate the locations of 4 Pacific hot spots (Hawaii, Easter Island, Macdonald, Juan de Fuca). Note only one of these hot spots (Hawaii) is located on a geoid anomaly high. The crosses indicate other hot spots included in Crough & Jurdy (1980). The map is dominated by a belt of positive geoid anomalies trending WNW-ESE across the younger part of the Pacific plate.

Figure 12 and trend ENE-WSW across the younger part of the Pacific plate. We note that the geoid highs show no clear correlation with the location of hot spots in the Pacific (Figure 13).

A similar pattern of geoid anomalies in the Pacific ocean has been mapped recently by Marsh et al (1981) and McKenzie et al (1980). Marsh et al (1981) used satellite-to-satellite tracking data between the ATS-6 and GEOS-3 satellites and the GEOS-3 data set adjusted by Brace (1977). McKenzie et al (1980) utilized filtered values of the GEOS-3 data set adjusted by Rapp (1979).

McKenzie et al (1980) have shown that the pattern of geoid and topography anomalies in the north Pacific ocean trend generally parallel to the absolute motion directions of the Pacific plate, as defined by the hot spot frame of reference. They showed that the observed correlation between geoid and topography could be explained by simple models in which the plan form of convection consists not of rolls but of three-dimensional rising and sinking jets, elongated in the direction of the absolute motion of the Pacific plate. The spacing between the maxima and minima of geoid data is about 1500 km and McKenzie et al (1980) used this pattern of geoid and topography anomalies as strong support for a small scale of convection occurring in the mantle.

CONCLUSIONS

We have shown in this review that gravity and topography anomalies on the Earth's surface may provide useful new information on deep processes occurring in the Earth, such as those associated with mantle convection. There are two main reasons for this. First, there has been a steady improvement in the resolution of the long wavelength gravity field particularly in the wavelength range of a few hundred to a few thousand km, mainly due to increased coverage of terrestrial gravity measurements and the development of radar altimeters in orbiting satellites. Second, there have been an increase in the number and type of numerical and laboratory experiments of convection in the Earth, including some with deformable upper and lower boundaries and temperature-dependent viscosity.

The oceans appear to hold the most promise to determine long wavelength gravity and topography anomalies, since their evolution has been relatively simple compared to that of the continents. The oceanic lithosphere, like the continental lithosphere will, of course, obscure the effects of deep processes in the mantle due to its thermal and mechanical properties. However, simple models now exist for both the thermal (Parsons & Sclater 1977) and mechanical (Watts et al 1980) evolution of oceanic lithosphere. Thus, these effects of the oceanic lithosphere can now be quantified and, if necessary, isolated from surface observations of gravity and topography.

The distribution of long wavelength gravity and topography anomalies by themselves provides little information on, for example, the form of mantle convection. However, as argued most recently by McKenzie (1977), the correlation between long wavelength gravity and topography anomalies is a strong argument for convection.

We have shown in this review that a good correlation between long wavelength gravity and topography anomalies exists over some portions of the ocean floor. The most striking correlations occur in the North Atlantic and North Pacific oceans. In the North Atlantic north of about 30° N a broad region of relatively shallow sea floor is associated with a gravity anomaly high. In the Pacific an ENE-WSW trending pattern of gravity anomaly highs extends across the young part of the ocean basin.

These long wavelength gravity and topography anomalies in the Pacific ocean have recently been interpreted by McKenzie et al (1980) in terms of convective flow in the underlying mantle. There are, however, still a number of problems concerning the gravity anomaly patterns that exist in the Pacific. The ENE-WSW trending anomalies (Figure 13) do not extend to the East Pacific rise crest but seem to be confined to the young part of the lithosphere (about 30 to 110 m.y.B.P.). In addition, the ENE-WSW trending anomalies appear to be absent in the South Pacific, south of about 10° south.

Future work, during the next decade, should better define the pattern of long wavelength gravity and topography anomalies that exist in both the oceans and continents. In the oceans, the large amounts of surface ship bathymetry, seismic, and magnetic anomaly data that have been collected during the past few decades should be used to systematically correct observed sea-floor depths for the effects of age and sediment loading. In the continents, the large amounts of unclassified gravity data and elevation data should be examined to determine whether similar relationships between gravity and topography exist in the continents as do in the oceans. Better global gravity field models based on both terrestrial and satellite-derived measurements should be constructed. These largely observational studies, in conjunction with numerical and laboratory experiments of convection, appear to hold particular promise in the future of determining the depth and plan form of convection in the Earth's mantle.

ACKNOWLEDGMENTS

This work was supported by Office of Naval Research Contract N00014-80-C-0098 Scope I and National Aeronautics Space Administration grant NASA-NCC-5-3. The work was carried out while SFD was supported by a Lamont-Doherty Post-doctoral Fellowship. We are grateful to J. Marsh for providing a preprint of his paper on the gravity field of the central Pacific and to J. Cochran, W. Haxby, and G. Karner for critically reading the manuscript. Lamont-Doherty Geological Observatory Contribution N. 3080.

Literature Cited

Airy, G. B. 1855. On the computation of the effect of the attraction of mountain-masses, as disturbing the apparent astronomical latitude of stations in geodetic surveys. *Philos. Trans. R. Soc.* 145:101–4

Allan, R. R. 1972. Depth of origin of gravity anomalies. *Nature* 236:261

Anderson, D. L., Minster, J. B. 1980. Seismic velocity, attenuation and rheology of the upper mantle. *Phys. Earth Planet. Inter. Jean Coulomb Symp.* In press

Anderson, R. N., McKenzie, D., Sclater, J. G. 1973. Gravity, bathymetry and convection in the earth. *Earth Planet. Sci. Lett.* 18:391–407

Balmino, G., Lambeck, K., Kaula, W. M. 1973. A spherical harmonic analysis of the Earth's topography. *J. Geophys. Res.* 78:478–81

Barrell, J. 1914. The strength of the Earth's crust, Part VIII. Physical conditions controlling the nature of lithosphere and asthenosphere. *J. Geol.* 22:425–43

Bills, B. G. 1978. A harmonic and statistical analysis of the topography of the Earth, Moon and Mars. PhD thesis. Calif. Inst. Technol., Pasadena, Calif. 263 pp.

Bodine, J. H., Steckler, M. S., Watts, A. B. 1981. Observations of flexure and the rheology of the oceanic lithosphere. *J. Geophys. Res.* In press

Bott, M. H. P., Browitt, C. W. A., Stacey, A. P. 1971. The deep structure of the Iceland-Faeroe Ridge. *Mar. Geophys. Res.* 1:328–51

Brace, K. L. 1977. *Preliminary ocean-area geoid from GEOS-III satellite radar altimetry.* Paper presented at NASA GEOS-III Final Meet., New Orleans. 26 pp.

Chapman, M. E. 1978. Shape of the ocean surface and implications for the Earth's interior. PhD thesis. Columbia Univ., New York. 208 pp.

Chase, G. C. 1979. Subduction, the geoid, and lower mantle convection. *Nature* 282:464–68

Cochran, J. R. 1979. An analysis of isostasy in the world's oceans, Part 2, Mid-ocean ridge crests. *J. Geophys. Res.* 84:4713–29

Cochran, J. R., Talwani, M. 1977. Free-air gravity anomalies in the world's oceans and their relationship to residual elevation. *Geophys. J. R. Astron. Soc.* 50:495–552

Cook, A. H. 1963. Sources of harmonics of low order in the external gravity field of the Earth. *Nature* 198:1186

Crough, S. T. 1978. Thermal origin of mid-plate hot-spot swells. *Geophys. J. R.*

Astron. Soc. 55:451–70

Crough, S. T. 1979. Geoid anomalies across fracture zones and the thickness of the lithosphere. *Earth Planet. Sci. Lett.* 44:224–30

Crough, S. T., Jurdy, D. M. 1980. Subducted lithosphere, hot spots and the geoid. *Earth Planet. Sci. Lett.* 48:15–22

Curry, J. R., Emmel, F. J., Moore, D. G., Raitt, R. W. 1981. Structure, tectonics, and geological history of the northeastern Indian ocean. In *The Ocean Basins and Margins*, ed. A. S. Nairn, F. G. Stehli. New York: Plenum. In press

Cutler, S. T. 1977. Geophysical investigation of the Nacza Ridge. MSc thesis. Univ. Hawaii, Honolulu.

Davies, G. F. 1977. Whole mantle convection and plate tectonics. *Geophys. J. R. Astron. Soc.* 49:459–86

DeBremaecker, J.-C. 1976. Relief and gravity anomalies over a convecting mantle. *Geophys. J. R. Astron Soc.* 45:349

Detrick, R. S., Crough, S. T. 1978. Island subsidence, hot spots, and lithospheric thinning. *J. Geophys. Res.* 83:1236–44

Egyed, L. 1969. The slow expansion hypothesis. In *The Application of Modern Physics to the Planetary Interiors*, ed. S. K. Runcorn, pp. 65–76. Wiley

Elsasser, W. M. 1969. Convection and stress propagation in the upper mantle. In *The Application of Modern Physics to the Earth and Planetary Interiors*, ed. S. K. Runcorn pp. 223–46 Wiley

Francis, T. J. G., Shor, G. G. 1966. Seismic refraction measurements from the northwest Indian ocean. *J. Geophys. Res.* 71:427–49

Gaposchkin, E. M. 1974. Earth's gravity field to the eighteenth degree and geocentric coordinates for 104 stations from satellite and terrestrial data. *J. Geophys. Res.* 76:5377–5411

Gaposchkin, E. M., Lambeck, K. 1971. Earth's gravity field to the sixteenth degree and station coordinate from satellite and terrestrial data. *J. Geophys. Res.* 76:4855–83

Garrels, R. M., Mackenzie, F. T., Siever, R. 1972. Sedimentary cycling in relation to the history of the continents and oceans. In *The Nature of the Solid Earth*, ed E. C. Robertson, pp. 93–124. McGraw-Hill

Goetze, C., Evans, B. 1979. Stress and temperature in the bending lithosphere as constrained by experimental rock mechanics. *Geophys. J. R. Astron Soc.* 59:463–78

Goslin, J., Beuzart, P., Francheteau, J., Le Pichon, X. 1972. Thickening of the oceanic

446   WATTS & DALY

layer in the Pacific ocean. *Mar. Geophys. Res.* 1:418–27

Grow, J. A. 1973. Crustal and upper mantle structure of the central Aleutian arc. *Geol. Soc. Am. Bull.* 84:2169–92

Grow, J. A., Bowin, C. O. 1975. Evidence for high density mantle beneath the Chile Trench due to the descending lithosphere. *J. Geophys. Res.* 80:1449–58

Gunn, R. 1943. A quantitative evaluation of the influence of the lithosphere on the anomalies of gravity. *J. Franklin Inst.* 236:373–96

Haxby, W. F. 1981. Geoid anomalies across mid-ocean ridges: A new constraint on lithospheric thermal models. *Earth Planet. Sci. Lett.*

Haxby, W. F., Turcotte, D. L. 1978. On isostatic geoid anomalies. *J. Geophys. Res.* 83:5473–78

Henry, M., Karig, D. E., Shor, G. G. 1975. Two seismic refraction profiles in the West Philippine Sea. *Deep Sea Drilling Project* 31:611–14

Jeanloz, R., Richter, F. M. 1979. Convection composition and the thermal state of the lower mantle. *J. Geophys. Res.* 84:5497–5504

Jeffreys, H. 1959. *The Earth.* Cambridge Univ. Press. 420 pp.

Johnson, R. H., Mahahoff, A. 1971. Relation of Macdonald volcano to migration of volcanism along the Austral Chain. *J. Geophys. Res.* 76:3282–90

Kahle, H,.-G., Talwani, M. 1973. Gravimetric Indian Ocean geoid. *Z. Geophys.* 39:167–87

Kaula, W. M. 1969. Earth's gravity field: Relation to global tectonics. *Science* 169:982–85

Khan, M. A. 1977a. Satellite techniques in geophysics and their relationship to marine geodesy. *Proc. Int. Symp. Appli. Mar. Geodesy, Columbus, Ohio*, pp. 87–112

Khan, M. A. 1977b. Depth of sources of gravity anomalies. *Geophys. J. R. Astron. Soc.* 48:197–209

Lambeck, K. 1972. Gravity anomalies over ocean ridges. *Geophys. J. R. Astron. Soc.* 30:37–53

Lee, W. H. 1966. Analysis of the Earth's topography, Orbital pertubations from terrestrial gravity data. *Final Rep.*, contract A. F. (601)-4171 Sect. 2a. 213 pp. Inst. Geophys. Planet. Phys., Univ. Calif., Los Angeles

Leitao, C. D., McGoogan, J. T. 1975. SKYLAB radar altimeter; short-wavelength pertubations detected in ocean surface profiles. *Science* 186:1208–9

Lerch, F. J., Wagner, C. A., Richardson, J. A., Browand, J. E. 1974. Goddard Earth models (5 and 6). *Rep. X-921-74-145.* Goddard Space Flight Cent., Greenbelt, Md. 100 pp.

Lerch, F. J., Klosko, S. M., Laubscher, R. E., Wagner, C. A. 1979. Gravity model improvement using GEOS-3 (GEM 9 and GEM 10). *J. Geophys. Res.* 84:3897–3916

Louden, K. E., Forsyth, D. W. 1976. Thermal conduction across fracture zones and the gravitational edge effect. *J. Geophys. Res.* 81:4869–74

Macdonald, G. J. F. 1963. Deep structure of continents. *Rev. Geophys.* 1:587

Mammerickx, J., Herron, E., Dorman, L. 1980. Evidence for two fossil spreading ridges in the southeast Pacific. *Geol. Soc. Am. Bull.* 91:263–71

Marsh, B. D., Marsh, J. G. 1976. On global gravity anomalies and two-scale mantle convection. *J. Geophys. Res.* 81:5267–80

Marsh, J. G., Marsh, B. D., Williamson, R. G., Wells, W. T. 1981. The gravity field in the central Pacific from satellite to satellite tracking. *J. Geophys. Res.* In press

McKenzie, D. P. 1967. Some remarks on heat flow and gravity anomalies. *J. Geophys. Res.* 72:6261–73

McKenzie, D. P. 1977. Surface deformation, gravity anomalies and convection. *Geophys. J. R. Astron. Soc.* 48:211–38

McKenzie, D. P., Bowin, C. 1976. The relationship between bathymetry and gravity in the Atlantic ocean. *J. Geophys. Res.* 81:1903–15

McKenzie, D. P., Weiss, N. O. 1975. Speculations on the thermal and tectonic history of the earth. *Geophys. J. R. Astron. Soc.* 42:131–74

McKenzie, D. P., Roberts, J., Weiss, N. 1974 Numerical models of convection in the Earth's mantle. *Tectonophysics* 19:89–103

McKenzie, D. P., Watts, A. B., Parsons, B., Roufousse, M. 1980. The plan form of mantle convection beneath the Pacific ocean. *Nature.* 288:442–416

McNutt, M. 1980. Implications of regional gravity for state of stress in the Earth's crust and upper mantle. *J. Geophys. Res.* 85:6377–96

Menard, H. W. 1973. Depth anomlies and the bobbing motion of drifting islands. *J. Geophys. Res.* 78:5128–37

Menard, H. W., Dorman, L. M. 1977. Dependence of depth anomalies upon latitude and plate motion. *J. Geophys. Res.* 82:5329

Morgan, W. J. 1972. Deep mantle convection plumes and plate motions. *Am. Assoc. Petrol. Geol. Bull.* 56:203–13

Munk, W. H., Macdonald, G. J. F. 1960. *The Rotation of the Earth.* London: Univ. Press Cambridge

O'Connell, R. J. 1977. On the scale of mantle convection *Tectonophysics* 38:119–36

Parsons, B., Sclater, J. G. 1977. An analysis of the variation of ocean floor bathymetry and heat flow with age. *J. Geophys. Res.* 82:803–27

Peterson, J. J., Fox, P. J., Schreiber, E. 1974. Newfoundland ophiolites and the geology of the oceanic layer. *Nature* 247:194–96

Pratt, J. H. 1858. On the attraction of the Himalaya mountains, and of the elevated regions beyond them, upon the plumb-line in India. *Philos. Trans. R. Soc. London* 145:53–100

Prey, A. 1922. Darstellung der Höhen- und Tiefen- cerhältnisse der Erde durch eine Entwicklung nach Kugelfunktionen bis sur 16. *Ordnung. Nachr. Akad. Wiss. Gottingen, Math.-Physik. K1.* 11 (1):1–29

Rapp, R. H. 1979. GEOS-3 data processing for the recovery of geoid undulations and gravity anomalies. *J. Geophys. Res.* 84: 3784–92

Richter, F. M., 1973. Dynamical models for sea floor spreading. *Rev. Geophys. Space Phys.* 11:223–87

Richter, F. M., Daly, S. F. 1978. Convection models having a multiplicity of large horizontal scales. *J. Geophys. Res.* 83: 4951–56

Richter, F. M., McKenzie, D. 1978. Simple plate models of mantle convection. *J. Geophys.* 44:441–71

Richter, F. M., Parsons, B. 1975. On the interaction of two scales of convection in the mantle. *J. Geophys. Res.* 80:2529–41

Roufousse, M. C. 1978. Interpretation of altimeter data. *Proc. 9th GEOP Conf. An Int. Symp. on the applications of Geodesy to geodynamics, Dept. Geodetic Sci. Rep. 280,* pp. 261–66 Ohio State Univ.

Runcorn, S. K. 1963. Satellite gravity measurements and convection in the mantle. *Nature* 200:628–30

Runcorn, S. K. 1965. Changes in the convection pattern in the Earth's mantle and continental drift: Evidence for a cold origin of the Earth. *Philos. Trans. R. Soc. London.* 258:228

Sclater, J. G., Francheteau, J. 1970. The implications of terrestrial heat flow observations on current tectonic and geochemical models of the crust and upper mantle of the Earth. *Geophys. J. R. Astron. Soc.* 20:509–42

Sclater, J. G., Lawver, L. A., Parsons, B., 1975. Comparison of long wavelength residual elevation and free-air gravity anomalies in the North Atlantic and possible implications for the thickness of the lithospheric plate. *J. Geophys. Res.* 80:1031–52

Shor, G. G., Pollard, D. D. 1964. Mohole

site selection studies North of Maui. *J. Geophys. Res.* 69:1627

Sibuet, J. C., Le Pichon, X., Goslin, J. 1974. Thickness of the lithosphere deduced from gravity edge effects across the Mendocino Fault. *Nature* 252:676–79

Sibuet, J. C., Veyrat-Peinet, B. 1980. Gravimetric model of the Atlantic Equatorial Fracture Zones. *J. Geophys. Res.* 85: 943–54

Stanley, H.-R. 1979. The GEOS 3 project. *J. Geophys. Res.* 84:3861–71

Talwani, M., Poppe, H. R., Rabinowitz, P. D. 1972. Gravimetrically determined geoid in the western North Atlantic, Sea Surface Topography from Space, 2. *NOAA Tech. Rep. ERL-228-AOML 7-2,* pp. 1–34

Vening Meinesz, F. A. 1941. Gravity over the Hawaiian Archipelago and over the Madiera Area; Conclusions about the Earth's crust. *Proc. Kon. Ned. Akad. Wetensia.* 44 pp.

Vening Meinesz, F. A. 1964. *The Earth's and Mantle.* Amsterdam: Elsevier. 124 pp.

Wagner, C. A. 1979. The geoid spectrum from altimetry. *J. Geophys. Res.* 84: 3861–71

Wagner, C. A., Lerch, F. J., Brownd, J. E., Richardson, J. A. 1976. Improvement in the geopotential derived from satellite and surface data (GEM 7 and 8). *Rep. X-921-76-20.* Goddard Space Flight Center, Greenbelt, Md. 11 pp.

Walcott, R. I. 1970. Flexural rigidity, thickness and viscosity of the lithosphere. *J. Geophys. Res.* 75:3941–54

Walcott, R. I. 1976. Lithospheric flexure, analysis of gravity anomalies, and the propagation of seamount chains. *Int. Woolard Symp., AGU Monogr. 19,* pp. 431–38

Watts, A. B. 1976. Gravity and bathymetry in the Central Pacific Ocean. *J. Geophys. Res.* 81:1533–53

Watts, A. B. 1978a. An analysis of isostasy in the world's oceans, Part 1, Hawaiian-Emperor Seamount chain. *J. Geophys. Res* 83:5989–6004

Watts, A. B. 1978b. Comments on 'On global gravity anomalies and two-scale mantle convection' by Bruce D. Marsh and James G. Marsh. *J, Geophys. Res.* 83:3551–54

Watts, A. B. 1979. On geoid heights derived from GEOS 3 altimeter data along the Hawaiian-Emperor seamount chain. *J. Geophys. Res.* 84:3817–26

Watts, A. B., Cochran, J. R. 1974. Gravity anomalies and flexure of the lithosphere along the Hawaiian-Emperor Seamount Chain. *Geophys. J. R. Astron. Soc.* 38: 119–41

Watts, A. B., Leeds, A. R. 1977. Gravimetric geoid in the northwest Pacific ocean. *Geophys J. R. Astron Soc.* 50:249–278

Watts, A. B., Talwani, M. 1975a. Gravity effect of downgoing lithospheric slabs beneath island arcs. *Geol. Soc. Am. Bull.* 86:1–4

Watts, A. B., Talwani, M. 1975b. Gravity field of the northwest Pacific ocean basin and its margin: Hawaii and vicinity. *Geol. Soc. Am. Special Map and Chart Ser. MC-9.* 6 pp.

Watts, A. B., Bodine, J. H., Ribe, N. M. 1980. Observations of flexure and the geological evolution of the Pacific ocean basin. *Nature* 283:532–37

Weissel, J. K., Hayes, D. E. 1974. The Australian-Antarctic discordance: New results and implications. *J. Geophys. Res.* 79:2579–87

Woollard, G. P. 1972. Regional variations in gravity. in *The Nature of the Solid Earth*, ed. E. C. Robertson, pp. 463–505. New York: McGraw-Hill

Ann. Rev. Earth Planet. Sci. 1981. 9:449–86

THE BIOGEOCHEMISTRY OF THE AIR-SEA INTERFACE

✖ 10157

Leonard W. Lion and James O. Leckie

Environmental Engineering and Science, Department of Civil Engineering,
Stanford University, Stanford, California 94305

INTRODUCTION

In recent years scientific investigators have evidenced a steadily growing interest in the interface between the atmosphere and the oceans. This region, termed the air-sea interface or surface microlayer, has been shown to have unique properties that may influence the geochemical cycles of many environmental pollutants. It is intuitively obvious that the air-sea interface must be traversed in the process of ocean-atmosphere exchanges. If, as the preponderance of evidence indicates, the interface has physical, chemical, and biological characteristics different from those of bulk seawater, it follows that those characteristics may influence the form and fate of materials that are introduced into the microlayer.

In this review the physical, chemical, and biological characteristics of the air-sea interface are described with particular emphasis on those features that may impact upon the form and/or fate of pollutants in the sea. There have been three major reviews that treat the air-sea interface (MacIntyre 1974a, Liss 1975, Duce & Hoffman 1976). Although these reviews consider biological and chemical interactions of toxic trace metals and organic pollutants at the air-sea interface, they are principally concerned with the process of chemical fractionation, particularly as it affects the exchange of materials between the oceans and the atmosphere. Emphasis in this review is placed on the composition of the surface microlayer and on the chemical and biological interactions that may ensue.

The air-sea interface is an extremely complex matrix. The discussion that follows is intended to provide the reader with an indication of the complexity of the interfacial solution and a sense of the possible interactions between trace metals and the other components of the interfacial

449

matrix. The potential for such interactions is inherent in the occurrence of high interfacial concentrations of metals, organics, particulates, and microbiota.

TRACE METALS AT THE AIR-SEA INTERFACE

The surface microlayer of the ocean is a poorly defined region on both an operational and a conceptual basis. Operationally, the depth of the surface microlayer is often defined by the type of surface sampler employed. Consequently, the reported concentrations of trace metals at the air-sea interface reflect not only temporal and geographic variability but also— unfortunately—differences in the types of sampling devices used. The art of interfacial sampling is still developing and several devices have been applied, each of which varies in its efficiency or the effective depth sampled. Table 1 lists some of the interfacial samplers now employed and their reported sampling depths. The depths sampled have varied over more than four orders of magnitude, from 30 nm to 1 mm.

Table 1   Air-Sea interface sampling devices

Type	Depth sampled	Investigator
Screen	$\approx 150 \ \mu m$	Garrett 1965, 1972
Glass plate	$60-100 \ \mu m$	Harvey & Burzell 1972
Rotating drum	$\approx 60 \ \mu m$	Harvey 1966
V-shaped tube	qualitative for foams and collapsed films	Szekielda et al 1972
Teflon disk	?	Miget et al 1974
Teflon plate	$50-100 \ \mu m$	Larsson et al 1974
Bubble microtome	$0.5-10 \ \mu m$	MacIntyre 1968
Funnel	?	Morris 1974
Tray	1.0 mm	Parker & Wodehouse 1971
Prism	$\approx 30 \ nm$	Baier 1972

Conceptually, the "true" surface microlayer is often described as a monomolecular layer (Garrett 1967a, Duce et al 1972) which would extend to a few hundred angstroms in depth. However, calculations based on gas exchange and some thermal measurements indicate interfacial thicknesses of 30 to 100 μm (Liss 1975). The actual microlayer thickness may in fact vary with respect to the parameter of interest. That is, microlayer thicknesses of 100 μm for thermal gradients and 10 nm for chemical gradients are not necessarily mutually exclusive phenomena. These uncertainties in understanding microlayer thickness make equally uncertain the understanding of metal and ligand concentrations in the surface microlayer. Consequently, questions concerned with metal speciation, biological toxicity, and transport become very difficult to answer.

The relative sampling depths of some of the surface samplers in Table 1 have been compared by Roy et al (1970), Hatcher & Parker (1974), and Daumas et al (1976). The common feature of all interfacial samplers used, with the possible exception of those of MacIntyre (1968), Szekielda et al (1972), and Baier (1972), is that to varying degrees they dilute a sample from a monomolecular surface layer with bulk marine water. Thus, sample concentrations may need to be multiplied by a factor of up to 10^4 in order to reflect concentrations in a monomolecular microlayer.

Table 2 summarizes some of the reported trace metal measurements made in the surface microlayer of both marine and fresh waters. For ease of comparison, the tabulated data show total trace metal concentrations (the sum of particulate and dissolved organic and inorganic concentrations where reported separately) averaged over the indicated number of independent samples. Concentrations of metals measured at the interface are seen to vary over a large range and, on the average, virtually all metals are found to be partitioned at the surface microlayer. It is worth noting the differences in microlayer sampling techniques when examining the data in Table 2. Samplers that select for foams or collapsed surface films, i.e. those of Szekielda et al (1972) and Pellenbarg (1976), provide sample metal concentrations up to 10^4 higher than those obtained from other surface-sampling devices, a factor roughly equivalent to the previously discussed dilution correction factor.

In some instances the authors cited here do not provide sufficient information to permit evaluation of the adequacy of their analytical procedures. The measurement of trace metals in the environment is, in fact, far from straightforward. Obtaining accurate, contaminant-free analyses of trace metals in environmental samples is now recognized as a major stumbling block by field investigators (Batley & Gardner 1977, Bruland et al 1978). In cases where discussion of analytical techniques is insufficient, we have assumed that the reported trace metal concentrations at the air-sea interface and their partitioning from bulk solution accurately reflect actual environmental concentrations. Although the results vary, some common features link many of the investigations: (a) the occurrence of measurable partitioning (surface excess) of trace metals is not a consistent phenomenon; the frequency of detectable partitioning increases with the presence of observable surface organic slicks (Piotrowicz et al 1972, Elzerman 1976); (b) the relative amount of organically associated (extractable) trace metals is higher at the surface microlayer than in bulk waters (Duce et al 1972, Barker & Zeitlin 1972); (c) particulate trace

Table 2 Summary of reported microlayer trace metal concentrations

Location	Type of sampler	Metal species	No. of samples	Microlayer concentration[a] (range and/or mean), ug/l	Mean partitioning[b]	Remarks	References
Narragansett Bay, Rhode Island	Screen	Pb	2	4.1, 9	1.7	Reported data differentiate between particulate (>0.45 μm) dissolved organic ($CHCl_3$ extractable), and dissolved inorganic metal species.	Duce et al 1972
		Cu	2	4.4, 16.2	3.5		
		Fe	2	57.1, 826.5	15.0		
		Ni	2	26.9, 39	1.9		
Narragansett Bay, Rhode Island	Screen	Pb	8	2.6–14 ($\bar{x} = 6.7$)	1.9	Reported data differentiate between particulate (>0.45 μm), dissolved organic ($CHCl_3$ extractable), and dissolved inorganic metal species.	Piotrowicz et al 1972
		Cu	8	4.1–16 ($\bar{x} = 6.6$)	2.3		
		Fe	8	3.7–830 ($\bar{x} = 149$)	5.1		
		Ni	8	12–40 ($\bar{x} = 25.5$)	1.3		
Delaware Bay (convergence zone)	V-shaped tube	Pb	nr[c]	$10^5 - 5 \times 10^5$	$\simeq 10^7$	Microlayer values are for the analysis of the $CHCl_3$ extraction residue of a film sample. Enrichments are calculated from seawater values given by Goldberg (1965).	Szekielda et al 1972
		Cu		$10^5 - 5 \times 10^5$	$\simeq 10^5$		
		Fe		$2 \times 10^6 - 10^7$	$\simeq 5 \times 10^6$		
		Ni		$10^5 - 5 \times 10^5$	$\simeq 10^5$		
		Zn		$2 \times 10^5 - 10^6$	$\simeq 5 \times 10^4$		
		Hg		300	$\simeq 10^3$		
		Cr		$10^5 - 5 \times 10^5$	$\simeq 5 \times 10^6$		
		Ag		$10^4 - 5 \times 10^4$	$\simeq 10^5$		
		Co		$5 \times 10^4 - 2.5 \times 10^5$	$\simeq 10^5$		
		Mn		$1 \times 10^5 - 5 \times 10^5$	$\simeq 10^5$		

Location	Sampler	Metal	n	Surface microlayer concentration	Bulk water metal concentration	Comments	Reference
Swedish West Coast	Teflon disk	Zn Ni	nr	$\simeq 100$ ppm $\simeq 100$ ppm	not reported	Metal concentration is ppm of $CHCl_3$-MeOH (2:1 v/v) extractable lipid weight.	Larsson et al 1974
Canary Creek Salt Marsh, Delaware	V-shaped tube	Cu Fe Zn		$1.48 \times 10^4 \pm 3.4 \times 10^3$ $1.39 \times 10^7 \pm 1.8 \times 10^4$ $1.03 \times 10^5 \pm 6.1 \times 10^4$	10.1 15.5 17.5	Averages for flood tide over three seasons (Pilottown Station). Data also available for other stations and for ebb tides.	Pellenbarg 1976, Pellenbarg & Church 1979
Lake Michigan	Screen	Cu Pb Zn Cd	51 62 62 62	$0.74-85$ ($\bar{x}=4.85$) $0.9-190$ ($\bar{x}=11.0$) $1.0-88$ ($\bar{x}=16.8$) $0.05-4.7$ ($\bar{x}=0.44$)	2.6 4.8 2.8 3.0	Data also available for dissolved and particulate ($>0.4\ \mu m$) metals and for lakes Ontario and Mendota	Elzerman 1976, Elzerman & Armstrong 1979
Openocean	Glass plate	Cu Fe Zn	nr	11.5 ± 3.7 56.8 ± 8.0 39.7 ± 13.7	3.1 4.2 2.8	Reported data differentiate between total and $CHCl_3$ extractable trace metals.	Barker & Zeitlin 1972
Lake Mendota	Foam	Cu Fe Pb Zn Cd	30 30 30 30 30	$140-1410$ ($\bar{x}=900$) $7,500-12,900$ ($\bar{x}=11,300$) $20-5,910$ ($\bar{x}=2,210$) $20-2,260$ ($\bar{x}=880$) $0.1-340$ ($\bar{x}=110$)	448 240 1,110 293 544	Microlayer analysis performed on acid-digested foam samples.	Eisenreich et al 1978
Northwest Atlantic	Screen	Hg	6	0.005 to 0.01	no surface enrichment observed	Analysis for reactive and total Hg; range of values shown does not include analysis from coastal stations.	Fitzgerald 1976, Fitzgerald & Hunt 1974

Table 2 (*continued*)

Location	Type of sampler	Metal species	No. of samples	Microlayer concentration[a] (range and/or mean), ug/l	Mean partitioning[b]	Remarks	References
Open ocean, California current	Screen	Cd	13	0.015–0.060 ($\bar{x}$ = 0.030)	3.7	Partitioning calculated vs. 0.3 m bulkwater samples. Average values shown do not include analysis from coastal stations.	Martin et al 1976
Lake foam	Screen	Fe	1	1,920	190		Pojasek & Zajicek 1978
		Mn	1	190	3.8		
South San Francisco estuary—salt marsh border	Screen	Pb	18	1.1–48 ($\bar{x}$ = 13.1)	11.6	Data averaged over two separate tidal cycles. Total metal represents sum of dissolved and particulate (>0.4 μm) forms.	Lion 1980
		Cu	18	5.6–181 ($\bar{x}$ = 37.8)	4.8		
		Cd	18	0.36–2.31 ($\bar{x}$ = 0.80)	1.7		
North Sea coastal waters		Pb	15	1.5–10.7 ($\bar{x}$ = 3.8)	1.7	Data shown are for particulate trace metal concentrations.	Hunter 1980
		Cu	15	0.38–5.3 ($\bar{x}$ = 1.7)	2.3		
		Fe	15	156–854 ($\bar{x}$ = 415)	0.84		
		Ni	15	<0.35–36.0	—		
		Zn	15	2.0–13.2 ($\bar{x}$ = 5.3)	1.5		
		Cd	15	<0.06–0.92	—		
		Mn	15	3.0–20.3 ($\bar{x}$ = 8.1)	0.74		

[a] Uncorrected for dilution effects.

[b] Partitioning = $\dfrac{\text{microlayer concentration}}{\text{bulkwater concentration}}$. [Note: In some cases the mean partitioning indicated represents an average of observed values from different stations or over a time series of measurements which were reported separately by the cited authors.]

[c] nr = not reported.

metals (especially lead) occur at higher levels in the surface microlayer (Piotrowicz et al 1972, Elzerman et al 1979).

These general conclusions suggest several possible processes for trace metal transport to the air-sea interface. Processes to be considered would be those that would preferentially transport organically bound metals or particulate metals. Such processes include rising bubbles from bulk solution (Lemlich 1972a) or airborne deposition of particulates. An alternative (not necessarily exclusive of the other processes) is that metals delivered to the microlayer react there to form organic complexes or are adsorbed at the surface microlayer onto particulates. In the sections that follow, the chemical and biological nature of the surface microlayer and the possibilities for transport processes and reactions in the microlayer are considered in further detail.

ORGANICS AT THE AIR-SEA INTERFACE

In this section the enrichment of organic compounds at the air-sea interface is reviewed. Organic enrichment is seen to be predictable based on simple thermodynamic considerations. Classes of organics that may be partitioned are described qualitatively and compared to available field data. The possible effects of organic surface films are reviewed and shown to include damping of capillary waves, perturbations in gas transfer efficiencies, stabilization of bubbles, and generation of particulate organic aggregates. The formation of organic aggregates may influence the form and fate of trace metals. Metal complexation by surface-active organics is also discussed.

Thermodynamics of Surface Adsorption and Classes of Surface-Active Organics

There is a large body of literature on the physical-chemical nature of the air-solution interface for aqueous systems. The information available from that literature is of significant use in establishing a framework for discussion of the behavior of organic solutes at the surface microlayer and to give a basis for comparison.

The Gibbs adsorption isotherm, which describes the basic thermodynamic relationships for interfacial phenomena, can be represented as

$$-d\gamma = \sum_{i=1}^{n} \Gamma_i d\mu_i, \tag{1}$$

where γ is the surface tension, Γ_i is the surface excess (as moles/cm^2), and μ_i is the chemical potential of species i.

456   LION & LECKIE

Also,

$$\mu_i = \mu_i^0 + RT \ln a_i,$$

where a_i is the activity of species i, R is the universal gas constant, and T is the absolute temperature.

Equation (1) predicts that positive adsorption or creation of a surface excess will result in a decrease in the surface tension of a given bulk liquid. At constant temperature and pressure and considering the surface as occupying zero volume, the free energy, G, of a system and surface tension are related by

$$dG = \gamma \, dA + \sum_{i=1}^{n} \mu_i \, dn_i, \tag{2}$$

where n_i is the moles of species i at the surface, and dA is the differential change in surface area.

A solute species, i, which decreases surface tension, will positively adsorb at the air-solution interface [Equation (1)] and this reaction will occur spontaneously [Equation (2)]. Most organic solutes decrease the surface tension of water (National Research Council 1928) and would therefore be expected to positively adsorb at the air-solution interface. Conversely, most inorganic electrolytes raise the surface tension of water (Osipow 1977) and therefore will tend to be negatively adsorbed (excluded) at the surface.

These simple relationships lead to some rather important conclusions. The principal one of interest for this discussion is that all naturally occurring air-water interfaces will exist with chemical constituents that differ either in chemical nature or in concentration from those of the bulk liquid. Such predictions have been dramatically confirmed by results of field sampling and analysis performed by Garrett (1967a, 1970) and Baier (1972). Surface-active organic materials are consistently found at all natural air-water and air-sea interfaces sampled regardless of the absence of observable slicks.

The formation of aquatic surface films preferentially occurs with certain types of molecules, which have been categorized by Gaines (1966) as (a) nonpolymeric substances, which are essentially insoluble but whose molecules have sufficient attraction for the aqueous phase to permit dispersion, and (b) polymeric materials, including certain proteins. The long-chain fatty acids and alcohols, which have a nonpolar hydrophobic hydrocarbon tail and a polar hydrophilic functional "head" group, such as $-COOH$ or $-OH$, are a typical example of molecules in category (a).

The metabolic by-products of phytoplankton and algae are additional potential sources of surface-active materials in the marine environment.

Wilson & Collier (1972) have shown that marine diatoms produce organics capable of forming stable foams. Hoyt (1970) reports phytoplankton and algal exudates to be high molecular weight ($>50,000$) polysaccharides which can form surface slicks.

Within the first surface-active organic category (fatty acids, alcohols, and their esters) it is generally a safe simplification to consider that polar groups confer water solubility and hydrophobic groups tend to prevent it. In the absence of a polar group, long-chain hydrocarbon molecules such as decane ($C_{10}H_{22}$) will not form monolayers but merely float as drops or lenses on a water surface. As the length of a hydrocarbon chain increases, the solubility of molecules decreases. Solubility of a given chain length is controlled by its functional groups. The "effectiveness" of functional groups in providing attraction to water has been listed qualitatively by Gaines (1966). Alkylhalides and hydrocarbon functional groups do not form films. Typically, organic acids, alcohols, and amino groups all produce stable films with C_{16} hydrocarbon chains.

For some simple solutions with one adsorbate, the Langmuir isotherm [Equation (3)] may be used to predict adsorption behavior in a quantitative manner:

$$\Gamma_i = \frac{\Gamma_{max}Ka_i}{a_w + Ka_i},\tag{3}$$

where K is the equilibrium constant for the adsorption reaction $a_i(aq)$ + surface site $(S) = a_iS$, Γ_{max} is the maximum surface adsorption capacity, a_i is the activity of adsorbate, and a_w is the activity of water.

The free energy and the equilibrium constant for the adsorption reaction are related by

$$\Delta G^\circ = -RT \ln K.\tag{4}$$

(It should be noted that the symbol ΔG° may be misleading since an established convention for standard states for adsorption reactions has not been agreed upon.)

Szyszkowski (1908) described the surface adsorption reactions of unionized aliphatic acids using a variable that reflected the free energy contribution from Van der Waals interactions of the hydrophobic tails (which increase in proportion to chain length). Davies & Rideal (1963) show that these data may be expressed in the form of a Langmuir isotherm as

$$\Gamma = \frac{K(C/K')}{1 + C/K'} = \text{moles/cm}^2,\tag{5}$$

where C is the solute concentration (moles/liter), K' equals $1.9 \exp(-710\ m/RT)$, m is the number of CH_2 groups in the organic acid, and K is 5.28×10^{-10}.

From this approximation, qualitative estimations can be made of the relative surface excess for acids of differing chain lengths. However, such approximations must remain crude because in aquatic environments there exist a large number of surface-active materials that compete for the relatively few surface sites. Competitive adsorption phenomena and other complexities of the marine environment, such as ionic strength effects and turbulence (which minimizes the probability of adsorbates existing as simple monolayers), make realistic quantitative predictions of actual adsorption densities unlikely. Results by Garrett (1967b) indicate that competitive adsorption phenomena do indeed occur in the marine environment. In the presence of a compressed monolayer (sea slick), high molecular weight fatty acids and alcohols are thought to force lower molecular weight (more soluble) constituents out of the surface environment.

Field Observations of Surface-Active Organics

The results of analysis of field samples for surface organic constituents are summarized in Table 3. It is not surprising that the organics observed at the microlayer fall into the two surface-active categories described above, i.e. (a) fatty acids, alcohols, and lipids, and (b) proteinaceous materials. Also represented at the air-sea interface are carbohydrates and insoluble hydrocarbons from pollutant and natural sources and chlorinated hydrocarbon pollutants.

Measured organic concentrations at the air-sea interface are generally elevated above those of the bulk liquid in the sea. The reported enrichments for total organic carbon (TOC), dissolved organic carbon (DOC), and particulate organic carbon (POC) are summarized in Table 4. In general, the magnitude of POC partitioning is found to be greater than that for DOC at the surface microlayer. The interrelationship between DOC and POC at the air-sea interface is discussed later in this section.

Duce et al (1972) reported fatty acids, hydrocarbons, and chlorinated hydrocarbons (PCBs) to be enriched from 1.5 to 28 times in the surface layer (100 to 150 μm) relative to the bulk liquid 20 cm deep. Bidleman & Olney (1974) indicated PCB and DDT enrichment in the surface microlayer. Enrichment of pesticides in the surface microlayer has also been reported by Seba & Corcoran (1969) who measured total microlayer residue (DDT, DDE, dieldrin, and aldrin) concentrations ranging from 0.093 to 12.75 ppb vs 1 ppt in bulk waters. W. MacIntyre et al (1974) found higher DDT in microlayer samples from Chesapeake Bay than in bulkwater samples but reported uniform distribution of PCBs. It is interesting to note that the atmospheric origin of PCBs and DDT in the marine environment corresponds to their relatively high levels in surface films. Thus, surface films may play a role in the marine accumulation

Table 3 Characterization of organic molecules at the air-sea interface

Nature of organics determined	Reference(s)
Glycoproteins and proteoglycans	Baier et al 1974, Baier 1970, 1972, 1975, 1976
Triglicerides, free fatty acids, wax esters, cholesterol, DDT, DDD, DDE, PCBs, and phthalic acid esters	Larsson et al 1974
Fatty esters, free fatty acids, fatty alcohols, and hydrocarbons	Garrett 1967a, 1970, Jarvis et al 1967
Free and esterified fatty acids, pollutant and natural product hydrocarbons	Morris 1974, Morris & Culkin 1974
Free and esterified fatty acids	Marty & Saliot 1974
Fatty acids, hydrocarbons, and chlorinated hydrocarbons (PCBs)	Duce et al 1972
Normal, branched, and cyclo-alkanes	Ledet & Laseter 1974a,b
Normal alkanes	Marty & Saliot 1976
Normal alkanes	Hardy et al 1977a,b
Hydrocarbons (alkanes and aromatics)	Wade & Quinn 1975
PCBs, chlorinated hydrocarbon pesticides, free fatty acids and esters, and normal paraffinic hydrocarbons	W. MacIntyre et al 1974
PCBs and DDT	Bidleman & Olney 1974, Bidleman et al 1976
Dieldrin, aldrin, DDT, DDE, acetone, butyraldehyde, and 2-butanone	Seba &Corcoran 1969
Carbohydrates (mono- and polysaccharides)	Sieburth & Conover 1965 Sieburth et al 1976

and transport of these organic pollutants in addition to their apparently important role with respect to trace metals. Concentrations of pesticides in surface layers might have greater effects on insects, fishes, and birds that inhabit or associate with this zone than on organisms in subsurface waters (Parker & Barsom 1970).

The identity of specific dominant microlayer organic constituents is the subject of some controversy. Garrett (1967a, 1970) holds that an apparent competition exists for surface adsorption sites, with the high molecular weight and less soluble fatty acids and alcohols being the most surface-active and therefore predominant. These organics are typical of those previously described in category (a). Garrett (1971) has qualitatively identified proteins and carbohydrates in surface water but concludes that their contribution to surface effects is small because of their greater water solubility.

Table 4 Enrichment of total, dissolved, and particulate organic carbon at the air-sea interface.

Location	Type of sampling device	Form of carbon	Average microlayer concentration (range)	Mean Partitioning	Reference
Delaware Bay (convergence zone)	V-shaped tube	TOC	6.2%	n.r.[a]	Szekielda et al 1972
Open ocean	Glass plate	TOC	18.4 ± 16.5 mg/l	10.2	Barker & Zeitlin 1972
North Atlantic Ocean	Screen	DOC	1.73 mg/l (1.55–2.13)	1.6	Sieburth et al 1976
Bay of Brusc	Rotating drum	POC	0.94 mg/l	2.7	Daumas et al 1976
	Screen	POC	0.56 mg/l	1.6	
Pacific Ocean	Screen	DOC	2.7 mg/l (1.9–4.4)	2.63	Williams 1967
	Screen	POC	0.98 mg/l (0.17–2.5)	7.0	
Open ocean	Screen	POC	0.42	10.5	Nishizawa 1971
	Screen	DOC	1.42	1.71	
Strait of Georgia	Glass plate	DOC	6.24 (3.11–11.89)	4.4	Dietz et al 1976
		POC	2.79 (0.76–5.83)	10.24	

[a] nr = not reported.

Baier's (1970, 1972) experiments indicate a different nature for aquatic surface films. By dipping germanium prisms in water bodies, surface materials were non-destructively analyzed by multiple attenuated internal reflection spectroscopy (MAIR). After MAIR analysis of thousands of samples Baier concluded that the air-water interface in fresh water and the marine environment is dominated by glycoproteins and proteoglycans; both are typical of compounds previously described in category (b). These results differ strikingly from the conclusions of Garrett; dominance of liquid-like material was not confirmed by Baier's results. Baier has also characterized sea foams as being dominated by glycoproteins and proteoglycans and as stabilized by silica-dominated particulate debris, including diatom fragments (Baier et al 1974).

Liss (1975) and Hunter & Liss (1977) have argued that the extraction and analysis technique employed by Garrett (1965) selects for only a small portion of the total DOC present. The qualitative organic analysis performed by Baier may more accurately determine the dominant organic compounds at the air-sea interface. However, Hunter & Liss (1977) note that the IR spectra obtained by Baier are consistent with those for carbohydrates and that Sieburth et al (1976) found mono- and polysaccharides to constitute an average of 28 percent of the DOC at the surface microlayer.

Morris (1974) has indicated that greater than 85 percent of extractable microlayer organics in the Mediterranean were pollutant hydrocarbons; however, he also measured free and esterified fatty acids [major constituents were C 16:0 and C 18:1 similar to those described by Garrett (1970)]. Wade & Quinn (1975) measured average hydrocarbon concentrations of 155 μg/l in the Sargasso Sea microlayer vs 73 μg/l in bulk water. Detailed characterization of microlayer alkanes in the Gulf of Mexico has been performed by Ledet & Laseter (1974b). They indicate approximately 70 percent by weight of sampled alkanes were branched, 50 percent 3-methyl branched, 13 percent cycloalkanes, and 3 percent n-alkanes. W. MacIntyre et al (1974) measured 300 to 500 μg/l (total) of C 10 to C 24 n-paraffinic hydrocarbons in Chesapeake Bay. Also measured were 700 to 7800 μg/l (total) free and combined fatty acids with C 14:0, 16:0, 16:1, 18:0, 18:1, and 22:0 fatty acids predominant.

Organic Complexation of Trace Metals

Regardless of the nature of the specific organic constituents of the surface microlayer, it seems safe to conclude that the interface represents a chemically unique and relatively concentrated solution. The interfacial enrichment of organic matter offers an increased potential for the formation of metal-organic complexes at the air-sea interface.

Trace metal-organic compounds have been classified by Siegel (1971) as:

1. Those species existing in equilibrium with their free components whose reactions can be described by a reversible reaction and an appropriate equilibrium constant, and

2. Those compounds whose relatively constant concentration in marine waters is the result of a steady-state condition as opposed to an equilibrium process, and are thermodynamically unstable relative to their concentrations in the ocean. Such compounds are termed "non-labile," "inert," or "robust," i.e. exchange of added metal ion with incorporated metal ion will occur slowly over a long period of time (Dwyer & Mellor 1964, Siegel 1971).

The availability of organics to react with trace metals in seawater has been the subject of some debate. Magnesium and calcium, in high concentration in the oceans, may associate with organic chelators making them unavailable to trace metals (Siegel 1971). However, Duursma (1970) suggests that trace metals that bind strongly with organics may compete for available organic ligands and calculates that at low chelate (EDTA) concentrations Fe^{+3} and Cr^{+3} will be preferentially complexed. Although organics are known to be concentrated in surface films with respect to underlying waters, it is not known whether organic chelators are present in excess amounts or are limiting.

The relative order of organic binding strength for some divalent cations with bidentate ligands has been compared by Baes (1973) using cadmium as a baseline. The relative bond strength that can be inferred for bidentate bonds with oxygen and/or nitrogen is approximately $Mn^{+2} < Fe^{+2} < Cd^{+2} < Co^{+2} < Zn^{+2} < Ni^{+2} < Cu^{+2} < Hg^{+2}$, which corresponds closely to the "Irving Williams" (Williams 1953) order of complex stability for metal-activated enzymes: $Mn^{+2} < Fe^{+2} < Co^{+2} < Ni^{+2} < Cu^{+2} > Zn^{+2}$.

It is possible through the use of equilibrium models to calculate the effect of organic ligands on trace metal speciation in a simplified seawater matrix. Such calculations permit some clarification of metal competitive behavior given a reasonable internally consistent set of starting assumptions. These calculations have been performed by Lion (1980) for simulated surface microlayer solutions using the REDEQL computer program (Morel & Morgan 1972). The calculations indicate it is reasonable to expect that many trace metals (e.g. Cd, Ca, Cu, Ni, Pb, and Zn) may be complexed with surface-active organic ligands and/or adsorbed onto particulate surfaces at the surface microlayer. The calculated results also suggest that the aqueous chemistries of trace elements are sufficiently different so that the surface enrichment observed may vary from metal to metal.

Effects of Organic Surface Films

The presence of organics at the air-sea interface has ramifications other than those associated with binding of trace metals. Marine surface films have been observed to alter sea-surface temperatures, interrupting normal mass and thermal convection processes below the air-water interface (Jarvis et al 1962). Surface films may also damp capillary surface waves (Garrett & Bultman 1963, Garrett 1967c). Production of bubble film drops is reported to be suppressed by the presence of sea-surface films (Woodcock 1972), while insoluble monolayer films apparently can act to increase the production of condensation nuclei from bubbled seawater (Garrett 1968). Gas transport rates can be affected by the presence of compressed surface films which act as a molecular barrier and reduce flux of CO_2, H_2O, N_2, and O_2 in and out of the bulk liquid (Garrett 1971, Lou & Rasmussen 1973). Liss (1977) has argued that only a compressed surface film can substantially influence gas-exchange rates. In such situations, the transfer of gases whose exchange is controlled by gas-phase resistance (e.g. H_2O, SO_2, NO_2) would be affected to the greatest extent (Liss 1977, Hunter & Liss 1977). A thorough review of microlayer organic chemistry including a discussion of gas-exchange processes has been presented by Liss (1975).

The presence of surface-active materials in bulk solution and at the surface can extend the lifetime of bubbles at the air-sea interface. This phenomenon is illustrated by the presence of stable foams which may literally pile into windrows on the downwind side of lakes or in marine coastal areas and persist for long periods of time (Baier 1972). Bubble stability may be qualitatively predicted from the Gibbs adsorption isotherm [Equation (1)], which may be interpreted to indicate that a decrease in Γ (which would occur if a bubble were stretched) causes an increase in $d\gamma$ or in bubble strength. Such strengthening might be thought of as conferring "shock resistance" to bubbles with adsorbed monolayers. Resistance to decreases in bubble surface area may occur through repulsion of polar functional groups in the surface coating. Garrett (1967b) has shown that, in the sea, bubble stability is enhanced by soluble surface-active materials which "feed stabilizing molecules into a bubble-water interface and protect it against rupture." Compression of surface films was observed to decrease bubble lifetimes as less soluble rigid molecules replaced the soluble surface-active agents at the interface. Monolayers may also act to increase bubble lifetimes via viscosity effects, whereby drainage of liquid in bubble walls may be retarded. MacIntyre (1974b) has indicated that oceanic bubble film caps are more likely to be stabilized by wet surfactants (i.e. those that are relatively hydrophilic such as proteinaceous materials) than by dry surfactants (such as the more hydrophobic long-chain fatty acids and lipids).

Particulate Organics

One of the most interesting (and controversial) of interfacial phenomena is the reported conversion of DOC into POC. This phenomenon was first reported by Sutcliffe et al (1963) and stimulated considerable debate concerning its authenticity in nature. Experiments by Riley (1963), Baylor & Sutcliffe (1963), and Riley et al (1964, 1965) all indicated that bubbles could produce particulate aggregates from filtered seawater containing dissolved organic material. It was shown that the aggregates formed might serve as a source of particulate food for marine organisms (Baylor & Sutcliffe 1963) and that the particulate aggregates scavenged (adsorbed?) phosphate and organics from solution (Sutcliffe et al 1963, Riley et al 1965). Subsequent experiments by Menzel (1966) using a carefully filtered air supply implicated organic contamination as the source of particulates produced by the other investigators. Experiments by Barber (1966) indicated that bacteria must be present in order to obtain particles. Extensive investigations by Batoosingh et al (1969) support the conclusion that the formation of POC from DOC can occur under specific conditions (i.e. in the presence of small numbers of $0.22-1.2$ μm seed particles). Wangersky (1972, 1976) speculated that POC aggregates that form at the sea surface may be polymerized by ultraviolet light and altered into forms that are increasingly insoluble. Results by Johnson (1976) show that dissolution of small-sized rising bubbles can produce POC in seawater. Wheeler (1975) presented data that suggest POC can be formed from the collapse of surface films such as might result from wind, or wave forces, or possibly Langmuirian circulation. Trent et al (1978) have made in-situ and laboratory measurements on macroscopic particulate aggregates (marine snow) taken from seawater and indicate adsorption onto rising bubbles as one possible mechanism for their formation. Bubble aggregation of DOC combined with the UV energy input from high-intensity sunlight have been suggested as a possibility for the primevil origin of life (Goldacre 1958, Wangersky 1965).

Association of metals with colloidal or particulate organic material may range from weak attractive interactions (physical adsorption) to strong associations comparable to chemical bonding (chemisorption). Since sinking solids of various sorts have been implicated in trace metal transport in the marine environment (Martin 1970, Turekian et al 1973, Boyle et al 1977, Brewer & Hao 1979), it is interesting to consider the transport potential of the POC generated at the microlayer. Nishizawa & Nakajima (1971) have measured vertical variations in POC and calculate a downward flux of 260 mgC/m^2/day which they estimate to be the rate of POC formation at the microlayer. Wangersky & Gordon (1965) analyzed POC and particulate Mn in the North Atlantic and concluded that DOC and Mn are incorporated into bubble-produced organic particulate aggregates which can settle to the sediments. Hirsbrunner & Wangersky (1975) produced organic aggregates containing Na, K, Mg, Ca, and Fe by shaking filtered seawater samples. However, Siegel & Burke (1965) found trace metal (Zn and Mn) adsorption onto bubble-produced organic aggregates to be minimal, indicating that sinking particulates may not remove significant quantities of these metals from the microlayer by adsorptive processes. Wallace & Duce (1975) have shown that rising bubbles can effectively transport POC and particulate trace metals (PTM) to the surface microlayer under laboratory conditions. Wallace et al (1977) measured PTM and POC in a series of marine-surface water samples. The relative variations in PTM, POC, and particulate Al indicated sinking of POC may regulate PTM abundance in open-ocean surface waters. PTM removal rate by sinking POC was estimated to be of the same order of magnitude as the rate of supply of trace metals to surface waters through atmospheric aerosol deposition.

An important aspect of the interrelationship between DOC and POC at the surface microlayer is the special case of pollutant hydrocarbons. The fate of oil introduced into the marine environment through oil spills, shipping traffic, and other anthropogenic sources has received considerable study. The sequence of hydrocarbon weathering at the microlayer encompasses (a) loss of volatile components, e.g. aromatic and aliphatic acids (Garrett 1974, Hansen 1975, 1977, Harrison et al 1975), (b) UV oxidation resulting in the production of water-soluble carboxylic acids and oil in water emulsions (Hansen 1975, 1977), and (c) formation of a tar-like particulate residue (Garrett 1974). Baier (1972) has shown that bubble bursting is an efficient way of removing surface oil films, while Fontana (1976) has shown millimeter-thick oil slicks reduce the amount of spray produced by bursting bubbles. Particulate hydrocarbons have been found to constitute a significant portion of surface microlayer particulate phase in the Atlantic and Pacific Ocean and the Mediterranean (Horn et al 1970, Morris 1971, Wong et al 1974, Morris et al 1975, Wade & Quinn 1975). An additional anthropogenic source of microlayer organic particulates is plastic particles (Carpenter & Smith 1972, Colton et al 1974). Both tar and plastic particles apparently are used for attachment by sessile

marine organisms including bryozoans, diatoms, hydroids, barnacles, iso-
pods, algae, and bacterial films (Horn et al 1970, Carpenter & Smith
1972, Wong et al 1974). The fate of hydrocarbons in the marine environment has been reviewed by Hardy et al (1977a).

The literature summarized here shows that organic ligands will be partitioned at the air-sea interface. To a certain extent the types of organics
that will be surface enriched may be predicted *a priori*; such estimates
agree with observations made in the marine environment. Surface-active
organics at the air-sea interface may bind with trace metals and thereby
enhance metal partitioning. The formation of particulate organic aggregates at the sea surface has a potentially important role with respect to
metal transport. In addition, the presence of elevated organic concentrations at the surface microlayer has important implications for surface
microbiota. Possible interactions include the utilization of organics as a
microbial substrate and the amelioration of trace metal toxicity through
complexation with organic ligands. These latter topics are discussed below.

MICROBIAL ECOLOGY OF THE
AIR-SEA INTERFACE

It is far from obvious that viable organisms should exist at the marine
surface. The ecological conditions of the surface microlayer imply that it
is an extremely high-stress environment. For example, Zaitsev (1971) indicated that the upper 10 cm of the sea captures about half the total
entering sunlight and more than 75 percent of 254 mμ UV radiation. An
increase in surface-water temperature would be expected in response to
the surface absorption of light energy. Temperatures of 27.6°C and 26°C
have been reported at depths of 10 and 30 cm, respectively, in the Caspian
Sea (Zaitsev 1971). Other phenomena that may stress biological communities at the surface include evaporative cooling and salinity disturbances
caused by atmospheric precipitation.

In the previous sections the presence of high concentrations of toxic
trace metals and chlorinated hydrocarbon pollutants were documented at
the air-sea interface. Depending upon the degree of their surface concentration and the extent of organic chelation, metal concentrations at the
surface could prove quite toxic to organisms. The toxic effects of oil pollutants represent a further stress which occasionally can drastically alter
conditions in the surface layer.

Trace metal toxicity to marine organisms is not well understood. Association of metals with living organisms can occur via adsorption to the
general body surface or cell membrane in the case of bacteria, or in higher
order organisms adsorption through special areas such as gills or across

the walls of the gut (Bryan 1971). The following list of atoms, functional groups, and molecules which are probably bonded to various metals in living organisms (in decreasing order of occurrence) is excerpted from a review by Saxby (1969):

Fe: porphyrin, imidazole (a monocyclic nitrogen compound),
     $-NH_2, =NH, R_2S, -S^-, -COO^-, -O^-, =PO_4^-$

Mg: $-COO^-, =PO_4^-$, porphyrin, imidazole

Cu: $-NH_2, =NH, R_2S, -S^-$

Zn: $-NH_2, =NH, R_2S, -S^-$

Cd: $R_2S, -S^-$

Cr: $-COO^-, =PO_4^-$

Typically the growth rate of marine microbiota is accelerated by the presence of trace metals and a chelating agent (Johnston 1963). There are two interpretations for this phenomenon, the first being that organic chelators (i.e. EDTA) convert free essential trace metals (i.e. Fe) to organically bound trace metals, thereby improving their availability to microbiota and the second being that metal chelation reduces inhibition of toxic metals (e.g. Cu) to the organisms. Recent studies by Sunda & Guillard (1976) and Rueter et al (1979) have shown that complexation of cupric ion reduces its toxicity to aquatic organisms. Toxic effects were observed at free cupric ion concentrations as low as 10^{-10} M. Jackson & Morgan (1978) using an equilibrium chemical model to determine copper speciation, and kinetic calculations to estimate iron availability, have argued that the presence of chelators in solution *must* act to increase organism growth through reduction of copper toxicity. Whether the extent of organic binding of trace metals at the surface microlayer is increased relative to bulk seawater will depend on many factors including the relative magnitudes of ligand and metal enrichment and metal-ligand binding strengths.

In contrast to the array of stresses at the air-sea interface, there are some beneficial features that might be conducive to microbial life in this habitat. Drachev et al (1965) reported that the high content of surface-active organic compounds and suspensates at the marine surface may serve as a favorable substrate for the development of microorganisms. The iron, phosphorus, and nitrogen content of foams were determined to be 10 to 100 times concentrations in the bulk water; the biochemical oxygen demand of foams was said to be comparable to that of sewage. These results confirm the work of Wilson (1959) who found high nitrogen and phosphorus concentrations in sea foam. Further verification is provided by Szekielda et al (1972), who measured 6.2 percent organic carbon,

0.8 percent nitrogen, and 500 to 2500 ppm phosphorus in a condensed surface film taken from a convergence zone in Delaware Bay. Williams (1967) reported high surface-microlayer concentrations of particulate and dissolved nitrogen and phosphorus relative to concentrations in subsurface water. Concentration ratios for DOC, dissolved organic nitrogen (DON), and dissolved organic phosphorus (DOP) in surface films (screen samples) relative to concentrations at 15 to 20 cm depth ranged from 1.7 to 4.9. Typical surface DOC, DON, and DOP concentrations were 3,000, 300, and 20 μg/l, respectively. Barker & Zeitlin (1972) reported a four-fold enrichment of total phosphorus and 10-fold enrichment of nitrate nitrogen in open ocean microlayer samples taken with a glass-plate sampler. Nitrate enrichment was not observed in surface samples (3-mm depth) taken by Goering & Menzel (1965) but higher levels of NH_3, NO_2, and PO_4 were found. Nishizawa (1971) reported order-of-magnitude enrichment of particulate organic nitrogen in the surface microlayer as did Dietz et al (1976). Higher phosphorus concentrations for <0.2-mm surface samples were reported by Balashov et al (1974).

Zaitsev (1971) indicated that sea foam can stimulate plant root systems, extend the survival rate of shrimp larvae, accelerate hatching and prolong the life of eggs of the goby *Pomatoschistus* sp., and apparently increase the hatching rate of *Artemia salina* (brine shrimp) eggs. Baylor & Sutcliffe (1963) have demonstrated that particulate matter formed by bubble aggregation will support the growth of *Artemia*.

The available literature demonstrates that the air-sea interface is, from a biological point of view, a high-stress ecological niche with some ameliorating rewards for those organisms able to adapt to life there. It is interesting to observe that many members of the microlayer planktonic community or neuston have specific physiological adaptions to maintain them at the microlayer. Maynard (1968a) has observed that many periphytic "benthic" organisms are able to use the marine surface film as a habitat, and found high concentrations of diatoms as well as small numbers of dinoflagellates, green, and blue-green algae in sea foams. Surface-layer utilization by attached organisms has been observed by Norris (1965) who classified many of the neustonic choanoflagellates which attach themselves to the surface film by a protuberance of the protoplast or by appendages. Typical surface-film hangers described by Norris (1965) are extremely interesting; these organisms have apparently adapted to hanging upside down from the marine surface and feed upon the abundant marine bacteria using their flagellae to carry food-rich currents to the cell.

Parker & Barsom (1970) have observed that many bloom-producing blue-green algae use their gas vacuoles to produce bouyancy in order to concentrate at the marine surface. This feature may allow them to more effectively capture light, absorb gas, and to escape water-soluble algicides. David (1965) described a variety of surface-adaptive techniques, ranging from physiological adaptions, such as chitonons floats, to behavioral adaptions, such as the use of bubble rafts or attachment to flotsam.

Also dwelling at the air-sea interface are insects of the genus *Halobates*, commonly referred to as water striders. These insects support themselves on top of surface films via non-wetable appendages and use surface ripples as a means of communication (Milne & Milne 1978, Wilcox 1979).

The neustonic community apparently encompasses a diverse array of organisms. Zaitsev (1971) has listed the following members of the neuston in order of their appearance in the food chain: (*a*) microorganisms; (*b*) protozoa, the primary consumers; (*c*) small metazoans (invertebrates), including several species of rotifers, larvae of many species of polychaetes, gastropod and lamellibranch mollusks, copepods, cirripeds, echinoderms, some species of cladocerans, etc; (*d*) large metazoans, including representatives of the Polychaeta, Isopoda, Amphipoda, Cumacea, Mysidacea, and Decapoda; and (*e*) fish eggs, larvae, and fry.

Selective adaption for the air-sea interface also occurs in the microbial community. Young (1978) has observed web-like extracellular exudates produced by marine bacterioneuston which might act to improve organism surface activity. Protective pigmentation may be a physiological mechanism used by microorganisms to permit life in the microlayer. Tsyban (1971) has noted the brightest and most diverse shades of bacterial strains (yellow, yellow-green, orange, brown, and red) to be associated with cultures of marine bacterioneuston. Carlucci & Williams (1965) found a selective enrichment of pigmented vs non-pigmented bacteria in aquatic foam. Pigmentation was not thought to be a significant factor in fractionation, but to indicate a different type population in the foam. Blanchard & Syzdek (1978) presented evidence suggesting that pigmented cells of the marine bacterium *S. marcescens* have a greater tendency (relative to non-pigmented cells) to remain at bubble surfaces and thus may be more easily transported from bulk solution to the surface microlayer.

Microorganisms may also select for the microlayer by attachment to partitioned particulate materials. Recent field studies indicate that bacterial fractionation at the air-sea interface is closely correlated with degree of particulate fractionation (Harvey & Young 1980). Microscopic observation using acridine orange epifluorescence showed a high percentage of interfacial microorganisms to be associated with particulate surfaces. Carlucci & Williams (1965) reported that the presence of particulates in solution enhances the bubble transport of bacteria to the surface.

Bacteria and POC represent overlapping categories in the analysis of marine samples, and few investigators have simultaneously categorized both these entities in the same samples. The association of bacteria with particulate matter (both inorganic and organic) represents a chicken-and-egg type dilemma. It is not known whether bacteria attach to surfaces that are first partitioned at the air-sea interface or whether the particulates are partitioned because associated bacterial activity has resulted in a change in their surface activity.

There is some controversy over the activity of microbiota at the air-sea interface. Marumo et al (1971) reported order-of-magnitude levels of bacterial partitioning, but plate counts on different growth media indicated most bacterial cells in the microlayer to be dead or senescent. [The reader should note that bacterial enumeration by plate counting selects for a relatively small portion of metabolically active bacteria. Use of a range of different growth media is designed to help overcome this difficulty; however, alternate procedures, such as measurement of ATP or ^{14}C uptake, are currently preferred as indicators of biological activity.] Dietz et al (1976) reported two order-of-magnitude enrichment of microbial populations (colony-forming units) in the surface microlayer but lower ATP levels and heterotrophic activity (measured by ^{14}C uptake rate). Wallace et al (1972) found that rising bubbles most efficiently fractionate organisms in the post-stationary growth phase. However, the data of Sieburth et al (1976) show surface microlayer enrichment of ATP in seven out of nine stations sampled in the North Atlantic. Di Salvo (1973) measured up to a three order-of-magnitude increase in the number of bacteria that attached to glass surfaces at the air-sea interface. Since bacterial attachment may take place through cellular production of bridging polymers, stimulated air-sea interfacial bacterial activity is implied. Daumas et al (1976) measured greater enrichment of ATP in surface samples taken with a rotating drum (50 to 80 μm), than with screen (440 μm) suggesting stratification of living biomass at the upper surface microlayer. The opposite effect was observed for photosynthetic pigments indicating planktonic algae did not accumulate in the uppermost surface layer. Gallagher (1975), using a glass-plate sampler, measured greater than two order-of-magnitude enrichment of plant pigments in the surface microlayer of a Georgia salt marsh. Enrichment varied over a tidal cycle with highest levels on flood tide. An average of 37 percent of total net plankton photosynthesis was contributed by surface-film organisms.

Although the activity of surface-dwelling organisms may be in question, there is no dispute over the occurrence of elevated population numbers. Observations of organism densities in the field have been reported by Sieburth (1965, 1971), who found mean surface microlayer (150 μm) concentrations of 2398 organisms/ml vs 8, 26, 23, and 4 at 1, 10, 51, and 105 meters depth, respectively. Dominant film isolates were *Pseudomonas*; 86

percent of 200 isolates were able to oxidize glucose, 93 percent could hydrolyze lipids; 91.5 percent proteins and only 23.5 percent were able to attack starch. The bacterial proteolytic and lipolytic capabilities coincide with the previously discussed predominance of glycoproteins and/or fatty acid esters in the microlayer. Elevated field levels of bacteria and algae at the surface have been found by Harvey (1966) who detected changes in concentration with time of day. Parker & Barsom (1970) reported 5000 colorless flagellates, 31,270 dinoflagellates, 330 ciliates, and 930 diatoms per liter at the surface in contrast to 0, 3900, 370, and 3770 individuals per liter at a depth of 10 m.

Crow et al (1975) have measured bacterial concentrations of the order of 10^5 cell/ml in coastal surface interface waters using a membrane-adsorption sampling procedure. Microlayer bacterial concentrations were two orders of magnitude greater than those of the bulk (10 cm) water. Tsyban (1971), Marumo et al (1971), and Dietz et al (1976) all reported one to two orders-of-magnitude enrichment factors for marine bacterioneuston.

If the microlayer is indeed a stressful habitat it would follow from ecological principles that the diversity of organisms there should be reduced. Certain genera of bacteria would be expected to display relatively greater concentrations at the microlayer, and, as already seen, their concentrations may outstrip those in bulk solution. These predictions are confirmed by the results of Tsyban (1971) who observed the genera *Bacterium* and *Pseudomonas* to dominate the Black Sea microlayer vs *Bacterium* and *Bacillus* at 0.5–3 m depth and *Bacterium, Mycobacterium,* and *Pseudomonas* at 4–10 m depth. W. MacIntyre et al (1974) compared diversity index values for microlayer and one-meter plankton communities and showed increased diversity in the subsurface population. These results contradict those of Taguchi & Nakajima (1971) who reported increased diversity of plankton in surface microlayer samples. Most of the identified cells in microlayer samples were diatoms; fifteen species were found to occur only at the surface microlayer. The predominance of diatoms in the air-sea interfacial plankton has also been observed by R. Harvey (personal communication) in samples taken from south San Francisco Bay.

In summary, a variety of conflicting ecological pressures may be seen to operate at the air-sea interface. High intensity UV radiation and elevated concentrations of pollutant chemicals stress surface communities, while increased levels of organic substrate and nutrients reward organisms adapted for surface living. Although the activity and diversity of neustonic organisms are in dispute, there is general agreement over the presence of concentrated surface microlayer populations.

The occurrence of so many marine organisms in the surface microlayer invites speculation with respect to their interaction with the elevated trace metal and chlorinated hydrocarbon concentrations which are also present in the surface environment. The opportunity is presented for the direct incorporation of high pollutant concentrations into the marine food chain at many trophic levels. The importance of pollutant uptake and concentration occurring at the microlayer with respect to subsurface uptake by organisms has yet to be evaluated.

Microbiota at the air-sea interface might influence the fate of trace metals both directly through active or passive uptake or indirectly through production of extracellular metal-complexing organics and/or through modification of particulate surface activity (via their attachment). The alternatives for transport of microbiota, particulates, surface-active organics, and associated trace metals are discussed below.

TRANSPORT OF METALS, ORGANICS, AND ORGANISMS AT THE AIR-SEA INTERFACE

Interest in transport processes at the air-sea interface dates to early investigations which sought an explanation for the ionic content of rainwater and atmospheric aerosol (Blanchard 1963). The primary sea-to-air transport process indicated was jet drop ejection by bursting bubbles. When bubbles break at the air-water interface, current evidence indicates that they act as extremely effective microtomes, skimming material from the surface and ejecting it as jet drops into the atmosphere (MacIntyre 1968, 1970, 1974a). The bursting of bubbles has powerful implications for the transport and distribution of surface-concentrated materials. Zobell & Mathews (1936) have observed marine bacteria in atmospheric aerosols 30 miles inland and terrestrial bacterial species 130 miles at sea. Bubble bursting is also indicated as a source for the atmospheric occurrence of marine diatoms, microflagellates (Stevenson & Collier 1962), and algae (Maynard 1968b). Bacterial concentrations in jet drops from bursting bubbles may be 1000-fold greater than concentrations in the bulk liquid from which they emerge (Blanchard & Syzdek 1970, 1972, Bezdek & Carlucci 1972). The water-to-air transfer of virus through bubble jets and film drops has also been shown to occur (Baylor et al 1977a,b).

Baier (1970) has observed that oil films that built up on lakes due to recreational use can be eliminated within three days and that the principal mechanism of material removal is bubble bursting. Similar results have been reported by Bezdek & Carlucci (1974) who were able to remove stearic acid monolayers from seawater in the laboratory. MacIntyre (1970,

1974a) and Blanchard (1964) have indicated that jet drops from bursting bubbles at a solution surface coated with organic molecules may in turn be coated with compacted or multiple organic layers. Thus bubble breaking can serve as a source for the introduction of marine organic materials into the atmosphere. Barger & Garrett (1970, 1976) and Blanchard (1964) have found surface-active organics on marine aerosols. Baier (1975, 1976) has found surface-active organic materials with characteristics similar to those found on surface microlayers in sea fog samples. Wilson (1959) suggested that the ocean surface can also serve as a source for airborne nitrogenous materials and phosphorus. Bubble transport and aerosol enrichment of organic materials have been studied in laboratory systems by Blanchard & Syzdek (1974a), Hoffman & Duce (1976), and Pueschel & Van Valin (1974). Bubble path length and the presence of surface-active organics were shown to be important variables that control the transfer of organics from solution to aerosols.

The importance of bubble bursting with respect to removal of trace metals from the marine surface and subsequent redistribution is not well understood. Ion fractionation by bubble bursting and distribution of elements with aerosol particle size is currently the subject of intense investigation (Bloch et al 1966, Barker & Zeitlin 1972, Bloch & Leucke 1972, Chesselet et al 1972, Hoffman & Duce 1972, Seto & Duce 1972, Tsunogai et al 1972, Wilkniss & Bressan 1972, Glass & Matteson 1973, MacIntyre 1974b,c, Van Grieken et al 1974, Hsu & Whelan 1976). An excellent review of interface-atmosphere exchange of microorganisms, nutrients, trace metals, and major marine cations and anions has been prepared by Duce & Hoffman (1976).

It is interesting to note that aerosols of marine origin may act as a primary source for atmospheric Na, Cl, K, Mg, and SO_4 (Rancitelli & Perkins 1970, Buat-Menard et al 1974, Duce et al 1976) and that a significant fraction of atmospheric Cu may also originate from the seas (G. Hoffman et al 1972, Cattell & Scott 1978). In-situ sea-to-atmosphere transport of Fe, Zn, and Cu by bursting bubbles has been demonstrated by Piotrowicz et al (1979). The study was conducted using a bubble interfacial microlayer sampler developed by Fasching et al (1974) and showed that bubble scavenging from bulk solution was a factor influencing the degree of aerosol enrichment of Cu and Zn, while aerosol enrichment of Cu (in part) and Fe apparently resulted from their accumulation at the surface microlayer.

A wide array of studies has been published in which atmospheric elemental ratios are compared to those of crustal materials and seawater as a means of identifying their source (Rancitelli & Perkins 1970, G. Hoffman et al 1972, Buat-Menard et al 1974, Chester et al 1974, E. Hoffman et al

1974, Korzh 1974, Peirson et al 1974, Yano et al 1974, Zoller et al 1974, Duce et al 1976). Although the investigators do not totally agree, typically the elements Al, Ce, Co, Cr, Fe, Sc, Th, and possibly V are thought to have a continental origin, while Ag, As, Cu, Hg, Pb, Sb, Sd, Se, Sn, and Zn are enriched relative to reference elements suggesting an anthropogenic source. Elevated metal concentrations may not necessarily indicate man-made inputs, since plants have recently been shown to release Zn and Pb containing particles to the atmosphere (Beauford et al 1977).

If the sea surface can serve as a source of atmospheric trace metals, the reverse process is certainly also a possibility. Martin et al (1976) measured higher surface microlayer enrichments of cadmium in nearshore Pacific Ocean samples than in samples taken further out to sea. Eaton (1976) has measured higher Cd concentrations in Atlantic surface waters and suggests an anthropogenic origin. The atmosphere serves as a primary source of automobile-generated lead to marine surface waters (Chow et al 1969, Patterson & Settle 1974, Benninger et al 1975, Patterson et al 1976). Fitzgerald (1976) has estimated rainfall input of Hg to the oceans to be 30×10^8 g/year. In the Southern California Bight, Young & Jan (1977) estimated that the airborne inputs of Fe, Mn, and Pb from forest fires were 10 to 100 times the inputs from municipal wastewater discharge. Atmospheric input of Cd and Pb to the New York Bight has been estimated by Duce et al (1976) as respectively 4.1 metric tons/year and 220 metric tons/year from dry fallout and 8.2 metric tons/year and 400 metric tons/year from rainfall. Atmospheric deposition was estimated to account for up to 13, 8, 5, and 1 to 2 percent of the total Bight inputs of Pb, Zn, Fe, and Cd, respectively. Wallace et al (1977) estimated the atmospheric flux of Al, Mn, Fe, Ni, Cr, Cu, Zn, Pb, and Cd to the ocean; the calculated atmospheric input closely compared to the calculated removal rate of these metals via sinking of organic carbon particulates, which was estimated independently. Sedimentary records in both marine and fresh water environments indicate a recent anthropogenic input of Cd, Cu, Hg, Pb, and Zn which is thought to be deposited from the atmosphere (Winchester & Nifong 1971, Bruland et al 1974, Schell & Barnes 1976, Elzerman et al 1979, Galloway & Likens 1979, Hamilton-Taylor 1979).

Dusts of continental origin can transport geochemically significant quantities of materials to the marine environment (Goldberg 1971). One major dust source is the African Sahara Desert which Prospero & Carlson (1972) estimated may supply enough material to account for the present rate of pelagic sedimentation in the north equatorial Atlantic Ocean. The marine dust veil is reported to have high concentrations of Co, Cr, Mn, Ni, and V associated with it and may act as an important input source for Cd, Hg, Pb, and Zn (Chester 1971). G. L. Hoffman et al (1974) report microlayer concentrations of Fe near the African coast to be associated with dust loadings from the Sahara.

In addition to its role in inorganic transport, airborne deposition may act as a source of organics in the marine enviroment (Goldberg 1970). The importance of airborne inputs relative to those from continental runoff is not known with any precision, but primary production is clearly the principal source of marine organics (Duce & Duursma 1977, Handa 1977). Atmospheric deposition may, however, be a major source for input of pollutant chlorinated organics including DDT and PCBs (Bidleman et al 1976) and CCl_4 (Hunter & Liss 1977).

Aside from their important role in the transfer of materials from the oceans to the atmosphere, bubbles are also strongly implicated in the transport of materials from the bulk ocean waters to the air-sea interface. Use of bubbles to concentrate bacteria at surfaces has been applied in the laboratory to separate algae and bacteria from dilute suspensions (Boyles & Lincoln 1958, Gaudin et al 1960, Levin et al 1962, Rubin et al 1966, Grieves & Wang 1967, Wallace et al 1972). Laboratory researchers have discovered that flotation can be influenced by a wide range of parameters including bubble size, the presence of surfactants, gas flow rate, foam height, salt concentration, pH, age of bacterial components to be harvested, cell type, and cell surface functional groups.

Carlucci & Williams (1965) observed that dissolved organic material in seawater solutions may adsorb to rising bubbles and that bacteria may become attached to rising bubbles depending upon their size, shape, and individual affinity for the organic matrix. Rods were concentrated more effectively than cocci. Blanchard & Syzdek (1974a,b) have shown that the bactierum *Serratia marcescenes* attaches to bubbles at a rate of 3.1 cells per cm of bubble rise, giving a calculated collection efficiency of 0.12 percent. Transfer efficiency for bacteria to bubble jet drops was considerably higher and varied with the distance of bubble rise; efficiencies of 14 and 25 percent were given for 1.05-mm diameter bubbles rising 2.2 and 20.3 cm. Low bacterial transfer efficiencies from bulk solution to bubbles have been reported by Carlucci & Bezdek (1972).

Aside from the special case represented by bacterial particulates, rising bubbles have proven to be effective in transporting particulate materials in general. Separation of naturally occurring marine particulate organic carbon and nitrogen from bulk solution to the surface has been performed by Wallace et al (1972) and Wallace & Duce (1979). Quinn et al (1975) have used rising bubbles to transport latex particulate spheres.

The use of bubbles to scavenge both particulate and solute metals from solution in the presence of a surface-active "collector" is a well-established practice and various bubble separation processes have been used by the mineral processing industries since 1912 (Fuerstenau & Healy 1972). Bubble or foam separation of Cd, Cr, and Na (Kubota 1975, Kubota & Hayaski 1977), Cd (Chou & Okamoto 1976), Cd, Cu, and Pb (Huang 1974), and Cu and Zn (Jacobelli-Turi et al 1972) has been studied in laboratory systems.

Separation processes that may play an important role in trace metal removal from the bulk liquid and concentration at the surface have been categorized by Lemlich (1972b) and include:

1. Foam separation in which bubble-generated foams or froths are used to separate both particulate phases and solutes from solution. Separation of an individual species may be accomplished through a variety of processes, such as direct adsorption to bubble surfaces, induction of surface activity through addition of dissolved surface-active "collectors," and adsorption or coprecipitation of solutes onto surface-active solids.
2. Non-foam separations in which a foam or froth is not formed and materials are transported to the surface by rising bubbles.

It is not difficult to conceive that many of the separation processes listed by Lemlich (1972b) may occur in the marine environment. The possibility that trace metal separation occurs is strongly suggested by the presence of a wide array of potential chelating and surface-active agents in the sea. Nakashima (1979) and Kim & Zeitlin (1971a,b,c, 1972) have successfully separated dissolved Cu, Mo, Se(IV), U, and Zn from a seawater matrix through adsorption to or coprecipitation with $Fe(OH)_3(s)$ and bubble flotation in the presence of artificial surface-active organic collectors.

Wallace & Duce (1975) have demonstrated that rising bubbles can fractionate particulate trace metals (PTM) from seawater; for most metal species (except Cd) PTM transport to the surface correlated well with particulate organic transport. Based on their laboratory data, Wallace & Duce estimated the rate of bubble transport of trace metals to the sea surface from bulk solution via rising bubbles and compared this to the estimated atmospheric deposition rate of metals to the microlayer. Bubble transport was estimated to be a significant source for the elements Cd, Cr, Cu, Fe, Mn, Ni, Pb, V, and Zn.

Recent field results and model calculations by Hunter (1980) suggest that atmospheric deposition of trace metals in North Sea coastal waters competes effectively with bubble transport as a source for microlayer enrichment under calm weather conditions. The relative rates of metal transport from Brownian diffusion, gravitational settling, rising bubbles, and atmospheric deposition were estimated by Hunter (1980). These calculations suggested that rising bubbles may, in general, be a more effective mechanism for surface enrichment of Cu, Zn, and Cd in the coastal waters of the North Sea. Gravitational settling was indicated as a removal process to account for observed microlayer depletions of Fe and Mn.

The literature available makes it abundantly clear that rising bubbles can act as a powerful transport vehicle for the delivery of dissolved surface-active organics and/or particulate materials to the sea surface microlayer. It should also be clear that rising bubbles certainly do not constitute the only transport process that might bring about partitioning from bulk solution to the air-sea interface. Thermal convection cells, turbulent mixing, and diffusion processes also would contribute to molecular movement and aggregation. MacIntyre (1974a) provides an excellent review of surface transport and ion fractionation processes.

The data summarized here leave little doubt that trace metals associated with transported species, i.e. metals that are complexed with surface-active ligands or adsorbed onto particulate surfaces, are transported in concert with these materials to the sea surface. In the complex marine matrix trace metals will compete with one another and the major marine cations for the available organic ligands and adsorption sites. It follows that in situations in which the available concentrations of complexing organics and solid surfaces are limiting, those trace metals that have relatively stronger binding strengths may be preferentially enriched at the air-sea interface.

SUMMARY

The available data describing the chemical and biological nature of the air-sea interface are still very limited. Useful comparisons are hindered by diverse sampling techniques. There are, nevertheless, recognizable trends in reported results. The occurrence of elevated trace metal concentrations in the surface microlayer is widely reported to occur and appears to correlate well with the presence of observable condensed organic slicks. Particulate and organic forms of metals show the highest degrees of enrichment at the air-sea interface.

Both dissolved and particulate organic carbon are partitioned into the surface microlayer. There is some controversy over the specific nature of DOC, but evidence favors the predominance of proteinaceous or carbohyrate materials. Pollutant hydrocarbons and chlorinated hydrocarbons are also enriched at the microlayer and might be transferred there to surface microbiota.

In spite of stresses at the air-sea interface, surface biological populations are elevated above those in bulk solution. Adaptive behavior for dwelling at the air-sea interface is exhibited by some surface species. Activity of surface bacterial populations is uncertain; there is some indication that diversity of populations at the surface microlayer may be reduced.

Rising bubbles and atmospheric aerosol deposition are indicated as processes that supply trace metals and organics to the air-sea interface. Bubble droplet formation and sinking of particles are indicated to be important material-removal processes. DOC at the microlayer may be converted to POC (which then sinks) through the action of breaking bubbles. Bubble transport appears particularly effective at transferring surface-active organics and particulates to the surface microlayer. This corresponds to the observed accumulation of organically associated and particulate trace metals at the surface microlayer. The high interfacial concentrations of organic ligands and solids also favor complexation and adsorption reactions in situ.

Trace metal levels at the surface microlayer may be toxic to the microbial population there. However, the possibility of metal binding by a wide array of organic chelating agents or adsorbing surfaces obscures the picture. It is not known whether the metals at the interface are substantially available to the microbial population.

All in all, very little is known about the interrelationships of the trace metal, microbial, particulate, and organic constituents of the air-sea interface. Better definition of metal-organic, metal-microbial, and microbial-organic interdependence at the microlayer would not only offer the potential for delineating the important processes involved in the transport of the component species, but would indicate the importance of microlayer interactions in the fate and distribution of pollutants which mankind inflicts upon the seas.

Literature Cited

Baes, C. F. 1973. The properties of cadmium. In *Cadmium the Dissipated Element*, ed. W. Fulkerson, H. Goeller. *Oak Ridge Natl. Lab. Rep. ORNL NSF-EP-21*

Baier, R. E. 1970. Surface quality assessment of natural bodies of water. *Proc. 13th Conf. Great Lakes Res.*, Int. Assoc. Great Lakes Res., pp. 114–27

Baier, R. E. 1972. Organic films on natural waters: their retrieval, identification, and modes of elimination. *J. Geophys. Res.* 77(27):5062–75

Baier, R. E. 1975. Surface-active and infrared-detectable matter in North Atlantic aerosols, sea fogs, and sea-surface films. Summary report prepared for US Naval Res. Lab., Washington, DC. Contract N00173-76-C-005, Proj. No. VA-5788-M

Baier, R. E. 1976. Infrared spectroscopic analysis of sea fog water residues, ambient atmospheric aerosols and related samples collected during the USNS Hayes cruise off the coast of Nova Scotia, Canada, 29 July–12 August 1975. Prepared for *NRL-Edited Data Volume for the Cruise of the USNS Hayes, August 1975*, Calspan Corp. Rep. VA-5788-M-2

Baier, R. E., Goupil, D. W., Perlmutter, S., King, R. 1974. Dominant chemical composition of sea-surface films, natural slicks, and foams. *J. Rech. Atmos.* 8(3–4):571–600

Balashov, A. I., Zaitsev, Yu. P., Kogan, G. M., Mikhajlov, V. I. 1974. A study of some chemical components of sea water at the air-sea interface. *Okeanologiya* 14:817–22

Barber, R. T. 1966. Interaction of bubbles

and bacteria in the formation of organic aggregates in sea water. *Nature* 211:257

Barker, D. R., Zeitlin, H. 1972. Metal-ion concentrations in sea-surface microlayer and size-separated atmospheric aerosol samples in Hawaii. *J. Geophys. Res.* 77(27):5076–86

Barger, W. R., Garrett, W. D. 1970. Surface active organic material in the marine atmosphere. *J. Geophys. Res.* 75:4561–66

Barger, W. R., Garrett, W. D. 1976. Surface active organic material in air over the Mediterranean and over the Eastern Equatorial Pacific *J. Geophys. Res.* 81:3151–57

Batley, G. E., Gardner, D. 1977. Sampling and storage of natural waters for trace metal analysis. *Water Res.* 11:745–56

Batoosingh, E., Riley, G. A., Keshwar, B. 1969. An analysis of experimental methods for producing particulate organic matter in seawater by bubbling. *Deep-Sea Res.* 16:213–19

Baylor, E. R., Baylor, M. B., Blanchard, D. C., Syzdek L. D., Appel, C. 1977a. Virus transfer from surf to wind. *Science* 198:575

Baylor, E. R., Peters, V., Baylor, M. B. 1977b. Water-to-air transfer of virus. *Science* 197:763–64

Baylor, E. R., Sutcliffe, W. H. Jr. 1963. Dissolved organic matter in seawater as a source of particulate food. *Limnol. Oceanogr.* 8:369

Beauford, W., Barber, J., Barringer, A. R. 1977. Release of particles containing metals from vegetation into the atmosphere. *Science* 195:571–73

Benninger, L. K., Lewis, D. M., Turekian, K. K. 1975. The use of natural Pb-210 as a heavy metal tracer in the river-estuarine system. In *Marine Chemistry in the Coastal Enviroment*, ed. T. M. Church, p. 202. *ACS Symp. Ser. 18.* Washington, DC: Am. Chem. Soc.

Bezdek, H. F., Carlucci, A. F. 1972. Surface concentration of marine bacteria. *Limnol. Oceanogr.* 17(4):566–69

Bezdek, H. F., Carlucci, A. F. 1974. Concentration and removal of liquid microlayers from a sea water surface by bursting bubbles. *Limnol. Oceanogr.* 19(1):128

Bidleman, T. F., Olney, C. E. 1974. Chlorinated hydrocarbons in the Sargasso Sea atmosphere and surface water. *Science* 183:516-18

Bidleman, T. F., Rice, C. P., Olney, C. E. 1976. High molecular weight chlorinated hydrocarbons in the air and sea: rates

and mechanisms of air/sea transfer. In *Marine Pollutant Transfer*, ed. H. L. Windom, R. A. Duce, pp. 323–51. Lexington, MA: D. C. Heath

Blanchard, D. C. 1963. The electrification of the atmosphere by particles from bubbles in the sea. In *Progress in Oceanography*, ed. M. Sears, 1:73–202. New York: Macmillan

Blanchard, D. C.. 1964. Sea to air transport of surface active material. *Science* 146:396–97

Blanchard, D. C., Syzdek, L. D. 1970. Mechanism for the water-to-air transfer and concentration of bacteria. *Science* 170:626–28

Blanchard, D. C., Syzdek, L. D. 1972. Concentration of bacteria in jet drops from bursting bubbles. *J. Geophys. Res.* 77(27):5087–99

Blanchard, D. C., Syzdek, L. D. 1974a. Importance of bubble scavenging in the water-to-air transfer of organic material and bacteria. *J. Rech. Atmos.* 13:529

Blanchard, D. C., Syzek, L. D. 1974b. Bubble tube: apparatus for determining rate of collection of bacteria by an air bubble rising in water. *Limnol. Oceanogr.* 19(1):133

Blanchard, D. C., Syzdek, L. D. 1978. Seven problems in bubble and jet drop researches. *Limnol. Oceanogr.* 23(3):389

Bloch, M. R., Kaplan, D., Kertes, V., Schnerb, J. 1966. Ion separation in bursting air bubbles: an explanation for the irregular ion ratios in atmospheric precipitations. *Nature* 209:802–3

Bloch, M. R., Leucke, W. 1972. Geochemistry of ocean water bubble spray. *J. Geophys. Res.* 77(27):5100–5

Boyle, E., Sclater, F., Edmond, J. M. 1977. The distribution of dissolved copper in the Pacific. *Earth Planet. Sci. Lett.* 37:38

Boyles, W. A., Lincoln, R. E. 1958. Separation and concentration of bacterial spores and vegetative cells by foam flotation. *Appl. Microbiol.* 6:327

Brewer, P. G., Hao, W. M. 1979. Oceanic elemental scavenging. In *Chemical Modeling in Aqueous Systems*, ed. E. A. Jenne, p. 261. *ACS Symp. Ser. 93.* Washington, DC: Am. Chem. Soc.

Bruland, K., Bertine, K., Koide, M., Goldberg, E. D. 1974. A history of metal pollution in the Southern California coastal zone. *Environ. Sci. Technol.* 8:425–32

Bruland, K. W., Knauer, G. A., Martin, J. H. 1978. Zinc in Northeast Pacific water. *Nature* 271:741–43

Bryan, G. W.. 1971. The effects of heavy metals (other than mercury) on marine and estuarine organisms. *Proc. R. Soc. London, Ser.* 177:389–410

Buat-Menard, P., Morelli, J., Chesselet, R. 1974. Water-soluble elements in atmospheric particulate matter over tropical and equatorial Atlantic. *J. Rech. Atmos.* 13:661

Carlucci, A. F., Bezdek, H. F. 1972. On the effectiveness of a bubble for scavenging bacteria from seawater. *J. Geophys. Res.* 77:6608–10

Carlucci, A. F., Williams, P. M. 1965. Concentration of bacteria from seawater by bubble scavenging. *J. Cons. Perm. Int. Explor. Mer* 30:28–33

Carpenter, E. J., Smith, K. L. 1972. Plastics on the Sargasso Sea surface. *Science* 175:1240–41

Cattell, F. C. R., Scott, W. D. 1978. Copper in aerosol particles produced by the ocean. *Science* 202(4366):429

Chesselet, R., Morelli, J., Buat-Menard, P. 1972. Variations in ionic ratios between reference sea water and marine aerosols. *J. Geophys. Res.* 77(27):5116–31

Chester, R. 1971. Geological, geochemical and environmental implications of the marine dust veil. In *Nobel Symp. 20; The Changing Chemistry of the Oceans*, ed. D. Dyrssen, D. Jagner, p. 291. New York:Wiley

Chester, R., Aston, S. R., Stoner, J. H., Bruty, D. 1974. Trace metals in soil-sized particles from the lower troposphere over the world ocean. *J. Rech. Atmos.* 13:777

Chou, C. E. J., Okamoto, Y. 1976. Removal of cadmium ion from aqueous solution. *J. Water Pollut. Control Fed.* 48(12):2747

Chow, T. J., Earl, J., Bennett, C. F. 1969. Lead aerosols in the marine atmosphere. *Environ. Sci. Technol.* 3:737–40

Colton, J. B. Jr., Knapp, F. D., Burns, B. R. 1974. Plastic particles in surface waters of the Northwestern Atlantic. *Science* 185:491–97

Crow, S. A., Ahearn, D. G., Cook, W. L., Bourquin, A. W. 1975. Densities of bacteria and fungi in coastal surface films as determined by a membrane adsorption procedure. *Limnol. Oceanogr.* 20(4):644

Daumas, R. A., Laborde, P. L., Marty, J. C., Saliot, A. 1976. Influence of sampling method on the chemical composition of water surface film. *Limnol. Oceanogr.* 21(2):319

David, P. M. 1965. The surface fauna of the ocean. *Endeavor* 24(92):95–100

Davies, J. T., Rideal, E. K. 1963. *Interfacial Phenomena*, p. 185. London:Academic

Dietz, A. S., Albright, L. J., Tuominen, T. 1976. Heterotrophic activities of bacterioneuston and bacterioplankton. *Can. J. Microbiol.* 22(12):1699

Di Salvo, L. H. 1973. Contamination of surfaces by bacterial neuston. *Limnol. Oceanogr.* 18(1):165–68

Drachev, S. M., Bylinkina, V. A., Sosunova, I. N. 1965. The importance of adsorption and surface phenomena in the self cleaning of lakes and ponds. *Biol. Anstr. 46* 101500 (1965) (Engl. trans.)

Duce, R. A., Duursma, E. K. 1977. Inputs of organic matter to the ocean. *Mar. Chem.* 5:319

Duce, R. A., Hoffman, E. J. 1976. Chemical fractionation at the air/sea interface. *Ann. Rev. Earth Planet. Sci.* 4:187–228

Duce, R. A., Quinn, J. G., Olney, C. E., Piotrowicz, S. R., Ray, B. J., Wade, T. L. 1972. Enrichment of heavy metals and organic compounds in the surface microlayer of Narragansett Bay, Rhode Island. *Science* 176:161–63

Duce, R. A., Wallace, G. T. Jr., Ray, B. J. 1976. Atmospheric trace metals over the New York Bight. Final report to MESA Proj. Off., NOAA. NOAA Grant No. 04-4-022-36

Duce, R. A., Hoffman, G. L., Ray, B. J., Fletcher, I. S., Wallace, G. T., Fasching, J. L., Piotrowicz, S. R., Walsh, P. R., Hoffman, E. J., Miller, J. M., Heffter, J. L. 1976. Trace metals in the marine atmosphere: sources and fluxes. In *Marine Pollutant Transfer*. See Bidleman et al 1976, p. 77

Duursma, E. K.. 1970. Organic chelation of ^{60}Co and ^{65}Zn by leucine in relation to sorption by sediments. *Symp. Organic Matter in Natural Waters*, ed. D. W. Hood, p. 387. Univ. Alaska Inst. Sci. *Occasional Publ. No. 1.*

Dwyer, F. P., Mellor, D. P., eds. 1964. *Chelating Agents and Metal Chelates.* New York:Academic

Eaton, A. 1976. Marine geochemistry of cadmium. *Mar. Chem.* 4:141

Eisenreich, S. J., Elzerman, A. W., Armstrong, D. E. 1978. Enrichment of micronutrients, heavy metals and chlorinated hydrocarbons in wind-generated lake foam. *Environ Sci. Technol.* 12(4):413

Elzerman, A. W. 1976. *Surface microlayer-microcontaminant interactions in freshwater lakes.* PhD thesis. Univ. Wis., Madison

Elzerman, A. W., Armstrong, D. E. 1979. Enrichment of Zn, Cd, Pb and Cu in the surface microlayer of Lakes Michigan, Ontario and Mendota. *Limnol. Oceanogr.* 24(1):133

Elzerman, A. W., Armstrong, D. E., Andren, A. W. 1979. Particulate zinc, cadmium, lead and copper in the surface microlayer of southern Lake Michigan. *Environ. Sci. Technol.* 13(6):720

Fasching, J. L., Courant, R. A., Duce, R. A., Piotrowicz, S. R. 1974. A new surface microlayer sampler utilizing the bubble microtome. *J. Rech. Atmos.* 8:649–52

Fitzgerald, W. F. 1976. Mercury studies of seawater and rain: geochemical flux implications. See Bidleman et al 1976, p. 121

Fitzgerald, W. F., Hunt, C. D. 1974. Distribution of mercury in the surface microlayer and in subsurface waters of the Northwest Atlantic Ocean. *J. Rech. Atmos.* 13:629

Fontana, M. 1976. An aspect of coastal pollution—the combined effect of detergent and oil at sea on sea spray composition. *Water, Air & Soil Pollut.* 5:269–80

Fuerstenau, D. W., Healy, T. W. 1972. Principles of mineral flotation. In *Adsorptive Bubble Separation Techniques*, ed. R. Lemlich, p. 91. New York: Academic

Gaines, G. L. Jr. 1966. *Insoluble Monolayers at Liquid-Gas Interfaces.* New York:Interscience

Galloway, J. N., Likens, G. E. 1979. Atmospheric enhancement of metal deposition in Adirondack Lake sediments. *Limnol. Oceanogr.* 24(3):427

Gallagher, J. L. 1975. The significance of the surface film in salt marsh plankton metabolism. *Limnol. Oceanogr.* 20(1):120

Garrett, W. D. 1965. Collection of slick forming materials from the sea surface. *Limnol. Oceanogr.* 10:602–5

Garrett, W. D. 1967a. The organic chemical composition of the ocean surface. *Deep-Sea Res.* 14:221–27

Garrett, W. D. 1967b. Stabilization of air bubbles at the air-sea interface by surface active material. *Deep-Sea Res.* 14:661–72

Garrett, W. D. 1967c. Damping of capillary waves at the air-sea interface by oceanic surface active material. *J. Mar. Res.* 25:279–91

Garrett, W. D. 1968. The influence of monomolecular surface films on the production of condensation nuclei from bubbled seawater. *J. Geophys. Res.* 73:5145–50

Garrett, W. D. 1970. Organic chemistry of natural seasurface films. See Duursma 1970, p. 469

Garrett, W. D. 1971. Impact of natural and man-made surface films on the properties of the air-sea interface. See Chester 1971, p. 75

Garrett, W. D. 1972. Collection of film-forming materials from the sea surface. In *A Guide to Marine Pollution*, ed. E. D. Goldberg, p. 76. New York, London, Paris: Gordon & Breach Science Publishers

Garrett, W. D. 1974. The surface activity of petroleum and its influence on the spreading and weathering of oil films at sea. *J. Rech. Atmos.* 13:55–62

Garrett, W. D., Bultman, J. D. 1963. Capillary-wave damping by insoluble organic monolayers. *J. Colloid Sci.* 18:798–801

Gaudin, A. M., Mular, A. L., O'Connor, R. R. 1960. Separation of microorganisms by flotation. 2. Flotation of spores of *Bacillus subtilus* var. *niger*. *Appl. Microbiol.* 8:91

Glass, S. J., Matteson, M. J. 1973. Ion enrichment in aerosols dispersed from bursting bubbles in aqueous salt solutions. *Tellus* 25:272–80

Goering, J. J., Menzel, D. W. 1965. The nutrient chemistry of the sea surface. *Deep-Sea Res.* 12:839–43

Goldacre, R. J. 1958. Surface films, their collapse on compression, the shapes and sizes of cells and the origin of life. In *Surface Phenomena in Chemistry and Biology*, ed. J. F. Danielle, H. G. A. Parkhurst, A. C. Riddiford, pp. 278–98. New York: Pergamon

Goldberg, E. D. 1965. Minor elements in seawater. In *Chemical Oceanography*, ed. J. P. Rilley, G. Skirrow, p. 163. London/New York:Academic

Goldberg, E. D. 1970. Air transport of organic contaminants to the marine environment. See Duursma 1970, p. 351

Goldberg, E. D. 1971. Atmospheric transport. In *Impingement of Man on the Oceans*, ed. D. W. Hood, p. 75. New York: Wiley

Grieves, R. B., Wang, S. L. 1967. Foam separation of *Pseudomonas fluorescens* and *Bacillus subtilus* var. *niger*. *Appl. Microbiol.* 15:76

Hamilton-Taylor, J. 1979. Enrichments of zinc, lead, and copper in recent sediments of Windermere, England. *Environ. Sci. Technol.* 13(6):693

Handa, N. 1977. Land sources of marine organic matter. *Mar. Chem.* 5:341

Hansen, H. P. 1975. Photochemical degradation of petroleum surface films on sea water. *Mar. Chem.* 3:183–85

Hansen, H. P. 1977. Photodegradation of hydrocarbon surface films. In *Petroleum Hydrocarbons in the Marine Environment*, ed. A. D. McIntyre, K. J. Whittle, pp. 101–6. (*Cons. Int. Explor. Mer, Rapp. p.-v. réun. 171*)

Hardy, R., Mackie, P. R., Whittle, K. J. 1977a. Hydrocarbons in the marine ecosystem—a review. See Hansen 1977, pp. 17–26

Hardy, R., Mackie, P. R., Whittle, K. J., McIntyre, A. D., Blackman, R. A. A. 1977b. Occurrence of hydrocarbons in the surface film; subsurface water and sediment in the waters around the United Kingdom. See Hansen 1977, pp. 61–65

Harrison, W., Winnik, M. A., Kwong, P. T. Y., Mackay, D. 1975. Crude oil spills. Disappearance of aromatic and aliphatic components from small sea-surface slicks. *Environ. Sci. Technol.* 9:231–34

Harvey, G. W. 1966. Micro-layer collection from the sea surface. A new method and initial results. *Limnol. Oceanogr.* 11:608–14

Harvey, G. W., Burzell, L. A. 1972. Simple microlayer method for small samples. *Limnol. Oceanogr.* 17:156–57

Harvey, R. W., Young, L. Y. 1980. The enrichment and association of bacteria and particulates in the salt marsh surface water. *Appl. Environ. Microbiol.* 39(4):894–99

Hatcher, R. F., Parker, B. C. 1974. Laboratory comparisons of four surface microlayer samplers. *Limnol. Oceanogr.* 19(1):162

Hirsbrunner, W. R., Wangersky, P. J. 1975. Determination of Na, K, Mg, Ca, Fe and Mn in organic particulate matter in sea water by atomic absorption. *Mar. Chem.* 3:55

Hoffman, G. L., Duce, R. A. 1972. Consideration of the chemical fractionation of alkali and alkaline earth metals in the Hawaiian marine atmosphere. *J. Geophys. Res.* 77(27):5161–69

Hoffman, E. J., Duce, R. A. 1976. Factors influencing the organic carbon content of marine aerosols: a laboratory study. *J. Geophys. Res.* 81(21):3667

Hoffman, E. J., Hoffman, G. L., Duce, R. A. 1974. Chemical fractionation of alkali and alkaline earth metals in atmospheric particulate matter over the North Atlantic. *J. Rech. Atmos.* 13:675

Hoffman, G. L., Duce, R. A., Hoffman, E. J. 1972. Trace metals in the Hawaiian atmosphere. *J. Geophys. Res.* 77(27):5322

Hoffman, G. L., Duce, R. A., Walsh, P. R., Hoffman, E. J., Ray, B. J., Fasching, J. L. 1974. Residence time of some particulate trace metals in the oceanic surface microlayer: significance of atmospheric deposition. *J. Rech. Atmos.* 8:745–59

Horn, J. M., Teal, J. M., Backus, R. H. 1970. Petroleum lumps on the surface of the sea. *Science* 168:245–46

Hoyt, J. W. 1970. High molecular weight algal substances in the sea. *Mar. Biol.* 7:93–9

Hsu, S. A., Whelan, T. III. 1976. Transport of atmospheric sea salt in the coastal zone. *Environ. Sci. Technol.* 10(3):281–83

Huang, R. C. H. 1974. *The removal of copper, cadmium, and lead ions from aqueous solutions using foam fractionation.* PhD thesis. Univ. Ottawa, Ottawa, Canada

Hunter, K. A. 1980. Processes affecting particulate trace metals in the sea surface microlayer. *Mar. Chem.* 9(1):49

Hunter, K. A., Liss, P. S. 1977. The input of organic material to the oceans: air-sea interactions and the organic chemical composition of the sea surface. *Mar. Chem.* 5:361

Jackson, G. A., Morgan, J. J. 1978. Trace metal-chelator interactions and phytoplankton growth in seawater media: theoretical analysis and comparison with reported observations. *Limnol. Oceanogr.* 23(2):268

Jacobelli-Turi, C., Maracci, M. A., Palmera, M. 1972. Separation of metallic ions by foaming: studies in Italy. See Fuerstenau & Healy 1972, p. 265

Jarvis, N. L., Timmons, C. O., Zisman, W. A. 1962. The effect of monomolecular film on the surface temperature of water. In *Retardation of Evaporation by Monolayers*, ed. V. K. LaMer, pp. 41–58. New York: Academic

Jarvis, N. L., Garrett, W. D., Scheiman, M. A., Timmons, C. O. 1967. Surface chemical characterization of surface-active material in seawater. *Limnol. Oceanogr.* 12:88–96

Johnson, B. D. 1976. Nonliving organic particle formation from bubble dissolution. *Limnol. Oceanogr.* 21(3):444

Johnston, R. 1963. Seawater, the natural medium of phytoplankton. I. General features. *J. Mar. Biol. Assoc. U. K.* 43:427

Kim, Y. S., Zeitlin, H. 1971a. Separation of trace-metal ions from seawater by adsorptive colloid flotation. *Chem. Commun.*, p. 672

Kim, Y. S., Zeitlin, H. 1971b. Separation of uranium from seawater by adsorbing colloid flotation. *Anal. Chem.* 43:1390

Kim, Y. S., Zeitlin, H. 1971c. A rapid adsorbing colloid flotation method for the separation of molybdenum from seawater. *Sep. Sci.* 6(4):505–13

Kim, Y. S., Zeitlin, H. 1972. The separation of zinc and copper from seawater by adsorption colloid flotation. *Sep. Sci.* 7(1):1–12

Korzh, V. D. 1974. Some general laws governing the turnover of substance within the ocean-atmosphere-continent-ocean cycle. *J. Rech. Atmos.* 13:653

Kubota, K. 1975. Non-foaming adsorptive bubble separation of DBS Na and foam separation of Cd. *Can. J. Chem. Eng.* 53(6):706

Kubota, K., Hayaski, S. 1977. Removal of sodium, cadmium and chromium ions from dilute aqueous solutions using foam fractionation. *Can. J. Chem. Eng.* 55:286–92

Larsson, K., Odham, G., Södergren, A. 1974. On lipid surface films on the sea. I. A simple method for sampling and studies on composition. *Mar. Chem.* 2:49.

Ledet, E. J., Laseter, J. L. 1974a. A comparison of two sampling devices for the recovery of organics from aqueous surface films. *Anal. Lett.* 7:553

Ledet, E. J., Laseter, J. L. 1974b. Alkanes at the air-sea interface from offshore Louisiana and Florida. *Science* 186:261–63

Lemlich, R. 1972a. Adsubble processes: foam fractionation and bubble fractionation. *J. Geophys. Res.* 77(27):5204–9

Lemlich, R., ed. 1972b. *Adsorptive Bubble Separation Techniques.* New York: Academic

Levin, G. V., Clendenning, J. R., Gibor, A., Bogar, F. D. 1962. Harvesting of algae by froth flotation. *Appl. Microbiol.* 10:169

Lion, L. W. 1980. *Cadmium, copper and lead in estuarine salt marsh microlayers: accumulation, speciation, and transport.* PhD thesis. Stanford Univ. Stanford, Calif.

Liss, P. S. 1975. Chemistry of the sea surface microlayer. In *Chemical Oceanography,* ed. J. P. Riley, G. Skirrow, 2:193. London, New York, San Francisco: Academic 2nd ed.

Liss, P. S. 1977. Effect of surface films on gas exchange across the air-sea interface. See Hansen 1977, pp. 120–24

Lou, Y. S., Rasmussen, G. P. 1973. Evaporation retardation by molecular layers. *Water Res.* 9(5):1258

MacIntyre, F. 1968. Bubbles: a boundary-layer 'microtome' for micron thick samples of a liquid surface. *J. Phys. Chem.* 72:589

MacIntyre, F. 1970. Geochemical fractionation during mass transfer from sea to air by breaking bubbles. *Tellus* 22:451–61

MacIntyre, F. 1974a. Chemical fractionation and sea-surface microlayer processes. In *The Sea, Vol. 5, Marine Chemistry,* ed. E. D. Goldberg, pp. 245–99. New York: Wiley

MacIntyre, F. 1974b. Non-lipid-related possibilities for chemical fractionation in bubble film caps. *J. Rech. Atmos.* 8:515–27

MacIntyre, F. 1974c. The top millimeter of the ocean. *Sci. Am.* May 1974, p. 62

MacIntyre, W. G., Smith, C. L., Munday, J. C., Gibson, V. M., Lake, J. L., Windsor, J. G., Dupuy, J. L., Harrison, W., Oberholtzer, J. D. 1974. Investigation of surface films—Chesapeake Bay entrance. *US Environ. Prot. Agency Ser.* #EPA-670/2-73-099

Martin, J. H. 1970. The possible transport of trace metals via moulted copepod exoskeletons. *Limnol. Oceanogr.* 15:756

Martin, J. H., Bruland, K. W., Broenkow, W. W. 1976. Cadmium transport in the California current. See Bidleman et al 1976, p. 159

Marty, J. C., Saliot, A. 1974. Etude chimique comparée du film de surface et de l'eau de mer sous-jacent: acide gras. *J. Rech. Atmos.* 8:563–70

Marty, J. C., Saliot, A. 1976. Hydrocarbons (normal alkanes) in the surface microlayer of sea water. *Deep-Sea Res.* 23:863–73

Marumo, R., Taga, N., Nakai, T. 1971. Neustonic bacteria and plankton in surface microlayers of the Equatorial waters. *Bull. Plankton Soc. J.* 18(2):36–41

Maynard, N. G. 1968a. Aquatic foams as an ecological habitat. *Z. Allg. Mikrobiol.* 8:119

Maynard, N. G. 1968b. Significance of airborne algae. *Z. Allg. Mikrobiol.* 8:225–26

Menzel, D. W. 1966. Bubbling of seawater and the production of organic particles, a re-evaluation. *Deep-Sea Res.* 13:963

Miget, R., Kator, H., Oppenheimer, C., Laseter, J. L., Ledet, E. J. 1974. New sampling device for the recovery of petroleum hydrocarbons and fatty acids from aqueous surface films. *Anal. Chem.* 46:1154–57

Milne, L. J., Milne, M. 1978. Insects of the water surface. *Sci. Am.* 238(4):134

Morel, F., Morgan, J. 1972. A numerical method for computing equilibria in aqueous chemical systems. *Environ. Sci. Technol.* 6:58–67

Morris, B. F. 1971. Petroleum: tar quantities floating in the Northwestern Atlantic taken with a quantitative neuston net. *Science* 173:430–32

Morris, B. F., Butler, J. N., Zsolnay, A. 1975. Pelagic tar in the Mediterranean Sea, 1974–1975. *Environ. Conserv.* 2:275–81

Morris, R. J. 1974. Lipid composition of surface films and zooplankton from the Eastern Mediterranean. *Mar. Pollut. Bull.* 5:105–8

Morris, R. J., Culkin, F. 1974. Lipid chemistry of Eastern Mediterranean surface layers. *Nature* 250:640

Nakashima, S. 1979. Flotation separation and atomic absorption spectrometric determination of selenium(IV) in water. *Anal. Chem.* 51:654

National Research Council. 1928. *International Critical Tables of Numerical Data, Physics, Chemistry and Technology,* Vol. IV, p. 466. New York: McGraw-Hill

Nishizawa, S. 1971. Concentration of organic and inorganic material in the surface skin at the equator, 155° W. *Bull. Plankton Soc. Jpn.* 18(2):42–4

Nishizawa, S., Nakajima, K. 1971. Concentration of particulate organic material in the sea surface layer. *Bull. Plankton Soc. Jpn.* 18(2):12–19

Norris, R. E. 1965. Neustonic marine Creaspedomonadales (Choanoflagellates) from Washington and California. *J. Protozool.* 12(4):589–602

Osipow, L. I. 1977. *Surface Chemistry: Theory and Industrial Applications.* Huntington, NY:R. E. Krieger

Parker, B., Barsom, G. 1970. Biological and chemical significance of surface microlayers in aquatic ecosystems. *Bioscience* 20(2):87–93

Parker, B. C., Wodehouse, E. B. 1971. Ecology and water quality criteria. In *Water for Texas,* pp. 114–134. *Proc. Water Resour. Inst., 15th Ann. Conf. Texas A&M Univ.*

Patterson, C., Settle, D. 1974. Contribution of lead via aerosol deposition to the Southern California Bight. *J. Rech. Atmos.* 13:957

Patterson, C., Settle, D., Glover, B. 1976. Analysis of lead in polluted coastal seawater. *Mar. Chem.* 4:305–19

Pellenbarg, R. E. 1976. *The aqueous surface microlayer and trace metals in the salt marsh.* PhD thesis. Univ. Del., Newark, Del.

Pellenbarg, R. E., Church, T. M. 1979. The estuarine surface microlayer and trace metal cycling in the salt marsh. *Science* 203(4384):1010

Peirson, D. H., Cawse, P. A., Cambray, R. S. 1974. Chemical uniformity of airborne particulate material, and a maritime effect. *Nature* 251:675–79

Piotrowicz, S. R., Ray, B. J., Hoffman, G. L., Duce, R. A. 1972. Trace metal enrichment in the sea-surface microlayer. *J. Geophys. Res.* 77(27):5243

Piotrowicz, S. R., Duce, R. A., Fasching, J. L., Weisel, C. P. 1979. Bursting bubbles and their effect on the sea-to-air transport of Fe, Cu, and Zn. *Mar. Chem.* 7(4):307–24

Pojasek, R. B., Zajicek, O. T. 1978. Surface microlayers and foams—source and metal transport in aquatic systems. *Water Res.* 12:7–11

Prospero, J. M., Carlson, T. N. 1972. Vertical and area distribution of Saharan dust over the Western Equatorial N. Atlantic Ocean. *J. Geophys. Res.* 77(27):5255

Pueschel, R. F., Van Valin, C. C. 1974. The mixed nature of laboratory-produced aerosols from seawater. *J. Rech. Atmos.* 13:601

Quinn, J. A., Steinbrook, R. A., Anderson, J. L. 1975. Breaking bubbles and the water-to-air transport of particulate matter. *Chem. Eng. Sci.* 30:1177–84

Rancitelli, L. A., Perkins, R. W. 1970. Trace element concentrations in the troposphere and stratosphere. *J. Geophys. Res.* 75:3055–64

Riley, G. A. 1963. Organic aggregates in seawater and the dynamics of their formation and utilization. *Limnol. Oceanogr.* 8:372

Riley, G. A., Wangersky, P. J., Van Hemert, D. 1964. Organic aggregates in tropical and subtropical surface waters of the North Atlantic Ocean. *Limnol. Oceanogr.* 9:546–50

Riley, G. A., Van Hemert, D., Wangersky, P. J. 1965. Organic aggregates in surface and deep waters of the Sargasso Sea. *Limnol. Oceanogr.* 10:354–63

Roy, V. M., Dupuy, J. L., MacIntyre, W. G., Harrison, W. 1970. Abundance of marine phytoplankton in surface films: a method of sampling. In *Hydrobiology. Bioresources of Shallow Water Environments,* ed. W. G. Weist, Jr., P. E. G. Reeson, pp. 371–80. Urbana, IL: Am. Water Resour. Assoc.

Rubin, L. B., Cassel, E. A., Henderson, O., Johnson, J. D., Lamb J. C. III. 1966. Microflotation: new low gas-flow rate foam separation technique for bacteria and algae. *Biotechnol. Bioeng.* 8:135

Rueter, J. G., McCarthy, J. J., Carpenter, E. J. 1979. The toxic effect of copper on *Oscillatoria* (*Trichodesmium*) *theibautii.* *Limnol. Oceanogr.* 24(3):558

Saxby, J. D. 1969. Metal-organic chemistry of the geochemical cycle. *Rev. Pure Appl. Chem.* 19:131

Schell, W. R., Barnes, R. S. 1976. Lead and mercury in the aquatic environment of Western Washington State. In *Aqueous-Environmental Chemistry of Metals,* ed. A. J. Rubin, p. 129. Ann Arbor, MI: Ann Arbor Sci. Publishers

Seba, D. B., Corcoran, E. F. 1969. Surface slicks as concentrators of pesticides in the marine environment. *Pestic. Monit. J.* 3:190

Seto, F. Y. B., Duce, R. A. 1972. A laboratory study of iodine enrichment on atmospheric sea-salt particles produced by bubbles. *J. Geophys. Res.* 77(27):5339–49

Sieburth, J. McN. 1965. Bacteriological samplers for air-water and water-sediment interfaces. In *Ocean Sciences Ocean Engineering,* p. 1064. Washington, D.C: Mar. Technol. Soc. Am. Soc. Limnol. Oceanogr.

Sieburth, J. McN. 1971. Distribution and activity of oceanic bacteria. *Deep-Sea Res.* 18:1111–21

Sieburth, J. McN., Conover, J. T. 1965. Slicks associated with *Trichodesmium* blooms in the Sargasso Sea. *Nature* 205:830–31

Sieburth, J. McN., Willis, P. J., Johnson, F. M., Burney, C. M., Lavoie, D. M., Hinga, K. R., Caron, D. A., French, F. W., III, Johnson, P. W., Davis, B. G. 1976. Dissolved organic matter and heterotrophic microneuston in the surface microlayers of the North Atlantic. *Science* 194:1415

Siegel, A. 1971. Metal-organic interactions in the marine environment. In *Organic Compounds in Aquatic Environments*, ed. S. D. Faust, J. V. Hunter, p. 265. New York:Marcel Dekker

Siegel, A., Burke, B. 1965. Sorption studies of cations on 'bubble produced organic aggregates' in sea water. *Deep-Sea Res.* 12:789

Stevenson, R. E., Collier, A. 1962. Preliminary observations on the occurrence of air-borne marine phytoplankton. *Lloydia* 25:89

Sunda, W., Guillard, R. R. L. 1976. The relationship between cupric ion activity and the toxicity of copper to phytoplankton. *J. Mar. Res.* 34:511

Sutcliffe, W. H., Baylor, E. R., Menzel, D. W. 1963. Sea surface chemistry and Langmuir circulation. *Deep-Sea Res.* 10:233

Szekielda, K. H., Kupferman, S. I., Klemas, V., Polis, D. F. 1972. Element enrichment in organic films and foam associated with aquatic frontal systems. *J. Geophys. Res.* 77(27):5278

Szyszkowski, B. 1908. Experimentelle Studien über kapillare Eigenschaften der wässerigen Lösungen von Fettsäuren. *Z. Physik. Chem.* 64:385

Taguchi, S., Nakajima, K. 1971. Plankton and seston in the sea surface of three inlets of Japan. *Bull. Plankton Soc. Jp.* 18(2):20–36

Trent, S. D., Shanks, A. L., Silver, M. W. 1978. In situ and laboratory measurements on macroscopic aggregates in Monterey Bay, California. *Limnol. Oceanogr.* 23(4):626

Tsunogai, S., Saito, O., Yamada, K., Nakaya, S. 1972. Chemical composition of oceanic aerosol. *J. Geophys. Res.* 77(27):5283

Tsyban, A. V. 1971. Marine bacterioneuston. *J. Oceanogr. Soc. Jpn.* 27:56–66

Turekian, K. K., Katz, A., Chan, L. 1973. Trace element trapping in Pteropod tests. *Limnol. Oceanogr.* 18(2):240–49

Van Grieken, R. E., Johansson, T. B., Winchester, J. W. 1974. Trace metal fractionation effects between sea water and aerosols from bubble bursting. *J. Rech. Atmos.* 13:611

Wade, T. L., Quinn, J. G. 1975. Hydrocarbons in the Sargasso Sea microlayer. *Mar. Pollut. Bull.* 6:54–57

Wallace, G. T., Duce, R. A. 1975. Concentration of particulate trace metals and particulate organic carbon in marine surface waters by a bubble flotation mechanism. *Mar. Chem.* 3:157–81

Wallace, G. T. Jr., Duce, R. A. 1979. Transport of particulate organic matter by bubbles in marine waters. *Limnol. Oceanogr.* 23(6):1155

Wallace, G. T. Jr., Loeb, G. I., Wilson, D. F. 1972. On the flotation of particulates in sea water by rising bubbles. *J. Geophys. Res.* 77(27):5293

Wallace, G. T. Jr., Hoffman, G. L., Duce, R. A. 1977. The influence of organic matter and atmospheric deposition on the particulate trace metal concentration of Northwest Atlantic surface seawater. *Mar. Chem.* 5:143

Wangersky, P. J. 1965. The organic chemistry of seawater. *Am. Sci.* 53:358

Wangersky, P. J. 1972. The cycle of organic carbon in seawater. *Chimia* 26(11):559–64

Wangersky, P. J. 1976. The surface film as a physical environment. *Ann. Rev. Ecol. Syst.* 7:161–76

Wangersky, P. J., Gordon, D. C. 1965. Particulate carbonate, organic carbon and Mn^{++} in the open ocean. *Limnol. Oceanogr.* 10:544

Wheeler, J. R. 1975. Formation and collapse of surface films. *Limnol. Oceanogr.* 20(3):338

Wilcox, R. S. 1979. Sex discrimination in *Gerris remigis*: role of a surface wave signal. *Science* 206(4424):1325

Wilkniss, P. E., Bressan, D. J. 1972. Fractionation of the elements F, Cl, and K at the sea-air interface. *J. Geophys. Res.* 77(27):5307

Williams, P. M. 1967. Sea surface chemistry: organic carbon and organic and inorganic nitrogen and phosphorus in surface films and subsurface waters. *Deep-Sea Res.* 14:791–800

Williams, R. J. P. 1953. Metal ions in biological systems. *Biol. Rev.* 28:381

Wilson, A. T. 1959. Surface of the ocean as a source of air-borne nitrogenous material and other plant nutrients. *Nature* 184:99–101

Wilson, W. B., Collier, A. 1972. The production of surface-active material by marine phytoplankton cultures. *J. Mar. Res.* 30:15–26

Winchester, J., Nifong, G. 1971. Water pollution in Lake Michigan by trace elements from pollution aerosol fallout. *Water, Air & Soil Pollut.* 1:50

Woodcock, A. H. 1972. Smaller salt particles in oceanic air and bubble behavior in the sea. *J. Geophys. Res.* 77(27):5316

Wong, C. S., Green, D. R., Cretney, W. J. 1974. Quantitative tar and plastic waste distributions in the Pacific Ocean. *Nature* 247:30–32

Yano, N., Katsuragawa, H., Maebashi, K. 1974. Assessment of sources of maritime aerosols by neutron activation analysis. *J. Rech. Atmos.* 13:807

Young, D. R., Jan, T. K. 1977. Fire fallout of metals off California. *Mar. Pollut. Bull.* 8:109

Young, L. Y. 1978. Bacterioneuston examined with critical point drying and transmission electron microscopy. *Microb. Ecol.* 4:267–77

Zaitsev, Yu. P. 1971. *Marine Neustonology.* Translated from Russian, Natl. Mar. Fisheries Serv., Natl. Oceanic and Atmos. Admin., US Dept. Comm. and Natl. Sci. Found.

Zobell, C. E., Mathews, H. M. 1936. A qualitative study of the bacterial flora of sea and land breezes. *Proc. Natl. Acad. Sci. US* 22:567

Zoller, W. H., Gladney, E. S., Duce, R. A. 1974. Atmospheric concentrations and sources of trace metals at the South Pole. *Science* 183(4121):198

AUTHOR INDEX

(Names appearing in capital letters indicate authors of chapters in this volume.)

A

Abe, K., 398
Abraham, F. F., 235
Abraham, K., 189
Abramowitz, M., 93
Abu-Eid, R. M., 366, 371, 373-75
Adams, J. B., 377
Aggarwal, Y. P., 98
Agnew, D. C., 395, 402
Ahearn, D. G., 471
A'Hearn, M. F., 114, 125
Ahmad, S. N., 318
Airy, G. B., 415
Aki, K., 91, 104, 397, 399, 404
Akimoto, S., 161, 169
Alabi, A. O., 163
Albarede, F., 333
Albright, L. J., 460, 468, 470, 471
Albritton, D. L., 239
Aldrich, L. T., 4, 12, 170
Aldridge, L. P., 366
Allan, R. R., 419
Alldredge, L. R., 152
Allegre, C. J., 312, 321, 333
Allen, C. R., 94
Allen, G. C., 350
Allen, J. L., 377
Alsop, L. E., 392, 401
Alterman, Z., 391, 392
Altschuler, Z. S., 256, 270, 271
Amitai, K., 465
Amthauer, G., 361, 362, 367, 368, 370
Anderson, A. T., 315
Anderson, D. J., 188
Anderson, D. L., 95-97, 101, 201, 321, 324, 394, 399, 400, 403, 404, 429
Anderson, D. M., 14
Anderson, J. L., 475
Anderson, R. N., 258, 416, 433, 436
Anderson, W. L., 157
Anderssen, R. S., 385
Ando, M., 95
Andren, A. W., 455, 474
Andrew-Jones, D. A., 176
Andrews, D. J., 82, 104
Andrews, J. T., 199, 210, 214, 216, 221-23
Annersten, H., 366-68, 370
Appleman, D. E., 352
Araki, T., 353, 362, 367
Archambeau, C. B., 394, 400, 403
Arinder, G., 38, 39

Armstrong, D. E., 453, 474
Armstrong, R. L., 325
Arnold, F., 235, 239
Arth, J. G., 335
Arthur, M. A., 256, 259, 274, 277, 279
Ashby, M. F., 96
Aston, S. R., 473
Atlas, E. L., 255, 262, 263
Ausloos, P., 244

B

Baadsgaard, H., 13
Backus, G. E., 391, 393, 399
Backus, R. H., 465, 466
Baes, C. F., 462
Baier, R. E., 450, 451, 456, 458, 461, 463, 465, 472, 473
Bailey, R. C., 151, 155, 157, 158, 164, 165
Bainbridge, A., 317
Bainbridge, K. T., 9
Bakun, W. H., 105
Balashov, A. I., 468
Balazs, E. I., 94, 98, 102, 103
Baldwin, A. C., 241
Baldwin, B., 29
Ballard, R. D., 316
Balling, N., 201
Balmino, G., 422, 423
Bambach, R. K., 276
Banbury, D. St. P., 362, 367
Bancroft, G. M., 364, 366-68
Banerjee, S. K., 377
Banks, R. J., 150, 151, 164
Bannister, J. R., 170
Banno, S., 188
Barber, J., 474
Barber, N. F., 9
Barber, R. T., 464
Bardeen, J., 9
Barger, W. R., 473
Barker, D. R., 451, 453, 460, 468, 473
Barker, T., 101
Barnes, R. A., 41, 239
Barnes, R. S., 474
Barrell, J., 415, 429
Barrett, D. R., 285, 304
Barringer, A. R., 474
Barry, R. G., 199
Barsom, G., 459, 468, 471
Barton, J. M. Jr., 176-81, 183, 185, 186, 189, 191-94
Batley, G. E., 451
Batoosingh, E., 464

Baturin, G. N., 252, 256, 258, 261-63, 265, 269, 270, 272, 273, 280
Baylor, E. R., 464, 468, 472
Baylor, M. B., 472
Beach, L., 170
Beamish, D., 155
Beauford, W., 474
Beblo, M., 166
Beeson, D. E., 31-33, 50, 51
Bell, J. A., 243
Bell, P. M., 346, 366, 371, 373-75, 377
Bender, M. L., 258
Benioff, H., 392, 398
Ben-Menahem, A., 385, 398
Bennett, C. F., 474
Benninger, L. K., 474
BENSON, R. H., 59-80; 59, 65, 66, 69, 71, 72
Beran, A., 353
Berdichevski, M. N., 170, 171
Berger, J., 391, 395, 402
Berger, W. H., 277
Berggren, W. A., 274
Bertine, K., 474
Besancon, J. R., 377
Bessel, F. W., 133
Beuzart, P., 426
Beynon, J. H., 12
Bezdek, H. F., 472, 475
Bezrukov, P. L., 272, 273
Bidleman, T. F., 458, 459, 475
Biemann, K., 14
Bigg, E. K., 23, 24, 37, 227, 228
Biggs, W. D., 60
Bilby, B. A., 85
Biller, J. E., 14
Bills, B. G., 422, 423
Biot, M. A., 98
Birch, P. V., 125
Bischke, R. E., 103
BISHOP, F. C., 175-98; 184-90
Bittner, H., 353
Black, L. P., 177
Blackman, R. A. A., 458
Blackwelder, E., 253, 256
Blaha, J. J., 22
Blake, N. J., 264
Blake, L., 377
Blanchard, D. C., 469, 472, 473, 475
Blanchard, M. B., 39
Blaumont, J. B., 30, 32-34
Blifford, I. H., 242
Bloch, M. R., 473
Block, B., 395
Blohm, E. K., 157, 158, 165

CUMULATIVE INDEXES

CONTRIBUTING AUTHORS VOLUMES 5–9

CHAPTER TITLES VOLUMES 5–9

500

ORDER FORM   ANNUAL REVIEWS INC.

Please list on the order blank on the reverse side the volumes you wish to order and whether you wish a standing order (the latest volume sent to you automatically each year). Volumes not yet published will be shipped in month and year indicated. Prices subject to change without notice. Out of print volumes subject to special order.

NEW TITLES FOR 1981

ANNUAL REVIEW OF NUTRITION   ISSN 0199-9885
    Vol. 1 (avail. July 1981): $20.00 (USA), $21.00 (elsewhere) per copy

INTELLIGENCE AND AFFECTIVITY: Their Relationship During Child Development
    A monograph, translated from a course of lectures by Jean Piaget   ISBN 0-8243-2901-5
    Avail. Feb. 1981   Hard cover: $8.00 (USA), $9.00 (elsewhere) per copy

SPECIAL PUBLICATIONS

ANNUAL REVIEWS REPRINTS: CELL MEMBRANES, 1975–1977   ISBN 0-8243-2501-X
    A collection of articles reprinted from recent *Annual Review* series
    Published 1978   Soft cover: $12.00 (USA), $12.50 (elsewhere) per copy

ANNUAL REVIEWS REPRINTS: IMMUNOLOGY, 1977–1979   ISBN 0-8243-2502-8
    A collection of articles reprinted from recent *Annual Review* series
    Published 1980   Soft cover: $12.00 (USA), $12.50 (elsewhere) per copy

THE EXCITEMENT AND FASCINATION OF SCIENCE, VOLUME 1   ISBN 0-8243-1602-9
    A collection of autobiographical and philosophical articles by leading scientists
    Published 1965   Clothbound: $6.50 (USA), $7.00 (elsewhere) per copy

THE EXCITEMENT AND FASCINATION OF SCIENCE, VOLUME 2: Reflections by Eminent Scientists
    Published 1978   Hard cover: $12.00 (USA), $12.50 (elsewhere) per copy   ISBN 0-8243-2601-6
                        Soft cover: $10.00 (USA), $10.50 (elsewhere) per copy   ISBN 0-8243-2602-4

THE HISTORY OF ENTOMOLOGY   ISBN 0-8243-2101-7
    A special supplement to the *Annual Review of Entomology* series
    Published 1973   Clothbound: $10.00 (USA), $10.50 (elsewhere) per copy

ANNUAL REVIEW SERIES

Annual Review of ANTHROPOLOGY   ISSN 0084-6570
    Vols. 1–8 (1972–79): $17.00 (USA), $17.50 (elsewhere) per copy
    Vol. 9 (1980): $20.00 (USA), $21.00 (elsewhere) per copy
    Vol. 10 (avail. Oct. 1981): $20.00 (USA), $21.00 (elsewhere) per copy

Annual Review of ASTRONOMY AND ASTROPHYSICS   ISSN 0066-4146
    Vols. 1–17 (1963–79): $17.00 (USA), $17.50 (elsewhere) per copy
    Vol. 18 (1980): $20.00 (USA), $21.00 (elsewhere) per copy
    Vol. 19 (avail. Sept. 1981): $20.00 (USA), $21.00 (elsewhere) per copy

Annual Review of BIOCHEMISTRY   ISSN 0066-4154
    Vols. 28–48 (1959–79): $18.00 (USA), $18.50 (elsewhere) per copy
    Vol. 49 (1980): $21.00 (USA), $22.00 (elsewhere) per copy
    Vol. 50 (avail. July 1981): $21.00 (USA), $22.00 (elsewhere) per copy

Annual Review of BIOPHYSICS AND BIOENGINEERING   ISSN 0084-6589
    Vols. 1–9 (1972–80): $17.00 (USA), $17.50 (elsewhere) per copy
    Vol. 10 (avail. June 1981): $20.00 (USA), $21.00 (elsewhere) per copy

Annual Review of EARTH AND PLANETARY SCIENCES   ISSN 0084-6597
    Vols. 1–8 (1973–80): $17.00 (USA), $17.50 (elsewhere) per copy
    Vol. 9 (avail. May 1981): $20.00 (USA), $21.00 (elsewhere) per copy

Annual Review of ECOLOGY AND SYSTEMATICS   ISSN 0066-4162
    Vols. 1–10 (1970–79): $17.00 (USA), $17.50 (elsewhere) per copy
    Vol. 11 (1980): $20.00 (USA), $21.00 (elsewhere) per copy
    Vol. 12 (avail. Nov. 1981): $20.00 (USA), $21.00 (elsewhere) per copy

Annual Review of ENERGY   ISSN 0362-1626
    Vols. 1–4 (1976–79): $17.00 (USA), $17.50 (elsewhere) per copy
    Vol. 5 (1980): $20.00 (USA), $21.00 (elsewhere) per copy
    Vol. 6 (avail. Oct. 1981): $20.00 (USA), $21.00 (elsewhere) per copy

Annual Review of ENTOMOLOGY   ISSN 0066-4170
    Vols. 7–25 (1962–80): $17.00 (USA), $17.50 (elsewhere) per copy
    Vol. 26 (avail. Jan. 1981): $20.00 (USA), $21.00 (elsewhere) per copy

Annual Review of FLUID MECHANICS   ISSN 0066-4189
    Vols. 1–12 (1969–80): $17.00 (USA), $17.50 (elsewhere) per copy
    Vol. 13 (avail. Jan. 1981): $20.00 (USA), $21.00 (elsewhere) per copy

Annual Review of GENETICS   ISSN 0066-4197
    Vols. 1–13 (1967–79): $17.00 (USA), $17.50 (elsewhere) per copy
    Vol. 14 (1980): $20.00 (USA), $21.00 (elsewhere) per copy
    Vol. 15 (avail. Dec. 1981): $20.00 (USA), $21.00 (elsewhere) per copy

(continued on reverse)

Annual Review of MATERIALS SCIENCE   ISSN 0084-6600
  Vols. 1–9 (1971–79): $17.00 (USA), $17.50 (elsewhere) per copy
  Vol. 10 (1980): $20.00 (USA), $21.00 (elsewhere) per copy
  Vol. 11 (avail. Aug. 1981): $20.00 (USA), $21.00 (elsewhere) per copy

Annual Review of MEDICINE: Selected Topics in the Clinical Sciences   ISSN 0066-4219
  Vols. 1–3, 5–15, 17–31 (1950–52, 1954–64, 1966–80): $17.00 (USA), $17.50 (elsewhere) per copy
  Vol. 32 (avail. Apr. 1981): $20.00 (USA), $21.00 (elsewhere) per copy

Annual Review of MICROBIOLOGY   ISSN 0066-4227
  Vols. 15–33 (1961–79): $17.00 (USA), $17.50 (elsewhere) per copy
  Vol. 34 (1980): $20.00 (USA), $21.00 (elsewhere) per copy
  Vol. 35 (avail. Oct. 1981): $20.00 (USA), $21.00 (elsewhere) per copy

Annual Review of NEUROSCIENCE   ISSN 0147-006X
  Vols. 1–3 (1978–80): $17.00 (USA), $17.50 (elsewhere) per copy
  Vol. 4 (avail. Mar. 1981): $20.00 (USA), $21.00 (elsewhere) per copy

Annual Review of NUCLEAR AND PARTICLE SCIENCE   ISSN 0066-4243
  Vols. 10–29 (1960–79): $19.50 (USA), $20.00 (elsewhere) per copy
  Vol. 30 (1980): $22.50 (USA), $23.50 (elsewhere) per copy
  Vol. 31 (avail. Dec. 1981): $22.50 (USA), $23.50 (elsewhere) per copy

Annual Review of PHARMACOLOGY AND TOXICOLOGY   ISSN 0362-1642
  Vols. 1–3, 5–20 (1961–63, 1965–80): $17.00 (USA), $17.50 (elsewhere) per copy
  Vol. 21 (avail. Apr. 1981): $20.00 (USA), $21.00 (elsewhere) per copy

Annual Review of PHYSICAL CHEMISTRY   ISSN 0066-426X
  Vols. 10–21, 23–30 (1959–70, 1972–79): $17.00 (USA), $17.50 (elsewhere) per copy
  Vol. 31 (1980): $20.00 (USA), $21.00 (elsewhere) per copy
  Vol. 32 (avail. Nov. 1981): $20.00 (USA), $21.00 (elsewhere) per copy

Annual Review of PHYSIOLOGY   ISSN 0066-4278
  Vols. 18–42 (1956–80): $17.00 (USA), $17.50 (elsewhere) per copy
  Vol. 43 (avail. Mar. 1981): $20.00 (USA), $21.00 (elsewhere) per copy

Annual Review of PHYTOPATHOLOGY   ISSN 0066-4286
  Vols. 1–17 (1963–79): $17.00 (USA), $17.50 (elsewhere) per copy
  Vol. 18 (1980): $20.00 (USA), $21.00 (elsewhere) per copy
  Vol. 19 (avail. Sept. 1981): $20.00 (USA), $21.00 (elsewhere) per copy

Annual Review of PLANT PHYSIOLOGY   ISSN 0066-4294
  Vols. 10–31 (1959–80): $17.00 (USA), $17.50 (elsewhere) per copy
  Vol. 32 (avail. June 1981): $20.00 (USA), $21.00 (elsewhere) per copy

Annual Review of PSYCHOLOGY   ISSN 0066-4308
  Vols. 4, 5, 8, 10–31 (1953, 1954, 1957, 1959–80): $17.00 (USA), $17.50 (elsewhere) per copy
  Vol. 32 (avail. Feb. 1981): $20.00 (USA), $21.00 (elsewhere) per copy

Annual Review of PUBLIC HEALTH   ISSN 0163-7525
  Vol. 1 (1980): $17.00 (USA), $17.50 (elsewhere) per copy
  Vol. 2 (avail. May 1981): $20.00 (USA), $21.00 (elsewhere) per copy

Annual Review of SOCIOLOGY   ISSN 0360-0572
  Vols. 1–5 (1975–79): $17.00 (USA), $17.50 (elsewhere) per copy
  Vol. 6 (1980): $20.00 (USA), $21.00 (elsewhere) per copy
  Vol. 7 (avail. Aug. 1981): $20.00 (USA), $21.00 (elsewhere) per copy

To ANNUAL REVIEWS INC., 4139 El Camino Way, Palo Alto, CA 94306 USA   (Tel. 415-493-4400)

Please enter my order for the following publications:
(Standing orders: indicate which volume you wish order to begin with)

_____, Vol(s). _____ Standing order ☐

_____, Vol(s). _____ Standing order ☐

_____, Vol(s). _____ Standing order ☐

_____, Vol(s). _____ Standing order ☐

Amount of remittance enclosed $_____   California residents please add applicable sales tax.
Please bill me ☐   Prices subject to change without notice.

SHIP TO (include institutional purchase order if billing address is different)

Name _____

Address _____

_____

_____ Zip Code _____

Signed _____ Date _____

☐ Please add my name to your mailing list to receive a free copy of the current Prospectus each year.
☐ Send free brochure listing contents of recent back volumes for *Annual Review(s)* of _____